Evolutionary Biology: Past, Present and Future

Evolutionary Biology: Past, Present and Future

Editor: Jesse Santos

R CALLISTO REFERENCE

www.callistoreference.com

Callisto Reference,
118-35 Queens Blvd., Suite 400,
Forest Hills, NY 11375, USA

Visit us on the World Wide Web at:
www.callistoreference.com

ISBN: 978-1-64116-247-0 (Hardback)

Cataloging-in-Publication Data

Evolutionary biology : past, present and future / edited by Jesse Santos.
    p. cm.
Includes bibliographical references and index.
ISBN 978-1-64116-247-0
1. Evolution (Biology). 2. Evolution. 3. Biology. I. Santos, Jesse.
QH366.2 .E96 2020
576.8--dc23

Table of Contents

Preface VII

Chapter 1   **Population genetics of mouse lemur vomeronasal receptors: current versus past selection and demographic inference** 1
Philipp Hohenbrink, Nicholas I. Mundy and Ute Radespiel

Chapter 2   **Correlated duplications and losses in the evolution of palmitoylation writer and eraser families** 12
Stijn Wittouck and Vera van Noort

Chapter 3   **Discovery and evolution of novel hemerythrin genes in annelid worms** 24
Elisa M. Costa-Paiva, Nathan V. Whelan, Damien S. Waits, Scott R. Santos,
Carlos G. Schrago and Kenneth M. Halanych

Chapter 4   **Exploring the evolutionary origin of floral organs of *Erycina pusilla*, an emerging orchid model system** 35
Anita Dirks-Mulder, Roland Butôt, Peter van Schaik, Jan Willem P. M. Wijnands,
Roel van den Berg, Louie Krol, Sadhana Doebar, Kelly van Kooperen,
Hugo de Boer, Elena M. Kramer, Erik F. Smets, Rutger A. Vos,
Alexander Vrijdaghs and Barbara Gravendeel

Chapter 5   **Evolutionary origin and function of NOX4-art, an arthropod specific NADPH oxidase** 53
Ana Caroline Paiva Gandara, André Torres, Ana Cristina Bahia,
Pedro L. Oliveira and Renata Schama

Chapter 6   **Reconstruction of the evolution of microbial defense systems** 69
Pere Puigbò, Kira S. Makarova, David M. Kristensen, Yuri I. Wolf and
Eugene V. Koonin

Chapter 7   **Natural selection drove metabolic specialization of the chromatophore in *Paulinella chromatophora*** 82
Cecilio Valadez-Cano, Roberto Olivares-Hernández, Osbaldo Resendis-Antonio,
Alexander DeLuna and Luis Delaye

Chapter 8   **168 million years old "marine lice" and the evolution of parasitism within isopods** 100
Christina Nagler, Matúš Hyžný and Joachim T. Haug

Chapter 9   **Identification of constraints influencing the bacterial genomes evolution in the PVC super-phylum** 114
Sandrine Pinos, Pierre Pontarotti, Didier Raoult and Vicky Merhej

Chapter 10   **Karyotypic evolution of the *Medicago* complex: *sativa-caerulea-falcata* inferred from comparative cytogenetic analysis** 128
Feng Yu, Haiqing Wang, Yanyan Zhao, Ruijuan Liu, Quanwen Dou,
Jiangli Dong and Tao Wang

Chapter 11   **A passive mutualistic interaction promotes the evolution of spatial structure within microbial populations** 140
Marie Marchal, Felix Goldschmidt, Selina N. Derksen-Müller, Sven Panke,
Martin Ackermann and David R. Johnson

Chapter 12   **Evolution of group I introns in Porifera: new evidence for intron mobility and implications for DNA barcoding** 154
Astrid Schuster, Jose V. Lopez, Leontine E. Becking, Michelle Kelly,
Shirley A. Pomponi, Gert Wörheide, Dirk Erpenbeck and Paco Cárdenas

Chapter 13   **The protein subunit of telomerase displays patterns of dynamic evolution and conservation across different metazoan taxa** 175
Alvina G. Lai, Natalia Pouchkina-Stantcheva, Alessia Di Donfrancesco,
Gerda Kildisiute, Sounak Sahu and A. Aziz Aboobaker

Chapter 14   **Duplication and concerted evolution of MiSp-encoding genes underlie the material properties of minor ampullate silks of cobweb weaving spiders** 196
Jannelle M. Vienneau-Hathaway, Elizabeth R. Brassfield, Amanda Kelly Lane,
Matthew A. Collin, Sandra M. Correa-Garhwal, Thomas H. Clarke,
Evelyn E. Schwager, Jessica E. Garb, Cheryl Y. Hayashi and Nadia A. Ayoub

Permissions

List of Contributors

Index

Preface

Evolutionary biology is a subfield of biology concerned with the study of evolutionary processes, which are at the core of the emergence of complex and diverse life on Earth. Among these processes are the mechanisms of common descent, natural selection and speciation. The genetic architecture of adaptation, molecular evolution, and the forces of sexual selection, biogeography and genetic drift are explored and studied under the domain of evolutionary biology. Research in evolutionary biology covers diverse themes from molecular genetics to computer science. One dimension of research strives to explain the phenomena of speciation and evolvability, and the evolution of aging, cooperation and sexual reproduction. This book is a valuable compilation of topics, ranging from the basic to the most complex advancements in the field of evolutionary biology. It presents this complex subject in the most comprehensible and easy to understand language. For someone with an interest and eye for detail, this book covers the most significant topics in this field.

This book unites the global concepts and researches in an organized manner for a comprehensive understanding of the subject. It is a ripe text for all researchers, students, scientists or anyone else who is interested in acquiring a better knowledge of this dynamic field.

I extend my sincere thanks to the contributors for such eloquent research chapters. Finally, I thank my family for being a source of support and help.

Editor

Population genetics of mouse lemur vomeronasal receptors: current versus past selection and demographic inference

Philipp Hohenbrink[1,2], Nicholas I. Mundy[2] and Ute Radespiel[1*] [iD]

Abstract

Background: A major effort is underway to use population genetic approaches to identify loci involved in adaptation. One issue that has so far received limited attention is whether loci that show a phylogenetic signal of positive selection in the past also show evidence of ongoing positive selection at the population level. We address this issue using vomeronasal receptors (VRs), a diverse gene family in mammals involved in intraspecific communication and predator detection. In mouse lemurs, we previously demonstrated that both subfamilies of VRs (V1Rs and V2Rs) show a strong signal of directional selection in interspecific analyses. We predicted that ongoing sexual selection and/or co-evolution with predators may lead to current directional or balancing selection on VRs. Here, we re-sequence 17 VRs and perform a suite of selection and demographic analyses in sympatric populations of two species of mouse lemurs (*Microcebus murinus* and *M. ravelobensis*) in northwestern Madagascar.

Results: *M. ravelobensis* had consistently higher genetic diversity at VRs than *M. murinus*. In general, we find little evidence for positive selection, with most loci evolving under purifying selection and one locus even showing evidence of functional loss in *M. ravelobensis*. However, a few loci in *M. ravelobensis* show potential evidence of positive selection. Using mismatch distributions and expansion models, we infer a more recent colonisation of the habitat by *M. murinus* than by *M. ravelobensis*, which most likely speciated in this region earlier on.

Conclusions: These findings suggest that the analysis of VR variation is useful in inferring demographic and phylogeographic history of mouse lemurs. In conclusion, this study reveals a substantial heterogeneity over time in selection on VR loci, suggesting that VR evolution is episodic.

Keywords: VNO, Evolution, Madagascar, *Microcebus murinus*, *Microcebus ravelobensis*, V1R, V2R, Genetic diversity, Demography, Selection

Background

Adaptation leaves its mark in the genome. These genomic signatures of adaptation can be investigated using phylogenetic and population genetic approaches, which address the different evolutionary timescales of interspecific and intraspecific variation. A rapidly growing body of literature is documenting the loci involved in adaptation at one of these levels [1]. However, there are few studies which directly evaluate the relationship between the two, or separate studies examining the same loci at different timescales (for example, genes involved in brain development in primates [2, 3]). Hence it is an open question whether positive selection that occurs at loci among species is generally reflected in positive selection at the same loci within species.

Different selection regimes make different predictions for the relationship between past and ongoing selection. In cases where there is an ongoing co-evolutionary dynamic, such as in host-parasite or Red Queen systems, selection on key loci may be more or less continuous and a signal of selection is expected both in the past and present. Under balancing selection, for example, multiple

* Correspondence: ute.radespiel@tiho-hannover.de
[1]Institute of Zoology, University of Veterinary Medicine Hannover, Buenteweg 17, 30559 Hannover, Germany
Full list of author information is available at the end of the article

alleles may be under on-going frequency-dependent selection across speciation events, and similar signals of selection can therefore be expected to act from the past to the present. Another possibility is that directional selection during a certain evolutionary period leads to the fixation of novel and advantageous variants. After this first and potentially episodic period of adaptation, purifying selection may then follow to stabilize and maintain the acquired adaptations. Under these conditions, there will be discordance between inferences of past and ongoing selection. Many other patterns are also possible: for example, changes in the effective population size (N_e) will affect fixation of mildly deleterious alleles by drift and selection on compensatory alleles, so changes in N_e alone may lead to different signatures of selection in different timeframes [4].

In mammals, G-protein coupled chemoreceptors form one of the gene families that are most consistently found to be under positive selection in phylogenetic analyses [5]. One of the most diverse classes of chemoreceptors in terrestrial mammals is vomeronasal receptors (VRs) that function in intraspecific communication and predator detection. There is evidence for positive selection on both types of VR (V1R and V2R) in mammals [6, 7]. This is consistent with the possibility of arms races between the scent of predators and the VRs of their prey, or the action of sexual selection on VR evolution for intraspecific communication.

Mouse lemurs (*Microcebus*) are nocturnal strepsirrhine primates with an elaborated olfactory behavioural repertoire [8–10]. Olfactory stimuli play an important role in their intraspecific communication as well as in predator detection [11–14]. They have the largest repertoire of VR genes (~200 *V1R*s and 2 *V2R*s) of any primate. We previously showed a high level of positive selection acting on multiple families of *V1R* genes, as well as *V2R* genes, during mouse lemur evolution [15, 16]. We hypothesised that this positive selection may relate to predator detection and/or intraspecific communication. If this is the case then one would also expect current selection on mouse lemur VR loci. To the best of our knowledge, population genetics of VRs has not previously been investigated in the wild.

Functional loci have been relatively neglected in studies of demography, which favour neutral markers (e.g., [17–21]). The phylogeography of many taxa has been influenced by Pleistocene glaciation cycles, and this has been demonstrated in temperate regions as well as in the tropics (e.g., [22, 23]). For Madagascar with its highly complex ecogeography of central-eastern highlands, a dry western and a humid eastern zone and various seasonal forest types, various biogeographic hypotheses have been proposed that take into account Pleistocene climatic changes, topographic barriers such as

mountains and large rivers [24, 25]. It has been suggested that various recent endemic radiations in lemurs and other vertebrates are causally linked to these climatic fluctuations that led to cyclic vegetation changes within and across isolated centres of endemism [26].

Mouse lemurs (*Microcebus* spp.) are a highly diverse genus with currently 24 described species [27] that are mostly confined to rather local or regional geographic ranges that are separated from each other by large rivers. Allopatric speciation has therefore been proposed as predominant mechanism of diversification within this genus [28, 29]. However, one species, the grey mouse lemur (*Microcebus murinus*), occurs from southern Madagascar up to northwestern Madagascar and can be found in partial sympatry with at least five other local or regional mouse lemur species, *M. griseorufus*, *M. berthae*, *M. myoxinus*, *M. ravelobensis* and *M. bongolavensis* (reviewed in [30]). Two recent molecular studies suggest that this large geographic range is the result of a rather recent expansion from southern to northwestern Madagascar at some time point in the late Pleistocene [21, 31]. Schneider et al. [21] inferred this expansion based on modelling and simulating the diversity and gene genealogy of mitochondrial sequence data obtained from several populations in northwestern Madagascar which also included samples from the same study site as chosen for this study. The haplotype diversity sampled in that study could be best explained by a succession of two spatial expansions that could be dated to the late Pleistocene (younger than 350,000 years old). It was hypothesised that an ancestral population of *M. murinus* colonized this region (=inter-river-system, IRS) before the last glacial maximum (LGM) and then contracted into riverine refugia during the dry period that coincided with the LGM and presumably with the preceding glacial periods in Madagascar [26]. From there it may subsequently have expanded again in parallel with forest expansion [21]. In the second study, Blair et al. [31] inferred the rather recent expansion of *M. murinus* by modelling the split between *M. murinus* and its sister species, *M. griseorufus*, in southwestern Madagascar. Coalescent methods were employed on 55–124 sequences for four molecular markers (alpha enolase intron, alpha fibrinogen intron, von Willebrand factor intron, cytochrome B concatenated with cytochrome c oxidase subunit II), respectively, stemming from localities in southern to western Madagascar. Results were concordant with allopatric speciation from a narrowly distributed common ancestor that lived in southwest Madagascar. Only *M. murinus* underwent a subsequent range expansion to the north and experienced severe population dynamics during the Pleistocene [31]. Given these scenarios, the demographic history of *M. murinus* populations in northwestern Madagascar, the region which it last colonized, should be substantially different from those of their sympatric congeners that most likely evolved in those areas over longer timescales [28, 32].

Here we assay sequence variation in 15 *V1R* and two *V2R* loci in populations of the two sympatric mouse lemur species (*M. murinus* and *M. ravelobensis*) from northwest Madagascar. We perform various tests of neutrality and investigate the correlation between past and present selective pressures. In addition, we perform demographic analyses to assess the suitability of VR loci for making demographic inferences.

Methods

Data collection

We extracted DNA from 20 grey mouse lemurs, *M. murinus* (10 male, 10 female), and 20 golden-brown mouse lemurs, *M. ravelobensis* (10 male, 10 female, see Additional file 1 for details), using a phenol-chloroform protocol and a REPLI-g WGA kit (Qiagen). Small ear biopsies were collected between May and October 2008 in the Ankarafantsika National Park in northwestern Madagascar by S. Thorén (authorisation no. 062/08/ MEEFT/ SG/DGEF/DSA/SSE). All animals were live-trapped in the 30.6 ha study site JBA (46°48′E, 16°19′S; for details about the trapping procedure see [33]), and we selected individuals from different trapping locations to minimise the risk of sampling several individuals from the same sleeping group who are most likely related [34, 35]. All sampled animals of the same species are defined to belong to the same population.

We designed locus-specific primer pairs for 15 *V1R* and both *V2R* loci using the online software Primer3-Plus [36]. All loci (see Table 1 for details) are expressed in the VNO of the grey mouse lemur [37]. Previous work on these loci in mouse lemurs provided no evidence for paralogues or recent duplicates [15, 16]. We used twelve loci from seven of the nine monophyletic *V1R* clusters and three unclustered loci [16]. Clusters I and VIII were not analysed because locus-specific primers could not be successfully designed for these. The PCR amplicons of intronless *V1R* loci covered the whole locus as all primers bind outside of the coding sequence. *VN2R1* consists of six exons and because of long intron sequences between the exons we designed exon-specific external primers. In contrast to our first description of *V2R*s [15], *VN2R2* consists of only five exons. Previously, we had used cDNA of extracted RNA without introns and sequenced a fragment spanning exon 3 to 5 according to closely related *V2R*s of family D in mice. A short fragment (22 bp) between these exons was assigned to exon 4. However, the present study which uses external primers reveals that the 22 bp of exon 4 rather

Table 1 *V1R* and *V2R* loci analysed with corresponding gene cluster according to Hohenbrink et al. [16] and total length

Locus	Cluster	Length
VN1R Mmur001	IV	909 bp
VN1R Mmur011	VI*	930 bp
VN1R Mmur031	V*	909 bp
VN1R Mmur033*	uncl	942 bp
VN1R Mmur040	II*	948 bp
VN1R Mmur041	V*	906 bp
VN1R Mmur043	VI*	1005 bp
VN1R Mmur048*	VI*	957 bp
VN1R Mmur049	VII*	879 bp
VN1R Mmur060*	V*	906 bp
VN1R Mmur065	uncl	1008 bp
VN1R Mmur066*	uncl	897 bp
VN1R Mmur067	III	921 bp
VN1R Mmur074*	IX*	918 bp
VN1R Mmur075	IV	909 bp
VN2R1	V2R	2739 bp
VN2R2*	V2R	2310 of 2418 bp[a]

[a] = locus 3 bp longer in *M. ravelobensis*, *: Loci and clusters under significant positive selection [15, 16], *uncl* unclustered

belongs to exon 3 which is longer in mouse lemurs than in mice. The analogous exon 4 of mice is missing in mouse lemurs, which is in concordance with genomic data of the two strepsirrhines *Daubentonia madagascariensis* and *Otolemur garnettii* where no "exon 4" was detected previously (see [15]), but a similar 22 bp insertion at the end of exon 3 was in fact present. In the present study we could not design external primers for exon 2 of *VN2R2* because of missing genomic data. Due to internal primers we are missing 108 bp (=13.2%) of the sequence of exon 2 and therefore results on *VN2R2* are based on 95.7% of the whole coding sequence.

We used MyTaq DNA polymerase for amplification (Bioline; 25 μl total volume containing 5.0 μl MyTaq Reaction Buffer, 1 μl of each primer [10 μM stock concentration], 0.1 μl Taq DNA polymerase [5 U/μl] and 1 μl of DNA) with the following PCR conditions: 94 °C for 2 min, 40 times (94 °C for 30 s, 60 °C for 45 s, 72 °C for 90 s), 72 °C for 5 min. PCR products were sequenced on both strands using BigDye Terminator 3.1 (Applied Biosystems) under standard conditions and run on an Applied Biosystems 3500 capillary sequencing machine. Consensus sequences were built with SeqMan 5.05 (DNASTAR Inc., Madison, WI, USA). Sequences were aligned and analysed using MEGA 5 [38]. Before data analyses we concatenated the exons of the *V2R*s.

Data analyses

We used DnaSP 5.10 [39] to unphase the two alleles of each diploid sequence and to identify the different haplotypes. Nucleotide diversity and haplotype diversity (expected heterozygosity, or gene diversity, [40]) were calculated with DnaSP 5.10 to estimate the genetic variation within the population. DnaSP 5.10 was used to estimate the number of polymorphic synonymous (p_S) and nonsynonymous substitutions (p_N) per site [41, 42] and p_N/p_S ratios were calculated. McDonald-Kreitman tests (= MKT, [43]) were conducted online [44] to calculate the ratio of fixed differences to polymorphic differences for synonymous and nonsynonymous SNPs. Here, all haplotypes of *M. murinus* and the most closely related haplotype sequence of *M. ravelobensis* were entered to test for positive selection in *M. murinus*, and vice versa to test in *M. ravelobensis*.

Arlequin 3.5 [45] was used to calculate Tajima's D and Fu's Fs with 1000 simulated samples. Population contractions or balancing selection can result in significant positive values of Tajima's D, whereas negative values indicate population growth [46, 47]. Fu's Fs show negative values if the data contains an excess of rare haplotypes also indicating population growth and/or positive selection [48]. In combination, positive D and positive Fs values indicate an excess of intermediate-frequency alleles after population subdivision or balancing selection, whereas negative D and negative Fs values reflect relative excess of rare variants and reveal population growth [49]. We used False Discovery Rate (FDR) to correct for multiple testing across both species for each analysis, setting FDR to 0.05 [50].

We analysed the distribution of nonsynonymous SNPs along the *V1R* protein [transmembrane, extra- or intracellular region; for details see 16]. Observed vs. expected χ^2-tests were used to compare the observed distribution of nonsynonymous SNPs in the *V1R* protein with the expected distribution using Statistica 6.1 (StatSoft, Inc., Tulsa, OK). A previous study on *V1Rs* in strepsirrhines reported that the ligand binding site of the *V1R* protein is potentially formed by about half of the 4 and 5th transmembrane region and the in-between 2nd extracellular loop (= 3rd extracellular region) [51]. This estimation was based on VR sequence data of cluster I only, but assuming structural similarities between clusters, we also tested if nonsynonymous substitutions were concentrated on the binding site proposed by Yoder et al. [51]. All statistical comparisons of dependent data between the two species were conducted with the Wilcoxon Matched Pairs test in Statistica. Here, the sample size was large enough to ignore *p*-value corrections that would have been necessary for smaller sample sizes [52]. No such analyses were performed for V2Rs since there is relatively little structural information available for them.

Arlequin 3.5 was also used to test two expansion models (demographic and spatial expansion) on the mismatch distributions of the haplotypes within this single population of each species that span the same spatial scale. A mismatch distribution is a distribution of the number of nucleotide mismatches between all pairs of nucleotide sequences of one locus within a given sample. For this study each individual (homozygous or heterozygous) always entered two sequences into the data pool. Mismatch distributions can be directly compared between loci or species in this study, because of the same sampling regime (see above), same spatial spread of the samples, and the same sample size of 40 nucleotide sequences per locus and species. The shape of a mismatch distribution has been shown to be influenced by demographic events like past expansions or population bottlenecks [53]. Mismatch distributions are bell-shape (= unimodal) in populations having increased in the past as a consequence of one demographic [53] or spatial expansion [54, 55] or L-shaped in case of a very recent size reduction [53]. A previous modelling approach [21] revealed that two successive expansions can generate a tri-modal mismatch distribution and the position of these modes corresponds to the time of the expansion. In contrast, populations at demographic equilibrium show a more ragged distribution [53] (= multimodal). Demographic expansions usually result from past genetic bottlenecks, whereas spatial expansions usually follow a colonisation event by relatively few founder individuals. We tested the expansion models available in Arlequin 3.5 to evaluate the evidence for a preceding colonisation event [54, 56]. The models also calculate τ-(*Tau*-) values that reflect the time of the expansion (in mutation units, $\tau = 2T\mu$), although exact time points are difficult to infer as reliable mutation rates are often not known. However, higher values of τ indicate that the expansion happened further in the past.

Results

Comparison of genetic diversity

For *V1R*, almost all comparisons showed substantially higher genetic diversity in *M. ravelobensis* than *M. murinus* (Table 2). Across all loci, *M. ravelobensis* possessed significantly more polymorphic sites than *M. murinus* (Wilcoxon-Test, $n = 15$, $Z = 3.04$, $p < 0.01$), more haplotypes (Wilcoxon-Test, $n = 15$, $Z = 3.15$, $p = 0.001$), higher nucleotide diversity (Wilcoxon-Test, $n = 15$, $Z = 2.78$, $p = 0.005$), higher haplotype diversity (Wilcoxon-Test, $n = 15$, $Z = 2.44$, $p = 0.015$) and higher numbers of protein alleles (Wilcoxon-Test, $n = 15$, $Z = 2.76$, $p = 0.006$, Table 2). All alleles at all loci had open reading frames of the expected length except locus *Mmur040* in *M. ravelobensis*, where an allele with an internal stop codon,

Table 2 Measures of genetic diversity for each locus and species

Locus	Number of haplotypes		No. diff. AA sequences		Nucleotide diversity		Haplotype diversity		No. of poly-morphic sites	
	Mmur	Mrav	Mmur	Mrav	Mmur	Mrav	Mmur	Mrav	Mmur	Mrav
001	7	10	3	8	.00073	.00493	.528	.787	4	13
011	3	4	3	3	.00194	.00061	.600	.442	4	3
031	5	9	5	5	.00141	.00194	.686	.708	6	9
033	3	9	2	9	.00089	.00445	.472	.877	3	15
040	6	10	5	10	.00074	.00420	.592	.622	5	21
041	4	9	2	4	.00141	.00202	.727	.745	4	10
043	4	5	3	3	.00048	.00070	.377	.503	4	4
048	11	9	6	3	.00170	.00072	.785	.474	10	9
049	2	9	2	8	.00033	.00180	.296	.819	1	9
060	3	9	2	6	.00016	.00265	.145	.732	2	14
065	1	2	1	1	.00000	.00014	.000	.142	0	1
066	4	17	3	14	.00139	.00560	.558	.954	5	28
067	2	14	1	9	.00015	.00417	.142	.859	1	15
074	6	12	6	10	.00093	.00475	.487	.777	8	25
075	7	9	1	6	.00098	.00506	.668	.709	7	21
Ø V1R	4.5	9.1	3.0	6.6	.00088	.00292	.471	.677	4.3	13.1
VN2R1	25	28	24	25	.00332	.00268	.973	.976	29	32
VN2R2	14	27	8	20	.00053	.00220	.838	.967	16	28
Ø V2R	19.5	27.5	16	22.5	.00193	0.0024	.906	.972	22.5	30

diff. AA sequences: number of different amino acid sequences, *Mmur*: *M. murinus*, *Mrav*: *M. ravelobensis*, Ø V1R/V2R: mean

indicating pseudogenization, occurred at a frequency of 6/40 = 0.15.

For both species, there were more haplotypes at V2R loci than V1R loci (Table 2). However, the number of polymorphic sites in V2Rs was increased more dramatically in *M. murinus* than in *M. ravelobensis*. The number of unique amino acid sequences was considerably lower than the number of haplotypes in VN2R2, but this was not the case in VN2R1 indicating that here most haplotypes differed by at least one nonsynonymous substitution. Overall, the genetic diversity in V1Rs and VN2R2 differs between the species, whereas it was equally high in VN2R1 of both species.

Tests of neutrality

Several loci in both species had significantly negative neutrality tests using uncorrected p-values (Table 3). However, using FDR, only three loci remained significant, all for Fu's Fs in *M. ravelobensis* (Mmur048, VN2R1 and VN2R2). No significantly positive values were found and test values did not differ significantly between the species (Tajima's D: $n = 15$, $Z = 0$, $p = 1.000$; Fu's F_s: $n = 15$, $Z = 0.68$, $p = 0.500$).

Tests for selection and distribution of mutations across VR proteins

The p_S of V1Rs was significantly higher than p_N in both species (*M. murinus*: $n = 15$, $Z = 2.92$, $p = 0.004$; *M.*

ravelobensis: $n = 15$, $Z = 2.39$, $p = 0.017$, Table 3). The p_N/p_S ratios of V1Rs did not differ significantly between species (Table 3; $n = 13$, $Z = 1.01$, $p = 0.311$). Using uncorrected p-values, McDonald-Kreitman Tests (MKT) were only significant for VN1R Mmur066 in *M. ravelobensis* ($p = 0.026$, Table 4), but this result was not robust to multiple testing correction using FDR. There was no significant correlation between the d_N/d_S of loci estimated across *Microcebus* species [16] and the p_N/p_S ratios determined intraspecifically in the current study (*M. murinus*: $n = 5$ loci, $r_s = 0.5$, n.s.; *M. ravelobensis*: $n = 5$ loci, $r_s = 0.0$, n.s.).

The distribution of nonsynonymous SNPs in the domains of the V1R protein did not differ significantly from the expected distribution in any species when looking at the data of all V1R loci combined (*M. murinus*: $\chi^2 = 3.2$, df = 2, $p = 0.200$; *M. ravelobensis*: $\chi^2 = 1.0$, df = 2, $p = 0.606$). Similarly, no single locus showed a significant deviation from the expected distributions (for example, VN1R Mmur066, the only locus with significant MKT: $\chi^2 = 2.6$, df = 2, $p = 0.268$; all other loci also with $p > 0.05$). A proposed odorant binding site [51] comprised about 20% of the V1R protein. There was no evidence for concentration of NS mutations in this region: across all loci, 21.9% of *M. murinus* NS substitutions and 19.7% of *M. ravelobensis* NS substitutions occurred here, and no single

Table 3 Results of the neutrality tests, and p_N/p_S ratios for each locus and species

Locus	Tajima's D				Fu's Fs				p_N/p_S	
	Mmur	p	Mrav	p	Mmur	p	Mrav	p	Mmur	Mrav
001	−0.70	0.287	1.46	0.939	−3.83	0.005	0.78	0.649	0.19	0.18
011	2.18	0.978	−0.42	0.398	3.96	0.941	−0.70	0.286	0.32	0.87
031	−0.24	0.469	−0.48	0.341	0.36	0.605	−2.05	0.147	0.91	0.11
033	0.42	0.700	0.60	0.763	1.42	0.806	1.21	0.743	0.20	1.13
040	−1.02	0.170	0.76	0.828	−2.38	0.055	2.13	0.815	0.29	2.53
041	0.86	0.804	−0.66	0.294	1.37	0.794	−1.91	0.160	0.11	0.35
043	−1.25	0.107	−0.74	0.272	−1.25	0.167	−1.49	0.134	2.43	0.05
048	−1.13	0.137	−1.98	0.007	−4.48	0.011	6.69	*0.000	0.39	0.08
049	0.37	0.807	−0.70	0.283	0.84	0.497	−2.44	0.103	—	0.53
060	−1.30	0.072	−0.86	0.222	−2.03	0.012	−0.89	0.362	0.15	0.14
065	0.00	1.000	−0.56	0.252	0.00	1.000	−0.22	0.209	—	0.00
066	0.16	0.608	−0.50	0.358	1.29	0.755	−2.34	0.218	0.31	1.62
067	−0.56	0.252	0.28	0.673	−0.22	0.218	−2.63	0.152	0.00	0.59
074	−1.56	0.024	−0.87	0.226	−1.77	0.102	−0.63	0.443	0.17	0.45
075	−1.27	0.094	−0.23	0.471	−2.69	0.051	1.61	0.776	0.00	0.24
VN2R1	1.15	0.914	−0.08	0.526	−6.78	0.022	−13.58	*0.000	0.19	0.49
VN2R2	−1.39	0.072	−0.78	0.249	−6.33	0.004	−17.25	*0.000	0.28	0.36

*: $p < 0.05$ with FDR of q = 0.05; Mmur: *M. murinus*, Mrav: *M. ravelobensis*, —: p_S was zero

Table 4 Results of MK tests and mismatch distributions for each locus and species

Locus	MK		τ_{demo}				τ_{spat}			
	Mmur p	Mrav p	Mmur	p	Mrav	p	Mmur	p	Mrav	p
001	0.538	0.379	0.75	0.47	8.63	0.09	0.75	0.35	6.28	0.33
011	0.599	0.998	5.02	0.11	0.58	0.84	3.85	0.13	0.56	0.68
031	0.272	0.597	1.88	0.69	3.82	0.88	1.65	0.81	0.18	0.79
033	0.051	0.821	0.00	*0.00	6.47	0.20	1.77	0.33	4.86	0.38
040	0.140	0.393	0.88	0.10	11.69	0.01	0.87	0.08	9.11	0.43
041	0.711	0.873	1.50	0.82	2.61	0.02	1.51	0.82	2.49	0.12
043	0.090	0.397	0.48	0.70	0.74	0.84	0.48	0.47	0.73	0.69
048	0.135	0.779	1.77	0.86	0.98	0.91	1.01	0.84	0.83	0.86
049	0.531	0.731	2.98	0.18	1.48	0.19	0.38	0.08	1.49	0.10
060	0.995	0.869	3.00	0.36	3.95	0.27	0.10	0.33	3.06	0.85
065	-	0.334	0.00	-	0.21	0.39	0.00	-	0.16	0.35
066	0.704	0.026	2.82	0.11	8.28	0.57	2.24	0.10	5.69	0.18
067	0.493	0.107	0.21	0.52	7.27	0.61	0.16	0.33	3.80	0.39
074	0.449	0.142	3.73	0.49	6.63	0.05	0.05	0.41	5.37	0.22
075	0.263	0.399	1.00	0.10	8.00	0.02	1.01	0.04	6.80	0.53
VN2R1	0.935	0.225	0.48	0.27	0.74	0.69	0.48	0.16	0.73	0.39
VN2R2	0.887	0.344	2.54	0.36	4.46	0.74	2.46	0.56	3.83	0.81

*: $p < 0.05$ with FDR of q = 0.05; Mmur: *M. murinus*, Mrav: *M. ravelobensis*, τ_{demo}: τ-values for demographic expansion model, τ_{spat}: τ-values for spatial expansion model, —: incalculable because of absence of variation

locus showed an excess of NS substitutions in this region.

Mismatch distributions

The mismatch distributions showed huge variation between loci and species (Fig. 1 and Additional file 2). In *M. murinus* most *V1R* loci ($n = 11$) had half-bell shaped distributions close to zero pairwise differences or the peak was at zero (Fig. 1a, Table 5, Additional file 2). Furthermore, *M. murinus* had four loci with unimodal distributions (*VN1R Mmur011, 033, 066* and *074*) but no locus with multimodal or ragged distributions (Fig. 1c). In contrast, three loci in *M. ravelobensis* showed ragged mismatch distributions (*VN1R Mmur001, 040,* and *067*, see Fig. 1b) and six had mostly broad unimodal distributions (*VN1R Mmur033, 041, 060, 066, 074* and *075,* see Fig. 1d). The remaining six *V1Rs* showed half-bell shaped distributions similar to *M. murinus*. Notably, unimodal distributions in *M. murinus* were still close to zero pairwise differences with a high peak, whereas they were generally broader and flat in *M. ravelobensis* (compare Fig. 1c + d).

The patterns of occurrence of half-bell shaped, unimodal and ragged distributions in the two species are summarized in Table 5. About half of the loci (8 of 15) had similar types of distributions in both species, but six of the remaining mismatch distributions showed higher variation in *M. ravelobensis* than in *M. murinus*. In *V2Rs*

Table 5 Pairwise occurrence of the three types of observed mismatch distributions for the *V1R* loci in both species

		M. murinus			
		HB	UM	RG	Σ
M. ravelobensis	**HB**	5	1	0	6
	UM	3	3	0	6
	RG	3	0	0	3
	Σ	11	4	0	15

HB: half-bell (L-) shape distribution, *UM*: unimodal distribution, *RG*: ragged distribution, *Σ*: Sum

(Additional file 2), the mismatch distributions in *M. murinus* showed a ragged distribution for *VN2R1* and a unimodal distribution for *VN2R2*. The distributions in *M. ravelobensis* were unimodal for both *V2Rs*.

Using uncorrected p-values, tests of demographic expansion were significant in one locus of *M. murinus* (*VN1R Mmur033*, $p = 0.00$) and three *V1R* loci of *M. ravelobensis* (*VN1R Mmur040*: $p = 0.01$; *VN1R Mmur041*: $p = 0.02$; *Mmur075*: $p = 0.02$), and tests of spatial expansion were significant in one locus of *M. murinus* (*Mmur075*: $p = 0.04$, Table 4). However, only one of these results remained significant after FDR correction – demographic expansion for *Mmur033* in *M. murinus*. The τ-values were significantly larger in *M. ravelobensis* compared to *M. murinus* using both models (Table 4; demographic expansion: $n = 15$, $Z = 2.38$, $p =$

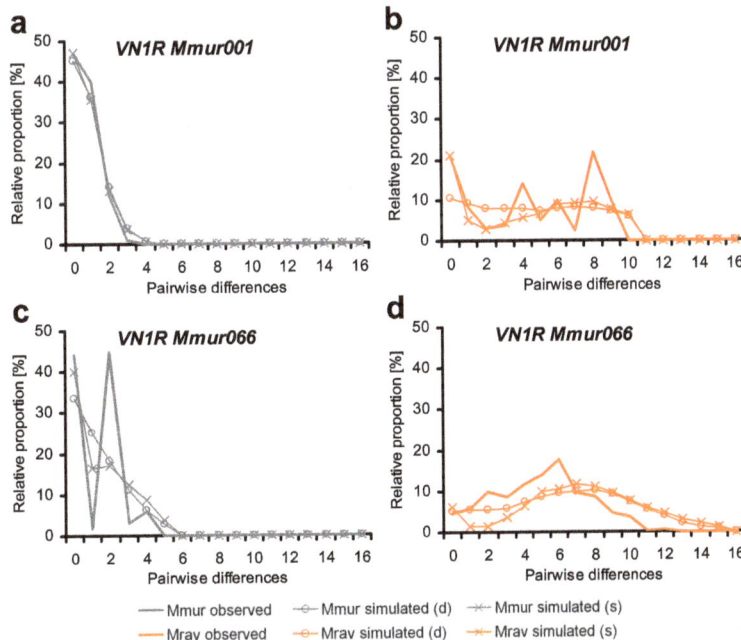

Fig. 1 Observed and simulated mismatch distributions of *VN1R Mmur001* and *066* in *M. murinus* (Mmur, left side, *grey*) and *M. ravelobensis* (Mrav, right side, *orange*); the two loci were selected to show the three observed types of distributions: half-bell shaped (**a**), unimodal (**c + d**) and ragged (**b**); mismatch distributions of the remaining loci are shown in Additional file 2; simulated (d) = simulated under demographic expansion model (line with *circles*), simulated (s) = simulated under spatial expansion model (line with *crosses*)

0.017; spatial expansion: $n = 15$, $Z = 2.44$, $p = 0.015$) indicating an older starting point of the putative expansion in *M. ravelobensis* compared to *M. murinus*.

Discussion

Selection in the recent evolutionary history of vomeronasal receptor genes

It was previously shown that the majority of *V1R* gene clusters in mouse lemurs evolved under strong positive selection and repeated gene duplication led to the evolution of a large *V1R* repertoire [16]. Positive selection still acted on *V1Rs* during the diversification of mouse lemurs as indicated by analyses of single *V1R* loci across different mouse lemur species [16] or of a *V1R* subfamily across different lemur species [51]. Positive selection, which was probably involved in generating this high *V1R* diversity in evolutionary timescales, could in principle be ongoing in present-day populations. In contrast to this expectation, the *VRs* we studied seem to be mostly evolving under purifying selection. Several results support the presence of purifying selection at the population level: 1) McDonald-Kreitman tests were mostly non-significant (with one possible exception, see below). 2) Neutrality tests were non-significant in most cases (see below for further discussion). 3) The p_N/p_S ratio was less than one in most loci of both species. 4) Nonsynonymous substitutions were randomly distributed within the *V1R* protein indicating neutral evolution rather than positive selection. This contrasts with our previous phylogenetic study where replacement substitutions were significantly concentrated at particular parts of the protein, consistent with odorant binding sites [16]. For one locus, *Mmur040* in *M. ravelobensis*, there is even evidence for ongoing loss of function, with an allele encoding a pseudogene segregating at appreciable frequency.

Only few loci showed patterns consistent with current positive selection. *VN1R Mmur066* in *M. ravelobensis* was the only example of a significant MK test with an excess of non-synonymous SNPs segregating in the population, although this was not robust to FDR correction. However, the high haplotype diversity and p_N/p_S ratio greater than one suggest that the possibility of balancing selection at this locus warrants further investigation. Three loci in *M. ravelobensis* showed significant evidence for departure from neutrality, with a negative Fu's Fs that was robust to FDR correction. The cause of this non-neutrality however is unclear – it may be due to directional selection or population expansion, which was not rejected for these loci from the mismatch distribution tests.

In a previous study we argued that diversification and positive selection on VR loci may be involved in reproductive isolation and speciation of mouse lemurs [16]. If

indeed functionally linked to reproduction, it would also be possible that the loci that are potentially under positive selection in the present study could be involved in olfactory mate choice or pheromonal communication in the context of finding a suitable mate (reviewed in [57]). Selection may thus act upon the VRs in the VNO and partly the main olfactory epithelium (MOE) [37]. Although some information is available on the function of certain *V1R* clusters in mice (e.g. [58, 59]), knowledge of the biological function of certain *V1Rs* or *V2Rs* are lacking completely for primates [16, 51]. In view of the rich and functional *V1R* repertoires of nocturnal strepsirrhines, future studies are urgently needed that shed light on their biological functions. Fruitful future experimental approaches may include behavioural assays involving individuals with polymorphisms at individual loci, and tests of the effect of odorants on the activity of VRs expressed in vitro. By contrast, the use of VNO tissue slices for immunohistochemistry or of anesthetised individuals for electrophysiological recordings, which has been a useful method in rodents [48], is not a viable option for these endangered species.

The higher haplotype diversity for *V2R* genes than *V1R* loci is interesting. The repertoire of *V1R* genes is much more diverse in mouse lemurs than their repertoire of *V2R* genes, with ~200 *V1R* loci [60] and only 2 *V2R* loci currently known in mouse lemurs [15]. It is possible that the paucity of *V2R* genes in the VNO is partly compensated by a high allelic diversity which also translates into a relatively high number of amino acid sequences and could therefore lead to a further increase of the olfactory sensory resolution in the VNO.

The presence of only weak evidence for ongoing selection at VR loci in these two mouse lemur species is in strong contrast to the prevalence of positive selection at the same loci in a comparative study [12]. This is one of the first studies to explicitly compare patterns of positive selection at these two levels, although there have been a few studies in humans [2, 3]. This obvious discordance between the results of the two studies suggests that fixation of beneficial mutations at VRs during mouse lemur evolution may have been highly episodic, i.e. occurring over short periods of time. However, one issue for further consideration is whether the power to detect positive selection is the same in the two approaches. Future studies would be needed to address this issue, for example using simulations.

Demographic history of two sympatric mouse lemur species

Under the assumption of a divergent phylogeographic history of both mouse lemur species in northwestern Madagascar, we predicted to find differences in the genetic diversity of the VRs between the two species. This prediction was confirmed by several datasets. The species with the longer phylogeographic history in the region, *M. ravelobensis*, possessed significantly more polymorphic sites, a higher number of haplotypes, a higher haplotype and nucleotide diversity, as well as a higher number of different amino acid sequences in its *V1R* loci than its congener with the supposedly shorter phylogeographic history in the region, *M. murinus*. Despite some degree of variability between loci, the results from these functional loci therefore support the conclusion of previous studies about the relatively recent expansion of *M. murinus* into northwestern Madagascar [21, 31] and suggest a distinct founder effect in this species. A similar interspecific difference was visible in the diversity of the two *V2R* loci, although these were generally more diverse in *M. murinus* than the *V1R*s.

Based on the scenario that *M. murinus* colonized northwestern Madagascar only sometime in the late Pleistocene [21, 31], we predicted to find signals of a stronger and/or more recent bottleneck in *M. murinus* than in *M. ravelobensis* which most likely evolved somewhere in this region earlier on [28] and may have maintained a larger ancient effective population size than the expanding founder population of *M. murinus*. As already noted above, although the majority of loci showed negative values in the summary statistics (Tajima's *D*, Fu's *Fs*) of both species, which would indicate population expansion under neutrality and/or positive selection, only very few loci deviated significantly from mutation-drift equilibrium. On the other hand, only *M. ravelobensis* showed significantly negative values of Fu's *Fs* in both *V2R* loci. However, the overall similarity in the patterns of these summary statistics does not support the hypotheses of a largely different demographic history of both species. There may be several reasons for these findings: first, it is possible that these results mostly reflect the most recent demographic history of both species that may have been rather similar in the late Pleistocene forests of northwestern Madagascar. The extent of the forest surface most likely contracted in all western lowland areas of Madagascar towards the last glacial maximum (LGM) and expanded again only afterwards [24–26]. During the LGM, all species, independent on whether they had a long or short phylogeographic history in these lowland forests, probably underwent population contractions and expanded again afterwards together with the forests [21, 26]. In addition, both tested mouse lemur species have most likely been equally affected by the anthropogenic habitat loss that started in large parts of Madagascar after the arrival of man within the last few thousand years and continues until today [61–63]. Second, it is possible that similar selection regimes in the two species acting across multiple loci could affect the allelic

diversity of VRs. However, this is unlikely, since there was little evidence for widespread directional selection across VR loci (see above).

In addition to the summary statistics discussed above, mismatch distributions were used to compare the distribution of haplotype diversity within both species. Three types of mismatch distributions were identified in both mouse lemur species, which differed in relative frequency: 1) Half-bell shaped or L-shaped distributions showing a lack of variation which indicates purifying selection or recent bottlenecks [53]; 2) Unimodal distributions that are seen after one demographic or spatial expansion [53]; 3) Ragged distributions that are typical for populations at demographic equilibrium when colonisation events are very old and diversification is not constrained. Although the two species shared the half-bell shape distribution in five loci, *M. ravelobensis* showed the more diverse types of distributions (*n* = 9) more often than *M. murinus* (*n* = 4). In accordance with the results on genetic diversity presented above, these findings are in agreement with the hypothesis of a larger ancient effective population size in *M. ravelobensis* that was able to maintain a larger degree of genetic variability across time than its congener *M. murinus*. The mismatch distributions of both species did not differ significantly from the simulated distribution after one demographic or spatial expansion in most loci (exception: one locus in *M. murinus*). However, the τ-values were significantly higher in *M. ravelobensis* indicating that the putative expansion of *M. ravelobensis* is older than that of *M. murinus*. A more precise estimation of the time since expansion is not possible, since reasonable mutation rates for *VR* loci are not available and evolutionary rates were shown to vary across the entire gene family [16]. Therefore, these species differences cannot be easily reconciled with the present knowledge on certain historic vegetation changes.

Conclusions

The current VR diversity of *M. murinus* and *M. ravelobensis* in northwestern Madagascar appears to be shaped by various processes such as divergent scenarios of population expansions, purifying selection, loss of function and a potential contribution from positive selection. Whereas strong positive selection, found in the whole *VR* repertoire (both *V1R*s and *V2R*s) and within individual gene clusters occurred in the past [16], ongoing selection may have shifted towards purifying selection in the majority of *V1R* loci to maintain the adaptive function of individual receptors, e.g. in the context of olfactory reproductive isolation between species as well as sex or kairomone recognition. This study only analysed a small subset of the large VR repertoire of mouse lemurs but gives important insights into the recent evolution of VRs and suggests a previously unknown shift in selection pressures acting on these functional genes that are probably of highest relevance for nocturnal solitary foragers that rely heavily on olfactory communication. Functional VR loci may not be best-suited for demographic modelling considering the difficulty of differentiating between signals of purifying selection and recent population bottlenecks. However, the simultaneous analysis of synonymous and nonsynonymous substitutions can help to disentangle these different processes. Future studies on the functional diversity and molecular evolution of olfactory receptor genes will certainly add substantially to our understanding of adaptive radiations, local adaptation and reproductive strategies in various mammalian clades whose life styles, social systems and reproductive strategies rely on pheromonal communication.

Additional files

Additional file 1: IDs of the sampled individuals; the table contains the IDs of all males and females that were used for sequencing 17 VR loci in the two mouse lemur species *Microcebus murinus* (n = 20) and *M. ravelobensis* (n = 20) in the study site JBA. (DOCX 14 kb)

Additional file 2: Observed and simulated mismatch distributions of *M. murinus* and *M. ravelobensis* under the demographic and the spatial expansion model; Graphic representation of the 15 mismatch distributions per species that were not included as examples in the main manuscript. The two species-specific distributions for each locus are displayed side by side (Mmur, left side, grey; Mrav, right side, orange). simulated (d) = simulated after demographic expansion model (line with circles), simulated (s) = simulated after spatial expansion model (line with crosses). (PDF 90 kb)

Acknowledgements
We thank the Institute of Animal Breeding and Genetics at the University of Veterinary Medicine Hannover for their technical help and Dr. Lounès Chikhi for commenting on an earlier version of this manuscript. We are particularly grateful to Sandra Thorén, Pia Eichmüller and Sarah Hohenbrink who collected the tissue samples in Madagascar.

Funding
This work was funded by the Volkswagen Foundation (grant number I/84 798).

Authors' contributions
Conception and design of the study: UR, PH, NIM. Genetic data collection: PH. Data analysis: PH, NIM, UR. Manuscript drafting: UR, NIM, PH. All authors contributed to the finalization of the manuscript and approved the final version.

Competing interests
The authors declare that they have no competing interests.

Author details
[1]Institute of Zoology, University of Veterinary Medicine Hannover, Buenteweg 17, 30559 Hannover, Germany. [2]Department of Zoology, University of Cambridge, Downing St, Cambridge CB2 3EJ, UK.

References
1. Vitti JJ, Grossman SR, Sabeti PC. Detecting natural selection in genomic data. Annu Rev Genet. 2013;47:97–120.
2. Evans PD, Gilbert SL, Mekel-Bobrov N, Vallender EJ, Anderson JR, Vaez-Azizi LM, Tishkoff SA, Hudson RR, Lahn BT. Microcephalin, a gene regulating brain size, continues to evolve adaptively in humans. Science. 2005;309(5741):1717–20.
3. Montgomery SH, Capellini I, Venditti C, Barton RA, Mundy NI. Adaptive evolution of four microcephaly genes and the evolution of brain size in anthropoid primates. Mol Biol Evol. 2011;28(1):625–38.
4. Mustonen V, Lassig M. From fitness landscapes to seascapes: non-equilibrium dynamics of selection and adaptation. Trends Genet. 2009;25(3):111–9.
5. Kosiol C, Vinar T, da Fonseca RR, Hubisz MJ, Bustamante CD, Nielsen R, Siepel A. Patterns of positive selection in six Mammalian genomes. Plos Genet. 2008;4(8):e1000144.
6. Mundy NI, Cook S. Positive selection during the diversification of class I vomeronasal receptor-like (V1RL) genes, putative pheromone receptor genes, in human and primate evolution. Mol Biol Evol. 2003;20(11):1805–10.
7. Shi P, Bielawski JP, Yang H, Zhang YP. Adaptive diversification of vomeronasal receptor 1 genes in rodents. J Mol Evol. 2005;60(5):566–76.
8. Glatston AR. Olfactory communication in the lesser mouse lemur (*Microcebus murinus*). In: Seth PK, editor. Perspectives in primate biology. New Delhi: Today and Tomorrow's Printers and Publishers; 1983. p. 63–73.
9. Perret M. Chemocommunication in the reproductive function of mouse lemurs. In: Alterman L, Doyle GA, Izard MK, editors. Creatures of the dark the nocturnal prosimians. New York: Plenum Press; 1995. p. 377–92.
10. Schilling A. Olfactoriy communication in prosimians. In: Doyle GA, Martin RD, editors. The study of prosimian behavior. New York: Academic Press; 1979. p. 461–542.
11. Buesching CD, Heistermann M, Hodges JK, Zimmermann E. Multimodal oestrus advertisement in a small nocturnal prosimian, *Microcebus murinus*. Folia Primatol. 1998;69(Suppl1):295–308.
12. Kappel P, Hohenbrink S, Radespiel U. Experimental evidence for olfactory predator recognition in wild mouse lemurs. Am J Primatol. 2011;73(9):928–38.
13. Perret M, Schilling A. Sexual responses to urinary chemosignals depend on photoperiod in a male primate. Physiol Behav. 1995;58:633–9.
14. Sündermann D, Scheumann M, Zimmermann E. Olfactory predator recognition in predator-naïve gray mouse lemurs (*Microcebus murinus*). J Comp Psychol. 2008;122(2):146–55.
15. Hohenbrink P, Mundy NI, Zimmermann E, Radespiel U. First evidence for functional vomeronasal 2 receptor genes in primates. Biol Lett. 2013;9(1):2012.1006.
16. Hohenbrink P, Radespiel U, Mundy NI. Pervasive and ongoing positive selection in the Vomeronasal-1 Receptor (V1R) repertoire of mouse lemurs. Mol Biol Evol. 2012;29(12):3807–16.
17. Excoffier L. Human demographic history: refining the recent African origin model. Curr Opin Genet Dev. 2002;12(6):675–82.
18. Kaessmann H, Heissig F, von Haeseler A, Paabo S. DNA sequence variation in a non-coding region of low recombination on the human X chromosome. Nat Genet. 1999;22(1):78–81.
19. Kawamoto Y, Takemoto H, Higuchi S, Sakamaki T, Hart JA, Hart TB, Tokuyama N, Reinartz GE, Guislain P, Dupain J, et al. Genetic structure of wild bonobo populations: diversity of mitochondrial DNA and geographical distribution. Plos One. 2013;8(3):e59660.
20. Quach H, Wilson D, Laval G, Patin E, Manry J, Guibert J, Barreiro LB, Nerrienet E, Verschoor E, Gessain A, et al. Different selective pressures shape the evolution of Toll-like receptors in human and African great ape populations. Hum Mol Genet. 2013;22(23):4829–40.
21. Schneider N, Chikhi L, Currat M, Radespiel U. Signals of recent spatial expansions in the grey mouse lemur (*Microcebus murinus*). BMC Evol Biol. 2010;10:105.
22. Burney DA. Madagascar's prehistoric ecosystem. In: Goodman SM, Benstead JP, editors. The natural history of Madagascar. Chicago: University of Chicago Press; 2003. p. 47–51.
23. Hewitt GM. Post-glacial re-colonization of European biota. Biol J Linn Soc. 1999;68(1-2):87–112.
24. Vences M, Wollenberg KC, Vieites DR, Lees DC. Madagascar as a model region of species diversification. Trends Ecol Evol. 2009;24(8):456–65.

25. Wilmé L, Goodman SM, Ganzhorn JU. Biogeographic evolution of Madagascar's microendemic biota. Science. 2006;312:1063–5.

26. Mercier JL, Wilmé L. The Eco-Geo-Clim model: explaining Madagascar's endemism. Madagascar Conserv Dev. 2013;8(2):63–8.

27. Hotaling S, Foley ME, Lawrence NM, Bocanegra J, Blanco MB, Rasoloarison R, Kappeler PM, Barrett MA, Yoder AD, Weisrock DW. Species discovery and validation in a cryptic radiation of endangered primates: coalescent-based species delimitation in Madagascar's mouse lemurs. Mol Ecol. 2016;25(9):2029–45.

28. Olivieri G, Zimmermann E, Randrianambinina B, Rasoloharijaona S, Rakotondravony D, Guschanski K, Radespiel U. The ever-increasing diversity in mouse lemurs: three new species in north and northwestern Madagascar. Mol Phylogenet Evol. 2007;43(1):309–27.

29. Weisrock DW, Rasoloarison RM, Fiorentino I, Ralison JM, Goodman SM, Kappeler PM, Yoder AD. Delimiting species without nuclear monophyly in Madagascar's mouse lemurs. Plos One. 2010;5(3):e9883.

30. Radespiel U. Can behavioral ecology help to understand the divergent geographic range sizes of mouse lemurs. In: Lehman SM, Radespiel U, Zimmermann E, editors. The dwarf and mouse lemurs of Madagascar: biology, behavior and conservation biogeography of the cheirogaleidae. Cambridge: Cambridge University Press; 2016.

31. Blair C, Heckman KL, Russell AL, Yoder AD. Multilocus coalescent analyses reveal the demographic history and speciation patterns of mouse lemur sister species. BMC Evol Biol. 2014;14(1):57.

32. Thiele D, Razafimahatratra E, Hapke A. Discrepant partitioning of genetic diversity in mouse lemurs and dwarf lemurs - Biological reality or taxonomic bias? Mol Phylogenet Evol. 2013;69(3):593–609.

33. Thorén S, Quietzsch F, Radespiel U. Leaf nest use and construction in the golden-brown mouse lemur (Microcebus ravelobensis) in the Ankarafantsika National Park. Am J Primatol. 2010;72(1):48–55.

34. Radespiel U, Jurić M, Zimmerman E. Sociogenetic structures, dispersal and the risk of inbreeding in a small nocturnal lemur, the golden-brown mouse lemur (Microcebus ravelobensis). Behaviour. 2009;146(4/5):607–28.

35. Radespiel U, Sarikaya Z, Zimmermann E, Bruford MW. Sociogenetic structure in a free-living nocturnal primate population: sex-specific differences in the grey mouse lemur (Microcebus murinus). Behav Ecol Sociobiol. 2001;50:493–502.

36. Untergasser A, Nijveen H, Rao X, Bisseling T, Geurts R, Leunissen JAM. Primer3Plus, an enhanced web interface to Primer3. Nucleic Acids Res. 2007;35:W71–4.

37. Hohenbrink P, Dempewolf S, Zimmermann E, Mundy NI, Radespiel U. Functional promiscuity in a mammalian chemosensory system: extensive expression of vomeronasal receptors in the main olfactory epithelium of mouse lemurs. Front Neuroanat. 2014;8:102.

38. Tamura K, Peterson D, Peterson N, Stecher G, Nei M, Kumar S. MEGA5: molecular evolutionary genetics analysis using maximum likelihood, evolutionary distance, and maximum parsimony methods. Mol Biol Evol. 2011;28(10):2731–9.

39. Librado P, Rozas J. DnaSP v5: a software for comprehensive analysis of DNA polymorphism data. Bioinformatics. 2009;25(11):1451–2.

40. Nei M. Molecular evolutionary genetics. New York: Columbia University Press; 1987.

41. Jukes T, Cantor C. Evolution of protein molecules. In: Munro HN, editor. Mammalian protein metabolism III. New York: Academic Press; 1969. p. 21–132.

42. Nei M, Gojobori T. Simple methods for estimating the numbers of synonymous and nonsynonymous nucleotide substitutions. Mol Biol Evol. 1986;3(5):418–26.

43. McDonald JH, Kreitman M. Adaptive protein evolution at the Adh locus in drosophila. Nature. 1991;351(6328):652–4.

44. Egea R, Casillas S, Barbadilla A. Standard and generalized McDonald-Kreitman test: a website to detect selection by comparing different classes of DNA sites. Nucleic Acids Res. 2008;36:W157–62.

45. Excoffier L, Lischer HEL. Arlequin suite ver 3.5: a new series of programs to perform population genetics analyses under Linux and Windows. Mol Ecol Resour. 2010;10(3):564–7.

46. Tajima F. The effect of change in population-size on DNA polymorphism. Genetics. 1989;123(3):597–601.

47. Tajima F. Statistical-method for testing the neutral mutation hypothesis by DNA polymorphism. Genetics. 1989;123(3):585–95.

48. Fu YX. Statistical tests of neutrality of mutations against population growth, hitchhiking and background selection. Genetics. 1997;147:915–25.

49. Bamshad MJ, Mummidi S, Gonzalez E, Ahuja SS, Dunn DM, Watkins WS, Wooding S, Stone AC, Jorde LB, Weiss RB, et al. A strong signature of balancing selection in the 5′ cis-regulatory region of CCR5. Proc Natl Acad Sci U S A. 2002;99(16):10539–44.

50. Benjamini Y, Hochberg Y. Controlling the false discovery rate: a practical and powerful approach to multiple testing. J R Stat Soc Series B Stat Methodol. 1995;57(1):289–300.

51. Yoder AD, Chan LM, dos Reis M, Larsen PA, Campbell CR, Rasoloarison R, Barrett M, Roos C, Kappeler P, Bielawski J, et al. Molecular evolutionary characterization of a V1R subfamily unique to strepsirrhine primates. Genome Biol Evol. 2014;6(1):213–27.

52. Mundry R, Fischer J. Use of statistical programs for nonparametric tests of small samples often leads to incorrect P values: examples from Animal Behaviour. Anim Behav. 1998;56:256–9.

53. Rogers AR, Harpending H. Population growth makes waves in the distribution of pairwise genetic differences. Mol Biol Evol. 1992;9(3):552–69.

54. Excoffier L. Patterns of DNA sequence diversity and genetic structure after a range expansion: lessons from the infinite-island model. Mol Ecol. 2004;13:853–64.

55. Ray N, Currat M, Excoffier L. Intra-deme molecular diversity in spatially expanding populations. Mol Biol Evol. 2003;20(1):76–86.

56. Schneider S, Excoffier L. Estimation of past demographic parameters from the distribution of pairwise differences when the mutation rates very among sites: Application to human mitochondrial DNA. Genetics. 1999; 152(3):1079–89.

57. Drea CM. D'scent of man: a comparative survey of primate chemosignaling in relation to sex. Horm Behav. 2015;68:117–33.

58. Isogai Y, Si S, Pont-Lezica L, Tan T, Kapoor V, Murthy VN, Dulac C. Molecular organization of vomeronasal chemoreception. Nature. 2011;478(7368):241–5.

59. Nodari F, Hsu F-F, Fu X, Holekamp TF, Kao L-F, Turk J, Holy TE. Sulfated steroids as natural ligands of mouse pheromone-sensing neurons. J Neurosci. 2008;28(25):6407–18.

60. Young JM, Massa HF, Hsu L, Trask BJ. Extreme variability among mammalian V1R gene families. Genome Res. 2010;20(1):10–8.

61. Burns SJ, Godfrey LR, Faina P, McGee D, Hardt B, Ranivoharimanana L, Randrianasy J. Rapid human-induced landscape transformation in Madagascar at the end of the first millennium of the Common Era. Quat Sci Rev. 2016;134:92–9.

62. Harper GJ, Steininger MK, Tucker CJ, Juhn D, Hawkins F. Fifty years of deforestation and forest fragmentation in Madagascar. Environ Conserv. 2007;34(4):325–33.

63. Zinner D, Wygoda C, Razafimanantsoa L, Rasoloarison R, Andrianandrasana HT, Ganzhorn JU, Torkler F. Analysis of deforestation patterns in the central Menabe, Madagascar, between 1973 and 2010. Reg Environ Change. 2013; 14:157–66.

2

Correlated duplications and losses in the evolution of palmitoylation writer and eraser families

Stijn Wittouck[1,2] and Vera van Noort[1]* iD

Abstract

Background: Protein post-translational modifications (PTMs) change protein properties. Each PTM type is associated with domain families that apply the modification (writers), remove the modification (erasers) and bind to the modified sites (readers) together called *toolkit domains*. The evolutionary origin and diversification remains largely understudied, except for tyrosine phosphorylation. Protein palmitoylation entails the addition of a palmitoyl fatty acid to a cysteine residue. This PTM functions as a membrane anchor and is involved in a range of cellular processes. One writer family and two erasers families are known for protein palmitoylation.

Results: In this work we unravel the evolutionary history of these writer and eraser families. We constructed a high-quality profile hidden Markov model (HMM) of each family, searched for protein family members in fully sequenced genomes and subsequently constructed phylogenetic distributions of the families. We constructed Maximum Likelihood phylogenetic trees and using gene tree rearrangement and tree reconciliation inferred their evolutionary histories in terms of duplication and loss events. We identified lineages where the families expanded or contracted and found that the evolutionary histories of the families are correlated. The results show that the erasers were invented first, before the origin of the eukaryotes. The writers first arose in the eukaryotic ancestor. The writers and erasers show co-expansions in several eukaryotic ancestral lineages. These expansions often seem to be followed by contractions in some or all of the lineages further in evolution.

Conclusions: A general pattern of correlated evolution appears between writer and eraser domains. These co-evolution patterns could be used in new methods for interaction prediction based on phylogenies.

Keywords: Post-translational modifications, Phylogenetic reconstruction, Tree reconciliation, Gene duplications, Gene losses, Correlated evolution

Background

Protein palmitoylation is a PTM that involves the addition of a 16-carbon saturated fatty acid, called palmitate, to a cysteine residue in a protein [7]. Due to the discovery of the palmitoylation writer enzyme family in the early 2000s and the recent developments in the application of large-scale MS to study various PTMs, including palmitoylation, this PTM has only recently been studied intensively. The primary function of the cysteine-attached palmitoyl group is to serve as a lipid anchor on soluble proteins, turning them into peripheral membrane proteins. Different classes

of lipids can function as lipid anchors [31], of which acyl groups and prenyl groups anchor proteins to the cytosolic side of a membrane. These acyl and prenyl groups often work together for stable and location-specific membrane attachment. Palmitoylation has a special position among them, as it is the only fully reversible lipid PTM, allowing for dynamics and regulation of protein-membrane interactions.

In addition to peripheral membrane proteins, integral membrane proteins are also frequently palmitoylated. Palmitoylated proteins are implicated in at least four classes of cellular processes. The first is the attachment of soluble proteins to the membrane and their localization to specific membrane compartments. The second function is the trafficking of membrane proteins between organelles and/or the plasma membrane (PM), and the third is the targeting

* Correspondence: vera.vannoort@kuleuven.be
[1]Centre of Microbial and Plant Genetics, KU Leuven, Leuven, Belgium
Full list of author information is available at the end of the article

of membrane proteins to specific parts of the PM such as postsynaptic clusters in neurons or lipid rafts. The fourth function is the stabilization of transmembrane proteins.

The principal protein family that is responsible for protein palmitoylation is the DHHC family of enzymes [18]. These are enzymes located in the membranes of the endomembrane system. They catalyze the transfer of a palmitoyl group from palmitoyl-CoA to a cysteine residue, forming a thioester bond. The first proteins of this family were discovered recently in yeast, and since then the family was found to be present in all eukaryotic species, with occurrences per genome ranging from less than ten in fungi to more than 20 in other eukaryotes [29]. The family is defined by its 51 residue DHHC domain, which is a variant of the C2H2 zinc finger motif. The DHHC domain containing proteins are characterized by i) a conserved sequence motif consisting of the residues aspartate, histidine, histidine and cysteine (DHHC) ii) six conserved cysteines iii) four to six transmembrane domains, the DHHC domain itself being located between two pairs of these on the cytosolic side of the membrane. This is compatible with the palmitate anchoring proteins to the intracellular side of the membrane in most cases.

Two small protein families are currently known to perform protein depalmitoylation: the acyl protein thioesterases (APTs) and protein palmitoyl thioesterases (PPTs) [7, 45]. In a structural/evolutionary classification of protein families, both are part of the alpha/beta hydrolase superfamily [32]. They make use of a nucleophile-acid-histidine catalytic triad. The APT and PPT families are sometimes situated within the serine hydrolase superfamily [38].

The APTs are cytosolic proteins [7]. Three of them have been found in humans: APT1, APTL1 and APT2. Originally they were identified as lysophospholipases (enzymes that deacylate monoacylated phospholipids), but recently they have all been found to perform depalmitoylation. APT1 is the best studied example [45]. It deacylates peripheral as well as integral membrane proteins, the classic example being signaling proteins of the Ras family. It shows some selectivity for substrates, and its efficiency varies across its substrates. It is strongly conserved among eukaryotes and, as opposed to proteins of the PPT and DHHC families, also present in prokaryotic species.

Enzymes of the PPT family have been shown to be targeted to the lysosomes [45]. They are conserved in eukaryotes. Humans have two of them: PPT1 and PPT2. They can both catalyze the depalmitoylation of palmitoyl-CoA, but only PPT1 is capable of depalmitoylating proteins. It has been shown that PPT2 is incapable of protein depalmitoylation [3]. PPT1 has been heavily studied because it has been identified as the causative gene of the disease infantile neuronal ceroid lipofuscinosis [24]. Neuronal ceroid lipofuscinoses (NCL) are a set of diseases characterized by an aggregation of so-called lipofuscin granules in neurons. These lipofuscin granules are composed of residues from lysosomal digestion. The nature of infantile NCL as lysosomal storage disorder seems very compatible with the function of PPT1 as a lysosomal enzyme. In non-diseased individuals, the protein is glycosylated at three places and transported to the lysosomes via the mannose 6-phosphate (M6P) pathway [24]. Nevertheless, many studies have shown that PPT1 is also implicated in processes outside of the lysosome. In neurons, the enzyme has sometimes been found in synaptic vesicles. It is often secreted (also in non-neuronal cells) and endocytosed again via the M6P receptor. However, much of the function of the protein remains to be discovered.

A lot less is known about protein depalmitoylation than is known about palmitoylation; it is very likely that other depalmitoylating enzymes remain to be discovered [7].

Explaining the seemingly irreducible complexity of writer-eraser-reader systems is an important challenge that needs to be addressed for every PTM type under investigation. In the case of tyrosine phosphorylation, it has been speculated that primitive PTPs in species without PTKs or reader domains are active to dephosphorylate Tyr residues that have been *accidentally* phosphorylated by promiscuous Ser/Thr kinases [30]. A prototypical example of this phenomenon is seen in *Saccharomyces cerevisiae*, where accidentally occurring pTyr residues exert unwanted allosteric effects, causing selective pressure for their removal [28].

Similarly to tyrosine phosphorylation, the very first form of lysine acetylation was also probably accidental. Acetylation has been shown to occur non-enzymatically in vitro and probably also in vivo [44]. In *E. coli*, acetylation levels reflect levels of acetyl-CoA present in the cell. This cofactor reflects, in turn, the nutrient status of the cell due to its central position in cellular metabolism. Deacetylation is performed by sirtuins, which allows for further regulation (i.e. in response to other signals than acetyl-CoA levels). The earliest acetylation system in the last universal common ancestor could have been similar to this system in *E. coli*. In this evolutionary scenario, the acetylation writer and reader domains have been later additions to this primitive but already functional PTM system using only eraser enzymes.

While for some PTM types like tyrosine phosphorylation and lysine acetylation there have been speculations on their stepwise origin, this is not the case for most of the PTM types, including palmitoylation. In this work we study the origin and evolution of the palmitoylation toolkit enzymes. The main aims are to identify lineage-specific family expansions and contractions and relate those to protein palmitoylation functionality and secondly to identify if the evolutionary histories of palmitoylation toolkit enzymes are correlated. If more studies like this accumulate it will become clear if the evolution of PTM systems on the long timescale follows general trends, such starting out as an accidental modification that gives rise to erasers as first of the toolkit domains.

Results

Profiles of (de-)palmitoylating protein families

We first searched the literature for proteins for which there is experimental evidence for palmitoylating or depalmitoylating enzymatic activity (Fig. 1a). We found five members in three species of the APT family with depalmitoylating activity (Table 1), four members in four species of the PPT family with depalmitoylating activity and 30 members in two species of the DHHC family with palmitoylating activity. The APT1 of *Saccharomyces cerevisiae* and *Toxoplasma gondii* are homologous to human APT1. For one of the palmitoylating enzymes ZDHHC13 the activity is still doubtful; the protein has

been shown to be autopalmitoylated but there is no direct evidence for its palmitoylating activity. There is one more human copy of the DHHC family ZDHHC11B but its activity has not been studied.

BLAST searches in proteomes of completely sequenced organisms resulted in 1576 DHHC sequences, 144 APT sequences and 136 PPT sequences. Based on a subset of these homologs, we generated multiple sequence alignments and manually corrected those. These seed alignments contained 159 DHHC sequences, 134 APT sequences and 128 PPT sequences. We used the corrected alignments to construct Hidden Markov Models (HMMs) for each of the three families (Fig. 1b–d). The DHHC

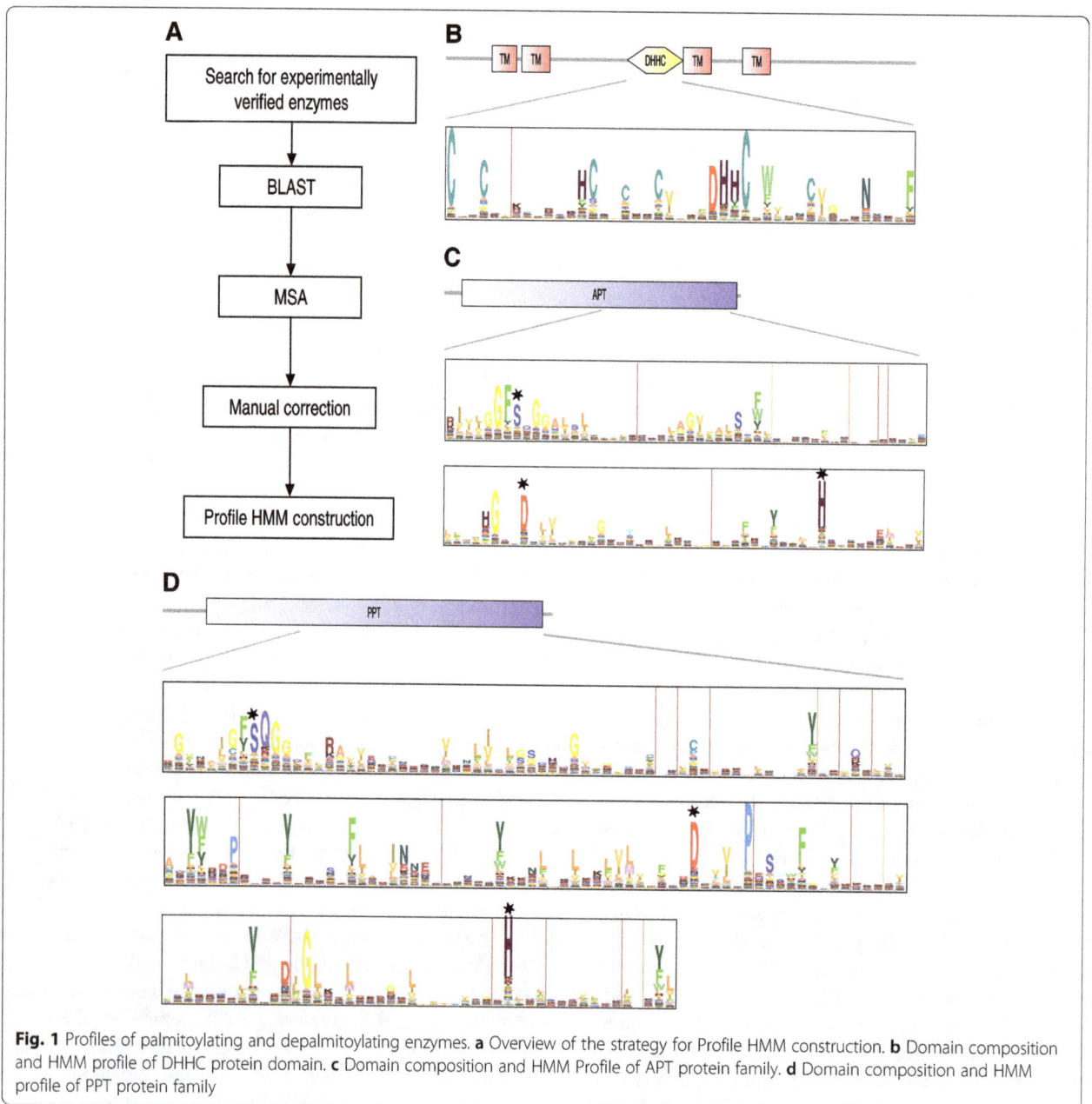

Fig. 1 Profiles of palmitoylating and depalmitoylating enzymes. **a** Overview of the strategy for Profile HMM construction. **b** Domain composition and HMM profile of DHHC protein domain. **c** Domain composition and HMM Profile of APT protein family. **d** Domain composition and HMM profile of PPT protein family

Table 1 Palmitoylation writer and eraser proteins of the APT, PPT and DHHC families of (de-)palmitoylating enzymes whose activity has been experimentally demonstrated

Family	species	gene	Reference
APT	Saccharomyces cerevisiae	APT1	[13]
	Homo sapiens	APT1	[11]
		APTL1	[41]
		APT2	[42]
	Toxoplasma gondii	APT1	[21]
PPT	Bos taurus	PPT1	[5]
	Homo sapiens	PPT1	[9]
	Drosophila melanogaster	Ppt1	[17]
	Caenorhabditis elegans	ppt-1	[35]
DHHC	Saccharomyces cerevisiae	AKR1	[23]
		ERF2	[1]
		SWF1	[43]
		PFA3	[39]
		PFA4	[26]
		PFA5	[19]
	Homo sapiens	ZDHHC1	[41]
		ZDHHC2	
		ZDHHC3	
		ZDHHC4	
		ZDHHC5	
		ZDHHC6	
		ZDHHC7	
		ZDHHC8	
		ZDHHC9	
		ZDHHC11	
		ZDHHC12	
		ZDHHC13	
		ZDHHC14	
		ZDHHC15	
		ZDHHC16	
		ZDHHC17	
		ZDHHC18	
		ZDHHC19	
		ZDHHC20	
		ZDHHC21	
		ZDHHC22	
		ZDHHC23	
		ZDHHC24	

residues as well as six conserved cysteines can clearly be observed in the profile of the DHHC family (Fig. 1b) that has 43 match states. Other conserved residues are a histidine at position 13, an aromatic residue at position 30,

asparagine at 39 and phenylalanine at 43. The APT and PPT profiles are much longer: 207 and 246 match states respectively (Fig. 1c–d). The actual domains are even slightly longer, because there are some insertion states that span multiple positions. Apart from the catalytic triads, the two profiles show other common characteristics. For example, the catalytic serine is located in an area of conserved hydrophobic residues. They also both contain many sites with conserved aromatic residues (green residues in the logos), especially the PPT profile that has more than 10 of these.

The length of the DHHC family being short is consistent with the multi-domain nature of these proteins (Fig. 1b) as opposed to the single domains of which the APT and PPT family consist. Palmitoyltransferases contain next to the DHHC domain, multiple Transmembrane domains that anchor them to the membrane.

Phylogenetic distribution
We used the HMM-profiles to search for protein family members in completely sequenced genomes of 119 eukaryotic and 1008 prokaryotic species. A length-score distribution was used to determine a threshold for inclusion and proteins not containing essential residues were removed. We found that metazoan genomes contain a large number of DHHC encoding genes, although there is considerable variation between species. Most of them have 10 to 25 DHHC genes, whereas vertebrate genomes contain at least 15 DHHCs. All eukaryotic species contain between 1 and 4 APTs (with one exception of zero) and zero to four PPTS (with one exception of six). Fungal genomes contain the smallest number of DHHCs; only three to seven.

Within the green plants (Viridiplantae) there is a clear distinction between two classes; the Chloryphyta have few DHHCs, one APT and one PPT whereas the Embryophyta (land plants) have many DHHCs, four or five APTs and two or three PPTs (Fig. 2). Only the APT family was found outside of eukaryotes with some proteobacteria having one or two copies of this family.

From the phylogenetic distribution (Figs. 2 and 3) it is already clear that the number of copies in a genome of each of the enzyme families is correlated, the correlation being 0.58 between DHHC and APT; 0.43 between DHHC and PPT and 0.53 between APT and PPT. However, such correlation could be due to few phylogenetic events; the distributions of DHHC, APT and PPT genes in extant genomes are not phylogenetically independent.

Duplications and losses in gene family trees
We used phylogenetic reconstructions and tree reconciliation to identify phylogenetically independent duplication and loss events and mapped these to a species tree. Bias in inferred duplication and losses can be introduced by small errors in gene trees. To reduce this bias, we performed gene tree rearrangement such that branches that are

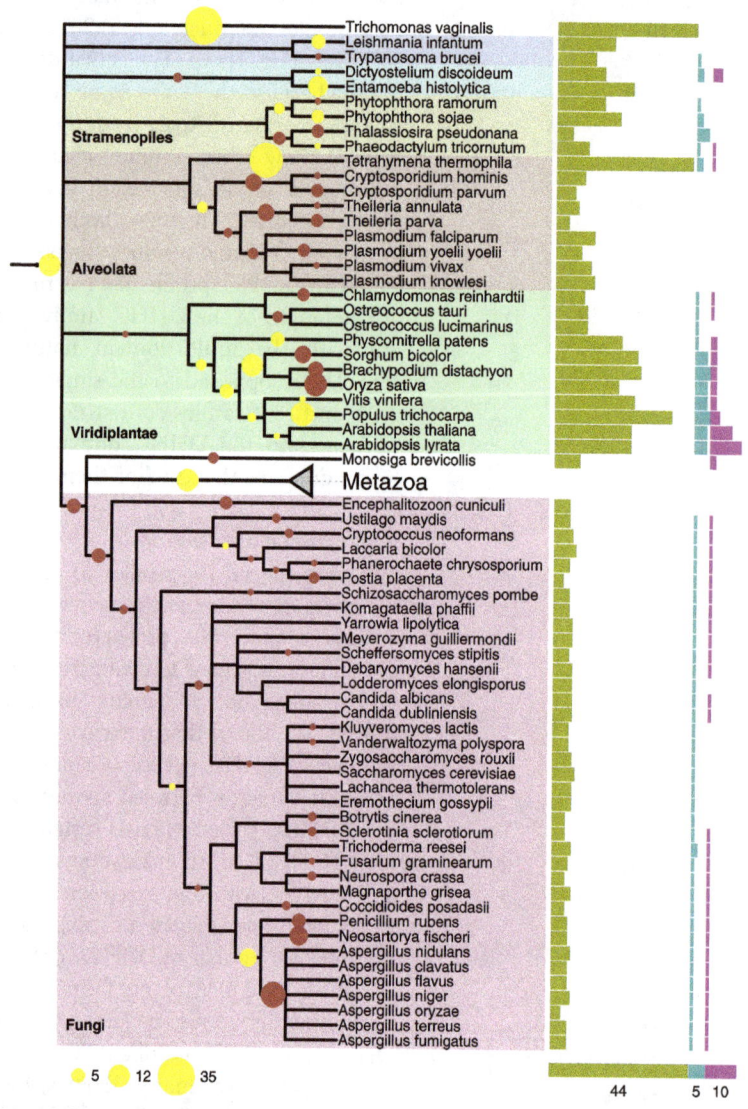

Fig. 2 Evolutionary history of DHHC family proteins. Tree reconciliation of the DHHC rearranged tree. *Green bars* indicate number of DHHC protein in the extant genome, cyan APT proteins, and magenta PPT proteins. *Yellow circles* indicate inferred increases in copy numbers of the DHHC family, *red circles* indicate inferred decreases in copy numbers of the DHHC family. The tree topology is extracted from NCBI taxonomy. Tree continues in Fig. 3

uncertain are rearranged in order to follow the species tree (see methods). If there are many duplications in a specific branch, the gene family is expanding whereas many losses in one branch result in contraction of the family. We identify an expansion of DHHC at the last eukaryotic common ancestor. This means part of the diversity in present day DHHC enzymes arose already in this earliest stage of eukaryotic evolution. At the root of the metazoan, the DHHC diversity was shaped by an early expansion followed by contractions. These contractions continued in the non-chordate eukaryotic species and led to their low DHHC numbers. Some of these species have a slightly larger number of DHHCs due to small late expansions. In the Chordata, the early eukaryotic

contractions were followed by expansions. One expansion is visible in the lancelet lineage, leading to the species *Branchiostoma oridae* . A second one is actually a stretch of expansions, starting at the Euteleostomi (Vertebrata) and continuing in two lineages: via the Clupeocephala until the Percomorphaceae ancestor and via the Tetrapoda until the Boreoeutheria ancestor. This stretch of expansions was followed by late losses in all lineages. This leads to the conclusion that the copy number of DHHCs peaked in at least three ancestral species. First in the common eukaryotic ancestor, that appears to have had a larger number of DHHCs than the single cell eukaryotes and Ecdysozoa living today. After that in the common ancestors of the Clupeocephala and Boreoeutheria, that both

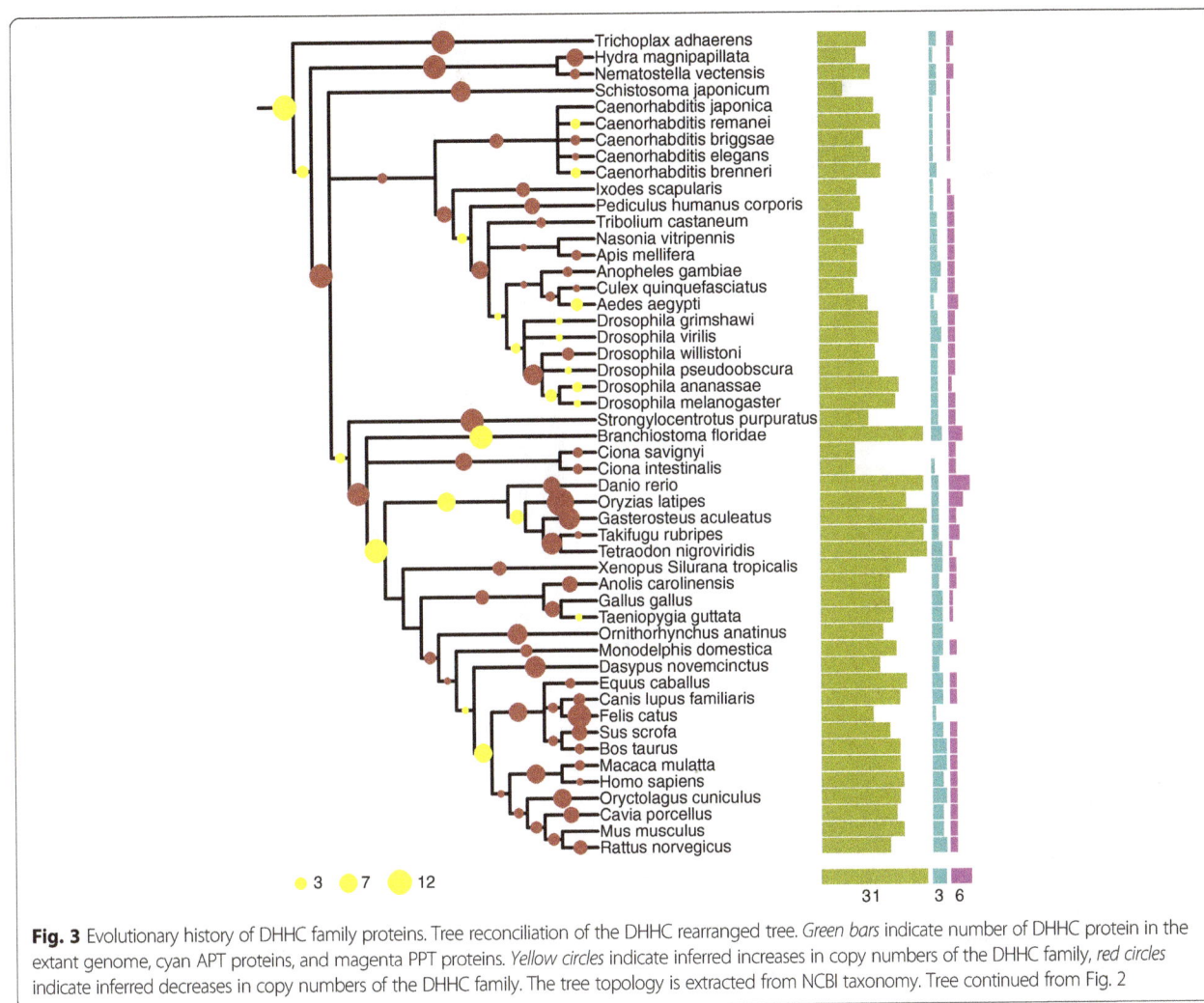

Fig. 3 Evolutionary history of DHHC family proteins. Tree reconciliation of the DHHC rearranged tree. *Green bars* indicate number of DHHC protein in the extant genome, cyan APT proteins, and magenta PPT proteins. *Yellow circles* indicate inferred increases in copy numbers of the DHHC family, *red circles* indicate inferred decreases in copy numbers of the DHHC family. The tree topology is extracted from NCBI taxonomy. Tree continued from Fig. 2

had larger DHHC counts than any eukaryotic species sequenced today. In the superphylum of the Alveolata, small contractions in the apicomplexan linages led to a small number of DHHCs.

In the APT and PPT families we also observe gains and losses but the copy number per genome is much smaller than the DHHC (Additional files 1 and 2: Figures S1 and S2). In the APT family we observe expansions at the last common eukaryotic ancestor, at the metazoan ancestor and the ancestor of the euteleostomi. The apicomplexa lost all APT family members, likely linked to their parasitic lifestyle.

Correlation between DHHC and APT/PPT evolution
The phylogenetic placing of evolutionary events of the DHHC, APT and PPT families are strikingly similar. At species tree branches with many DHHC duplication events and few losses, the APT family also often expanded (Fig. 4a). Conversely, at branches with few DHHC duplications and a lot of losses, the APT family often reduced (Fig. 4a). The association between the two families is only apparent for

species tree branches with a large number of duplications or losses; the pattern is not visible for species tree branches with less than five DHHC duplications and less than five losses.

We assessed statistical significance by considering the net expansions in the DHHC family for each internal branch and dividing them into three groups according to events in APT or PPT family; expansion, contraction or no change. We used the Wilcoxon rank-sum test for independent samples. The association between evolutionary events in the DHHC family with events in the APT family is significant (Fig. 4c, $p = 0.0011$). The association between DHHC and PPT evolutionary events (Fig. 4b) is even stronger (Fig. 4d, $p = 0.0001$). In addition, the association between the APT and PPT events is also significant (Fisher exact test; $p = 2e-7$).

For comparison, reconciliation of the original ML DHHC tree without rearrangement can be found in Additional file 3: Figure S3. As a negative control, we also inferred the evolutionary history of the histone deacetylase (HDAC) enzyme family that has 710 members in the genomes we analyzed. We tested its association with the APT and PPT families in

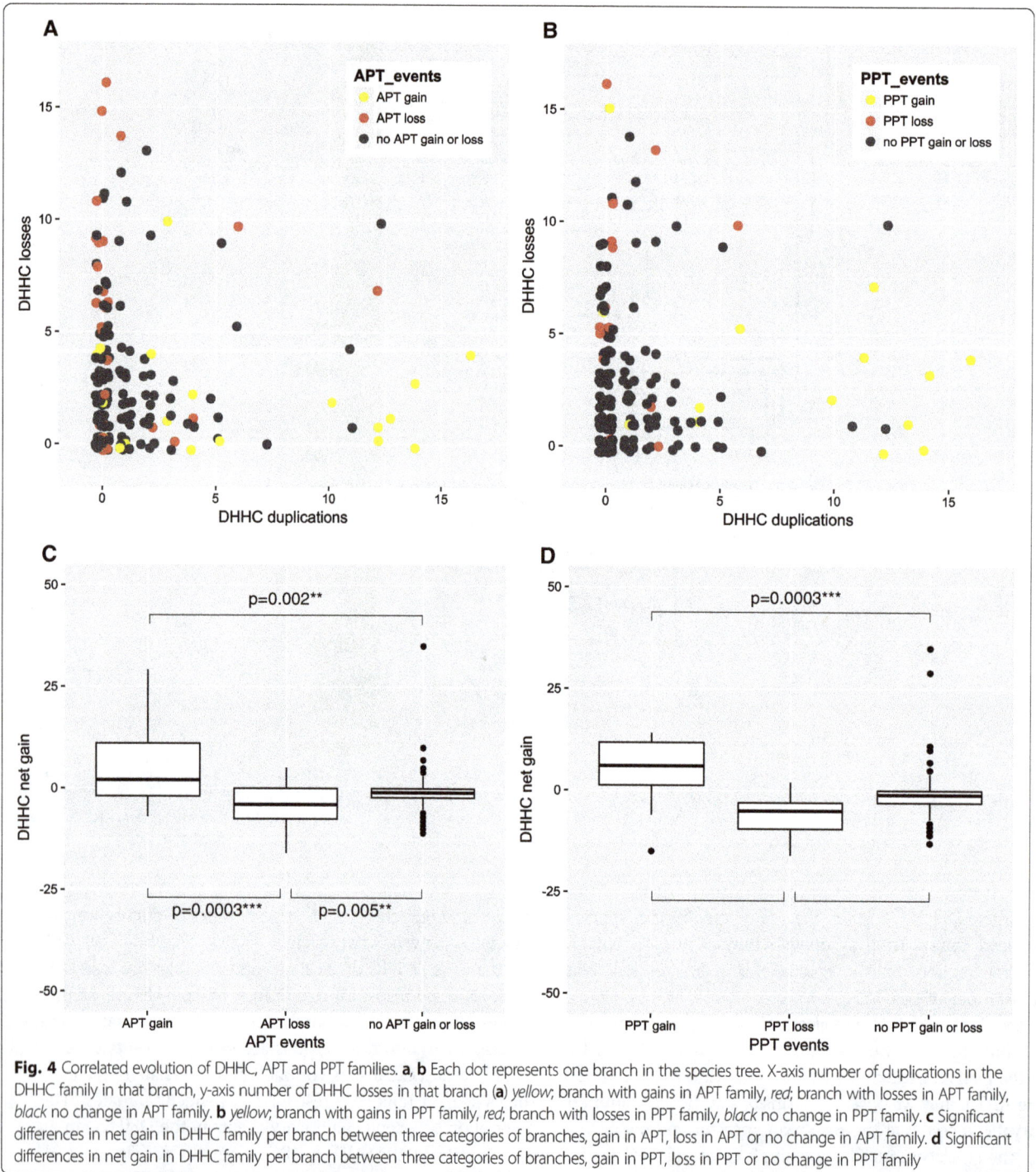

Fig. 4 Correlated evolution of DHHC, APT and PPT families. **a, b** Each dot represents one branch in the species tree. X-axis number of duplications in the DHHC family in that branch, y-axis number of DHHC losses in that branch (**a**) *yellow*; branch with gains in APT family, *red*; branch with losses in APT family, *black* no change in APT family. **b** *yellow*; branch with gains in PPT family, *red*; branch with losses in PPT family, *black* no change in PPT family. **c** Significant differences in net gain in DHHC family per branch between three categories of branches, gain in APT, loss in APT or no change in APT family. **d** Significant differences in net gain in DHHC family per branch between three categories of branches, gain in PPT, loss in PPT or no change in PPT family

the same manner as we did for the DHHC family. Although some level of association is observed, there is no significant correlation between HDAC net gain and changes in the APT and PPT families (Additional file 4: Figure S4).

Discussion

In this work we have analyzed the phylogenetic distribution of palmitoylating and depalmitoylating enzymes and show

that these families co-evolve. We find that the DHHC family is only present in eukaryotes. The APT family on the other hand is also present in bacteria. More specifically, we find two clusters of proteins matching the APT model: a high scoring cluster with exclusively proteins from the Proteobacteria, and a lower scoring cluster with proteins from Proteobacteria as well as other bacterial clades. The presence of APTs in bacteria strongly suggests that they arose before the DHHCs in the course of evolution. What could the function of this palmitoylation eraser family be without there being palmitoylation writer enzymes? A possibility is that they have evolved to remove accidental palmitoylation, as this modification has been shown to occur non-enzymatically. This removal of accidental modifications is similar to what has been speculated for tyrosine phosphorylation and lysine acetylation erasers [2].

In the DHHC family, we find consecutive periods of expansions and contractions. A first hypothesis to explain the expansion-contraction patterns is a temporary selection pressure for a larger number of DHHCs. In general, the adaptive expansion of a gene family can occur for two reasons: a dosage increase of the proteins or a functional diversification of the family [12]. Contractions of gene families are thought to be mainly the result of neutral selection; in other words, the loss alleles are fixated by random drift because they are not deleterious.

In this light, the question becomes why contractions of the DHHC family after its expansion are not disadvantageous. Actually, Hogeweg and co-worker have shown that in an evolving system, lineages with whole genome duplication are better able to adapt to a changing environment [10]. In case of expansion for dosage increase, it could be speculated that gene expansion is the fastest way to achieve this dosage increase. Amplification of an, initially, low-efficiency enzyme can result in adaptive mutations to arise in the enzymatic function or the regulation in any of the gene duplicates. As the occurrence of adaptive mutations is limited by the per base mutation rate, a duplication increases the options to adapt. Over time, the expression levels or the enzymatic efficiency of the individual genes is optimized to the new function, rendering the extra copies superfluous. In case of functional diversification, the explanation for gene loss might be that after new and improved types of DHHCs have evolved from duplicated genes, they partly replace the functions of the conserved types, also rendering them obsolete.

The co-occurrence of DHHC and APT expansions fits in this selection hypothesis in two possible ways. First, the APT expansions likely comprised only one or two duplications. The evolution of a new palmitoylation eraser enzyme might have created opportunities for more extensive use of palmitoylation in general, leading to selection for more DHHCs. Alternatively, another molecular invention or a change in some environmental factor might have created a selection pressure for the palmitoylation machinary in general, including both writers and erasers.

In contrast to the APT family, the strong co-evolution of the PPT family with the DHHC and APT families was rather unexpected, for a number of reasons. Firstly, it is fairly certain that PPTs reside in the lysosomes. Therefore, they are unlikely to participate in any signaling processes

and it is unclear why their diversification in function might be useful in case of increased use of protein palmitoylation, although selection pressure for duplications for dosage increase is conceivable. And secondly, very few PPT enzymes are experimentally confirmed as palmitoylation erasers, and for some of them it has even been shown that they are not capable of this function, such as human PPT2. If the selection scenario is correct, this might be an indication that protein depalmitoylation is in fact the main function of the PPT family, and that these non-depalmitoylating PPTs are rather the exception.

A pattern of expansion followed by contractions is often seen at whole-genome duplications (WGDs). A possible pitfall of this analysis is that we simply observe the effect of WGDs. In Metazoa, known WGDs occurred in the common ancestor of the Vertebrata and in the common ancestor of the teleost fishes [34]. Also in the history of land plant evolution, one or more WGDs are known to have occurred at the origin of multiple clades or species that are present in our data: the Poaceae, eudicots, *Arabidopsis thaliana*, *Populus trichocarpa* and *Physcomitrella patens* [34]. Although some overlap is visible with DHHC expansions and WGD events, DHHC expansions also occurred when no WGD took place and gene family expansions continue in clades after WGD events.

Conclusions

This study and previous studies suggests that the functional link between writer and eraser domains is reflected in correlated evolution on the level of duplications and losses. Conversely, this information could be used to infer functional relationships. So far, prediction of functional interaction based on phylogenetic information has been based on correlated sequence evolution or correlated presence-absence profiles (for a review see [22]) but not on duplications and losses. These methods work well specifically for bacteria and archaea but not as well in eukaryotes [15]. The co-evolution patterns we find here, could be employed to further develop functional interaction prediction methods specific to eukaryotes.

Methods
Experimentally validated enzymes
The literature resources Scopus and KU Leuven LIMO were searched for research articles describing DHHCs, APTs and PPTs using the search terms protein acylation; protein depamitoylation; protein acyl thioesterase and palmitoyl protein thioesterase . We read the articles and stored protein and species names of characterized enzymes in a table.

Protein sequences
We downloaded all protein sequences from the STRING database (version 9.1) [16]. Locally installed BLAST (version

2.2.30) [4] was used to find homologs of the experimentally validated APT and PPT proteins, with an *e*-value of 10^{-50}. To find homologs of the DHHC family an *e*-value cut-off of 10^{-10} was used. Other search paramaters were default; gap opening penalty 11, gap extension penalty 1, word size 6 and the BLOSUM62 substitution matrix. For the APT and PPT families, the hydrolase domains make up the largest part of the proteins. Thus, whole protein sequences were used as queries. For the DHHC family only the DHHC domain was used as query. Non-redundant BLAST results were collected together for each set of experimentally confirmed enzymes with the same function.

Construction of seed alignments

The MAFFT package version 7 [20] contains three algorithms: the FFT-NS-i, L-INS-i and G-INS-i. All of these are based on a progressive alignment using a guide tree, followed by iterative refinement. FFT-NS-i is the fastest of the three methods. L-INS-i and G-INS-i are slower but more accurate.

The BLAST search results of the DHHC family resulted in a large number of protein sequences.

Not all of these sequences are needed to capture the common characteristics and the diversity of the family. Taking a subset has the advantage of being able to create a more accurate alignment, more atypical sequences are excluded and it is easier to remove problematic sequences.

We implemented a subset function in R and makes use of the R package Phangorn [37]. The function starts with reading the MSA and the construction of a distance matrix, making use of maximum likelihood distance estimation and the LG model of amino acid substitution. Then follows an iterative process of two steps. First, the two sequences with the smallest distance between them are identified. And then, of these two, the sequence with the largest total distance to all other sequences is removed. These two steps are repeated until the number of sequences is reduced to the required number. For the DHHC family, the BLAST search results were first aligned with the fast MAFFT FFT-NS-I method. Then, a subset of 200 sequences was extracted using the subset method.

The sequences in this subset were then aligned with the accurate L-INS-i algorithm. Next, the alignment was inspected and doubtful sequences were removed. The criteria for inclusion were: the presence of multiple cysteine residues in the DHHC domain, the DHHC motif itself and the 2x2 transmembrane structure of the proteins. For the prediction of the transmembrane structures, we used the TMHMM server, version 2.0 [25].

For the alignments of the APT and PPT families, the G-INS-i algorithm was used. While for the DHHC family, L-INS-i seemed to give better results than G-INS-i, the opposite was true for the PPT and APT families. The reason for this is that for these families, a much larger portion of

the sequences was alignable, and thus it is appropriate to include global instead of local pairwise alignment information in the iterative optimization process. We made accurate alignments for the APT and PPT families by making use of the multithreading option implemented in MAFFT. First, we aligned the full set of BLAST results with an extended G-INS-i algorithm, using 10,000 optimization cycles. Then we manually inspected the resulting MSA. We made small corrections to the alignment and excluded some sequences based on the knowledge that the catalytic triad is an essential feature of the protein family.

Profile HMMs

For the construction of profile HMMs and the subsequent database searches using these profiles, we used the HMMER software, version 3.1b1 [14]. By default, the hmmbuild command will select all columns of the alignment that contain less than 50% gaps and model these as the match states. To increase the specificity of the search, columns with many gaps or low conservation were excluded. A strict non-gap percentage threshold of 80% and a conservation threshold of $5*10^{-6}$ were applied. These selection criteria were implemented by the trimAl software, version 1.2rev59 [6]. The boundaries of the family domain were determined by visual inspection of the multiple sequence alignments and columns outside of these boundaries were excluded. Skylign was used to visualize the profile HMMs.

HMMER searches

The constructed profile HMMs were used to search the protein sequence database. The domain list output of hmmsearch was used for further analysis. Inclusion thresholds were set based on i) visual inspection of the length-score plot, ii) sequence characteristics of the results iii) functional annotations. After setting the inclusion thresholds domain hits were removed that did not contain essential parts of the domain, the DHHC motif or the catalytic triad.

Phylogenetic reconstruction

To construct a multiple alignment of the DHHC family, first complete domain hits without large insertions or deletions were aligned with each other with the accurate L-INS-i algorithm of MAFFT. Other sequences were added with the −add option. Prealigned L-INS-i DHXC sequences were added with the −addprofile option. Alignment columns with less than 0.5% residues (99.5% gaps) were removed to save computation time. The APT and PPT sequences were aligned in one step with the MAFFT G-INS-i algorithm with 10,000 optimization cycles.

For the construction of all phylogenetic trees, we used version 8.1.21 of RAxML [40]. The Pthreads implementation was used for parameter optimization, while we used the hybrid implementation for the actual tree inferences.

We determined the optimal protein substitution model using a script provided by A. Stamatakis on the RAxML website. The script determines the substitution model that results in the highest likelihood value of a fixed maximum parsimony (MP) tree. The initial rearrangement parameter determines the depth of the tree search in each iteration of the search algorithm. The RAxML software contains an option to determine this parameter automatically, but the manual advises to test this automatic option versus a fixed setting of 10 for a couple of trees. The option (fixed or automatic) that results in the tree with highest likelihood value should be used for the final tree inferences. We generated five MP starting trees and tested both the automatic and fixed options on each of the trees. A fixed rearrangement setting of 10 gave the best results for all families.

For the construction of phylogenetic trees for fewer than 1000 sequences, the RAxML manual advises the use of the analyses invoked by the -f a option, which combines rapid bootstrapping with an extensive search for the ML tree. We used this strategy for the APT and PPT families. The algorithm starts by computing bootstrap trees with RBS enabled. It then uses these bootstrap trees as starting points to explore the tree space in three steps, increasing the depth of the search but decreasing the number of trees in each step. First, it does a fast search, using every fifth bootstrap tree as a starting tree. In the second step, the ten most promising results of the fast searches are further improved by doing slower optimizations. Finally, the best of these resulting trees is thoroughly optimized. This last step always uses the gamma model of rate heterogeneity, even when the CAT option is specified. The approach above is less attractive for large trees. One reason for this is that it takes a lot of computation time, and another that the required time is impossible to predict. Therefore we used a slightly different strategy for the DHHC family. We started by computing bootstrap trees in batches of 100 (with RBS enabled). After each batch, we combined the bootstraps of all batches and tested the MRE criterion. We stopped when the criterion converged.

To find the ML tree, we did 20 searches starting from independent parsimony starting trees. To estimate the irregularity of the likelihood surface, we calculated the average Robinson-Foulds distance (RF) as well as the average WRF between all trees. This resulted in a RF of 10.9% and a WRF of 2.8%. These values indicate that while the trees differ quite substantially in general, they are very similar at their highly supported branches. For this reason, we did not perform extra ML tree searches. To obtain the final tree, we picked the ML tree with the highest likelihood between these 20 trees and the trees obtained in the process of tuning the rearrangement setting. We then used the -f b option to draw the bootstrap confidence values on this tree.

RAxML tree construction algorithms always produce unrooted trees. We used a rooting algorithm built into RAxML. It uses a variant of midpoint rooting; the tree is rooted in such a way that the sums of the branches of both subtrees of the root are equal.

Tree rearrangement, reconciliation and mapping

The first step was the preprocessing of the ML gene tree using R. We rooted the tree with bootstrap values on the root given by the RAxML rooted tree. We gave names to the internal nodes of the tree to make later tracking easier and uniform. Tree rearrangement was carried out with NOTUNG [8]. NOTUNG is a gene tree-species reconciliation software package that supports duplication-loss event models with a parsimony-based optimization criterion. It thus identifies the smallest (weighted) number of independent evolutionary events that explain the phylogenetic gene tree. NOTUNG functions include rearranging of a rooted gene tree in areas of weak sequence support, thus avoiding overestimating duplications in gene trees that are incongruent with the species-tree. We used the standard parsimony weight parameters of the software: 1.5 for a duplication, 0.0 for a conditional duplication and 1.0 for a loss. For the bootstrap cut-off value to identify weak branches, the value of 90 was used. This is a relatively strict value. For the rearrangement procedure, a binary species tree is needed. We obtained binary species trees for each gene family by extracting the NCBI taxonomy species identifiers of all species present in the gene tree and uploading them to the phyloT online tree generator (biobyte solutions GmbH, 2014) to generate a binary species tree [36]. We used the following options in phyloT: NCBI taxonomy IDs as identifiers, collapsed internal nodes, no polytomies (this option randomly resolves the polytomies of the underlying non-binary species tree from NCBI), newick format.

After the tree rearrangements we performed the reconciliation, both on the raw ML gene tree and on the rearranged gene tree. For this step we needed non-binary species trees; these were generated using phyloT with the same options as described in the previous paragraph, except that the polytomies were retained. For the inferred duplications, NOTUNG outputs a lower and an upper bound. The lower bound represents the oldest species in which the duplication was present; the upper bound is the youngest species in which the duplication was not present. The losses are written to a table with the species node names and the number of losses.

The fact that the gene trees were rearranged using a species tree with randomly resolved polytomies, means that some random rearrangements were introduced in the gene tree. This is however no problem, because in the reconciliation process the non-binary species tree was used. The random rearrangements then corresponded again to polytomies in the species tree, meaning that they can only have led to conditional duplication inferences. These were not retained in the results.

The inferred duplication and loss events were mapped on a species tree and visualized with the online tool iTol [27]. The conversion of the event data to the suitable format for iTol was done in R. We used the DHHC species tree for data visualization since it contained all eukaryotic species in the database that we used. In both the event data and the species tree, the NCBI taxonomy identifiers of the species were converted to the species names. To handle phylogenetic trees in R, we made use of the APE package [33].

Evolutionary history of the HDAC family

An HMM of the histone deacetylase domain was retrieved from Pfam (PF00850). The HMMER search, construction of phylogenetic tree, rearrangement and reconciliation steps were performed in the same way as for the APT and PPT families. In total 710 HDAC sequences were included in the phylogenetic tree. We set a fixed value of 200 bootstraps in the tree inference process.

Additional files

Additional file 1: Figure S1. Tree reconciliation of the APT ML and rearranged trees. The upper half of the circles represents the results using the ML tree; the lower half represents the results from the rearranged tree. Yellow semicircles indicate inferred gain in the APT family, red semicircles indicate inferred losses in the APT family. Black indicates no inferred copy-number changes. The tree topology is extracted from NCBI taxonomy. (PDF 41 kb)

Additional file 2: Figure S2. Tree reconciliation of the PPT ML and rearranged trees. The upper half of the circles represents the results using the ML tree; the lower half represents the results from the rearranged tree. Yellow semicircles indicate inferred gains of the PPT family, red semicircles indicate inferred gene losses of the PPT family. Black indicates no inferred copy-number changes. The tree topology is extracted from NCBI taxonomy. (PDF 38 kb)

Additional file 3: Figure S3. Tree reconciliation of the DHHC ML tree (without rearrangement). Yellow circles indicate inferred increases in copy numbers of the DHHC family, red circles indicate inferred decreases in copy numbers of the DHHC family. The tree topology is extracted from NCBI taxonomy. (PDF 49 kb)

Additional file 4: Figure S4. Histone deacetylase family as a negative control. A) Differences in net gain in HDAC family per branch between three categories of branches, gain in APT, loss in APT or no change in APT family B) Differences in net gain in HDAC family per branch between three categories of branches, gain in PPT, loss in PPT or no change in PPT family (PDF 41 kb)

Funding
This work was supported by the KU Leuven research fund.

Authors' contributions
VvN has provided the idea for the study. SW has carried out analyses. SW and VvN wrote the manuscript. Both authors read and approved the final manuscript.

Competing interests
The authors declare that they have no competing interests.

Author details
[1]Centre of Microbial and Plant Genetics, KU Leuven, Leuven, Belgium. [2]Department of Bioscience Engineering, University of Antwerp, Antwerp, Belgium.

References
1. Bartels DJ, Mitchell DA, Dong X, Deschenes RJ. Erf2, a novel gene product that affects the localization and palmitoylation of Ras2 in Saccharomyces cerevisiae. Mol Cell Biol. 1999;19:6775–87.
2. Beltrao P, Bork P, Krogan NJ, van Noort V. Evolution and functional cross-talk of protein post-translational modifications. Mol Syst Biol. 2013;9:714.
3. Calero G, Gupta P, Nonato MC, Tandel S, Biehl ER, Hofmann SL, Clardy J. The crystal structure of palmitoyl protein thioesterase-2 (PPT2) reveals the basis for divergent substrate specificities of the two lysosomal thioesterases, PPT1 and PPT2. J Biol Chem. 2003;278:37957–64.
4. Camacho C, Coulouris G, Avagyan V, Ma N, Papadopoulos J, Bealer K, Madden TL. BLAST+: architecture and applications. BMC Bioinformatics. 2009;10:421.
5. Camp LA, Hofmann SL. Purification and properties of a palmitoyl-protein thioesterase that cleaves palmitate from H-Ras. J Biol Chem. 1993;268: 22566–74.
6. Capella-Gutiérrez S, Silla-Martínez JM, Gabaldón T. trimAl: a tool for automated alignment trimming in large-scale phylogenetic analyses. Bioinformatics. 2009;25:1972–3.
7. Chamberlain LH, Shipston MJ. The physiology of protein S-acylation. Physiol Rev. 2015;95:341–76.
8. Chen K, Durand D, Farach-Colton M. NOTUNG: A Program for Dating Gene Duplications and Optimizing Gene Family Trees. J Comput Biol. 2000;7:429–47.
9. Cho S, Dawson PE, Dawson G. In vitro depalmitoylation of neurospecific peptides: implication for infantile neuronal ceroid lipofuscinosis. J Neurosci Res. 2000;59:32–8.
10. Cuypers TD, Hogeweg P. Virtual Genomes in Flux: An Interplay of Neutrality and Adaptability Explains Genome Expansion and Streamlining. Genome Biol Evol. 2012;4:212–29.
11. Dekker FJ, Rocks O, Vartak N, Menninger S, Hedberg C, Balamurugan R, Wetzel S, Renner S, Gerauer M, Schölermann B, et al. Small-molecule inhibition of APT1 affects Ras localization and signaling. Nat Chem Biol. 2010;6:449–56.
12. Demuth JP, Hahn MW. The life and death of gene families. Bioessays. 2009; 31:29–39.
13. Duncan JA, Gilman AG. Characterization of Saccharomyces cerevisiae acyl-protein thioesterase 1, the enzyme responsible for G protein alpha subunit deacylation in vivo. J Biol Chem. 2002;277:31740–52.
14. Eddy SR. Accelerated Profile HMM Searches. PLoS Comput Biol. 2011;7:e1002195.
15. Franceschini A, Lin J, von Mering C, Jensen LJ. SVD-phy: improved prediction of protein functional associations through singular value decomposition of phylogenetic profiles. Bioinformatics. 2016;32(7):1085-7.
16. Franceschini A, Szklarczyk D, Frankild S, Kuhn M, Simonovic M, Roth A, Lin J, Minguez P, Bork P, von Mering C, et al. STRING v9.1: protein-protein interaction networks, with increased coverage and integration. Nucleic Acids Res. 2013;41:D808–15.
17. Glaser RL, Hickey AJ, Chotkowski HL, Chu-LaGraff Q. Characterization of Drosophila palmitoyl-protein thioesterase 1. Gene. 2003;312:271–9.
18. Greaves J, Chamberlain LH. DHHC palmitoyl transferases: substrate interactions and (patho)physiology. Trends Biochem Sci. 2011;36:245–53.
19. Hou H, John Peter AT, Meiringer C, Subramanian K, Ungermann C. Analysis of DHHC acyltransferases implies overlapping substrate specificity and a two-step reaction mechanism. Traffic. 2009;10:1061–73.
20. Katoh K, Standley DM. MAFFT multiple sequence alignment software version 7: improvements in performance and usability. Mol Biol Evol. 2013;30:772–80.
21. Kemp LE, Rusch M, Adibekian A, Bullen HE, Graindorge A, Freymond C, Rottmann M, Braun-Breton C, Baumeister S, Porfetye AT, et al. Characterization of a serine hydrolase targeted by acyl-protein thioesterase inhibitors in Toxoplasma gondii. J Biol Chem. 2013;288:27002–18.

22. Kensche PR, van Noort V, Dutilh BE, Huynen MA. Practical and theoretical advances in predicting the function of a protein by its phylogenetic distribution. J R Soc Interface. 2008;5:151–70.

23. Kihara A, Kurotsu F, Sano T, Iwaki S, Igarashi Y. Long-chain base kinase Lcb4 Is anchored to the membrane through its palmitoylation by Akr1. Mol Cell Biol. 2005;25:9189–97.

24. Kollmann K, Uusi-Rauva K, Scifo E, Tyynelä J, Jalanko A, Braulke T. Cell biology and function of neuronal ceroid lipofuscinosis-related proteins. Biochim Biophys Acta. 2013;1832:1866–81.

25. Krogh A, Larsson B, von Heijne G, Sonnhammer EL. Predicting transmembrane protein topology with a hidden Markov model: application to complete genomes. J Mol Biol. 2001;305:567–80.

26. Lam KKY, Davey M, Sun B, Roth AF, Davis NG, Conibear E. Palmitoylation by the DHHC protein Pfa4 regulates the ER exit of Chs3. J Cell Biol. 2006;174:19–25.

27. Letunic I, Bork P. Interactive Tree Of Life v2: online annotation and display of phylogenetic trees made easy. Nucleic Acids Res. 2011;39:W475–8.

28. Lim WA, Pawson T. Phosphotyrosine Signaling: Evolving a New Cellular Communication System. Cell. 2010;142:661–7.

29. Mitchell DA, Vasudevan A, Linder ME, Deschenes RJ. Protein palmitoylation by a family of DHHC protein S-acyltransferases. J Lipid Res. 2006;47:1118–27.

30. Moorhead GBG, De Wever V, Templeton G, Kerk D. Evolution of protein phosphatases in plants and animals. Biochem J. 2009;417:401–9.

31. Nadolski MJ, Linder ME. Protein lipidation. FEBS J. 2007;274:5202–10.

32. Nardini M, Dijkstra BW. Alpha/beta hydrolase fold enzymes: the family keeps growing. Curr Opin Struct Biol. 1999;9:732–7.

33. Paradis E, Claude J, Strimmer K. APE: Analyses of Phylogenetics and Evolution in R language. Bioinformatics. 2004;20:289–90.

34. Van de Peer Y, Maere S, Meyer A. The evolutionary significance of ancient genome duplications. Nat Rev Genet. 2009;10:725–32.

35. Porter MY, Turmaine M, Mole SE. Identification and characterization of Caenorhabditis elegans palmitoyl protein thioesterase1. J Neurosci Res. 2005;79:836–48.

36. Sayers EW, Barrett T, Benson DA, Bryant SH, Canese K, Chetvernin V, Church DM, DiCuccio M, Edgar R, Federhen S, et al. Database resources of the National Center for Biotechnology Information. Nucleic Acids Res. 2009;37:D5–D15.

37. Schliep KP. phangorn: phylogenetic analysis in R. Bioinformatics. 2011;27:592–3.

38. Simon GM, Cravatt BF. Activity-based proteomics of enzyme superfamilies: serine hydrolases as a case study. J Biol Chem. 2010;285:11051–5.

39. Smotrys JE, Schoenfish MJ, Stutz MA, Linder ME. The vacuolar DHHC-CRD protein Pfa3p is a protein acyltransferase for Vac8p. J Cell Biol. 2005;170:1091–9.

40. Stamatakis A. RAxML version 8: a tool for phylogenetic analysis and post-analysis of large phylogenies. Bioinformatics. 2014;30:1312–3.

41. Tian L, McClafferty H, Knaus H-G, Ruth P, Shipston MJ. Distinct acyl protein transferases and thioesterases control surface expression of calcium-activated potassium channels. J Biol Chem. 2012;287:14718–25.

42. Tomatis VM, Trenchi A, Gomez GA, Daniotti JL. Acyl-protein thioesterase 2 catalyzes the deacylation of peripheral membrane-associated GAP-43. PLoS One. 2010;5:e15045.

43. Valdez-Taubas J, Pelham H. Swf1-dependent palmitoylation of the SNARE Tlg1 prevents its ubiquitination and degradation. EMBO J. 2005;24:2524–32.

44. Weinert BT, Iesmantavicius V, Wagner SA, Schölz C, Gummesson B, Beli P, Nyström T, Choudhary C. Acetyl-phosphate is a critical determinant of lysine acetylation in E. coli. Mol Cell. 2013;51:265–72.

45. Zeidman R, Jackson CS, Magee AI. Protein acyl thioesterases (Review). Mol Membr Biol. 2009;26:32–41.

3

Discovery and evolution of novel hemerythrin genes in annelid worms

Elisa M. Costa-Paiva[1,2], Nathan V. Whelan[2,3], Damien S. Waits[2], Scott R. Santos[2], Carlos G. Schrago[1] and Kenneth M. Halanych[2*]

Abstract

Background: Despite extensive study on hemoglobins and hemocyanins, little is known about hemerythrin (Hr) evolutionary history. Four subgroups of Hrs have been documented, including: circulating Hr (cHr), myohemerythrin (myoHr), ovohemerythrin (ovoHr), and neurohemerythrin (nHr). Annelids have the greatest diversity of oxygen carrying proteins among animals and are the only phylum in which all Hr subgroups have been documented. To examine Hr diversity in annelids and to further understand evolution of Hrs, we employed approaches to survey annelid transcriptomes *in silico*.

Results: Sequences of 214 putative Hr genes were identified from 44 annelid species in 40 different families and Bayesian inference revealed two major clades with strong statistical support. Notably, the topology of the Hr gene tree did not mirror the phylogeny of Annelida as presently understood, and we found evidence of extensive Hr gene duplication and loss in annelids. Gene tree topology supported monophyly of cHrs and a myoHr clade that included nHrs sequences, indicating these designations are functional rather than evolutionary.

Conclusions: The presence of several cHrs in early branching taxa suggests that a variety of Hrs were present in the common ancestor of extant annelids. Although our analysis was limited to expressed-coding regions, our findings demonstrate a greater diversity of Hrs among annelids than previously reported.

Keywords: Blood pigments, Respiratory proteins, Transcriptome, Annelida

Background

Metabolism in metazoans requires oxidation of organic molecules. Thus natural selection has presumably favored proteins that can reversibly bind and transport oxygen [1]. Such oxygen-binding proteins likely originated from enzymes whose primary function was to protect the body from oxygen toxicity, and, secondarily, these enzymes acquired the ability to carry oxygen molecules [2]. Several different classes of oxygen-carrying proteins, or respiratory pigments, are found across animal life. Although these molecules can reversibly bind oxygen, their binding affinities and evolutionary origins differ. In animals, oxygen-binding proteins are usually divided into two main groups: proteins that use iron to bind oxygen, including hemoglobins and hemerythrins,

and hemocyanins that use copper [3]. Although hemoglobins and hemocyanins have been extensively investigated [4–8], knowledge on the evolutionary history of hemerythrins is limited [9]. Interestingly, medical sciences have increasingly been taking advantage of oxygen-binding proteins as blood substitutes [10, 11] or as carrier proteins for synthetic vaccines, (e.g., cancer vaccines; [12, 13]) making further study of oxygen binding protein diversity and evolution appealing.

Hemerythrins (Hrs) are a non-heme oligomeric protein family within the 'four-helical up-and-down bundle' fold and 'all alpha proteins' class according to the Structural Classification of Proteins database (SCOP) [14]. Oxygen-binding Hr proteins contain approximately 120 amino acid residues in a single domain and transport oxygen with the aid of two iron ions that bind directly to the polypeptide chain. Residues involved in iron binding include histidines (His) in positions 26 56, 75, 79, and 108, glutamic acid residue (Glu) in position 60 and aspartic acid residue (Asp) in position 113 (position

---

* Correspondence: ken@auburn.edu
[2]Department of Biological Sciences, Molette Biology Laboratory for Environmental and Climate Change Studies, Auburn University, Auburn, AL 36849, USA
Full list of author information is available at the end of the article

numbers from *Themiste zostericola* in [15]). Presence of these signature residues indicates putative respiratory function for Hrs [16]. Functional Hr subunits usually form a homooctamer, although dimeric, trimeric, or tetrameric Hrs have been observed in some sipunculid species, including *Phascolosoma arcuatum*, *P. agasizii*, and *Siphonosoma funafuti* [17–19]. The crystal structure of Hrs consists of a bundle of four antiparallel α-helices (A, B, C, and D) formed by polypeptides: an A α-helix formed by 19 amino acid residues from position 19 to 38, B α-helix with 23 amino acids residues from position 43 to 65, C α-helix formed by 16 amino acids residues from position 72 to 88, and D α-helix formed by 20 amino acids residues from position 98 to 118, using *T. zostericola* as the reference sequence [9]. The core of active sites contains two iron atoms bridged by two carboxylate groups from aspartate and glutamate residues and an oxygen-containing ligand [20, 21]. Binding of oxygen apparently requires other currently unknown cellular factors since purified Hr, by itself, usually does not bind oxygen [22, 23]. Observed oxygen binding capacity is about 25% greater in Hrs than heme-based proteins, including hemoglobins [15].

Although Hr-like proteins have also been reported in prokaryotes [24–26] oxygen binding Hrs have only been reported from marine invertebrates belonging to Annelida (which include sipunculids; [27]) Brachiopoda, Priapulida, Bryozoa, and a single species of both Cnidaria (*Nematostella vectensis*) and Arthropoda (*Calanus finmarchicus*) [9, 18, 28, 29]. Given this phylogenetic breadth of animals, whether all metazoan Hrs share a common origin is debated [28, 29]. Overall, Hr proteins exhibit variation in their quaternary structure, and four groups have been reported based mainly on their primary structure and location within animal bodies [9]. Specifically, hemerythrins found in vascular tissue, referred to here as circulating hemerythrins (cHrs), and muscle-specific myohemerythrins (myoHr), have been better characterized compared to the other two hemerythrin groups, ovohemerythrins (ovoHr) and neurohemerythrins (nHr) [28, 30]. cHrs are polymeric intracellular proteins that occur inside nucleated cells, hemerythrocytes or pink blood cells located in coelomic fluid or vascular systems of Hr-bearing organisms [15]. In contrast, myoHrs are monomeric cytoplasmic proteins present in muscle cells of annelids [31]. The main difference between these groups is the presence of a five-codon insertion found in myoHr immediately before the D α-helix. Expression of myoHrs seems not to be restricted to cHrs-bearing organisms, considering that some annelid species possess both myoHrs and hemoglobins [32]. The other two groups of Hrs, ovohemerythrin (ovoHr) and neurohemerythrin (nHr), are also intracellular and non-circulating. ovoHr was identified

in oocytes of the leech *Theromyzon tessulatum* and its presence during oogenesis possibly suggests a complex function in iron storage and detoxification [32, 33]. On the other hand, nHr was recently discovered in neural and non-neuronal tissues from the body wall of the leech *Hirudo medicinalis*, and it exhibits upregulation in response to septic injury [34]. Nevertheless, Vanin et al. suggested that nHr of leech may in fact be a myoHr [9]. Such diversification in Hr function may have involved gene duplications resulting in new proteins via neo- or subfunctionalization [32]. Moreover, ovoHrs and nHrs have only been reported in the literature a few times, and more studies are required to understand their function and evolution.

Annelids have the greatest diversity of oxygen-binding proteins among metazoans [35] and it is the only phylum from which all subtypes of Hr proteins have been documented [28]. While Hrs of annelids have been studied since the middle of the 20th century, until the 1990s, Hrs were recorded only from sipunculids and from a single polychaete family, Magelonidae [18, 36]. Later, Vanin et al. [9] found Hrs in a nereid and a leech and Bailly et al. [28] discovered Hrs genes in seven annelid species, suggesting Hrs are broadly distributed in annelids. Given the diversity of lifestyles among annelids known to have Hrs [29, 37], and the lack of information about Hrs in general [9], the occurrence and diversity of these molecules may be higher than currently recognized. Thus, to examine a wide diversity of annelid taxa for Hrs and to further understand how different forms of Hrs are evolutionarily related to each other, we employed approaches to survey Hrs from a diverse array of annelid transcriptomes *in silico*. We identified Hrs in 44 taxa and further describe the molecular diversity and evolution of Hrs in the light of annelid phylogeny [27, 38, 39]. Along with this, we assess whether described Hr subtypes consist of evolutionary lineages or result of independent adaptations to different organismal tissues.

Results

Our bioinformatic analyses (Additional file 1) recovered a total of 415 unique nucleotide sequences of hemerythrin-like genes. Following translation Pfam domain evaluation and manual removal of sequences with less than 100 amino acid residues, 214 putative novel Hr genes were retained from all taxa examined in this study, representing 44 annelid species in 40 different families (Table 1). Novel Hr genes accession numbers for each species is available in Additional file 2. The number of expressed Hrs in a given species ranged from one in *Alciopa* sp., *Cossura longocirrata*, *Enchytraeus albidus*, *Schizobranchia insignis*, and *Syllis* cf. *hyaline* to 11 in *Magelona berkeleyi* and *Phascolosoma agassizii*.

```
A. japonica myoHr        TTC AAG GAG CAC AAG GAG ATC CAC GAC GGA TTC ATT GCC CAT CTC TCC --- AAG
A. japonica cHr          TTT GCT GAA CAC AAG AAG GCT CAT GAC GAT TTC ATC CAT AGG CTG GAC --- ACC
H. filiformis myoHr      GTT GTG CCC CAC AAA AAG CTC CAC AAG GAC TTT CTG GAC AAG CTT GCA --- GGA
H. filiformis cHr        TAT GAT GAA CAC AAG GCG GCT CAT ACT GAA TTC ATT CGC AGG CTT GAT --- AGC
M. infundibulum myoHr    TAT GAT TCA CAC AAG AAG TTA CAT GCG GAG TTC GTG GGA AAA CTT GCG --- GCT
M. infundibulum cHr      TAT GAT GAA CAC AAG GCG GCT CAT ACT GAA TTC ATT CGC AGG CTT GAT --- AGC
P. agassizii myoHr       GTA GTA CCA CAC AAA GCA GCT CAC AAA GAT TTT CTT GGC AAG ATC AGA --- GCA
P. agassizii cHr         TAT GAT GAA CAC AAG GCG GCT CAT ACT GAA TTC ATT CGC AGG CTT GAT --- AGC
T. pyroides myoHr        TTT TCG ACC CAC TGT GCC GCC CAC AAA GGA TTC CTG GAC AAG ATA CGT --- GGT
T. pyroides cHr          TTT GCT GAA CAC AAG AAG GCT CAT GAC GAT TTC ATC CAT AGG CTG GAC --- ACC

A. japonica myoHr        CTG ACC ACC --- --- --- --- CCG GTG TCC GAC GAG GAA GTA TTC TAC TGC AAG
A. japonica cHr          TGG AAT GGT --- --- --- --- --- --- --- --- --- GAT ATC CAG TAT GCC AAG
H. filiformis myoHr      CTG TCG GCA --- --- --- --- CCT GTG CCT GCG AAA GAC GTC GAC TAT TGC AAG
H. filiformis cHr        TGG GAC GGT --- --- --- --- --- --- --- --- --- GAT CTC ACC TAT GCC AAG
M. infundibulum myoHr    CTC AGT GCA --- --- --- --- CCC CTA GAT GGC CCC AGC GTC TAC TAC TCC AAG
M. infundibulum cHr      TGG GAC GGT --- --- --- --- --- --- --- --- --- GAT CCC ACC TAT GCC AAG
P. agassizii myoHr       CTC AAA GCA --- --- --- --- CCC CTG GAT GAT GGC ACC ATA GAC TAC GCA AAG
P. agassizii cHr         TGG GAC GGT --- --- --- --- --- --- --- --- --- GAT CTC ACC TAT GCC AAG
T. pyroides myoHr        CTC AAG GCA --- --- --- --- CCC GTT GAT GAT GCC ACT ATC AAG TTC GCC AAA
T. pyroides cHr          TGG AAT GGT --- --- --- --- --- --- --- --- --- GAT ATC CAG TAT GCC AAG

A. japonica myoHr        AAC TGG CTG GCG CAG CAC ATC AAG AAC ATT GAC TTC GCG TAC AGA GGC AAG CTG
A. japonica cHr          GAG TGG CTG GTT AAC CAC ATC AAG ACC ATT GAC TTC AAA TAC AAG CAC AGG ATC
H. filiformis myoHr      GAA TGG TTG GTG AAT CAC ATC AAG GGC ACA GAC TTC AAA TAC AAA GGA AAA TTG
H. filiformis cHr        AAA TGG CTG GGC GAC CAC ATT AAG ACC ATT GAC TTC AAG TAC AGG GGC AAG ATC
M. infundibulum myoHr    AAC TGG TTG GTC AAC CAC ATC AAA GGA ATA GAT TTC AAA TAC AAG GGA AAG CTT
M. infundibulum cHr      AAA TGG CTG GTC GAC CAC ATC AAG ACC ATT GAC TTC AAG TAC AGG GGC AAG ATC
P. agassizii myoHr       GAC TGG TTG GTG AAT CAC ATT AAA GGA ACT GAC TTC AAA TAC AAG GGA AAG CTG
P. agassizii cHr         AAA TGG CTG GTC GAC CAC ATC AAG ACC ATT GAC TTC AAG TAC AGG GGC AAG ATC
T. pyroides myoHr        GAA TGG CTG GTG AAC CAT ATC AAA GGT ACT GAT TTC AAG TAC AAG GAA AAG CTG
T. pyroides cHr          GAG TGG CTG GTT AAC CAC ATC AAG ACC ATT GAC TTC AAA TAC AAG CAC AGG ATC
```

Fig. 1 Alignment of nucleotide dataset of few species. Showing the five-codon insertion (yellow background) between C α-helix (blue background) and D α-helix (green background)

Following trimming alignment of translated transcripts possessed 132 residue positions, with nearly all sequences, the exception being *Aphrodita japonica*, starting with a methionine residue. We decided to keep the apparently incomplete sequence from *A. japonica* due to its high similarity with the remaining sequences. All sequences in the alignment contained signature residues involved in iron binding, indicating putative respiratory function for these putative Hrs [16]. For the 214 sequences, 100 were unique and 114 identical for at least two species at the amino acid level.

Sequences were assembled into a final dataset containing 225 sequences being 214 new, two Hr sequences from *Lingula reevii*, a brachiopod, and nine annelid sequences previous used as "queries", with 396 aligned nucleotides positions. Of these, 209 sequences contained a five-codon insertion between the C and D α-helices described for myoHr, but not cHrs (Fig. 1) [9]. Datasets supporting conclusions of this article are available in the Figshare repository, under DOIs: https://doi.org/10.6084/m9.figshare.3505883.v1 and https://doi.org/10.6084/m9.figshare.3505886.v1.

Every species analyzed possessed at least one copy of myoHr gene. Bayesian inference rooted using two brachiopod Hr sequences revealed two major clades with one corresponding to cHrs clade (p.p. = 0.93; Fig. 2 blue clade) and the other supported monophyly for a myoHr clade (p.p. = 0.95; Fig. 2 black clade). The leech nHr sequence included in the dataset was found inside the myoHr clade (Fig. 2; orange). Within the myoHr clade low nodal support values and polytomies were found. However, such results are not uncommon for single protein or protein family trees [40].

Our analysis found 13 putative cHr sequences lacking the characteristic five-codon insertion before the D α-helix that define myoHrs, distributed across nine different families. Besides well-known records of cHrs in sipunculids, such as *Themiste pyroides* and *Phascolosoma agassizii*, we discovered cHrs in the sipunculid *Thysanocardia nigra* as well as six annelid families; Amphinomidae Aproditidae, Capitellidae, Oweniidae, Sabellidae, and Spionidae (Fig. 2; blue clade).

The topology of the Hr gene tree did not mirror recent phylogenies of Annelida based on phylogenomic datasets [27, 38, 39]. For example, we found 10 Hr sequences identical at the nucleotide level (Fig. 2, purple clade) belonging to distant annelid families indicating a strong conservation among those orthologs. Several of these sequences were prepared and sequenced at different times, making cross contamination unlikely. Those 10 identical sequences differed 28.54% (nucleotide level) from the consensus of all others myoHr sequences and the majority of nonsynonymous substitutions are concentrated in A and B α-helices.

Regarding paralogs multiple copies of Hr genes were found for several species, including two paralogs from both *Chaetopterus variopedatus* and *Mesochaetopterus taylori*, with these paralogs forming a monophyletic clade (p.p. = 0.85; Fig. 2; green clade). Both species were from Chaetopteridae suggesting a recent paralogous

Table 1 List of taxa, including collection site, total number of base pairs sequenced, total number of contigs after assembly, number of putative Hr genes, and GenBank accession numbers. GenBank accession numbers for each Hr copy is indicated in Additional file 2

Species	Collection site	Total bp	Total contigs number	Hr genes number	Accession number
ALCIOPIDAE					
Alciopa sp.	N 33° 07.' W 076° 06.4'	157,869,560	233,051	1	KY007275
ALVINELLIDAE					
Paralvinella palmiformis Desbruyères & Laubier, 1986	N 47° 56.9' W 129° 05.9'	59,602,987	85,363	1	KY007423
AMPHINOMIDAE					
Paramphinome jeffreysii (McIntosh, 1868)	N 63° 30.8' E 10° 25.0'	104,449,511	165,337	8	KY007424 to KY007431
APHRODITIDAE					
Aphrodita japonica Marenzeller, 1879	N 48° 28.6' W 122° 58.7'	84,662,357	120,025	9	KY007279 to KY007287
ASPIDOSIPHONIDAE					
Aspidosiphon laevis de Quatrefages, 1865	N 09° 22.6' W 82° 18.1'	120,601,137	168,072	4	KY007297 to KY007300
CAPITELLIDAE					
Heteromastus filiformis (Claparede, 1864)	N 41° 41.5' W 070° 37.6'	94,824,555	148,196	8	KY007364 to KY007371
CHAETOPTERIDAE					
Chaetopterus variopedatus (Renier, 1804)	N 41° 41.5' W 070° 43.5'	166,610,386	147,132	2	KY007307 and KY007309
Mesochaetopterus taylori Potts, 1914	N 48° 29.0' W 123° 04.3'	86,521,966	83,209	3	KY007383 to KY007386
CHRYSOPETALIDAE					
Arichlidon gathofi Watson Russell, 2000	N 33° 59.6' W 76° 42.1'	105,115,966	140,980	8	KY007288 to KY007296
CIRRATULIDAE					
Chaetozone sp.	N 66° 33.2' E 33° 06.7'	85,730,053	143,587	4	KY007310 to KY007313
COSSURIDAE					
Cossura longocirrata Webster & Benedict, 1887	N 33° 29.4' W 074° 48.0'	45,732,505	75,079	1	KY007319
ENCHYTRAEIDAE					
Enchytraeus albidus Henle, 1837	N 59° 5.2' E 16° 03.3'	13,345,974	22,776	1	KY007331
EUNICIDAE					
Eunice pennata (Müller, 1776)	N 39° 47.2' W 70° 46.3'	59,429,144	93,814	5	KY007332 to KY007336
EUPHROSINIDAE					
Euphrosine capensis Kinberg, 1857	S 34° 09.9' E 18° 26.0'	27,221,777	72,220	3	KY007337 to KY007339
FLABELLIGERIDAE					
Poeobius meseres Heath, 1930	N 36° 41.2' W 122° 02.0'	25,964,726	70,078	2	KY007441 and KY007442
GLYCERIDAE					
Glycera dibranchiata Ehlers, 1868	N 41° 54.1' W 070° 00.4'	51,282,233	101,455	3	KY007350 to KY007352
GOLFINGIIDAE					
Thysanocardia nigra (Ikeda, 1904)	N 48° 28.6' W 122° 58.7'	57,399,340	58,011	6	KY007480 to KY007485
GONIADIDAE					
Glycinde armigera Moore, 1911	N 36° 23.0' W 121° 57.9'	32,178,692	79,528	4	KY007353 to KY007356
HAPLOTAXIDAE					
Delaya leruthi (Hrabe, 1958)	N 43° 0.8' E 01° 2.5'	93,863,431	118,020	7	KY007320 to KY007326
HESIONIDAE					
Oxydromus pugettensis (Johnson, 1901)	N 48° 34.3' W 123° 10.1'	45,242,396	92,341	2	KY007421 and KY007422
LUMBRINERIDAE					
Ninoe sp.	N 35° 29.4' W 074° 48.0'	120,256,564	151,183	5	KY007409 to KY007413
MAGELONIDAE					
Magelona berkeleyi Jones, 1971	N 36° 22.8' W 121° 58.1'	16,339,407	50,123	10	KY007372 to KY007382

Table 1 List of taxa, including collection site, total number of base pairs sequenced, total number of contigs after assembly, number of putative Hr genes, and GenBank accession numbers. GenBank accession numbers for each Hr copy is indicated in Additional file 2 *(Continued)*

MALDANIDAE					
Clymenella torquata (Leidy, 1855)	N 41° 42.7' W 070° 19.7'	62,661,529	111,567	5	KY007314 to KY007314
Nicomache venticola Blake & Hilbig, 1990	N 47° 57.0' W 129° 05.9'	64,130,139	124,708	4	KY007405 to KY007408
NEPHTYIDAE					
Nephtys incisa Malmgren, 1865	N 40° 53.0' W 070° 25.0'	126,720,409	188,338	6	KY007396 to KY007401
NEREIDIDAE					
Alitta succinea (Leuckart, 1847)	N 41° 54.0' W 070° 00.3'	105,821,565	153,011	3	KY007402 to KY007404
OENONIDAE					
Drilonereis sp.	N 39° 54.1' W 070° 35.1'	3,490,940	12,598	4	KY007327 to KY007330
Oenone fulgida (Savigny in Lamarck, 1818)	S 34° 37.0' E 19° 21.4'	92973167	144,726	3	KY007414 to KY007416
ORBINIIDAE					
Naineris laevigata (Grube, 1855)	S 34° 35.0' E 19° 20.9'	123,970,343	218,272	4	KY007392 to KY007395
OWENIIDAE					
Galathowenia oculata (Zachs, 1923)	N 66° 33.2' E 33° 6.7'	128,195,375	179,612	10	KY007340 to KY007349
PECTINARIIDAE					
Pectinaria gouldii (Verrill, 1874)	N 41° 37.9' W 070° 53.3'	63,132,019	92,091	9	KY007432 to KY007440
PHASCOLOSOMATIDAE					
Phascolosoma agassizii Keferstein, 1866	N 48° 31.2' W 123° 01.0'	78,749,017	87,403	11	KY007443 to KY007453
PILARGIDAE					
Ancistrosyllis groenlandica McIntosh, 1879	N 40° 27.3' W 070° 47.6'	39,327,753	94,924	3	KY007276 to KY007278
POLYNOIDAE					
Halosydna brevisetosa Kinberg, 1856	N 48° 29.5' W 123° 01.1'	61,671,140	118,418	7	KY007357 to KY007363
RANDIELLIDAE					
Randiella sp.	S 14° 40.0' E 145° 27.1'	139,189,396	151,934	6	KY007454 to KY007459
SABELLIDAE					
Myxicola infundibulum (Montagu, 1808)	N 48° 28.6' W 122° 58.7'	156,042,620	217,996	5	KY007387 to KY007391
Schizobranchia insignis Bush, 1905	N 48° 33.3' W 122° 56.5'	55,085,979	102,002	1	KY007460
SPARGANOPHILIDAE					
Sparganophilus sp.	N 40° 50.3' W 92° 5.3'	117,343,038	123,905	8	KY007461 to KY007468
SPIONIDAE					
Boccardia proboscidea Hartman, 1940	N 48° 29.2' W 123° 04.1'	78,374,988	117,570	7	KY007301 to KY007307
STERNASPIDAE					
Sternaspis scutata Ranzani, 1817	N 48° 29.1' W 123° 04.3'	81,147,455	115,096	3	KY007469 to KY007471
SYLLIDAE					
Syllis cf. *hyalina* Grube, 1863	S 34° 37.0' E 19° 21.4'	76,801,405	106,283	1	KY007472
THEMISTIDAE					
Themiste pyroides (Chamberlin, 1919)	N 48° 21.4' W 123° 43.4'	75,495,745	88,157	7	KY007473 to KY007479
TOMOPTERIDAE					
Tomopteris sp.	N 36° 41.2' W 122° 02.0'	30,525,410	66,655	3	KY007486 to KY007488
Oligochaeta gen. sp. (unidentified Crassiclitellata - Place Kabary 2)	S 12° 59.0' E 49° 17.4'	107,638,847	146,018	4	KY007417 to KY007420

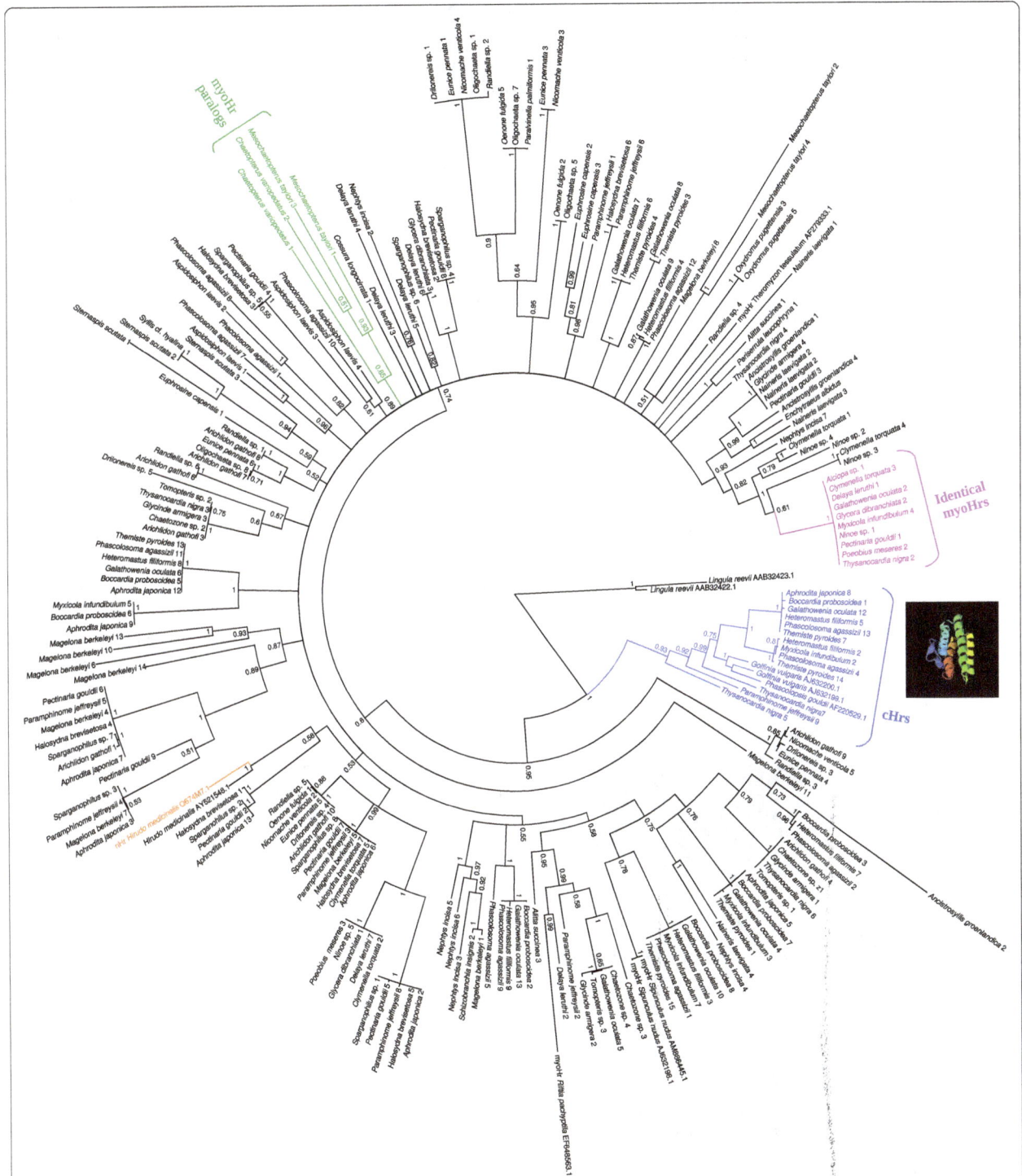

Fig. 2 Bayesian tree using MrBayes 3.2.1 [51] rooted with two Hr sequences from *Lingula reevii*, a brachiopod. The blue clade represents the cHrs otologs sequences with a prediction of the consensus cHr sequence quaternary structure using I-TASSER 4.4 [61]. Black, purple, and green sequences represent the myoHr sequences and orange sequence is the only nHr from *Hirudo medicinalis*. The green clade shows the paralogs sequences from *C. variopedatus* and *M. taylori* and the purple clade represents the identical orthologs myoHr sequences. The number after the name of each sequence indicates the GenBank accession numbers for each Hr gene and it is indicated in Additional file 2: Table S1

duplication [26]. Given the presence of multiple Hrs, apparently early annelids already contained several copies of Hr genes, with some paralogs arising later (as in *C. variopedatus* and *M. taylori*).

Differences in evolutionary rates between cHrs and myoHrs sequences was accessed and relative rates of change in different positions were calculated using two different approaches. Sites with high variation were found not to have significantly different rates among inter-helices sites using DIVERGE [41]. For helix regions A and C α-helices had one site each with high evolutionary rate, while D α-helix did not have any sites with a high evolutionary rate, indicating a highly conservative region. At the same time, B α-helix had five sites, suggesting that this helix is evolving faster than others (Fig. 3). RELAX [42] was also used to assess differences in selection on cHrs relative to myoHrs while accounting for lineage-specific rate differences. Similar to the DIVERGE analyses, no significant differences ($P = 0.218$) were found between the two sets of genes. Thus, the gene duplication event leading to cHrs versus myoHrs does not seem to have been accompanied by a significant change in substitution rate or selection differences on the two gene lineages. Alternatively, any evidence of such a rate or selection difference could have been lost during the 500+ MY since these genes diverged.

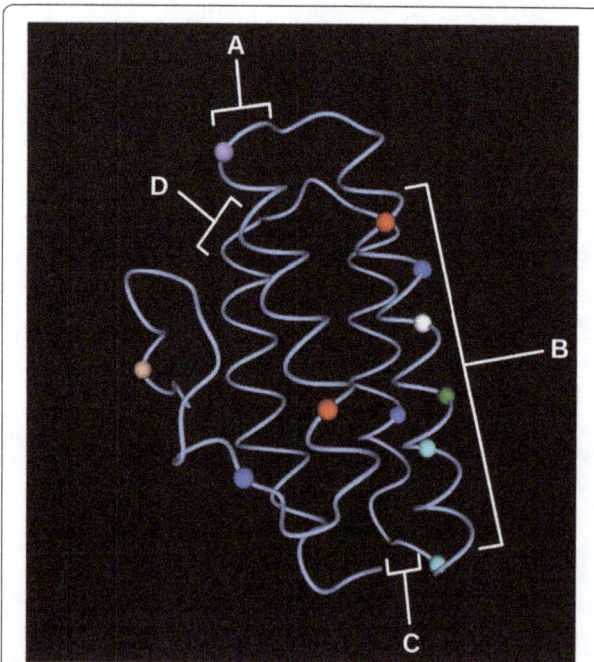

Fig. 3 Hr showing differences in evolutionary rates between cHrs and myoHrs sequences calculated with DIVERGE [41] using a 0.7 cutoff. A and C α-helices presented one site each above the cutoff value, D α-helix did not present any sites, and B α-helix presented five sites. *Colored dots* represent sites above cutoff value, and different colors are only to aid illustration

Discussion

As demonstrated here expression of Hr genes among annelids is much more common than previously reported, revealing unrecognized diversity of these genes in this phylum. All 44 of the examined species possessed actively transcribed myoHrs, while cHrs were less frequently recovered from these transcriptomes. Although this diversity and wide distribution of Hrs in annelids could, in part, be explained by the need to carry oxygen (Hr have approximately 25% greater oxygen affinity than hemoglobins; [15]) secondary functional specializations could also be important for driving diversification. For example, Hrs participate in iron storage, metal detoxification, and immunity in some annelids (e.g., *Theromyzon tessulatum*, *Hirudo medicinalis* and *Neanthes diversicolor*) [33, 34, 43]. Our findings build on recent publications demonstrating that sequence diversity among Hr-bearing species was larger than traditionally suspected [9, 28, 29]. However, those studies were based predominately on genomic data. In this case, use of transcriptomes shows that Hrs genes were present and expressed.

Bayesian phylogenetic reconstruction recovered monophyletic clades for annelid cHrs sequences as well as myoHrs sequences with strong support corroborating Vanin et al.'s [9] previous findings. The presence of multiple copies of myoHr genes across the annelid phylogeny implies these proteins have undergone several instances of gene duplication during their evolution, as previously reported for other bacterial, archeal, and eukaryotic taxa [26]. Moreover, the unexpected diversity of myoHrs could be associated with functional diversification of this gene, as observed for myoglobins [44] and also for Hrs involved in heavy metal detoxification and aspects of innate immunity [45].

Classification of specific Hr subtypes [28, 32–34] was not validated by the gene genealogy. Although our analyses used whole organisms (including reproductive and nerve tissues), our results failed to recover Hr proteins that corresponded to ovoHrs or nHrs. These categories, however, were described based on limited differences in amino acid sequence and do not reflect distinct monophyletic subgroups within the Hr gene family. Given this, recognizing only two primary types of Hrs, circulating Hrs (cHr) and non-circulating Hrs (ncHrs), is perhaps more appropriate. Although tentative, this reinterpretation deserves further consideration.

Incongruence between our gene genealogy relative to current knowledge of annelid evolutionary history indicates that Hrs have a complex history which possibly involved events of gene losses and duplications, paralogs replacements and lateral gene transfer [26]. Although previous work supported the idea of a monomeric protein as an ancestral myoHr within annelids [9, 15], the presence of cHrs in *Paramphinome jeffreysii* (Amphinomidae),

Galathowenia oculata (Oweniidae), *Phascolosoma agassizii*, *Themiste pyroides*, and *Thysanocardia nigra* (the last three belonging to Sipuncula), all members of lineages near the base of the annelid tree [27, 38, 39] (Fig. 4), indicates that both cHrs and myoHrs were likely present in the ancestor of annelids, which date back to the Cambrian [46]. Interestingly, Hrs from multiple species, representing hundreds of millions of years of evolution, possessed identical amino acid, and in some cases nucleotide, sequences (Fig. 2, purple clade), suggesting a level of conservation and selection orders of magnitude greater than most of the genome. Additional studies across metazoan together with studies of the gene structure of Hr proteins and physiological aspects of organisms are the next important steps toward a better understanding of the evolutionary patterns involved in this family of oxygen carrying proteins.

Conclusions

Our findings demonstrate that sequence diversity among Hr-bearing annelid species is much greater than traditionally suspected and that many of these Hrs are actively expressed. There are two primary types of Hrs circulating Hrs (cHr) and non-circulating Hrs (ncHrs), instead of the four subtypes reported in the literature. Incongruence between our gene genealogy relative to current knowledge of annelid evolutionary history indicates that Hrs have a complex history. Our findings indicate that both cHrs and myoHrs were likely present in the ancestor of annelids, as both subtypes occur in all lineages near the base of the annelid tree.

Methods

Sample collection

Information on species employed herein is provided in Table 1. Transcriptomes of these species were collected as part of the WormNet II project to resolve annelid phylogeny and were collected with a variety of

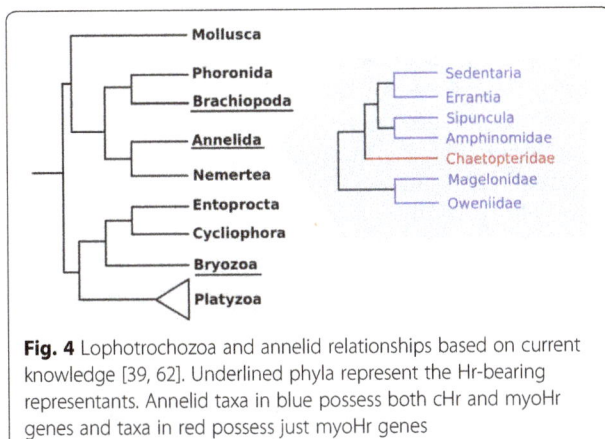

Fig. 4 Lophotrochozoa and annelid relationships based on current knowledge [39, 62]. Underlined phyla represent the Hr-bearing representants. Annelid taxa in blue possess both cHr and myoHr genes and taxa in red possess just myoHr genes

techniques, including intertidal sampling, dredge and box cores. Upon collection, all samples were either preserved in RNALater or frozen at −80 °C.

Data collection & sequence assembly

RNA extraction, cDNA preparation and high-throughput sequencing generally followed Kocot et al. [47] and Whelan et al. [48]. Briefly, total RNA was extracted from either whole animals (for small specimens) or the body wall and coelomic region (for larger specimens). After extraction, RNAs were purified using TRIzol (Invitrogen) or the RNeasy kit (Qiagen) with on-column DNase digestion, respectively. The SMART cDNA Library Construction Kit (Clonetech) was utilized to reverse transcribe single stranded RNA template. Double stranded cDNA synthesis was completed with The Advantage 2 PCR system (Clontech). Libraries were barcoded and sequenced with Illumina technology by The Genomic Services Lab at the Hudson Alpha Institute (Huntsville, Alabama, USA). Because transcriptomic sequencing was conducted from 2012–2015, Paired End (PE) runs were of 100 bp or 125 bp lengths, utilizing either v3 or v4 chemistry on Illumina HiSeq 2000 or 2500 platforms (San Diego, California). To facilitate sequence assembly, paired-end transcriptome data were digitally normalized to an average k-mer coverage of 30 using normalize-by-median.py [49] and assembled using Trinity r2013-02-25 with default settings [50].

Data mining and gene identification

Two approaches were utilized in searching transcriptomic data for putative Hr genes *in silico*. The first approach employed BLASTX [51] at an e-value cutoff of 10^{-6} to compare each assembled transcriptome contig ('queries') to a protein database composed of 21 protein sequence from the National Center for Biotechnology (NCBI) database (Additional file 3: Table S2) of at least 110 amino acid residues and previously identified as annelid Hrs ($n = 11$), myoHrs ($n = 9$), or nHr ($n = 1$). Sequences of ovoHrs were not included since there were only two relatively short (18 amino acid residues each) sequences available from NCBI. The BLASTX approach, rather than a tBLASTn, assured that any transcriptome contig with a significant 'hit' to an Hr would be further evaluated in the pipeline. Contigs recovered from initial BLAST searches were then utilized in BLASTX searches against the NCBI protein database (minimum e-value of 10^{-10}) and only top hits longer than 300 nucleotides retained. These were considered putative cHr, myoHr or nHr genes, as appropriate.

The second approach processed the transcriptomic data through the Trinotate annotation pipeline (http://trinotate.github.io/) [50], which utilizes a BLAST-based approach to provide, among others, GO (The Gene

Ontology Consortium) annotation [52]. Recovered sequences were verified by the functional annotation they received. Transcripts annotated as hemerythrins, using the 10^{-6} e-value cutoff obtained by using BLASTX, were also considered putative hemerythrin-like gene orthologs.

Contigs putatively identified as Hr genes by the two approaches were subsequently translated into amino acids using TransDecoder with default settings [53]. Since TransDecoder can produce multiple open reading frames (ORFs), all translations were additionally subject to a Pfam Domain evaluation using the EMBL-EBI database with an e-value cutoff of 10^{-5}. Translations returning an Hr-like Pfam domain and that were longer than 100 amino acids residues were retained for subsequent analyses. Moreover, we manually evaluated the presence of residues involved in iron binding, which are: histidine residues (His) in positions 26, 56, 75, 79, and 108; glutamic acid residue (Glu) in position 60; and aspartic acid residue (Asp) in position 113, numbered by reference sequence *T. zostericola*. Presence of these signature residues indicates putative respiratory function for Hrs. Transcripts passing the criteria described above were considered Hr genes (Table 1). ovoHr has been just reported once for a single species and there is no complete sequence available at GenBank, so we are not able to investigate if ovoHr is present in our dataset.

Sequence alignment

The protein dataset consisted of 225 sequences, including two Hr sequences from *Lingula reevii*, a brachiopod, nine annelid sequences previous used as "queries" (Additional file 3: Table S2), and a remaining 214 sequences from translated transcripts (Additional file 4). All sequences were initially aligned with MAFFT using the "accurate E-INS-i" algorithm [54], followed by visual inspection and manual curation in order to remove spuriously aligned sequences based on similarity to the protein alignment as a whole. Subsequently, ends of aligned sequences were manually trimmed in Geneious 8.1.6 [55] to exclude 5'residues leading to the putative start codon and 3' residues following the first seven amino acids subsequent to the end of the D α-helix.

In order to employ nucleotide sequences in the phylogenetic analysis, an alignment from corresponding aligned protein sequences (Additional file 4) was performed using PAL2NAL [56]. The resulting nucleotide alignment was used for all subsequent analyses (Additional file 5).

Phylogenetic analysis

JModelTest2 was applied to carry out statistical selection of best-fit models of nucleotide substitution for the dataset using the Akaike and Bayesian Information Criteria (AIC and BIC, respectively) methods [57]. Bayesian phylogenetic inference was performed with MrBayes 3.2.1 [58] using the GTR + G substitution model. Two independent runs with four Metropolis-coupled chains were run for 10^7 generations, sampling the posterior distribution every 500 generations. In order to confirm if chains achieved stationary and determine an appropriate burn-in, we evaluated trace plots of all MrBayes parameter output in Tracer v1.6 [59]. The first 25% of samples were discarded as burn-in and a majority rule consensus tree generated using MrBayes. Bayesian posterior probabilities were used for assessing statistical support of each bipartition.

Two alternative approaches were used to root the Hr gene genealogy. Firstly, the tree inferred in MrBayes was rooted using two Hr sequences from *Lingula reevii* (AAB32422.1 and AAB32423.1) as outgroup. We also inferred the root of the tree using using BEAST 1.8.3 [60] to infer a rooted tree of Hrs under the strict molecular clock. This was done because we were unable to decisively rule out the possibility that Hr sequences from *Lingula reevii* were closely evolutionarily related to one of the annelid Hr lineages. The strategy using BEAST is similar to midpoint rooting, although the Bayesian implementation in BEAST allows for a more flexible treatment of the evolutionary rate via a normal prior with mean and standard deviation equal to 1. Moreover, tree topology was jointly estimated. In BEAST, we adopted the same substitution model settings used in MrBayes and the MCMC algorithm was run for 50,000,000 generations and sampled every 1,000th generation, with 50% of the run discarded as burn-in. Trees was summarized in TreeAnnotator 1.8.3 [60] and Markov chain stationarity was assessed in Tracer by ESS values > 1,000. Since the inferred root node separated *Lingula reevii* sequences from annelid sequences with 0.86 posterior probability, the results are reported using the gene genealogy rooted with outgroup.

Evolutionary rate analyses

The protein alignment (Additional file 4) was used in DIVERGE [41] to examine site-specific shifted evolutionary rates and assesses whether there has been a significant change in evolution rate after duplication or speciation events by calculating the coefficient of divergence (θD) and determining if the null hypothesis of no functional divergence between myoHrs and cHrs could be statistically rejected. We employed a cutoff of 0.7 for detection of site-specific shifted evolutionary rates.

Additionally, we used RELAX [42] to detect changes in selection intensity between cHRs and myoHrs based on aligned nucleotide data (Additional file 5). Rather than just looking in *dN/dS* ratios averaged across branches using a 'branch-site' model approach, RELAX examines dN/dS ratios along each branch at a given site by drawing values from a discrete distribution of dN/dS ratios independent of other branches. In addition to assessing values in a branch independent manner, RELAX results are analyzed within a phylogenetic framework.

Additional files

Additional file 1: Flow chart of bioinformatics pipeline. Rounded purple rectangles represent input/output files, orange ovals represent software or scripts, and the green hexagon represents a step which involving manual evaluation. Nine annelid Hrs sequences previous used as query and two *Lingula* (Brachipoda) sequences from Genbank (Additional file 3) were also included in the dataset. (DOC 1870 kb)

Additional file 2: Hr genes with their respective accession numbers. Novel Hr genes accession numbers for each species. From the manuscript of Costa-Paiva et al. BMC Evolutionary Biology. (DOC 217 kb)

Additional file 3: Outgroup and query sequences used to search assembled translated transcriptomes. Query seqeunces used to search transcriptomes. The Hr sequence for H. medicinalis possesses the myoHr five codon insertion between the C and D α-helix, so in this work we considered it a myoHr. Sequences in bold was also included in the dataset previous to the alignment. From the manuscript of Costa-Paiva et al. BMC Evolutionary Biology (DOC 53 kb)

Additional file 4: The amino acid alignment used in analyses. The nucleotide alignment used for all subsequent analyses. (TXT 31 kb)

Additional file 5: The nucleotide alignment used for analyses. (TXT 89 kb)

Abbreviations
cHr: Circulating hemerythrin; Hr: Hemerythrin; myoHr: Myohemerythrin; ncHr: Non-circulating hemerythrin; nHr: Neurohemerythrin; ovoHr: Ovohemerythrin

Acknowledgments
We thank Fernando Avila Queiroz for valuable help with the figures and Antonio Sole-Cava, Joana Zanol, and Francisco Prosdocimi for critically reading this manuscript. We are grateful to Christer Erséus and Sam James for oligochaete samples used in transcriptomes. A Fellowship for the first author was provided by CAPES (Coordenação de Aperfeiçoamento de Pessoal de Nível Superior, Brazil). The findings and conclusions in this article are those of the authors and do not necessarily represent the views of the U.S. Fish and Wildlife Service. Use of SkyNet computational resources at Auburn University is acknowledged. This paper is part of the D. Sc. Requirements of Elisa Maria Costa-Paiva at Biodiversity and Evolutionary Biology Graduate Program of the Federal University of Rio de Janeiro. This is Molette Biology Laboratory contribution 61 and Auburn University Marine Biology Program contribution 153.

Funding
This work was funded by National Science Foundation grant DEB-1036537 to K. M. Halanych and Scott R. Santos and OCE-1155188 to K. M. Halanych.

Authors' contributions
EMC and KMH conceived this study. EMC performed the *in silico* analysis helped by DSW. EMC, NVW and CGS participate in phylogenetic analysis. EMC, CGS and KMH interpreted the data. EMC and KMH drafted the manuscript and all others revised critically. All the authors read and approved the version to be published.

Competing interests
The authors declare that they have no competing interests.

Author details
[1]Departamento de Genética, Laboratório de Biologia Evolutiva Teórica e Aplicada, Universidade Federal do Rio de Janeiro, Rio de Janeiro, RJ, Brazil. [2]Department of Biological Sciences, Molette Biology Laboratory for Environmental and Climate Change Studies, Auburn University, Auburn, AL 36849, USA. [3]Warm Springs Fish Technology Center, U.S. Fish and Wildlife Service, 5308 Spring ST, Warm Springs, GA 31830, USA.

References
1. Schmidt-Rhaesa A. The Evolution of Organs Systems. New York: Oxford University Press; 2007.
2. Terwilliger NB. Functional adaptations of oxygen-transport proteins. J Exp Biol. 1998;201:1085–98.
3. Terwilliger RC, Terwilliger NB, Schabtach E. Comparison of chlorocruorin and annelid hemoglobin quaternary structures. Comp Biochem Phys A. 1976;55(1):51–55.
4. Burmester T. Origin and evolution of arthropod hemocyanins and related proteins. J Comp Physiol B. 2002;172(2):95–107.
5. Burmester T. Evolution of respiratory proteins across the Pancrustacea. Integr Comp Biol. 2015;55(5):765–70.
6. Lecomte JT, Vuletich DA, Lesk AM. Structural divergence and distant relationships in proteins: evolution of the globins. Curr Opin Struc Biol. 2005;15(3):290–301.
7. Decker H, Hellmann N, Jaenicke E, Lieb B, Meissner U, Markl J. Minireview: Recent progress in hemocyanin research. Integr Comp Biol. 2007;47(4):631–44.
8. Vinogradov SN, Hoogewijs D, Bailly X, Arredondo-Peter R, Gough J, Dewilde S, Moens L, Vanfleteren JR. A phylogenomic profile of globins. BMC Evol Biol. 2006;6(1):31.
9. Vanin S, Negrisolo E, Bailly X, Bubacco L, Beltramini M, Salvato B. Molecular evolution and phylogeny of sipunculan hemerythrins. J Mol Evol. 2006;62:32–41.
10. Rousselot M, Delpy E, Rochelle CD, Lagente V, Pirow R, Rees JF, Hagege A, Guen D, Hourdez S, Zal F. *Arenicola marina* extracellular hemoglobin: A new promising blood substitute. Biotechnol J. 2006;1(3):333–45.
11. Zal F, Rousselot M. Use of a haemoglobin for the preparation of dressings and resulting dressings. U.S. Patent 9220929B2, December 28, 2015.
12. Jones L. Recent advances in the molecular design of synthetic vaccines. Nat Chem. 2015;7(12):952–60.
13. Zal, F. Use of haemoglobin of annelids for treating cancer. U.S. Patent 0374796A1, December 31, 2015.
14. Andreeva A, Howorth D, Brenner SE, Hubbard TJ, Chothia C, Murzin AG. SCOP database in 2004: refinements integrate structure and sequence family data. Nucleic Acids Res. 2004;32:226–9.
15. Mangum CP. Physiological function of the hemerythrins. In: Mangum CP, editor. Advances in Comparative & Environmental Physiology Vol 13 – Blood and Tissue Oxygen Carriers. Berlin: Springer; 1992. p. 173–92.
16. Thompson JW, Salahudeen AA, Chollangi S, Ruiz JC, Brautigam CA, Makris TM, Lipscomb JD, Tomchick DR, Bruick RK. Structural and molecular characterization of iron-sensing hemerythrin-like domain within F-box and leucine-rich repeat protein 5 (FBXL5). J Biol Chem. 2012;287(10):7357–65.
17. Addison AW, Bruce RE. Chemistry of *Phascolosoma lurco* hemerythrin. Arch Biochem Biophys. 1977;183:328–32.
18. Klippenstein GL. Structural aspects of hemerythrin and myohemerythrin. Am Zool. 1980;20:39–51.
19. Klotz IM, Kurtz Jr DM. Binuclear oxygen carriers: hemerythrin. Accounts Chem Res. 1984;17(9):16–22.
20. Stenkamp RE. Dioxygen and hemerythrin. Chem Rev. 1994;94:715–26.
21. Wirstam M, Lippard SJ, Friesner R. Reversible dioxygen binding to hemerythrin. J Am Chem Soc. 2003;125(13):3980–7.
22. Wells RMG. Respiratory characteristics of the blood pigments of three worms from an intertidal mudflat. New Zeal J Zool. 1982;9:243–8.
23. Kurtz Jr DM. Molecular structure/function of the hemerythrins. In: Mangum CP, editor. Advances in Comparative & Environmental Physiology Vol 13 – Blood and Tissue Oxygen Carriers. Berlin: Springer; 1992. p. 151–71.
24. French CE, Bell JML, Ward FB. Diversity and distribution of hemerythrin-like proteins in prokaryotes. FEMS Microbiol Lett. 2008;279(2):131–45.

25. Li X, Tao J, Hu X, Chan J, Xiao J, Mi K. A bacterial hemerythrin-like protein MsmHr inhibits the SigF-dependent hydrogen peroxide response in mycobacteria. Front Microbiol. 2015;5(Jan):1–11.

26. Alvarez-Carreño C, Becerra A, Lazcano A. Molecular evolution of the oxygen-binding hemerythrin domain. PLoS ONE. 2016;11(6):e0157904.

27. Weigert A, Helm C, Meyer M, Nickel B, Arendt D, Hausdorf B, Santos SR, Halanych KM, Purschke G, Bleidorn C, Struck TH. Illuminating the base of the annelid tree using transcriptomics. Mol Biol Evol. 2014;31(6):1391–401.

28. Bailly X, Vanin S, Chabasse C, Mizuguchi K, Vinogradov SN. A phylogenomic profile of hemerythrins, the nonheme diiron binding respiratory proteins. BMC Evol Biol. 2008;8:244.

29. Martín-Durán JM, De Mendoza A, Sebé-Pedrós A, Ruiz-Trillo I, Hejnol A. A broad genomic survey reveals multiple origins and frequent losses in the evolution of respiratory hemerythrins and hemocyanins. Genome Biol Evol. 2013;5:1435–42.

30. Klippenstein GL, Cote JL, Ludlam SE. The primary structure of myohemerythrin. Biochemistry. 1966;15(5):1128–36.

31. Ward KB, Hendrickson WA, Klippenstein GL. Quaternary and tertiary structure of hemerythrin. Nature. 1975;257:818–21.

32. Coutte L, Slomianny MC, Malecha J, Baert JL. Cloning and expression analysis of a cDNA that encodes a leech hemerythrin. Biochim Biophys Acta. 2001;1518(3):282–6.

33. Baert JL, Britel M, Sautière P, Malécha J. Ovohemerythrin, a major 14-kDa yolk protein distinct from vitellogenin in leech. Eur J Biochem. 1992;209:563–9.

34. Vergote D, Sautière PE, Vandenbulcke F, Vieau D, Mitta G, Macagno ER, Salzet M. Up-regulation of neurohemerythrin expression in the central nervous system of the medicinal leech, *Hirudo medicinalis*, following septic injury. J Biol Chem. 2004;279(42):43828–37.

35. Mangum CP. Major events in the evolution of the oxygen carriers. Amer Zool. 1998;38(1):1–13.

36. Manwell C, Baker CMA. *Magelona* haemerythrin: tissue specificity, molecular weights and oxygen equilibria. Comp Biochem Phys B. 1988;89(3):453–63.

37. Rouse G, Pleijel F. Polychaetes. New York: Oxford University Press; 2001.

38. Struck TH, Golombek A, Weigert A, Franke FA, Westheide W, Purschke G, Bleidorn C, Halanych KM. The evolution of annelids reveals two adaptive routes to the interstitial realm. Curr Biol. 2015;25(15):1993–9.

39. Weigert A, Bleidorn C. Current status of annelid phylogeny. Org Divers Evol. 2016;16:1–18.

40. DeSalle R. Can single protein and protein family phylogenies be resolved better? J Phylogenetics Evol Biol. 2015;3:e116.

41. Gu X, Zou Y, Su Z, Huang W, Zhou Z, Arendsee A, Zeng Y. An update of DIVERGE software for functional divergence analysis of protein family. Mol Biol Evol. 2013;30:1713–9.

42. Wertheim JO, Murrell R, Smith MD, Kosakovsky Pond SL, Scheffler K. RELAX: Detecting relaxed selection in a phylogenetic framework. Mol Biol Evol. 2014;32:820–32.

43. Demuynck S, Li KW, Van der Schors R, Dhainaut-Courtois N. Amino acid sequence of the small cadmium-binding protein (MP-II) from *Nereis diversicolor* (Annelida, Polychaeta) – evidence for a myohemerythrin structure. Eur J Biochem. 1993;217:151–6.

44. Koch J, Lüdemann J, Spies R, Last M, Amemiya CT, Burmester T. Unusual diversity of myoglobin genes in the lungfish. Mol Biol Evol. 2016;33(12):3033–41.

45. Coates CJ, Decker H. Immunological properties of oxygen-transport proteins: hemoglobin, hemocyanin and hemerythrin. Cell Mol Life Sci. 2017;74:293.

46. Liu J, Ou Q, Han J, Li J, Wu Y, Jiao G, He T. Lower Cambrian polychaete from China sheds light on early annelid evolution. Naturwissenschaften. 2015;102(5):1–7.

47. Kocot KM, Cannon JT, Todt C, Citarella MR, Kohn AB, Meyer A, Santos SR, Schander C, Moroz LL, Lieb B, Halanych KM. Phylogenomics reveals deep molluscan relationships. Nature. 2011;477:452–6.

48. Whelan NV, Kocot KM, Moroz LL, Halanych KM. Error, signal, and the placement of Ctenophora sister to all other animals. Proc Natl Acad Sci U S A. 2015;112:5773–8.

49. Brown CT, Howe A, Zhang Q, Pyrkosz AB, Brom TH. A reference-free algorithm for computational normalization of shotgun sequencing data. arXiv:1203.4802 [q-bio.GN]. 2012.

50. Grabherr MG, et al. Full-length transcriptome assembly from RNA-Seq data without a reference genome. Nat Biotechnol. 2011;29:644–52.

51. Altschul SF, Gish W, Miller W, Myers EW, Lipman DJ. Basic local alignment search tool. J Mol Biol. 1990;215:403–10.

52. Gene Ontology Consortium. The Gene Ontology (GO) database and informatics resource. Nucleic Acids Res. 2004;32(1):D258–61.

53. Haas BJ, Papanicolaou A, Yassour M, Grabherr M, Blood PD, Bowden J, Couger MB, Eccles D, Li B, Lieber M, MacManes MD, Ott M, Orvis J, Pochet N, Strozzi F, Weeks N, Westerman R, William T, Dewey CN, Henschel R, LeDuc RD, Friedman N, Regev A. De novo transcript sequence reconstruction from RNA-seq using the Trinity platform for reference generation and analysis. Nat Protoc. 2013;8(8):1494–512.

54. Katoh K, Standley DM. MAFFT multiple sequence alignment software version 7: improvements in performance and usability. Mol Biol Evol. 2013;30(4):772–80.

55. Kearse M, Moir R, Wilson A, Stones-Havas S, Cheung M, Sturrock S, Buxton S, Cooper A, Markowitz S, Duran C, Thierer T, Ashton B, Meintjes P, Drummond A. Geneious Basic: an integrated and extendable desktop software platform for the organization and analysis of sequence data. Bioinformatics. 2012;28(12):1647–9.

56. Suyama M, Torrents D, Bork P. PAL2NAL: robust conversion of protein sequence alignments into the corresponding codon alignments. Nucleic Acids Res. 2006;34 suppl 2:609–12.

57. Darriba D, Taboada GL, Doallo R, Posada D. jModelTest 2: more models, new heuristics and parallel computing. Nat Methods. 2012;9(8):772.

58. Ronquist F, Huelsenbeck JP. MrBayes 3: Bayesian phylogenetic inference under mixed models. Bioinformatics. 2003;19(12):1572–4.

59. Rambaut A, Suchard MA, Xie D, Drummond AJ. Tracer v1.6. Available from http://beast.bio.ed.ac.uk/Tracer. 2014. Accessed 20 July 2016.

60. Drummond AJ, Suchard MA, Xie D, Rambaut A. Bayesian phylogenetics with BEAUti and the BEAST 1.7. Mol Biol Evol. 2012;29:1969–73.

61. Yang J, Yan R, Roy A, Xu D, Poisson J, Zhang Y. The I-TASSER Suite: Protein structure and function prediction. Nat Methods. 2015;12:7–8.

62. Kocot KM, Struck TH, Merkel J, Waits DS, Todt C, Brannock PM, Weese DA, Cannon JT, Moroz LL, Lieb B, Halanych KM. Phylogenomics of Lophotrochozoa with consideration of systematic error. Syst Biol. 2016. doi:10.1093/sysbio/syw079.

4

Exploring the evolutionary origin of floral organs of *Erycina pusilla*, an emerging orchid model system

Anita Dirks-Mulder[1,2], Roland Butôt[1], Peter van Schaik[2], Jan Willem P. M. Wijnands[2], Roel van den Berg[2], Louie Krol[2], Sadhana Doebar[2], Kelly van Kooperen[2], Hugo de Boer[1,7,8], Elena M. Kramer[3], Erik F. Smets[1,6], Rutger A. Vos[1,4], Alexander Vrijdaghs[6] and Barbara Gravendeel[1,2,5*] (iD)

Abstract

Background: Thousands of flowering plant species attract pollinators without offering rewards, but the evolution of this deceit is poorly understood. Rewardless flowers of the orchid *Erycina pusilla* have an enlarged median sepal and incised median petal ('lip') to attract oil-collecting bees. These bees also forage on similar looking but rewarding Malpighiaceae flowers that have five unequally sized petals and gland-carrying sepals. The lip of *E. pusilla* has a 'callus' that, together with winged 'stelidia', mimics these glands. Different hypotheses exist about the evolutionary origin of the median sepal, callus and stelidia of orchid flowers.

Results: The evolutionary origin of these organs was investigated using a combination of morphological, molecular and phylogenetic techniques to a developmental series of floral buds of *E. pusilla*. The vascular bundle of the median sepal indicates it is a first whorl organ but its convex epidermal cells reflect convergence of petaloid features. Expression of *AGL6 EpMADS4* and *APETALA3 EpMADS14* is low in the median sepal, possibly correlating with its petaloid appearance. A vascular bundle indicating second whorl derivation leads to the lip. *AGL6 EpMADS5* and *APETALA3 EpMADS13* are most highly expressed in lip and callus, consistent with current models for lip identity. Six vascular bundles, indicating a stamen-derived origin, lead to the callus, stelidia and stamen. *AGAMOUS* is not expressed in the callus, consistent with its sterilization. Out of three copies of *AGAMOUS* and four copies of *SEPALLATA*, *EpMADS22* and *EpMADS6* are most highly expressed in the stamen. Another copy of *AGAMOUS*, *EpMADS20*, and the single copy of *SEEDSTICK*, *EpMADS23*, are most highly expressed in the stelidia, suggesting *EpMADS22* may be required for fertile stamens.

Conclusions: The median sepal, callus and stelidia of *E. pusilla* appear to be derived from a sepal, a stamen that gained petal identity, and stamens, respectively. Duplications, diversifying selection and changes in spatial expression of different MADS-box genes shaped these organs, enabling the rewardless flowers of *E. pusilla* to mimic an unrelated rewarding flower for pollinator attraction. These genetic changes are not incorporated in current models and urge for a rethinking of the evolution of deceptive flowers.

Keywords: Deceptive pollination, Floral development, MADS-box genes, Mimicry, Vascular bundles

* Correspondence: Barbara.Gravendeel@naturalis.nl
[1]Endless Forms group, Naturalis Biodiversity Center, Vondellaan 55, 2332 AA Leiden, The Netherlands
[2]Faculty of Science and Technology, University of Applied Sciences Leiden, Zernikedreef 11, 2333 CK Leiden, The Netherlands
Full list of author information is available at the end of the article

Background

Flowering plants interact with a wide range of other organisms including pollinators. Pollinators can either receive nectar, oil, pollen or shelter in return for pollen transfer in a rewarding relationship, or nothing at all in a deceptive relationship [1]. One of the deceptive strategies is mimicry, defined as the close resemblance of one living organism, 'the mimic', to another, 'the model', leading to misidentification by a third organism, 'the operator'. Essential for mimicry is the production of a false signal (visual, olfactory and/or tactile) that is used to mislead the operator, resulting in a gain in fitness of the mimic [1]. Mimicry in plants generally serves the purpose of attraction of pollinators to facilitate fertilization. In these cases, an unrewarding plant species mimics traits typical for co-flowering models, such as a specific floral shape, coloration, and presence of nectar guides, glands, trichomes or spurs. In this way, pollinators, that are unable to distinguish the two types of flowers from each other, are fooled [1, 2]. Despite the fact that deceptive pollination evolved in thousands of plant species, most notably orchids [3], the mechanisms by which this deceit evolved are still poorly understood.

Flowers are the main attractors of the majority of angiosperms to gain attention of pollinators. The outer first whorl of a flower is usually made up of sepals that generally serve as protection covering the other floral parts until anthesis. The outer second whorl consists of often-showy petals mainly involved in pollinator attraction. The sepals and petals together enfold the male and female reproductive organs in the inner floral whorls. Over the past decades, evolutionary developmental (evo-devo) studies have yielded many new insights in the role of duplication and neo-functionalization of developmental genes in floral diversification and the evolution of sepals, petals and male and female reproductive organs. These studies helped redefine the evolutionary origin of such organs [4].

Theoretically, an orchid flower can be considered to consist of five whorls of floral organs. Three sepals and three petals are present in the outer two whorls. Three external and three internal stamens and three carpels are present in the three inner whorls (Fig. 5a). Studies of the genetic plant model species *Arabidopsis thaliana* have shown that genes only associated with petals in *A. thaliana* are also expressed in the first floral whorl of petaloid monocots including orchids. Expression of these genes in the first whorl of petaloid monocots plays an important role in the similarity of sepals and petals in lilies, gingers and orchids [5–7]. From an evolutionary perspective, retention of expression of genes associated with petals in the outer floral whorl is considered an ancestral character for angiosperms [8]. In orchid flowers, the median petal, or 'lip', is often enlarged and ornamented with a wart-like structure, or 'callus'. The lip mostly functions as main attractor and landing platform for pollinators. Many hypotheses have been put forward about the evolutionary origin of the lip and its ornaments [9]. Hsu et al. [10] showed that the lip is homologous with true petals but gained an additional function possibly due to the duplication of a complex of modified developmental genes that gained novel expression domains.

A stamen usually consists of a filament and an anther where the pollen are produced. Many lineages in plant families such as buttercups, orchids, penstemons and witch-hazels, not only have fertile stamens but also rudimentary, sterile or abortive stamen-like structures. These structures are generally called staminodes and are often positioned between the fertile stamens and carpels, although they can also occur in other positions [11]. Multiple hypotheses exist about the function of the morphologically very diverse staminodes. In *Aquilegia*, staminodes play a role in protecting the early developing fruits as they usually remain present after pollination long after the other organs have abscised [12]. In other plant genera, staminodes are assumed to mediate pollination. Comparative gene expression and silencing studies showed that staminode identity in *Aquilegia* evolved from a pre-existing stamen identity program. Of the genes involved, one lineage duplicated and one paralog became primarily expressed in the staminodia [11, 12].

Characteristic for orchids is that the male and female reproductive organs are incorporated in a so-called 'gynostemium'. This structure is thought to result from a fusion of a maximum of six fertile to (partly) sterile stamens and parts of the pistil, in particular the style and stigma. It is a complex organ and the evolutionary origin of its different parts is not yet clear [9, 13, 14]. During the evolution of the orchids over the past 100 million years a reduction in the number of fertile stamens and fusion with the carpels occurred [15–17]. Six fertile stamens, positioned in floral whorls three and four, are commonly present in the closest relatives of the orchids in Asparagales. In the Apostasioideae, the earliest diverging of the five subfamilies of orchids, the number of fertile stamens is reduced to three in the genus *Neuwiedia*, one in floral whorl three and two in whorl four. In the genus *Apostasia*, a staminode develops in floral whorl three or nothing resulting in two fertile stamens [18]. In subfamily Cypripedioideae only two fertile stamens are present. A further reduction into a single fertile stamen in floral whorl three evolved in subfamilies Vanilloideae, Orchidoideae and Epidendroideae [13]. Since the two subfamilies with either three or two fertile stamens are the least diverse, reduction to a single fertile stamen may have contributed to species diversification. The sterile stamens have evolved into many other structures. In the majority of the Epidendroid orchids with a single fertile stamen, the mature gynostemium evolved appendages projecting to the front or side, clearly differentiating from broadened or flattened tissue at the base, that help pollinators to position themselves in the correct way to remove or deposit pollinia, which ensures pollination. The shapes of these appendages differ greatly and different terms are used to describe them, e.g. column wings or 'stelidia' [19–21]. The oldest hypothesis postulates that the stelidia are remnants of male reproductive tissue [22, 23] and following this hypothesis, stelidia are interpreted as vestiges of the lateral stamens of the third and fourth floral whorls [24].

Current models explaining floral organ development

The genetic basis of floral organ formation can be explained with various genetic models of MADS-box transcription factors. The core eudicot 'ABCDE model' included the A-class gene *APETALA1* (*AP1*), B-class genes *APETALA3* (*AP3*) and *PISTILLATA* (*PI*), C-class gene *AGAMOUS* (*AG*), D-class gene *SEEDSTICK* (*STK*) and E-class gene *SEPALLATA* (*SEP*). This model has been revised for the monocots to reflect two key differences: (i) there are no *AP1* orthologs outside the core eudicots so *FRUITFULL* (*FUL*)-like genes are the closest homologs, and (ii) many monocots have entirely petaloid perianths. Class A + B + E genes specify petaloid sepals, A + B + E control petals, B + C + E determine stamens, C + E specify carpels, and D + E are necessary for ovule development [25–27] (Fig. 1a). As in the core eudicots, these genetic combinations are thought to function as protein complexes, as proposed by Theissen and Saedler [27] in the now well accepted 'floral quartet model' (Fig. 1b). For the highly specialized flowers of most orchid lineages, further elaborations have been proposed, including the 'orchid code' [28, 29], 'Homeotic Orchid Tepal' (HOT) model [30] and 'Perianth code' (P-code) [10].

The orchid code and HOT model (Fig. 1c) postulate that the four *AP3* lineages in orchids have experienced sub- and neo-functionalization to give rise to distinct petal and lip identity programs. In addition to original MADS-box genes incorporated in the ABCDE model, several *AGAMOUS-LIKE-6* (*AGL6*) gene copies were recently found to play an important role in orchid flower formation. According to the P-code model (Fig. 1d), there are two MADS-box protein complexes active in orchid flowers, one consisting of a set of *AP3/AGL6/PI* copies, specific for sepal/petal formation, and one consisting of another set of *AP3/AGL6/PI* copies, specific for the formation of the lip. When the ratio of these two complexes is skewed towards the latter, the lip is large. When the ratio is skewed towards the former, intermediate lip-structures are formed [10]. The P-code model has been functionally validated for wild-type *Oncidium* and *Phalaenopsis*, and also for *Oncidium* peloric mutants, in which the two petals are lip-like. The P-code model was also validated in orchids from other subfamilies than the Epidendroideae, to which *Oncidium* and *Phalaenopsis* belong, i.e. Cypripedioideae, Orchidoideae and Vanilloideae, and used to detect gene expression profiles in species with intermediate lip formation [10].

Erycina pusilla as an emergent orchid model: current resources and terminology

MADS-box genes have now been identified for several commercially important orchid genera (e.g. *Cymbidium, Dendrobium, Oncidium* and *Phalaenopsis*) [30–32] but long life cycles, large chromosome numbers and complex genomes of these genera hamper functional studies. DNA-mediated transformation can be used to study the function of orchid genes and *E. pusilla*, with its relatively short life cycle, functions as an emergent orchid model species for such studies [33, 34].

Erycina pusilla belongs to the Oncidiinae, which is a highly diverse subtribe of meso- and south-American epiphytic orchids in subfamily Epidendroideae [35]. It is a rapidly growing orchid species with a low chromosome number ($n = 6$) and a, for orchids, relatively small sized diploid genome of 1.475 Gb [36, 37]. It can be grown from seed to flowering stage in less than a year [33, 34] and plantlets can be grown without mycorrhizae in test tubes. Flowers develop in a few days in which five distinct floral developmental stages can be observed (Fig. 2a). The species produces deceptive flowers that are self-compatible but incapable of spontaneous self-pollination.

Oil-collecting *Centris* bees are the main pollinators [38]. The lateral sepals of *E. pusilla* are small and green. The median sepal is larger and more colorful than the lateral sepals. The lip is the largest part of the flower and very different in shape compared to the lateral petals and sepals. On the basal part of the lip or 'hypochile', a callus is present that guides pollinators towards the stamen and stigma to either remove or deposit pollinia effectively. The gynostemium is enveloped on both sides by two large, wing-shaped structures that we further refer to as stelidia. During floral visits, *Centris* bees cling to these stelidia and the callus with their forelegs while searching for oils (Fig. 2b). In *E. pusilla* however, these bees are fooled because the flowers employ food deception by Batesian mimicry by resembling flowers of rewarding species of the unrelated Malpighiaceae [38–40]. Flowers of this family have five clawed petals that are often unequal in size. The sepals carry oil glands. It is generally assumed that the enlarged median sepal, incised lip, callus and stelidia of Oncidiinae evolved to mimick the shape of the petals and oil glands of rewarding flowers of Malpighiaceae (Figs. 2b–d and 3) in order to attract oil-collecting bees for pollination [35, 38, 40, 41].

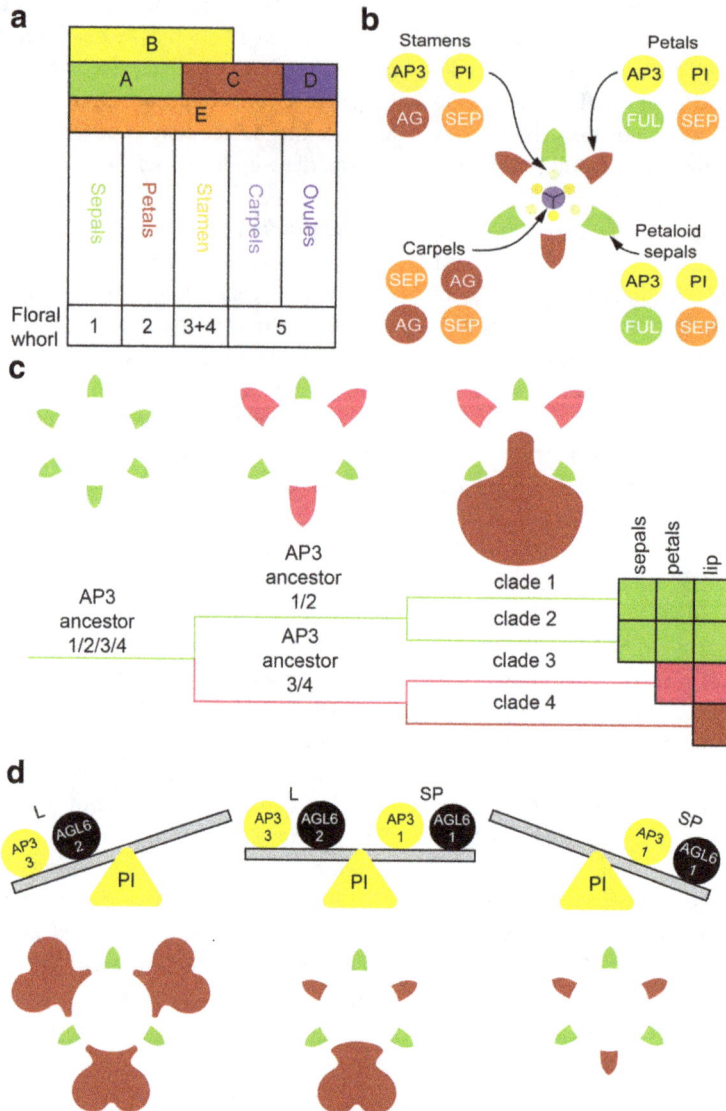

Fig. 1 Current models explaining floral organ development. **a** ABCDE model of floral development in petaloid monocots. **b** Floral quartet model. **c** Orchid code and HOT model. **d** Perianth code model [Illustrations by Bas Blankevoort]

Agrobacterium-mediated genetic transformation was recently developed for *E. pusilla* [33] and knockdown of genes is currently being optimized. It is expected that the entire genome will have been analyzed using a combination of next-generation sequencing techniques within the following years. Furthermore, transcriptome data of *E. pusilla* are included in the Orchidstra database [31]. Twenty-eight MADS-box genes from *E. pusilla* have been identified thus far including the most important floral developmental ones [34]. These resources make *E. pusilla* an ideal orchid model for evo-devo studies. Lin et al. [34] published expression data of MADS-box genes isolated from sepals, petals, lip, column and ovary of flowers of *E. pusilla* after anthesis together with a basic phenetic gene lineage analysis.

In this study, we employed a combination of micro-, macromorphological, molecular and phylogenetic techniques to assess the evolutionary origin of the median sepal, callus and stelidia of the flowers of *E. pusilla*. To accomplish this goal, we investigated early and late floral developmental stages with scanning electron microscopy (SEM), light microscopy (LM), 3D-Xray microscopy (micro-CT) and expression (RT-qPCR) of MADS-box genes belonging to six different lineages. In addition, we investigated gene duplication and putative neo-functionalization as indicated by inferred episodes of diversifying selection. Our aim was to test the hypotheses that the median sepal, callus and stelidia are derived from sepals, petals and stamens, respectively, to unravel the genetic basis of the evolution of deceptive flowers.

Fig. 2 General overview of *E. pusilla* flowers, pollinator and floral parts. **a** Five floral stages of *E. pusilla* [Photo by Rogier van Vugt]. **b** A female *Centris poecila* bee pollinating a flower of *Tolumnia guibertiana*, a close relative of *E. pusilla*, in Cuba [Photo by Angel Vale], showing the function of the stelidia and callus in freshly opened flowers of these orchids, i.e. attraction and providing a holdfast for the pollinator. **c** Frontal view of fully developed stelidia. **d** Adaxial side (with respect to the floral axis) of a flower. **e** Abaxial side (with respect to the floral axis). Abbreviations: s(cl) = callus; lse = lateral sepal; mse = median sepal; pe = petal; s(sl) = stelidium; fs = fertile stamen

Methods

Plant material and growth conditions

A more than 15 year old inbred line of *E. pusilla* originally collected in Surinam was grown in climate rooms under controlled conditions (7.00–23.00 h light regime), at a temperature of 20 °C and a relative humidity of 50%. The orchids were cultured in vitro under sterile conditions on Phytamax orchid medium with charcoal and banana powder (Sigma-Aldrich) mixed with 4 g/L Gelrite™ (Duchefa) culture medium. Pollinia of flowers from different plants were placed on each other's stigma after which ovaries developed into fruits. After 18–22 weeks, seeds were ripe and sown into containers with sterile fresh nutrient culture medium. The seeds developed into a new *E. pusilla* flowering plant within 20 weeks.

Fixation for micromorphology

Flowers and flower buds were fixed with standard formalin-aceto-alcohol (FAA: absolute ethanol, 90%;

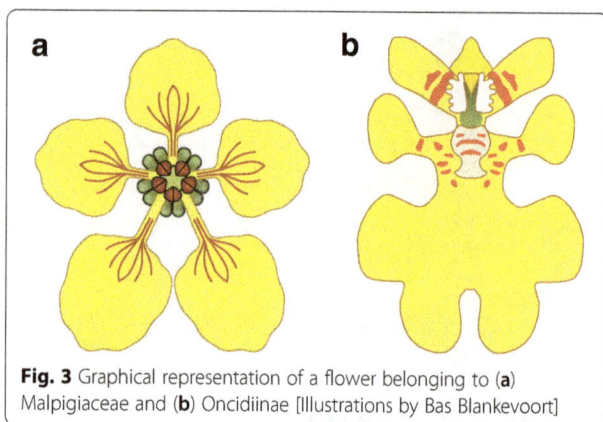

Fig. 3 Graphical representation of a flower belonging to (**a**) Malpigiaceae and (**b**) Oncidiinae [Illustrations by Bas Blankevoort]

glacial acetic acid, 5%, formalin; 5% acetic acid) for one hour under vacuum pressure at room temperature and for 16 h at 4 °C on a rotating platform. They were washed once and stored in 70% ethanol until further use.

SEM

Floral buds at different developmental stages were dissected in 70% ethanol under a Wild M3 stereo microscope (Leica Microsystems AG, Wetzlar, Germany) equipped with a cold-light source (Schott KL1500; Schott-Fostec LLC, Auburn, New York, USA). Subsequently, the material was washed with 70% ethanol and then placed in a mixture (1:1) of 70% ethanol and DMM (dimethoxymethane) for five minutes for dehydration. The material was then transferred to 100% DMM for 20 min and critical point dried using liquid CO_2 with a Leica EM CPD300 critical point dryer (Leica Microsystems, Wetzlar Germany). The dried samples were mounted on aluminium stubs using Leit-C carbon cement or double-sided carbon tape and coated with Platina-Palladium with a Quorum Q150TS sputtercoater (Quorum Technologies, Laughton, East Sussex, UK). Images were obtained with a JEOL JSM-7600 F Field Emission Scanning Electron Microscope (JEOL Ltd., Tokyo, Japan).

For the images presented in Fig. 4, fixed floral buds were critical point dried using liquid CO_2 with a CPD 030 critical point dryer (BAL-TEC AG, Balzers, Lichtenstein) and coated with gold with a SPI-ModuleTM Sputter Coater (SPI Supplies, West-Chester, Pennsylvania, USA). Scanning electron microscope (SEM) images were obtained with a Jeol JSM-6360 (JEOL Ltd., Tokyo) at the Laboratory of Plant Conservation and Population Biology (KU Leuven, Belgium).

3D-Xray microscopy

Fully grown flowers were infiltrated with 1% phosphotungstic acid (PTA) in 70% ethanol for 7 days in order to increase the contrast [42]. The PTA solution was changed every 1–2 days. The flowers were embedded in 1% low melting point agarose (Promega) prior to scanning. The scans were performed on a Zeiss Xradia 510 Versa 3D X-ray with a Sealed transmission 30–160 kV, max 10 W x-ray sources. Scanning was performed using the following settings: acceleration voltage/power 40 kV/ 3 W; source current 75 µA; exposure time 2 s; picture per sample 3201; camera binning 2; optical magnification 4 ×, with a pixel size of 3.5 µm. The total exposure time was approximately 3, 2 h. 3D images were stacked and processed with Avizo 3D software version 8.1.

RNA extraction

For organ dissection, floral buds of *E. pusilla* were collected from floral stages 2 and 4 (Fig. 2a). The earliest floral stage to dissect the different flower parts was at floral stage 2. The lateral sepals, median sepal, petals, lip, callus, stamen and the remaining part of the gynostemium with stelidia but excluding the ovary were dissected (Fig. 2c–e) and collected in individual tubes and immediately frozen on dry ice and stored at −80 °C until RNA extraction. Total RNA was extracted from seven different floral organs of *E. pusilla* using the RNeasy Plant Mini Kit (QIAGEN), following the manufacturer's protocol. A maximum of 100 mg plant material was placed in a 2.2 ml micro centrifuge tube with 7 mm glass bead. The TissueLyser II (QIAGEN) was used to grind the plant material. The amount of RNA was measured using the NanoVue Plus™ (GE Healthcare Life Sciences) and its integrity was assessed on an Agilent 2100 Bioanalyzer using the Plant RNA nano protocol. RNA samples with an RNA Integrity Number (RIN) < 7 were discarded. RNA was stored at −80 °C until further use. Extracted RNA was treated with DNase I, Amp Grade (Invitrogen 1U/µl) to digest single- and double-stranded DNA following the manufacturer's protocol.

cDNA synthesis

cDNA was synthesized with up to 1 µg of DNase-treated RNA using iScript™ cDNA Synthesis Kit (Bio-Rad Laboratories) following the manufacturer's protocol. A reaction mixture was prepared by addition of 1 µg of RNA, 4 µl 5x iScript reaction mix, 1 µl iScript reverse transcriptase to nuclease-free water up to a total volume of 20 µl. The reaction mixture was incubated at 25 °C for 5 min, 42 °C for 30 min and 85 °C for 5 min using a C1000 Touch™ thermal cycler machine (Bio-Rad). During this reaction, a positive control (CTRL) and no reverse transcriptase (NRT) control were included.

Primer design

DNA sequences were downloaded from NCBI Genbank and Orchidstra (http://orchidstra2.abrc.sinica.edu.tw).

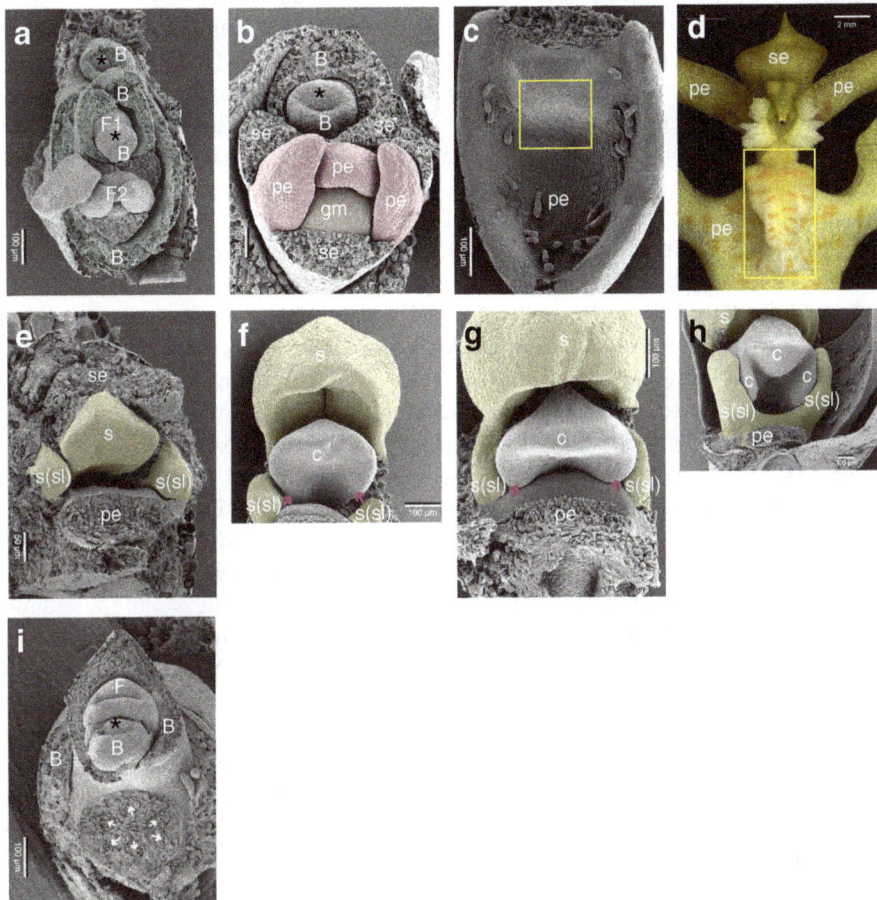

Fig. 4 Developing inflorescence of *E. pusilla*. **a** Apical view of a young developing inflorescence. A central meristem is present and below it two flowers are visible, each subtended by a bract. The distal flower (F1) is primordial and the next flower (F2) is somewhat more developed. **b** Apical view of a developing flower in an early developmental stage. The scars of the three removed sepals are visible, two are adaxially (lateral sepals) and one is abaxially (median sepal) situated. More central in the flower, two abaxial-lateral petals and one adaxial developing petal (lip) are present. Most central in the flower is the primordium of the gynostemium. **c–d** Developing adaxial petal (lip) with callus (*boxed*). **e–h** Successive stages of the development of the gynostemium with the developing fertile stamen central and stelidia laterally. In (**e**), the scar of the removed abaxial sepal is visible. Below the fertile stamen, the scar of the adaxial petal (lip) can be seen. In between the fertile stamen and the adaxial petal (lip), the stigmatic cavity is present. In (**f** and **g**), the two adaxial (lateral) carpels are visible (*arrowed*). In (**h**), the abaxial carpel is incorporated in the stigmatic cavity. **i** Apical view of an inflorescence axis with a removed developing flower. In the upper half of the micrograph, the apex of the axis is visible as well as a flower at very early developmental stage, subtended by a bract. In the lower half, in the scar of the removed developing flower, six vascular bundles are visible (arrowed). Abbreviations: Red asterisk = apical meristem; B = bract; F = flower (primordium); c = carpel; gm = gynostemium; pe = petal; se = sepal; s = fertile stamen; s(sl) = stelidium. Color codes: dark green = bract; red = petals; orange = gynostemium; yellow = androecium

For the MADS-box genes primers were designed on the C-terminal of the DNA sequences to avoid cross –amplification. Beacon Designer™ (Premier Biosoft, http://www.oligoarchitect.com) software was used to design primers (Additional file 1: Tables S1–S2). All primer pairs were screened for their specificity against the Orchidstra database and in a gradient PCR reaction. The reaction mixture (25 μl) contained: 2.5 ng cDNA, 0.2 μM of each primer, 0.1 mM dNTP's and 0.6 U *Taq* DNA polymerase (QIAGEN) in 1x Coral Load Buffer (QIAGEN). The amplification protocol was as follows: initial denaturation step of 5 min 94 °C followed by 40 cycles of [20 s 94 °C, 20 s <55–65 > °C, 20 s 72 °C], one final amplification step of 7 min 72 °C and ∞ 15 °C. Based on the results of the gradient PCR, the annealing temperature was set to 61.3 °C for the Quantitative Real-time PCR as this value gave the best results. Only when a specific product was detected was the primer pair used for subsequent quantification.

Reference genes and quantitative real-time PCR
Experimental and computational analyses with Lin-RegPCR (http://www.hartfaalcentrum.nl, v2015.1) [43, 44] indicate that *E. pusilla Ubiquitin-2, Actin*, and *F-box* were

stably expressed in the tissues of interest and these genes were chosen as reference genes for the expression assay. Expression of all MADS-box genes was normalized to the geometric mean of these three reference genes.

Quantitative real-time PCR was performed using the CFX384 Touch Real-Time PCR system (Bio-Rad Laboratories). The assays were performed using the iQ™ SYBR® Green Supermix (Bio-Rad Laboratories). The reaction mixture (7 μl) contained: 1x iQ™ SYBR® Green Supermix, 0.2 μM of each primer, 1 ng cDNA template from a specific floral organ (biological triplicate reactions) for each target gene and floral organ for two sets of isolated RNA (six reactions in total). All reactions were performed in Hard-Shell® Thin-Wall 384-Well Skirted PCR Plates (Bio-Rad Laboratories). For each amplicon group, a positive control was included (=CTRL, flower buds from floral stage 1 to 4), a negative control (=NTC, reaction mixture without cDNA) and a no reverse transcriptase treated sample (=NRT, control sample during the cDNA synthesis). For all the qPCR reactions, the amplification protocol was as follows: initial denaturation of 5 min 95 °C followed by; 20 s 95 °C; 30 s 61.3 °C; 30 s 72 °C; plate read, for 50 cycles; then followed by a melting curve analysis of 5 s, 65 °C to 95 °C with steps of 0.2 °C to confirm single amplified products (Additional file 2: Figure S2).

Normalization, data analysis and statistical analysis

The non-baseline corrected data were exported from the Bio-Rad CFX Manager™ (v3.1) to a spreadsheet. Quantification Amplification results (QAR) were used for analysis with LinRegPCR (v2015.1, dr. J.M. Ruijter). The calculated N_0-values represented the starting concentration of a sample in fluorescence units. Removal of between-run variation in the multi-plate qPCR experiments was done using Factor qPCR© (v2015.0) [45, 46]. Geometric means of the corrected N_0-values were calculated from the six samples together, i.e. two biological and three technical replicates. GraphPad Prism version 7.00 (http://www.graphpad.com) was used to perform a Two-Way ANOVA with Sidak's multiple comparison test to calculate significant differences between the two floral stages 2 and 4, and graphed with Standard Error of Measurement (SEM) error bars. Tukey's multiple comparisons test was used to compare the means between the floral organs. Variation for the two biological replicates was assessed by tests in triplicate.

Phylogenetic analyses

Nucleotide sequences of floral developmental genes were downloaded from NCBI GenBank® (Additional file 1: Table S1) and separate data sets were constructed for MADS-box gene classes *FUL-, AP3-, PI-, AG-, STK-, SEP-* and *AGL6-like*. For each gene class, protein-guided codon alignments were constructed by first performing multiple sequence alignments of the protein translations using MAFFT v.7.245 (with the algorithm most suited for proteins with multiple conserved domains, E-INS-I or "oldgenafpair" for backward compatibility), with a maximum of 1000 iterations [47] and then reconciling the nucleotide sequences with their aligned protein translations.

Gene trees were inferred from the codon alignments using PhyML v3.0_360-500M [48] under a GTR + G + I model with six rate classes and with base frequencies, proportion of invariant sites, and γ-shape parameter α estimated using maximum likelihood. Optimal topologies were selected from results obtained by traversing tree space with both nearest neighbor interchange (NNI) and subtree prune and regraft (SPR) branch swap algorithms, ie. PhyML's "BEST" option. Support values for nodes were computed using approximate likelihood ratio tests (SH-like aLRT, [49]).

To infer where on the gene trees duplications may have occurred the GSDI algorithm [50] was used as implemented in forester V1.038 (https://sites.google.com/site/cmzmasek/home/software/forester). Fully resolved species trees for GSDI testing were constructed based on the current understanding of the phylogeny of the species under study (Additional file 3: Figure S4).

Lastly, to detect lineage-specific excesses of non-synonymous substitutions, BranchSiteREL [51] analyses were performed as implemented in HyPhy [52] on the Datamonkey (http://datamonkey.org) cluster.

Results
Ontogeny, macro- and micromorphology of flowers of *E. pusilla*

Floral ontogeny in *E. pusilla* can be divided into two main phases: early and late. Early ontogeny starts from floral initiation (floral stage 1) up to the three-carpel-apex stage (floral stage 2) and late ontogeny starts from the three-carpel-apex stage (floral stage 2) until anthesis (floral stages 3, 4 and 5, Fig. 2a) [53].

The inflorescence of *E. pusilla* is branched and multiple flowers develop in succession (Fig. 2a). Up to floral stage 1, the perianth is formed following a classic monocot developmental pattern (Fig. 5a) [54] in which the sepals are among the first organs to become visible, followed by the petals. The position of the two abaxial petals is slightly shifted laterally (Fig. 4a). Stamen and carpel primordial are not visible in the course of the early phase, but instead a single massive primordium is present from which the gynostemium will develop (Fig. 4b).

On the hypochile of the lip a callus is formed from floral stage 2 onwards (Fig. 4c–d). The fertile stamen differentiates after floral stage 1. The stelidia appear at each

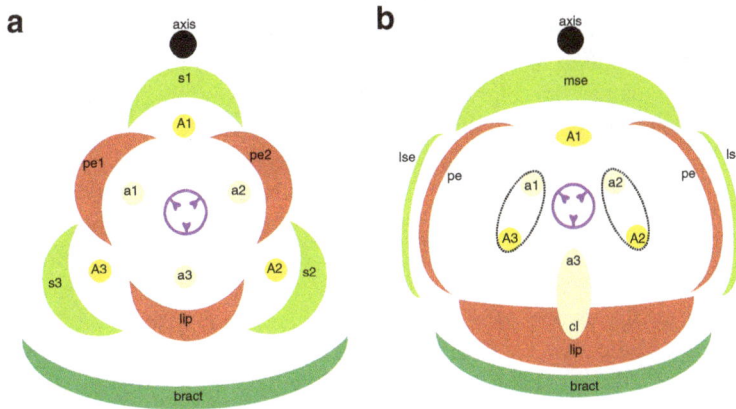

Fig. 5 Floral diagrams. **a** A typical monocot flower. **b** A resupinate flower of *E. pusilla*. Abbreviations: s_{1-3} = sepals; p_{1-3} = petals; A_{1-3} = anther in outer floral whorl; a_{1-3} = anther in inner floral whorl; lse = lateral sepal; mse = median sepal; pe = petal; cl = callus. Color codes: black interrupted = stelidia and callus on lip; purple = gynoecium [Illustrations by Erik-Jan Bosch]

side of the gynostemium (Fig. 4e–h) from where they elongate and start forming wing-like appendices (Fig. 2e). The abaxial carpel is incorporated in the stigmatic cavity, which forms a compound structure with the fertile stamen (Fig. 4h). The three-carpel-apex stage is clearly visible in floral stage 2. At this stage the six staminal vascular bundles can also be observed just above the inferior ovary (Fig. 4i). In floral stage 3, no new organs are formed, but in floral stage 4 (Fig. 2a) the mature flower becomes resupinate (Fig. 5b). The terms adaxial and abaxial are used here to indicate the position of the distinct floral parts with respect to the inflorescence axis (Fig. 4a–b), thereby taking the position of the primordia of the floral organs as a reference. For example, with

respect to the inflorescence axis, the lip is the adaxial petal, which by resupination becomes the lowermost part of the flower.

Using micro-CT scanning, vascular bundles were observed in a fully-grown floral stage 5 flower (Fig. 6a–f and Additional file 4: Movie S1). In the inferior ovary six vascular bundles could be discerned, indicated in purple. Three of these vascular bundles, indicated in green, run to the adaxial (median) sepal and abaxial (lateral) sepals, respectively. Three main groups of vascular bundles, indicated in red, run towards the petals including the lip, where they split up. Four vascular bundles (indicated in yellow) are present; one bundle, already split into two at the base, runs to the fertile stamen, where it splits up

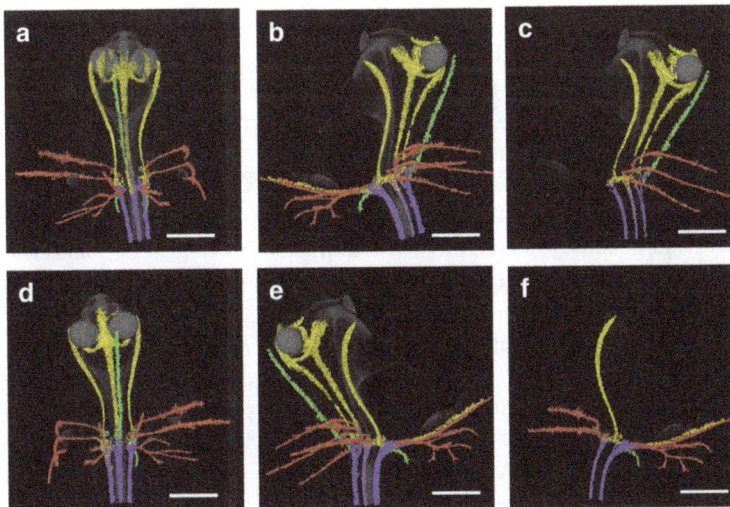

Fig. 6 Vascular bundle patterns of *E. pusilla*. **a** Frontal view of a 3D X-ray macroscopical reconstruction of the vascular bundle patterns in a mature flower of *E. pusilla*. **b** Successive clockwise turn of 45°. **c** Simplified version of (**b**). **d** Successive clockwise turn of 90°. **e** Successive clockwise turn of 135°. **f** Simplified version of (**e**). Color codes: green = vascular bundles in sepals; red = vascular bundles in petals; purple = vascular bundles in gynoecium; yellow = vascular bundles in androecium. Scale bar = 1 mm

further towards the two pollinia (Fig. 6a–e); two vascular bundles, originated from two pairs, run up into the stelidia (Fig. 6b–c; e–f) and one vascular bundle runs all the way up into the callus of the lip (Fig. 6b; e–f). When following the yellow vascular bundles downwards, they connect in a plexus situated on top of the inferior ovary with the rest of the vascular system of the flower.

Throughout late ontogeny, epidermal cells in all floral organs remained relatively undifferentiated and only expanded in size. Epidermal cells on the abaxial side of floral organs were mostly similar to the cells on the adaxial side, but more convex shaped (Additional file 5: Figure S1). Epidermal cells of the lateral sepals were irregular, flattened and rectangular shaped and longitudinally orientated from the base to the apex (Fig. 7a–c). Epidermal cells of the median sepal, as well as of the petals and the lip, develop from irregularly flattened shaped cells at floral stage 2, to a more convex shape in floral stage 5 (Fig. 7d–l). Epidermal cells of the callus develop from convex shaped cells in floral stage 2 to cells with a more conical shape in floral stage 5 (Fig. 7m–o). Epidermal cells of the stelidia become convex shaped during floral stage 2 and develop papillae on their apices during floral stage 5 (Fig. 7p–r).

Duplications, diversifying evolution and expression of eighteen MADS-box genes in selected floral organs of E. pusilla in two developmental stages

FUL-, SEP- and AGL6-like genes

The closest homologs of the *Arabidopsis* A class gene *APETALA1* in *E. pusilla* are the three *FUL*-like genes copies *EpMADS10, 11* and *12*. Our phylogenetic analyses reconstructed three orchid clades of *FUL*-like genes, containing the three copies present in the genome of *E. pusilla* (Additional file 6: Figure S5a), which was consistent with previous studies [55]. Diversifying selection was detected along the branch following the gene duplication leading to *EpMADS10*. The three *FUL*-like gene copies were expressed in all floral organs of *E. pusilla* but at low levels only (Additional file 7: Figure S3). During development, expression generally decreased in most floral organs for *EpMADS10* and *11* whereas it generally increased for the majority of floral organs for *EpMADS12* (Additional file 7: Figure S3 and Additional file 1: Table S3).

Four *SEP*-like orchid clades were retrieved (Additional file 6: Figure S5f), encompassing the four copies of *E. pusilla*, consistent with previous studies [55, 56]. The branch leading to the duplication that gave rise to *EpMADS6* and *EpMADS7* shows evidence of diversifying selection. *EpMADS6, 7, 8* and *9* were expressed in all floral organs at varying levels. *EpMADS6* was mainly expressed in the fertile stamen, a statistically

significant difference as compared to the other six floral organs (Additional file 7: Figure S3 and Additional file 1: Table S3).

Three *AGL6* orchid clades, also found by Hsu et al. [10] were retrieved, containing the three different copies present in the *E. pusilla* genome (Additional file 6: Figure S5g). Evidence for a moderate degree of diversifying selection could be detected on the branch leading to *EpMADS4*. The three different copies of *AGL6*-genes were not expressed in all floral organs and the level of expression also varied. *EpMADS3* was most highly expressed in the sepals and petals. *EpMADS4* was more highly expressed in the lateral sepals as compared with the median sepal, petals and lip. *EpMADS5* was mainly expressed in the lip and callus (Fig. 8).

AP3-like and PI-like genes

Initial phylogenetic analyses reconstructed the main duplication between the *AP3* and *PI* genes also found in many other studies [10, 30, 57] so two separate gene trees were retrieved for each lineage (Additional file 6: Figure S5b–c). Four orchid *AP3*-clades and three *PI*-clades were identified in these analyses. The three copies of *AP3* and a single copy of *PI* present in the genome of *E. pusilla* were placed in *AP3*-clades 1, 2 and 3 and *PI*-clade 2, respectively. No evidence for diversifying selection could be detected along the branches leading to the *PI*-clade containing *EpMADS16* but evidence for diversifying selection along the branch in the *AP3*-1 clade encompassing *EpMADS15* was found. *AP3*-like gene copy *EpMADS14* was most highly expressed in the lateral sepals. *AP3*-like gene copy *EpMADS13* was more highly expressed in the lip and callus than in the sepals and petals (Fig. 8). The *PI*-like gene *EpMADS16* was more highly expressed in the first four floral whorls in both floral stages (Figs. 8 and 9).

AG- and STK-like genes

Three orchid *AG*-clades and two *STK*-clades were identified in the phylogenetic analyses (Additional file 6: Figure S5d–e). *EpMADS20, 21 and 22* were placed in *AG*-clades 3, 1 and 2, respectively, and *EpMADS23* was placed in *STK*-clade 1, as also found by Lin et al. [34]. No evidence for diversifying selection in the branches supporting the three orchid *AG*-clades and *STK*-clade containing copies present in the genome of *E. pusilla* could be detected. *AG*-like gene copy *EpMADS20* was most highly expressed in the stelidia, whereas *EpMADS22* was most highly expressed in the stamen as compared with all other floral organs analyzed (Fig. 8). No expression of *AG*-like genes could be detected in the callus. *STK*-like gene copy

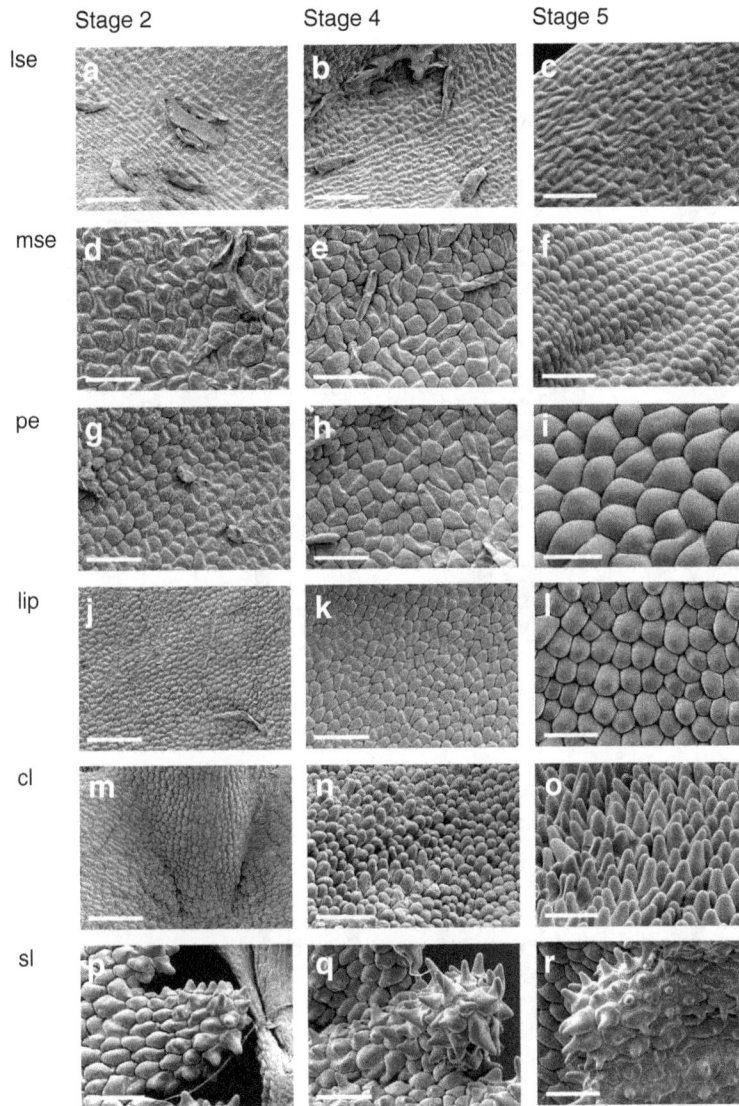

Fig. 7 Micromorphology of the epidermal cells on the adaxial side of a flower of *E. pusilla*. The three columns represent, from left to right, floral stage 2, 4 and 5 of the floral organs. Epidermal cells of (**a–c**) lateral sepal, (**d–f**) median sepal, (**g–i**) petal, (**j–l**) lip, (**m–o**) callus on lip and (**p–r**) stelidia. Scale bar = 100 μm. Abbreviations: lse = lateral sepal; mse = median sepal; pe = petal; cl = callus; sl = stelidia

EpMADS23 was most highly expressed in the stelidia as compared with all other floral organs analyzed (Figs. 8 and 9).

Discussion
Homology of the median sepal of *E. pusilla*
The floral ontogenetic observations and vascularization patterns indicate that the median sepal is derived from the first floral whorl. In contrast, the presence of convex epidermal cells suggests a petaloid origin [58]. The *AGL6* and *AP3* copies *EpMADS3* and *EpMADS15,* members of the sepal/petal-complex of the P-code model, were most highly expressed in the median sepal, lateral sepal and petal. A possible

correlation between expression and petaloidy was found for *AGL6* and *AP3* copies *EpMADS4* and *EpMADS14.* These two genes were lowly expressed in the median sepal, lip and petal as compared with the lateral sepal. Additional functional studies are needed to show whether loss of function of *EpMADS4* and *EpMADS14* is linked to sepal morphology in *E. pusilla* and other species that also possess a petaloid median sepal. The *AGL6* gene copy *EpMADS4* copy showed evidence of diversifying evolution. Lin et al. [34] identified fifteen motifs in the MIKC-type MADS-box proteins of *E. pusilla*. Two differences can be noticed within the K-region and C-terminal-region of *AP3* and *AGL6* genes of *E. pusilla*: (i) *AP3*

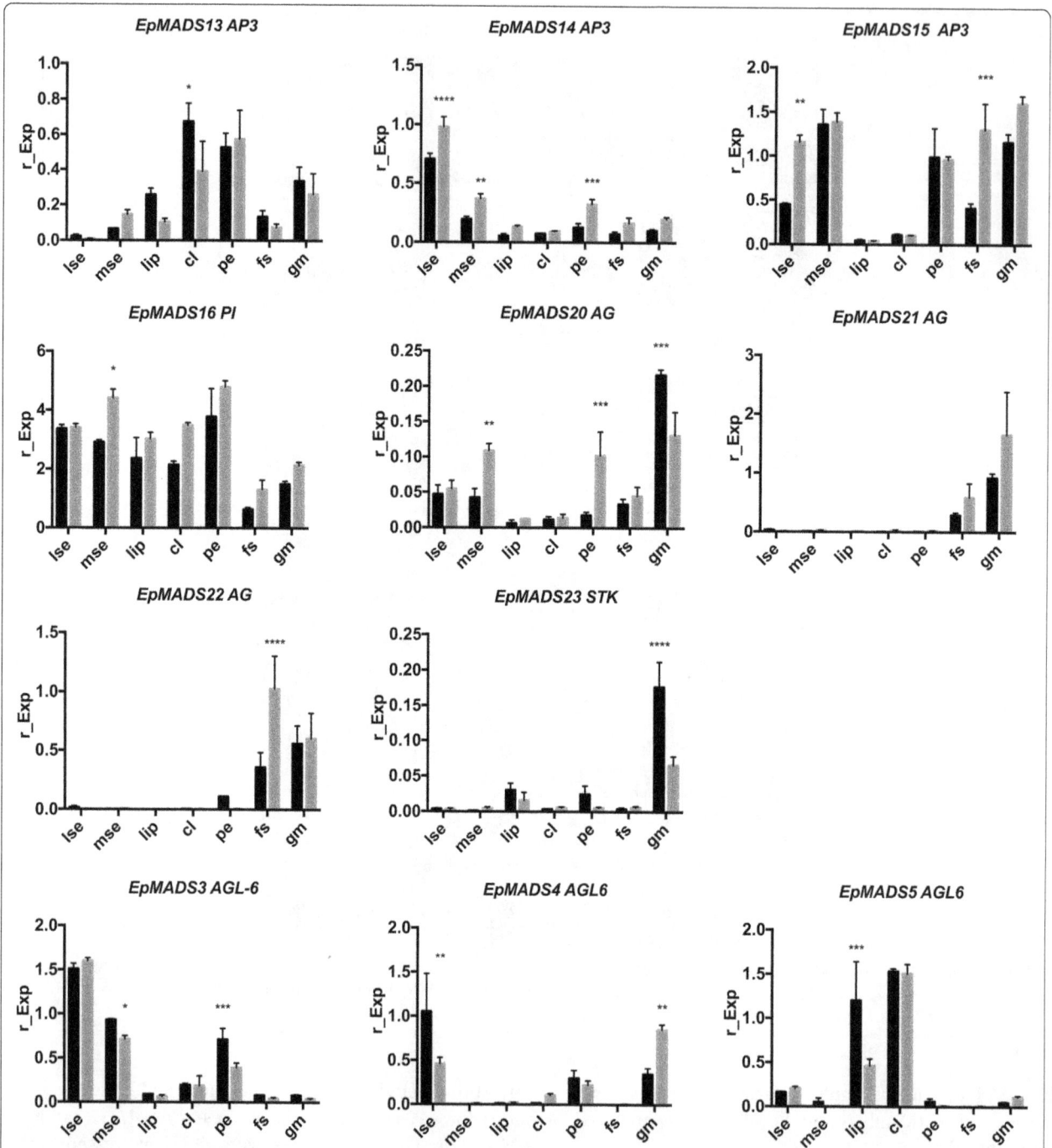

Fig. 8 Floral organ specific expression levels of selected MADS-box gene copies in *E. pusilla*. *AP3* (top row), *PI* (second row), *AG* (second and third row), *STK* (second row), *ALG6* (third row). RNA was extracted from seven different floral organs during two stages of development of *E. pusilla* and used for cDNA synthesis. Expression of the MADS-box genes was normalized to the geometric mean of three reference genes *Actin*, *UBI2* and *Fbox*. Each column shows the relative expression of 20 floral organs in two cDNA pools (10 floral organs per isolation), both tested in triplicate. Abbreviations: lse = lateral sepal; mse = median sepal; cl = callus; pe = petal; fs = fertile stamen; gm = gynostemium. Dark grey = floral stage 2 and light grey = floral stage 4. Y-axis: relative gene expression. The *error bars* represent the Standard Error of Mean. P-value style: GP: >0.05 (ns), <0.05 (*), <0.01 (**), <0.001 (***), <0.0001 (****)

EpMADS14 is missing motif 11, while the other B-class genes all contain motif 11. *AGL6 EpMADS4* also contains motif 11, while the other *AGL6* gene copies lack this motif; (ii) *AGL6 EpMADS4* is missing motif 6 whereas all the other *AGL6* gene copies contain motif 6. The differences found may contribute to the morphological differences between the median and lateral sepals *of E. pusilla*.

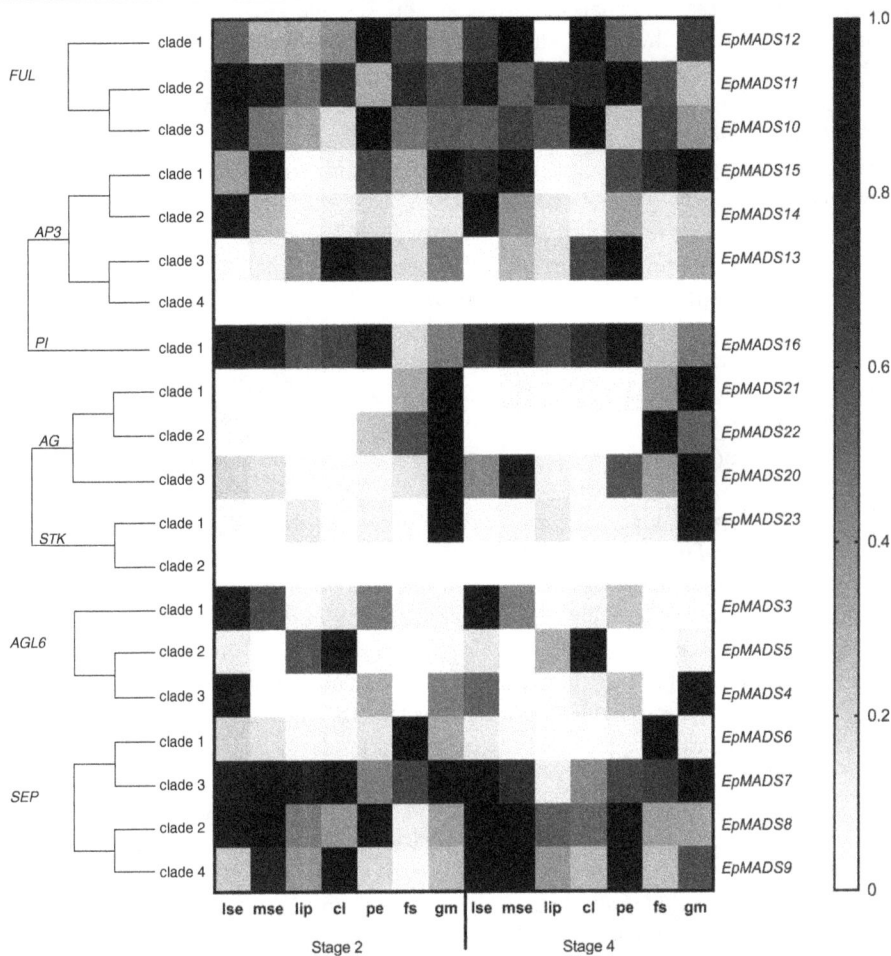

Fig. 9 Heat map representation of MADS-box gene expression in *E. pusilla*. The *FUL-, AP3-, PI-, AG-, STK-, SEP-* and *ALG6-* like copies were retrieved from different gene lineage clades during two stages of floral development. Expression of the MADS-box genes was normalised to the geometric mean of three reference genes *Actin, UBI2* and *Fbox*. The relative gene expression was normalised with the CTRL sample (= flower buds from floral stages 1-4). The scales for each gene and developmental stage are independent of each other and set to 1 for the highest value. Abbreviations: lse = lateral sepal; mse = median sepal; cl = callus; pe = petal; fs = fertile stamen; gm = gynostemium

Homology of the lip and callus of *E. pusilla*

The convex shaped epidermal cells on the lip and conical shaped epidermal cells on the callus are indicative of a petaloid function [58]. The *FUL*-like gene copy *EpMADS12*, *AP3*-like *EpMADS13* and *AGL6*-like *EpMADS5* are most highly expressed in lip and callus, further confirming a lip identity based on the ABCDE, floral quartet and P-code models, that dictate joint expression of A, B, E and *AGL6*-like genes in the petals and lip, respectively. According to these models, B, C and E class genes should be expressed in stamens but no evidence of expression of C class genes was found in the lip or callus. Notwithstanding, the possible staminal origin of the callus is supported by multiple lines of evidence. First of all, the ontogeny and function of the lip of *E. pusilla* are very different as compared with the ontogeny and function of the callus. The lip is formed from floral stage 1 onwards, mainly acts as a long

distance attraction and functions as a soft landing platform for pollinating bees. The callus is formed from floral stage 2 onwards and functions as short distance attraction by offering a sturdy holdfast to pollinators. This is in line with Carlquist [59], who states that different vascularization patterns are driven by different functional needs. Many Oncidiinae have a callus on the lip and in some of these species, the callus produces oil, making the functions of the lip and the callus even more distinct. Flowers with an oil-producing callus evolved twice in unrelated clades from species with non-rewarding flowers according to the molecular phylogeny of the Oncidiinae as presented in Pridgeon et al. [38]. One of the two rewarding clades, i.e. the one containing the genus *Gomesa*, is the sister group of the *Erycina* clade, showing that changes between an oil-producing and a non-rewarding callus occur quite easily in this group of orchids. This suggests that evolution towards oil production is correlated with increased venation as also stated by Carlquist [59]. We argue, however, that the venation in the callus is not only driven by functional needs but that the venation pattern is also informative regarding the evolutionary origin of the callus, as the callus of *E. pusilla* is connected with only one of the six original staminal bundles, physically distinct from the two adjacent vascular bundles leading to the lip. We consider this indicative of a possible staminal origin of the callus because of the occasional appearance of an infertile staminodial structure at this particular position, the inner adaxial stamen (a3), in teratologous orchid flowers [60]. Terata of monandrous orchids with both stelidia carrying an additional anther on their tip next to the anther on the apex of the gynostemium, such as *Bulbophyllum triandrum* and *Prostechea cochleata* var. *triandrum*, are commonly seen as support for a staminal origin of stelidia. Similarly, mutants in *Dactylorhiza* with a staminodial structure on their lip [60] could be interpreted as support for a staminal origin of the callus. Alternatively, these phenotypes could be caused by ectopic C gene expression that is transforming petal into stamen tissue. Homeotic transformation is not necessarily indicative of derivation. According to Carlquist [59] data from teratology are therefore not useful for studying the evolution of flowers. This publication was written at a time that experimental mutants could not yet be made though. Ongoing work on B- and C- class homeotic mutants in the established plant models *Arabidopsis*, *Antirrhinum* and *Petunia* shows how much can be gained from teratology. We hope that these mutants can be created in emerging orchid models such as *E. pusilla* in the future to provide more evidence for the evolutionary origin of the callus on the lip.

Homology of the stamen and stelidia of *E. pusilla*
Five vascular bundles, indicating a stamen-derived origin, lead to the stamen and stelidia. Our observations concur with those of Swamy [24] who showed that the ovary is traversed by multiple vascular bundles in monandrous orchids. He visualized 'compound' bundles of staminal origin in the ovary of a species of *Dendrobium* and discovered vascularizing bundles in the stelidia. In several other plant families, e.g. Brassicaceae (*Arabidopsis*), Commelinaceae (*Tradescantia*), and Cyperaceae (*Cyperus*), it has been shown that vascular bundles of different organs originate in the developing organs and grow towards the stele rather than being branched from the stele [61–64]. Based on Fig. 6 and Additional file 4: Movie S1, we hypothesize that especially the staminal vascular bundles are connected in a similar way to the rest of the vascular system. Of the three copies of *AG* and four copies of *SEP*, *EpMADS22* and *EpMADS6* were found to be highest expressed in the stamen. Another

copy of *AG*, *EpMADS20*, and the single copy of *STK*, *EpMADS23*, were found to be most highly expressed in the stelidia, suggesting that *EpMADS23* expression may be correlated with sterility.

Implications for current floral models
The ABCDE, orchid code, HOT and P-code models do not explain the morphological difference between median and lateral sepals as present in orchid species such as *E. pusilla*. Our results show that a differentiation between the sepaloid lateral sepals and petaloid median sepal of *E. pusilla* is correlated with a significant reduction of expression of AP3-like *EpMADS14* and ALG6-like *EpMADS4* in all petaloid organs (Fig. 10a).

The P-code model explains the development of the lip of *E. pusilla* as the SP-complex (AP3-like *EpMADS15*/AGL6-like *EpMADS3*/PI-like *EpMADS16*) was found to be most highly expressed in the sepals and petals, whereas the L-complex (AP3-like *EpMADS13*/AGL6-like *EpMADS5*/PI-like *EpMADS16*) was found to be most highly expressed in the lip (Fig. 10b). However, the model does not yet account for the development of the callus and the high expression of AGL6-like *EpMADS5* in this particular organ. To incorporate all new evidence found for the evolution and development of first and second floral whorl organs, we propose an Oncidiinae model (Fig. 11), summarizing the gene expression data presented in this study for *E. pusilla* and earlier studies carried out on *Oncidium* Gower Ramsey [10] [Illustrations by Bas Blankevoort].

All four MADS-box B class gene copies were found to be expressed in the fertile stamen of *E. pusilla*. In addition, AG-like *EpMADS22* and SEP-like *EpMADS6* were most highly expressed in this floral organ, confirming a stamen identity as predicted by the ABCDE model. The high expression of AG-like *EpMADS20* and STK-like *EpMADS23* in the stelidia cannot be explained with the ABCDE model. All current orchid floral models only describe evolution and development of the first and second whorl floral organs. We found evidence for differential gene expression in organs in the third and fourth floral whorl, i.e. the stamen and stelidia (Fig. 10c), and this argues for the development of additional models.

Conclusions
After examining vascularization, macro- and micromorpology, gene duplications, diversifying evolution and expression of different MADS-box genes in selected floral organs in two developmental stages, it can be concluded that: (i) the median sepal obtained a petal-identity, thus representing a particular character state of the character 'sepal', (ii) that the lip was derived from a petal but the callus from a stamen that gained petal identity, and (iii) the stelidia evolved from stamens. Duplications,

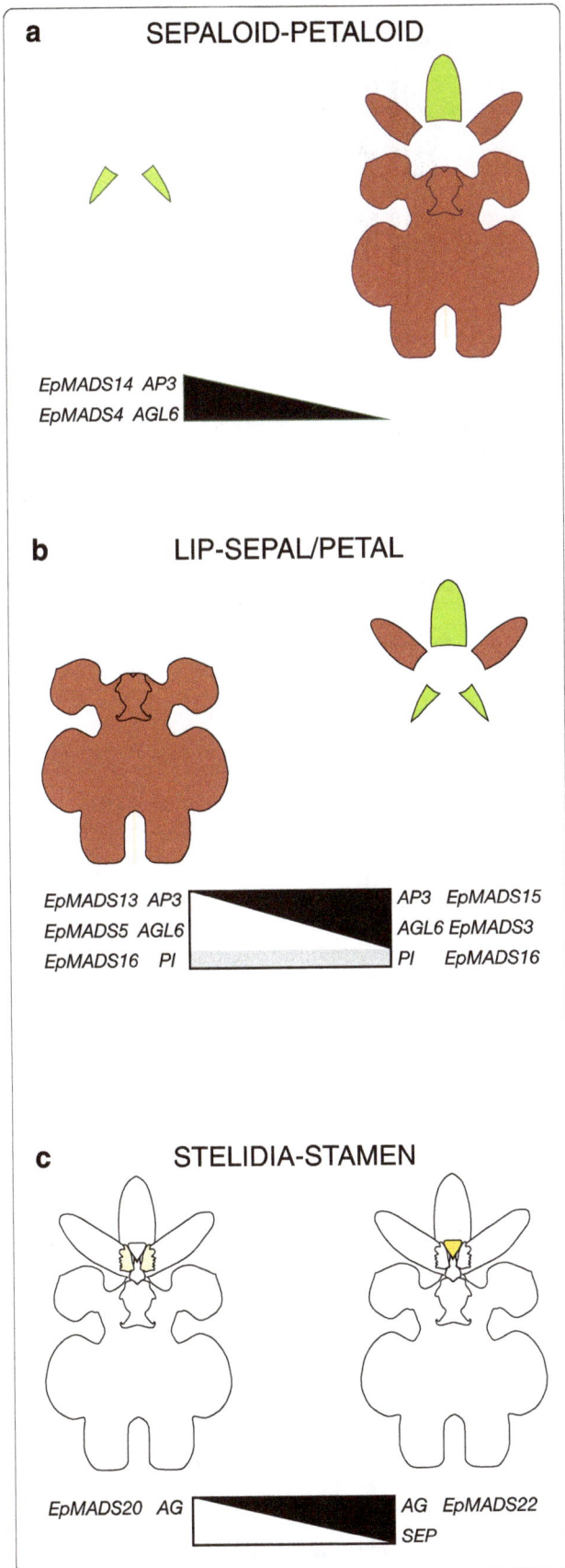

a SEPALOID-PETALOID

EpMADS14 AP3
EpMADS4 AGL6

b LIP-SEPAL/PETAL

EpMADS13 AP3   AP3 EpMADS15
EpMADS5 AGL6   AGL6 EpMADS3
EpMADS16 PI   PI   EpMADS16

c STELIDIA-STAMEN

EpMADS20 AG   AG EpMADS22
                       SEP

Fig. 10 Summary of expression of MADS-box genes involved in the differentiation of selected floral organs of *E. pusilla*. **a** Expression of *EpMADS4/14* (in *black*) correlating with a sepaloid-petaloid identity is high in the lateral sepals (*left side*) but low in the remainder of the perianth (*right side*), **b** Expression of the lip complex *EpMADS5/13/16* (in *white/grey*)) correlating with a lip identity is high in in the lip and callus (*left side*) but low in the remainder of the perianth (*right side*). Expression of the sepal/petal-complex *EpMADS3/15/16* (in *black/grey*) correlating with a sepal and petal identity is low in the lip (*left side*) but high in the sepals and petals (*right side*), **c** Expression of *EpMADS20/23* (in *white*) correlating with a stelidia-stamen identity is high in the stelidia (*left side*) but low in the stamen (*right side*). Expression of *EpMADS6/22* (in black) is low in the stelidia (left side) but high in the stamen (*right side*)

diversifying selection and changes in spatial expressions of *AP3 EpMADS14* and *AGL6 EpMADS4* may have contributed to an increase of petaloidy of the median sepal. The same can be applied to *AP3 EpMADS13* and *AGL6 EpMADS5* in the lip and callus. Differential expression of *AG* copies *EpMADS20* and *EpMADS22*, *STK* copy *EpMADS23* and *SEP* copy *EpMADS6* appear to be associated with the evolution of the stamen and stelidia, respectively.

The evolutionary origin of the median sepal, callus and stelidia of *E. pusilla* cannot be explained with any of the currently existing floral developmental models. Therefore, new models, like our Oncidiinae model, need to be developed to summarize MADS-box gene expression in more complex floral organs. Such models need validation by functional analyses. The genetic mechanisms discovered in this study ultimately contributed to

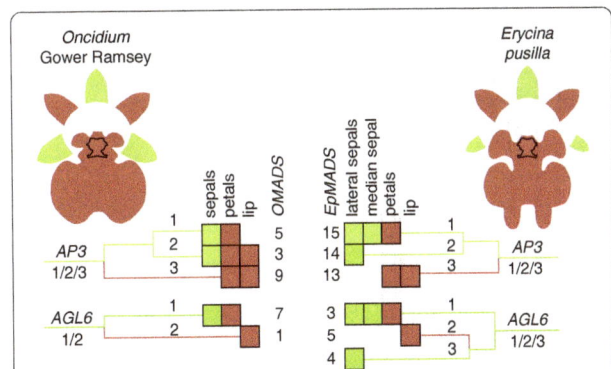

Fig. 11 Oncidiinae model summarizing expression of MADS-box genes involved in the differentiation of the perianth of *Oncidium* Gower Ramsey (*left*) and *E. pusilla* (*right*). Clade 1 *AP3*-like *OMADS5* and *EpMADS15* and clade 1 *AGL6*-like genes *OMADS7* and *EpMADS3* are expressed in the sepals and petals of both species. Clade 2 *AP3*-like *OMADS3* is expressed in the entire perianth of *O.* Gower Ramsey whereas *EpMADS14* is only expressed in the lateral sepals of *E. pusilla*. Clade 2 *AGL6*-like genes *OMADS1* and *EpMADS5* are expressed in the lip only of both species. Clade 3 *AP3*-like *OMADS9* and *EpMADS13* are expressed in the petals and lip of both species. Clade 3 *AGL6*-like gene *EpMADS4* is only expressed in the lateral sepals of *E. pusilla* [Illustrations by Bas Blankevoort]

the evolution of a deceptive orchid flower mimicking the morphologies of rewarding Malpighiaceae flowers. This mimicry enabled flowers of *E. pusilla,* and many other species in the highly diverse Oncidiinae, to successfully attract *Centris* bees for pollination, often, as is the case for *E. pusilla,* without offering a reward. Pollination by deceit is one of the most striking adaptations of orchids to pollinators. It is estimated that approximately a third of all orchid species employ deceit pollination, and that food mimicry is the most common type. Deceptive pollination is hypothesized to be correlated with species diversification as subtle changes in floral morphology can attract different pollinators and eventually lead to reproductive isolation. It was recently discovered that deceptive pollination augmented orchid diversity, not by accelerating speciation but by adding more species at roughly the same rate through time [17]. Ongoing research on the genomics of *E. pusilla* and other emergent plant models will shed more light on the role that key developmental genes played in the evolution of deceptive flowers.

Additional files

Additional file 1: Table S1. List of sequences used in the alignments and phylogenetic analyses. **Table S2.** Transcript primer sequences and amplicon characteristics used for quantitative real-time PCR validation of the expression profiles of eighteen MADS-box transcripts following MIQE guidelines [65]. **Table S3.** Difference in MADS-box gene expression between floral organs; variance analysis of measures using Tukey multicomparisons test. *P*-value style: GP: >0.05 (ns), <0.05 (*), <0.01 (**), <0.001 (***), <0.0001 (****). Abbreviations: lse = lateral sepal, mse = median sepal, cl = callus, pe = petal, fs = fertile stamen and gm = gynostemium. (DOCX 110 kb)

Additional file 2: Figure S2. Melting curve analysis of all primer pairs used in this study performed at the end of the PCR cycles to confirm the specificity of primer annealing. (PDF 9997 kb)

Additional file 3: Figure S4. Species phylogeny compiled based on Topik et al. [66], Biswal et al. [67], Takamiya et al. [68] and Chase et al. [69] for (a) *FUL*-, (b) *AP3*-, (c) *PI*- (d) *AG*- and *STK*-, (e) *SEP*- and (f) *AGL6*-like MADS-box gene lineage trees. (ZIP 584 kb)

Additional file 4: Movie S1. Animation of the 3D visualization as depicted in Fig. 6. (MPG 60302 kb)

Additional file 5: Figure S1. Scanning electron micrographs of epidermal cells on the abaxial side of an *E. pusilla* flower. The three columns represent, from left to right, stage 2, 4 and 5 floral organs. Epidermal cells of (a–c) lateral sepal, (d–f) median sepal, (g–i) petal and (j–l) lip. Scale bar = 100 μm. Abbreviations: lse = lateral sepal; mse = median sepal; pe = petal. (PDF 19591 kb)

Additional file 6: Figure S5. MADS-box gene lineage trees. (a) *FUL*-, (b) *AP3*-, (c) *PI*-, (d) *AG*-, (e) *STK*-, (f) *SEP*- and (g) *AGL6*-like trees. Color codes: green node = speciation event; red node = duplication event. Branches are colored along a gradient between blue and red, in proportion to the value of omega (dN/dS) for the third (i.e. the highest) rate class in the BranchSiteREL analysis. Hence, blue and red branches may be interpreted as suggesting, respectively, stabilizing and diversifying selection. Purple branches implicate a moderate level of diversifying selection. (ZIP 1121 kb)

Additional file 7: Figure S3. Floral organ specific expression levels of *FUL EpMADS10, EpMADS11* and *EpMADS12* and *SEP EpMADS6, EpMADS7, EpMADS8* and *EpMADS9*. RNA was extracted from seven different floral organs during two stages of development of *E. pusilla* and used for cDNA synthesis. Expression of the MADS-box genes was normalized to the geometric mean of three reference genes *Actin, UBI2* and *Fbox*. Each column shows the relative expression of 20 floral organs in two cDNA pools (10 floral organs per isolation), both tested in triplicate. Abbreviations: lse = lateral sepal; mse = median sepal; cl = callus; pe = petal; fs = fertile stamen; gm = gynostemium. Dark grey = floral stage 2 and light grey = floral stage 4. Y-axis: relative gene expression. The error bars represent the Standard Error of Mean. *P*-value style: GP: >0.05 (ns), <0.05 (*), <0.01 (**), <0.001 (***), <0.0001 (****). (TIFF 3032 kb)

Abbreviations
AG: *AGAMOUS*; *AGL6*: *AGAMOUS-LIKE-6*; *AP3*: *APETALA3*; cl: callus; fs: fertile stamen; *FUL*: *FRUITFULL*; gm: gynostemium; L-complex: Lip complex; LM: Light microscopy; lse: lateral sepal; micro-CT: 3D-Xray microscopy; mse: median sepal; P-code: Perianth code; pe: petal; *PI*: *PISTILLATA*; SEM: Scanning electron microscopy; *SEP*: *SEPALLATA*; SP-complex: Sepal/petal-complex; *STK*: *SEEDSTICK*

Acknowledgements
We thank Johan Keus for the culturing of *E. pusilla*, Bas Blankevoort, Erik-Jan Bosch, Rogier van Vugt (Hortus botanicus Leiden) and Angel Vale for the illustrations and photographs, Pieter van der Velden (LUMC), Stef Janson and Jan M. de Ruijter (UvA) for their support and input with the qPCR, Anneke de Wolf for support with the SEM, Øyvind Hammer for help with the Zeiss X radia and Marcel Lombaerts and Jan Oliehoek for help with the construction of the DNA sequence alignments.

Funding
This work was supported by grant 023.003.015 from the Netherlands Organization for Scientific Research (NWO) to AD and a Fulbright grant to BG.

Authors' contributions
ADM and BG designed the gene expression study and KvK, PvS and LK collected the expression data. RAV carried out the phylogenetic analyses with help of JWW. RvdB and SD collected the anatomical and micro-CT data. AV collected the electron microscope data and helped with the interpretation of the floral ontogeny. RB assisted with plant breeding. All authors contributed to the writing of the manuscript. All authors read and approved the final manuscript.

Competing interests
The authors declare that they have no competing interests.

Author details
[1]Endless Forms group, Naturalis Biodiversity Center, Vondellaan 55, 2332 AA Leiden, The Netherlands. [2]Faculty of Science and Technology, University of Applied Sciences Leiden, Zernikedreef 11, 2333 CK Leiden, The Netherlands. [3]Department of Organismic and Evolutionary Biology, Harvard University, 16 Divinity Ave, Cambridge, MA 02138, USA. [4]Institute for Biodiversity and Ecosystem Dynamics, University of Amsterdam, Science Park 904, 1098 XH Amsterdam, The Netherlands. [5]Institute Biology Leiden, Leiden University, Sylviusweg 72, 2333 BE Leiden, The Netherlands. [6]Ecology, Evolution and Biodiversity Conservation cluster, KU Leuven, Kasteelpark Arenberg 31, 3001 Leuven, Belgium. [7]The Natural History Museum, University of Oslo, P.O. Box 1172Blindern, 0318 Oslo, Norway. [8]Department of Organismal Biology, Evolutionary Biology Centre, Uppsala University, Norbyvägen 18D, Uppsala SE-75236, Sweden.

References

1. Cho SC, Jang SH, Chae SJ, Chung KM, Moon YH, An GH, Jang SK. Analysis of the C-terminal region of Arabidopsis thaliana APETALA1 as a transcription activation domain. Plant Mol Biol. 1999;40(3):419–29.
2. Roy BA, Widmer A. Floral mimicry: a fascinating yet poorly understood phenomenon. Trends Plant Sci. 1999;4(8):325–30.
3. Ackerman JD, Cuevas AA, Hof D. Are deception-pollinated species more variable than those offering a reward? Plant Syst Evol. 2011;293(1–4):91–9.
4. Preston JC, Hileman LC, Cubas P. Reduce, reuse, and recycle: developmental evolution of trait diversification. Am J Bot. 2011;98(3):397–403.
5. Kanno A, Nakada M, Akita Y, Hirai M. Class B Gene Expression and the Modified ABC Model in Nongrass Monocots. Sci World J. 2007;7:268–79.
6. Kanno A, Saeki H, Kameya T, Saedler H, Theissen G. Heterotopic expression of class B floral homeotic genes supports a modified ABC model for tulip (Tulipa gesneriana). Plant Mol Biol. 2003;52(4):831–41.
7. Nakamura T, Fukuda T, Nakano M, Hasebe M, Kameya T, Kanno A. The modified ABC model explains the development of the petaloid perianth of Agapanthus praecox ssp. orientalis (Agapanthaceae) flowers. Plant Mol Biol. 2005;58(3):435–45.
8. Soltis DE, Chanderbali AS, Kim S, Buzgo M, Soltis PS. The ABC model and its applicability to basal angiosperms. Ann Bot. 2007;100(2):155–63.
9. Endress PK. Development and evolution of extreme synorganization in angiosperm flowers and diversity: a comparison of Apocynaceae and Orchidaceae. Ann Bot. 2016;117(5):749–67.
10. Hsu HF, Hsu WH, Lee YI, Mao WT, Yang JY, Li JY, Yang CH. Model for perianth formation in orchids. Nat Plants. 2015;1(5):15046.
11. Decraene LPR, Smets EF. Staminodes: Their morphological and evolutionary significance. Bot Rev. 2001;67(3):351–402.
12. Kramer EM, Holappa L, Gould B, Jaramillo MA, Setnikov D, Santiago PM. Elaboration of B gene function to include the identity of novel floral organs in the lower eudicot Aquilegia. Plant Cell. 2007;19(3):750–66.
13. Rudall PJ, Bateman RM. Roles of synorganisation, zygomorphy and heterotopy in floral evolution: the gynostemium and labellum of orchids and other lilioid monocots. Biol Rev Camb Philos Soc. 2002;77(3):403–41.
14. Rudall PJ, Perl CD, Bateman RM. Organ homologies in orchid flowers re-interpreted using the Musk Orchid as a model. PeerJ. 2013;1:e26.
15. McKnight TD, Shippen DE. Plant telomere biology. Plant Cell. 2004;16(4):794–803.
16. Ramirez SR, Gravendeel B, Singer RB, Marshall CR, Pierce NE. Dating the origin of the Orchidaceae from a fossil orchid with its pollinator. Nature. 2007;448(7157):1042–5.
17. Givnish TJ, Spalink D, Ames M, Lyon SP, Hunter SJ, Zuluaga A, Iles WJ, Clements MA, Arroyo MT, Leebens-Mack J, et al. Orchid phylogenomics and multiple drivers of their extraordinary diversification. In: Proceedings Biological sciences/The Royal Society; 2015, 282(1814).
18. Kocyan A, Endress PK. Floral structure and development of Apostasia and Neuwiedia (Apostasioideae) and their relationships to other Orchidaceae. Int J Plant Sci. 2001;162(4):847–67.
19. Vermeulen P. The Different Structure of the Rostellum in Ophrydeae and Neottieae. Acta Bot Neerl. 1959;8(3):338–55.
20. Kurzweil H. Developmental studies in orchid flowers I: Epidendroid and vandoid species. Nord J Bot. 1987;7(4):427–42.
21. Kurzweil H, Kocyan A. Ontogeny of orchid flowers. In: Orchid Biology: Reviews and Perspectives, Viii. 2002. p. 83–138.
22. Brown R, Nees von Esenbeck CGD. Prodromus florae Novae Hollandiae et Insulae Van-Diemen : exhibens characteres plantarum. Norimbergae: Sumtibus L. Schrag; 1827.
23. Darwin C. The various contrivances by which orchids are fertilised by insects. London: John Murray; 1877.
24. Swamy BGL. Vascular anatomy of orchid flowers. Bot Mus Leafl Harv Univ. 1948;13(4):61–95.
25. Coen ES, Meyerowitz EM. The war of the whorls: genetic interactions controlling flower development. Nature. 1991;353(6339):31–7.
26. Theissen G. Development of floral organ identity: stories from the MADS house. Curr Opin Plant Biol. 2001;4(1):75–85.
27. Theissen G, Saedler H. Plant biology. Floral quartets. Nature. 2001;409(6819):469–71.
28. Mondragon-Palomino M, Theissen G. Why are orchid flowers so diverse? Reduction of evolutionary constraints by paralogues of class B floral homeotic genes. Ann Bot. 2009;104(3):583–94.
29. Mondragon-Palomino M, Theissen G. Conserved differential expression of paralogous DEFICIENS- and GLOBOSA-like MADS-box genes in the flowers of Orchidaceae: refining the 'orchid code'. Plant J. 2011;66(6):1008–19.
30. Pan ZJ, Cheng CC, Tsai WC, Chung MC, Chen WH, Hu JM, Chen HH. The duplicated B-class MADS-box genes display dualistic characters in orchid floral organ identity and growth. Plant Cell Physiol. 2011;52(9):1515–31.
31. Su CL, Chao YT, Yen SH, Chen CY, Chen WC, Chang YC, Shih MC. Orchidstra: an integrated orchid functional genomics database. Plant Cell Physiol. 2013;54(2):e11.
32. Cai J, Liu X, Vanneste K, Proost S, Tsai W-C, Liu K-W, Chen L-J, He Y, Xu Q, Bian J, et al. The genome sequence of the orchid Phalaenopsis equestris. Nat Genet. 2015;47(1):65–72.
33. Lee SH, Li CW, Liau CH, Chang PY, Liao LJ, Lin CS, Chan MT. Establishment of an Agrobacterium-mediated genetic transformation procedure for the experimental model orchid Erycina pusilla. Plant Cell Tissue Organ Cult. 2015;120(1):211–20.
34. Lin CS, Hsu CT, Liao DC, Chang WJ, Chou ML, Huang YT, Chen JJ, Ko SS, Chan MT, Shih MC. Transcriptome-wide analysis of the MADS-box gene family in the orchid Erycina pusilla. Plant Biotechnol J. 2016;14(1):284–98.
35. Neubig KM, Whitten WM, Williams NH, Blanco MA, Endara L, Burleigh JG, Silvera K, Cushman JC, Chase MW. Generic recircumscriptions of Oncidiinae (Orchidaceae: Cymbidieae) based on maximum likelihood analysis of combined DNA datasets. Bot J Linn Soc. 2012;168(2):117–46.
36. Felix LP, Guerra M. Chromosome analysis in Psygmorchis pusilla (L.) Dodson & Dressler: the smallest chromosome number known in Orchidaceae. Caryologia. 1999;52(3–4):165–8.
37. Chase MW, Hanson L, Albert VA, Whitten WM, Williams NH. Life history evolution and genome size in subtribe Oncidiinae (Orchidaceae). Ann Bot. 2005;95(1):191–9.
38. Pridgeon A. Genera Orchidacearum/Vol. 5, Epidendroideae (part two)/ed. by Alec M. Pridgeon … [et al.]. Oxford: Oxford University Press; 2009.
39. Vale A, Navarro L, Rojas D, Alvarez JC. Breeding system and pollination by mimicry of the orchid Tolumnia guibertiana in Western Cuba. Plant Species Biol. 2011;26(2):163–73.
40. Papadopulos AS, Powell MP, Pupulin F, Warner J, Hawkins JA, Salamin N, Chittka L, Williams NH, Whitten WM, Loader D, et al. Convergent evolution of floral signals underlies the success of Neotropical orchids. In: Proceedings Biological sciences/The Royal Society; 2013, 280(1765):20130960.
41. Carmona-Díaz G, García-Franco JG. Reproductive success in the Mexican rewardless Oncidium cosymbephorum (Orchidaceae) facilitated by the oil-rewarding Malpighia glabra (Malpighiaceae). Plant Ecol. 2008;203(2):253–61.
42. Staedler YM, Masson D, Schonenberger J. Plant tissues in 3D via X-ray tomography: simple contrasting methods allow high resolution imaging. PloS One. 2013;8(9):e75295.
43. Ruijter JM, Ramakers C, Hoogaars WM, Karlen Y, Bakker O, van den Hoff MJ, Moorman AF. Amplification efficiency: linking baseline and bias in the analysis of quantitative PCR data. Nucleic Acids Res. 2009;37(6):e45.
44. Tuomi JM, Voorbraak F, Jones DL, Ruijter JM. Bias in the Cq value observed with hydrolysis probe based quantitative PCR can be corrected with the estimated PCR efficiency value. Methods. 2010;50(4):313–22.
45. Ruijter JM, Thygesen HH, Schoneveld OJ, Das AT, Berkhout B, Lamers WH. Factor correction as a tool to eliminate between-session variation in replicate experiments: application to molecular biology and retrovirology. Retrovirology. 2006;3:2.
46. Ruijter JM, Ruiz Villalba A, Hellemans J, Untergasser A, van den Hoff MJ. Removal of between-run variation in a multi-plate qPCR experiment. Biomol Detect Quantif. 2015;5:10–4.
47. Katoh K, Standley DM. MAFFT multiple sequence alignment software version 7: improvements in performance and usability. Mol Biol Evol. 2013;30(4):772–80.
48. Guindon S, Dufayard JF, Lefort V, Anisimova M, Hordijk W, Gascuel O. New algorithms and methods to estimate maximum-likelihood phylogenies: assessing the performance of PhyML 3.0. Syst Biol. 2010;59(3):307–21.
49. Anisimova M, Gascuel O. Approximate likelihood-ratio test for branches: A fast, accurate, and powerful alternative. Syst Biol. 2006;55(4):539–52.
50. Zmasek CM, Eddy SR. A simple algorithm to infer gene duplication and speciation events on a gene tree. Bioinformatics. 2001;17(9):821–8.

51. Kosakovsky Pond SL, Murrell B, Fourment M, Frost SD, Delport W, Scheffler K. A random effects branch-site model for detecting episodic diversifying selection. Mol Biol Evol. 2011;28(11):3033–43.

52. Pond SL, Frost SD, Muse SV. HyPhy: hypothesis testing using phylogenies. Bioinformatics. 2005;21(5):676–9.

53. Kull T, Arditti J. Orchid Biology VIII: Reviews and Perspectives. Netherlands: Springer; 2013.

54. Rudall PJ, Bateman RM. Evolution of zygomorphy in monocot flowers: iterative patterns and developmental constraints. New Phytol. 2004;162(1): 25–44.

55. Acri-Nunes-Miranda R, Mondragon-Palomino M. Expression of paralogous SEP-, FUL-, AG- and STK-like MADS-box genes in wild-type and peloric Phalaenopsis flowers. Front Plant Sci. 2014;5:76.

56. Pan ZJ, Chen YY, Du JS, Chen YY, Chung MC, Tsai WC, Wang CN, Chen HH. Flower development of Phalaenopsis orchid involves functionally divergent SEPALLATA-like genes. New Phytol. 2014;202(3):1024–42.

57. Mondragon-Palomino M, Hiese L, Harter A, Koch MA, Theissen G. Positive selection and ancient duplications in the evolution of class B floral homeotic genes of orchids and grasses. BMC Evol Biol. 2009;9:81.

58. Whitney HM, Bennett KM, Dorling M, Sandbach L, Prince D, Chittka L, Glover BJ. Why do so many petals have conical epidermal cells? Ann Bot. 2011;108(4):609–16.

59. Carlquist S. Toward acceptable evolutionary interpretations of floral anatomy. Phytomorphology. 1969;19(4):332–62.

60. Bateman RM, Rudall PJ. The Good, the Bad, and the Ugly: using naturally occurring terata to distinguish the possible from the impossible in orchid floral evolution. Aliso. 2006;22:481–96.

61. Endress PK. Diversity and Evolutionary Biology of Tropical Flowers. Cambridge: Cambridge University Press; 1996.

62. Reynders M, Vrijdaghs A, Larridon I, Huygh W, Leroux O, Muasya AM, Goetghebeur P. Gynoecial anatomy and development in Cyperoideae (Cyperaceae, Poales): congenital fusion of carpels facilitates evolutionary modifications in pistil structure. Plant Ecol Evol. 2012;145(1):96–125.

63. Pizzolato TD. Procambial initiation for the vascular system in the aerial shoot of Costus (Costaceae, Zingiberales). Int J Plant Sci. 2007;168(4):393–413.

64. Scarpella E, Marcos D, Friml J, Berleth T. Control of leaf vascular patterning by polar auxin transport. Genes Dev. 2006;20(8):1015–27.

65. Bustin SA, Benes V, Garson JA, Hellemans J, Huggett J, Kubista M, Mueller R, Nolan T, Pfaffl MW, Shipley GL, et al. The MIQE guidelines: minimum information for publication of quantitative real-time PCR experiments. Clin Chem. 2009;55(4):611–22.

66. Topik H, Yukawa T, Ito M. Molecular phylogenetics of subtribe Aeridinae (Orchidaceae): insights from plastid matK and nuclear ribosomal ITS sequences. J Plant Res. 2005;118(4):271–84.

67. Biswal DK, Marbaniang JV, Tandon P. Age estimation for Asian Cymbidium (Orchidaceae: Epidendroideae) with implementation of fossil data calibration using molecular markers (ITS2 & matK) implying phylogeographic inference. PeerJ PrePrints. 2013;1:e94v91.

68. Takamiya T, Wongsawad P, Sathapattayanon A, Tajima N, Suzuki S, Kitamura S, Shioda N, Handa T, Kitanaka S, Iijima H, et al. Molecular phylogenetics and character evolution of morphologically diverse groups, Dendrobium section Dendrobium and allies. Aob Plants. 2014;6:plu045.

69. Chase MW, Cameron KM, Freudenstein JV, Pridgeon AM, Salazar G, Van den Berg C, Schuiteman A. An updated classification of Orchidaceae. Bot J Linn Soc. 2015;177(2):151–74.

Evolutionary origin and function of NOX4-art, an arthropod specific NADPH oxidase

Ana Caroline Paiva Gandara[1*†], André Torres[2†], Ana Cristina Bahia[3], Pedro L. Oliveira[1,4] and Renata Schama[2,4*] (iD)

Abstract

Background: NADPH oxidases (NOX) are ROS producing enzymes that perform essential roles in cell physiology, including cell signaling and antimicrobial defense. This gene family is present in most eukaryotes, suggesting a common ancestor. To date, only a limited number of phylogenetic studies of metazoan NOXes have been performed, with few arthropod genes. In arthropods, only NOX5 and DUOX genes have been found and a gene called NOXm was found in mosquitoes but its origin and function has not been examined. In this study, we analyzed the evolution of this gene family in arthropods. A thorough search of genomes and transcriptomes was performed enabling us to browse most branches of arthropod phylogeny.

Results: We have found that the subfamilies NOX5 and DUOX are present in all arthropod groups. We also show that a NOX gene, closely related to NOX4 and previously found only in mosquitoes (NOXm), can also be found in other taxonomic groups, leading us to rename it as NOX4-art. Although the accessory protein p22-*phox*, essential for NOX1-4 activation, was not found in any of the arthropods studied, NOX4-art of *Aedes aegypti* encodes an active protein that produces H_2O_2. Although NOX4-art has been lost in a number of arthropod lineages, it has all the domains and many signature residues and motifs necessary for ROS production and, when silenced, H_2O_2 production is considerably diminished in *A. aegypti* cells.

Conclusions: Combining all bioinformatic analyses and laboratory work we have reached interesting conclusions regarding arthropod NOX gene family evolution. NOX5 and DUOX are present in all arthropod lineages but it seems that a NOX2-like gene was lost in the ancestral lineage leading to Ecdysozoa. The NOX4-art gene originated from a NOX4-like ancestor and is functional. Although no p22-*phox* was observed in arthropods, there was no evidence of neo-functionalization and this gene probably produces H_2O_2 as in other metazoan NOX4 genes. Although functional and present in the genomes of many species, NOX4-art was lost in a number of arthropod lineages.

Keywords: NADPH oxidase, Gene loss, Gene family, Arthropods, ROS, Reactive oxygen species

Background

Reactive oxygen species (ROS) are generated by the partial reduction of oxygen, producing a number of short-lived and highly electrophilic molecules (eg. super-oxide anion ($\cdot O_2^-$), hydrogen peroxide (H_2O_2), among others). These molecules have originally been seen as deleterious but in the last decade a wide array of diverse biochemical functions were assigned to them in several organisms. ROS are implicated in important basic biological processes that include cell differentiation, development and motility, cytoskeletal reorganization, cell survival and apoptosis, stress response, gut homeostasis and defense, cell signaling and transcriptional regulation [1–3]. In *Drosophila melanogaster*, for example, the ROS producing dual oxidase gene (DUOX) seems to play a critical role in gut innate immunity [4]. ROS can be produced in different ways: (i) as a side reaction of common enzymatic activity, (ii) exogenous compound induction but also (iii) as a physiological response produced by specialized enzymes. The gene family known as NADPH oxidase (NOX) is responsible for the physiological production of ROS [2, 5]. This NOX-dependent generation of ROS is highly conserved across virtually all multicellular forms of life.

* Correspondence: acpgandara@gmail.com; schama@ioc.fiocruz.br
†Equal contributors
[1]Instituto de Bioquímica Médica Leopoldo de Meis, Universidade Federal do Rio de Janeiro, Rio de Janeiro, Brazil
[2]Laboratório de Biologia Computacional e Sistemas, Instituto Oswaldo Cruz, Fiocruz, Rio de Janeiro, Brazil
Full list of author information is available at the end of the article

Metazoan NOXes are divided in three subfamilies (NOX1-4, NOX5 and DUOXes). All proteins have two canonical domains that are also shared with ferric reductase enzymes: a heme-containing transmembrane domain and a C-terminal cytoplasmic dehydrogenase (DH) domain, which contains FAD and NADPH-binding sites [6, 7]. The ferric reductase domain is characterized by 6 transmembrane α-helices (TM1-6) where four conserved histidine residues (two on helix 3 and two on helix 5) bind two heme molecules. The electrons are transferred from NADPH to FAD, then to the heme molecules and, finally, to molecular oxygen (O_2) which becomes superoxide by partial reduction. Besides the canonical domains, members of the DUOX subfamily also have at least two EF-hand calcium-binding domains and one N-terminal peroxidase-like domain [8, 9] while NOX5 enzymes have four N-terminal EF-hand calcium binding domains [8, 10–13].

During the course of evolution, individual NOXes have acquired different regulatory systems [6, 14, 15] as separate cytosolic and membrane bound subunits [7]. The activation of NOX2 has been extensively studied since point mutations in this gene cause the X-linked chronic granulomatous disease (CGD) [16]. This was the first NOX gene described and, in mammals, it is expressed in phagocytic cells, producing superoxide [2, 17, 18]. However, ROS are also produced in a variety of other cell types and tissues [1, 19]. In the NOX1-4 subfamily, all proteins need to be associated with a non-glycosylated integral protein called p22-*phox* to be active [6, 12, 15]. This protein has a cytoplasmic proline-rich region (PRR) that helps stabilize the NOX enzymes at the membrane [20, 21]. NOX1-3 need other cytosolic proteins and NOX2 and NOX3 also need the small GTPase Rac to function [22–25]. NOX4 enzymes seem to require only p22-*phox* for basal ROS production [5, 14, 21, 26]. NOX4 is constitutively active in the presence of p22-*phox* [26] simply because the conformation of its DH domain seems to allow the transfer of electrons from NADPH to FAD [27]. In the other two subfamilies, NOX5 and DUOX1-2, apart from other proteins, calcium molecules are important for ROS production since they are needed for enzyme activation [10, 15, 28–31].

In mammals, the most widely studied group so far, there are seven genes (NOX1-4, NOX5 and DUOX1-2) that belong to the different subfamilies of NOXes. The presence of NOX genes in most eukaryotic groups suggests a common ancestor early in evolution with patterns of expansion and gene loss [6, 7]. In metazoa, NOX1-4 seem to have emerged from ancestral EF-hand containing subfamilies (NOX5 and DUOX) [7]. The relationship among the genes within the NOX1-4 subfamily is somewhat unclear. NOX2 is present in most groups and was suggested as the ancestral NOX1-4 in animals [32], however, it has been lost in Ecdysozoa

(phyla Nematoda and Arthropoda). Previous studies suggested that NOX4 appeared in the deuterostomes (although not in Echinodermata) but, recently, this gene was found in the genome of the sea anemone *Nematostella vectensis* which indicates an earlier divergence than previously thought [7]. NOX1 seems to be present only in vertebrates and NOX3 only in mammals and reptiles/birds [32]. In arthropods, only NOX5 and DUOX genes have been found [32]. Interestingly, a new gene called NOXm was found only in mosquitoes but its origin has not been carefully examined [7, 32].

To date, only a limited number of phylogenetic studies of metazoan NOXes have been performed where just a few available arthropod genes (from the species *Drosophila melanogaster*, *Apis mellifera* and *Anopheles gambiae*) were utilized [6, 7, 32]. The most recent study used a whole superfamily approach with deep branching nodes but again, only a few metazoan species were analyzed [7]. The NOXm gene described for mosquitoes, for example, has been linked to their hematophagous habit [32] and warrants attention. Despite its ecological importance and relevance to vector biology, its evolutionary history and functionality have never been looked at. We only know that *Wolbachia* infection in *Aedes aegypti* leads to increased NOXm and DUOX transcript levels and their silencing suppresses the expression of some antimicrobial peptides [33]. More recently, Park et al. 2015 also found that eicosanoids seem to mediate ROS production by a NOX4-like gene in the moth *Spodoptera exigua* [34].

Given the role of NOXes in insects and how little is known about their evolution in this wide and diverse group of animals, a deeper understanding of their phylogenetic relationship is needed. Furthermore, the discovery of the NOXm gene raises intriguing questions about its function and evolution in arthropods [32]. Here, we performed a more extensive analysis that profited by the availability of several recently sequenced arthropod genomes to characterize the evolutionary history of the NOX family in arthropods and the structural and functional features of NOXm. We performed a thorough search of genomes and transcriptomes, when available, and were able to browse most branches of arthropod phylogeny. Furthermore, through bioinformatics and, to some extent, experimental work we show that the functionality features necessary for ROS production are present in NOXm, a gene closely related to vertebrate NOX4. Finally, we determined that this gene is not restricted to mosquitoes and is present in a number of arthropod lineages although it has also been lost in many. Since NOXm is not limited to Culicidae (mosquitoes) but present in a number of arthropod lineages, we suggest renaming it to NOX4-art.

Results

Gene searches

Sequences encoding putative NOX genes were identified from the predicted protein set of 70 arthropods, one choanoflagellate, two cnidarians, one sponge, four mollusks, two annelids, six nematodes, one echinoderm, one cephalochordate, one urochordate, and eight vertebrate genomes [see Additional file 1]. This search greatly improves on previous searches of NOX genes in invertebrate genomes, especially nematodes, mollusks, annelids and arthropods; the latter being the main focus of this work. The searches and posterior gene structure analysis of all genes in all genomes (95 in total) revealed a total of 113 DUOX, 94 NOX5, 34 NOX4/NOX4-art, 6 NOX1, 3 NOX3 and 21 NOX2 genes. Additional file 1 summarizes the organisms analyzed and number of copies of NOX genes found in each genome and their source [see Additional file 1]. We found that in only four cases the automated genome predictions did not recover one or more of the NOX genes expected for the species and these were found with tBLASTn and Exonerate searches of the scaffolds. Four new genes were predicted using GeneWise: DUOX genes from the species *Eurytemora affinis* (Scaffold63, region 576882-539914) and *Blatella germanica* (Scaffold551, region 278933-220027) and NOX5 genes from the species *Phlebotomus papatasi* (scaffold:PpapN1:Scaffold23814, region 33-1651) and *Heliconius melpomene* (HE671948.1, region 63215-107001).

In public domain databases, arthropod genes that seem orthologous to vertebrate NOX4 have been identified in some species but their function and true evolutionary origin and how pervasive they are in the arthropod phylum have never been studied. For a better look of the distribution of NOX4-art genes in arthropod groups we performed tBLASTn searches against the Transcriptome Shotgun Assembly (TSA) sequence database [35] on NCBI website. This enabled us to find NOX4-art orthologs in species of arthropods without sequenced genomes. With NOX4-art searches against the TSA database, 55 sequences belonging to other species of arthropods than those for which we had full genomes were found [see Additional file 2]. Of these, 48 belonged to Hexapoda species, two to Chelicerata and five to Crustacea. Most sequences contained both the ferric reductase and the C-terminal cytoplasmic FAD and NADPH-binding sites. Nevertheless, to make sure the alignments were reliable and enough sites could be used, only sequences longer than 290 amino acids were used in the phylogenetic analysis.

In vertebrates, where these enzymes have been well studied, NOX genes 1 to 4 are regulated by a number of other proteins and the binding to the non-glycosylated integral protein p22-*phox* is essential to their function

[6, 21, 26]. For that reason, sequences encoding the putative p22-*phox* accessory protein were also searched in the genomes analyzed using the same approach described for the NOX genes. The search for the p22-*phox* gene in the 95 genomes studied yielded 20 genes in 18 genomes, with only the ascidian, *Ciona intestinalis*, and the cephalochordate, *Branchiostoma floridae*, with two copies [see Additional file 3]. The tBLASTn and Exonerate searches yielded one putative new gene for the species *Capitella teleta* that could only be partially predicted. Only for the two Annelida species (the new, partially predicted, gene from *Capitella teleta* and the gene from *Helobdella robusta*) the polybasic and proline-rich regions (PRR) could not be detected. It is important to highlight that orthologs of p22-*phox* were not found in any arthropod genome.

Phylogenetic analysis

Our phylogenetic analysis using the protein alignment of the ferric reductase and dehydrogenase (FAD and NADPH-binding sites) conserved domains of each sequence retrieved from the 95 metazoan genomes plus arthropod TSA sequences was able to divide the NOXes into four well-supported clades [see Additional file 4]. The DUOX subfamily (92% bootstrap [see Additional file 4]) is the most pervasive with almost all species having at least one gene (exceptions being the Annelida species *Helobdella robusta*, the two cnidarians *Nematostella vectensis* and *Hydra vulgaris* and the Choanoflagellida *Monosiga brevicollis*). These same species also lack a NOX5 gene (clade with 100% bootstrap support [see Additional file 4]). This could be due to genome assembly errors or this gene may have been lost in these lineages. Other Annelida species do have NOX5 and DUOX genes [6] and we did find these genes for the species *Capitella teleta* [see Additional file 1]. However, no NOX5 nor DUOX genes have ever been found in Cnidaria or Choanoflagellida [6, 7]. In agreement with the literature, we did not find a NOX5 gene in the Urochordate species *Ciona intestinalis* [7, 32, 36]. Of the five Arachnids, only in *Metaseiulus occidentalis* no NOX5 was found, suggesting an assembly error in this case. As expected, no NOX5 gene was found in Nematodes [6, 32]. Also, in agreement with Kawahara et al. 2007 NOX1 was only found in vertebrates and NOX3 in mammals and reptiles/birds (Fig. 1) [32]. The NOX1-3 clade also has a high bootstrap value (83%) but the relationships within the clade are not well defined with low bootstrap on most branches (Fig. 1). This could be due to saturation and loss of phylogenetic signal at these deep nodes or, less likely, because most substitutions have occurred within the different lineages sampled and not between them. In vertebrates, NOX1-3 forms a well-supported clade indicating that all three genes have diverged more recently from an ancestral vertebrate

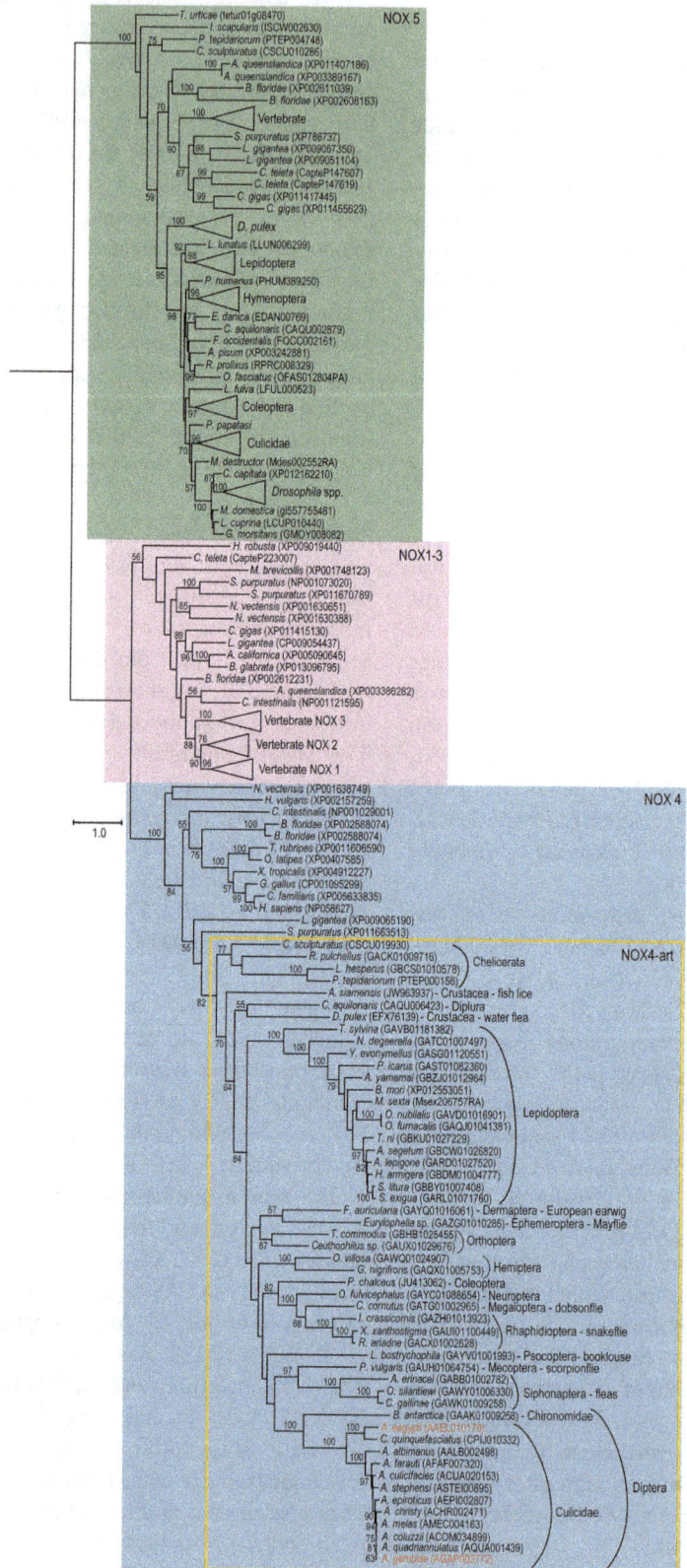

Fig. 1 (See legend on next page.)

(See figure on previous page.)
Fig. 1 Maximum Likelihood phylogeny of aligned NOX proteins identified in our searches, species names and their accession numbers. Three well-supported clades have been highlighted in different colors: NOX5, NOX1-3 and NOX4/NOX4-art. The yellow square within NOX4 clade depicts the arthropod specific genes (NOX4-art). The *Aedes aegypti* and *Anopheles gambiae* genes have the same sequence as the ones used in Kawahara et al. 2007 [32] and are highlighted in a different color. Numbers on branches are bootstrap support values from 1000 replicates; only numbers above 50% are shown. Scale bar is substitutions per site. The image was created using iTOL [100]

NOX2-like gene. Although Zhang et al. 2013 used the terminology NOX1-3 for NOX2 of non-vertebrate metazoans [7], we felt that it was confusing and, for the lack of a better name, we will refer to non-vertebrate NOX2 as NOX2-like. As expected, no NOX2-like gene was found in Ecdysozoa, pointing to a probable loss in their ancestral lineage. NOX4 forms a well-supported clade (100% bootstrap, Fig. 1). The divergence between the NOX1-3 and NOX4 clades probably happened in the ancestral eumetazoan as indicated by the presence of a NOX2-like in the choanoflagellate and Porifera species but of NOX4 only in Cnidaria (Fig. 1, [see also Additional file 1]). Although it was suggested that the sea urchin does not have a NOX4 gene [6], we found one that clusters with good support with other metazoan NOX4. A well-supported clade (82% bootstrap value) nested within the NOX4 genes is comprised only of arthropod genes. This clade includes the NOXm gene found in mosquitoes [32]. With our thorough search of the genomes and TSA database we found that this same gene is present in many other arthropod species. Therefore we suggest it to be renamed as NOX4-art instead of NOXm. The presence and absence of each NOX subfamily and p22-*phox* gene for each animal lineage is summarized in Additional file 1. When more than one sequence of NOX5, NOX2 or NOX4 was found for one particular species these copies seemed to be derived from duplications within the species genome and are therefore paralogs (Fig. 1).

Synteny and protein domain and motif analysis

For the synteny analysis, only species with complete genomes where NOX4-art was found were analyzed together in the orthoMCL searches. Although many species of *Anopheles* have their genomes sequenced, we chose to use only *Anopheles gambiae* since it is the best annotated and studied species. Within the species studied, synteny analyses revealed conservation of orthologous genes only among species of the same taxonomic group. Within the Culicidae (mosquitoes), NOX4-art genes are flanked by several orthologous groups [see Additional file 5]. Within three of the orthologous groups, paralogous genes were found indicating duplications within a species. Besides putative duplication events, we also found inversions and changes in gene orientation [see Additional file 5]. Taking *Anopheles gambiae* chromosome as reference, the whole syntenic block is inverted in *Aedes aegypti* and, in *Culex*

quinquefasciatus, inversions were detected in two pairs of genes, one upstream (CPIJ010328 and CPIJ010329) and one downstream (CPIJ010333 and CPIJ010334) of NOX4-art. In order to detect if this syntenic region was still present in *Drosophila melanogaster*, a species that lacks NOX4/NOX4-art, we used the genes found in this region in mosquitoes as queries for online BLASTp searches against *D. melanogaster*'s genome in GenBank's nr database. Although some orthologs were found in the fruit fly's genome, they were not found in the same gene cluster and even belonged to different chromosomes (2L and 3R; [see Additional file 5]). Synteny analysis in both lepidopteran scaffolds revealed two and five pairs of orthologous genes up and downstream of NOX4-art [see Additional file 5].

For all NOXes, specific structural features have been identified and many key residues and loop sizes were described as important for ROS production and structural stability of the enzymes [32, 37, 38]. Aligning the proteins of the NOX4/NOX4-art clade separately and using the human NOX4 as a guide, we identified previously described conserved amino acid residues, loops and segments. Figure 2 depicts a schematic representation of NOX4 structural features. As expected, all NOX4/NOX4-art proteins had the six transmembrane helixes, two FAD binding and four NADPH binding domains. All four histidine residues that are essential for heme binding and electron transport were present in most proteins. Only in those that did not seem complete, probably due to prediction problems, some important residues could not be found [see Additional file 6]. Loops A, C, D and the segment between NADPH3-4 were highly conserved in the number of residues they contain (Table 1). Apart from loop C, loops A and D and the segment between NADPH3-4 were also conserved in size among all NOX1-4 proteins, indicating that the size of the loops might be important for protein function [32]. In NOX4/NOX4-art, loop E varies in size among the different taxonomic groups (Table 1). Kawahara et al. 2007, identified this segment as being longer in all NOX4 genes when compared to NOX1-3 [32]. Nevertheless, in our study, with a higher number of taxa of different taxonomic groups, we can see that in some cases the loop is long but in other cases it is as small as in NOX1-3 (41-44 residues in [32], Table 1). One differentiating portion of NOX4/NOX4-art is loop C, that seems to be conserved in size in our

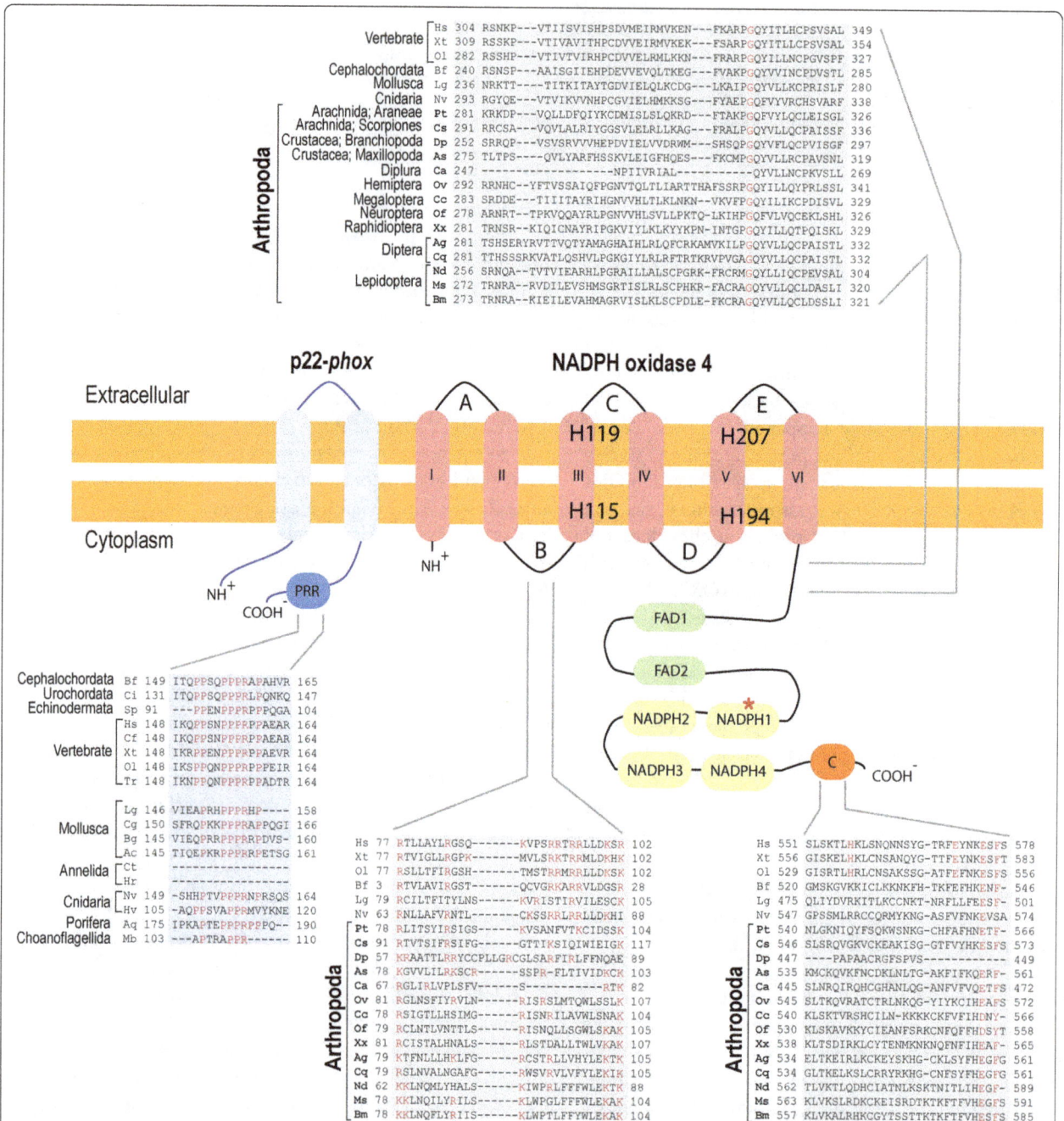

Fig. 2 Schematic representation of NOX4/NOX4-art and p22-*phox* proteins with partial regions of the alignment of important loops and segments. The six hydrophobic helixes of the ferric reductase domain are depicted in pink with the four histidine residues. The dehydrogenase domain is colored in *green* and *yellow* (FAD1-2 and NADPH1-4 respectively). The red asterisk in NADPH1 shows where the VXGPFG-motif is located. The C-terminal region, important for the interaction with p22-*phox*, is dark orange. Segments and loops are black with loops identified by capital letters. The protein p22-*phox* with its two transmembrane helixes and proline-rich region (PRR) is illustrated in blue. Partial alignments of important regions are highlighted in blue for p22-*phox* (where no arthropod species are present) and grey for NOX4/NOX4-art. Within the alignments important residues are colored in *red*. Hs - *Homo sapiens*, Xt - *Xenopus tropicalis*, Ol - *Oryzias latipes*, Bf - *Branchiostoma floridae*, Lg - *Lottia gigantea*, Nv - *Nematostella vectensis*, Pt - *Parasteatoda tepidariorum*, Cs - *Centruroides sculpturatus*, Dp - *Daphnia pulex*, As - *Argulus siamensis*, Ca - *Catajapyx aquilonaris*, Ov – *Okanagana villosa*, Cc - *Corydalus cornutus*, Of - *Osmylus fulvicephalus*, Xx - *Xanthostigma xanthostigma*, Ag - *Anopheles gambiae*, Cq - *Culex quinquefasciatus*, Nd - *Nemophora degeerella*, Ms. - *Manduca sexta*, Bm - *Bombyx mori*, Ac - *Aplysia californica*, Aq - *Amphimedon queenslandica*, Bg - *Biomphalaria glabrata*, Cf - *Canis familiaris*, Cg - *Crassostrea gigas*, Ci - *Ciona intestinalis*, Ct - *Capitella teleta*, Hr - *Helobdella robusta*, Hv - *Hydra vulgaris*, Mb - *Monosiga brevicollis*, Sp - *Strongylocentrotus purpuratus*, Tr - *Takifugu rubripes*

Table 1 Number of amino acids present in loops and segments joining transmembrane (TM) regions and canonical DH domains in NOX4/NOX4-art of the different taxonomic groups analyzed. # - Number of species in each taxonomic group

Group	#	loop A	loop B	loop C	loop D	loop E	TM6	-	FAD1-2	-	NADPH1	-	NADPH2	-	NADPH3	-	NADPH4	C-terminus
Cnidaria	2	16	26-27	25	13	41-73		46		49-69		19-20		49		20			28
Arachnida	5	16	27	25	13	45-76		46-51		43-57		15-16		51-52		20			27-29
Crustacea	6	16	12-33	15/25	13	12-40		27-46		36-58		7-11		48-52		20			27
Hexapoda[a]	62	16	21	25	13	11-69		46-52		36-52[b]		14-17		51-52[c]		20			27-30
Echinodermata	1	16	25	25	13	76		204		45		19		49		20			28
Urochordata	1	16	26	25	13	135		47		109		18		48		22			27
Cephalochordata	1	0	26	25	13	80-82		46		52-75		16		48		20			27
Vertebrates	6	16	26	25	13	48-75		46		48		15-16		48-49		20			28

[a]At least in 80% of the species analyzed; [b]varies among orders; [c]much bigger in Lepidoptera (median = 102.5)

analysis (Table 1) but longer in NOX1-2 (37-41 residues) and NOX3 (38 residues) [32]. In loop E, deletion of the THPPGC motif or mutation of the histidine (in position 222) and cysteine residues (in positions 226 and/or 270) switches hydrogen peroxide to superoxide production in NOX4 [39]. In arthropods, although the motif is different, the histidine and cysteine residues are present in most species [see Additional file 6]. Loop B (and arginine and lysine residues within), segment TM6-FAD (and a glycine residue within), the VXGPFG-motif (within NADPH1) and the extreme C-terminal regions (and glutamate and phenylalanine residues within) were identified as being important for NOX activity [32, 37, 38] (Fig. 2). All species have a polybasic-rich region in loop B, characterized by arginine and lysine residues [see Additional file 6]. In human NOX4, these residues are functionally important and bind to the C-terminal region of the DH (FAD/NADPH-binding) domain, providing proximity for the transmembrane heme-binding domain to interact with it [40]. Although many arginine residues are present in loop B of arthropod's NOX4-art genes, the RRXRR motif characteristic of vertebrates [37] is not present (Fig. 2 [see also Additional file 6]). In segment TM6-FAD, most NOX sequences have the glycine residue near position 336. In vertebrate NOX2, this residue and an arginine near position 80 in loop B, also present in most of our sequences (a lysine substitutes the arginine in Diptera), are thought to participate in the binding of p22-*phox* [41]. The expected glutamate (position 575) and phenylalanine (position 577) residues are present in the C-terminal region of all sequences but the histidine (position 557) and a second glutamate (position 571) residue are absent in most arthropod sequences (Fig. 4 [see also Additional file 6]). Since in the NOX4/p22-*phox* complex these residues are necessary for the constitutive production of ROS [37], their absence in NOX4-art sequences indicates that these genes are either not constitutively active or that the interaction of loop B and the C-terminal region might be different from that of vertebrate NOX4. All sequences have the VXGPFG-motif within NADPH1 binding site.

Silencing of NOX4-art

As p22-*phox* has been shown to be essential for enzymatic activity in the NOX4 proteins studied so far, the presence of the NOX4-art genes in arthropods, together with the lack of p22-*phox* could be interpreted as indicative of absence of catalytic activity. In order to test for the ROS producing activity of NOX4-art, we used dsRNA-mediated silencing of the NOX4-art gene in *Aedes aegypti* embryonic cells (Aag-2). NOX4-art silencing resulted in a significant decrease in hydrogen peroxide production (Fig. 3a and b), clearly demonstrating that the protein coded by the NOX4-art gene is

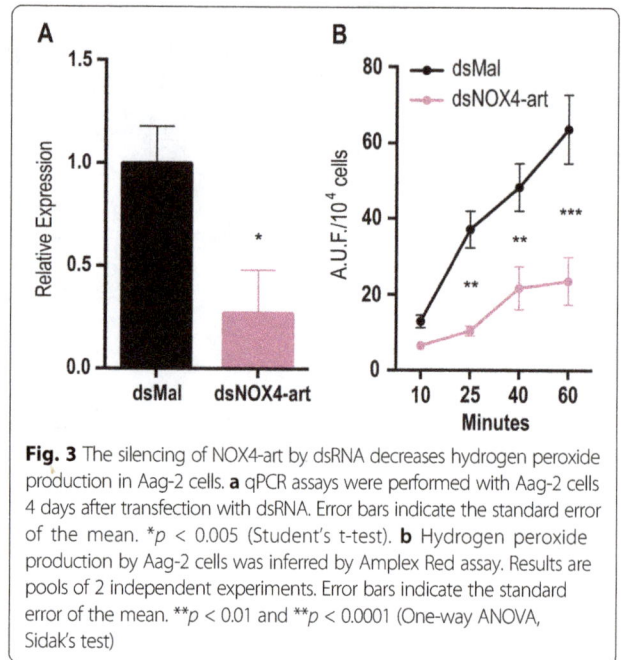

Fig. 3 The silencing of NOX4-art by dsRNA decreases hydrogen peroxide production in Aag-2 cells. **a** qPCR assays were performed with Aag-2 cells 4 days after transfection with dsRNA. Error bars indicate the standard error of the mean. *$p < 0.005$ (Student's t-test). **b** Hydrogen peroxide production by Aag-2 cells was inferred by Amplex Red assay. Results are pools of 2 independent experiments. Error bars indicate the standard error of the mean. **$p < 0.01$ and **$p < 0.0001$ (One-way ANOVA, Sidak's test)

catalytically active, possibly by direct production of hydrogen peroxide, as shown for vertebrate NOX4 and DUOX1-2 [2].

Discussion

In this study, we evaluated the diversity of NOXes in 70 arthropod genomes and investigated the origin and function of a NOX gene previously found only in the mosquitoes *Anopheles gambiae* and *Aedes aegypti*, here shown to be an ancestral trait of the arthropod lineage. The annotation of the complete NOX repertoire in arthropods allowed us to better understand gene divergence and the importance of deletion events in the evolution of this essential gene family. Combining the phylogenetic, domain and residue analysis and laboratory work, our data would suggest that: 1) Indeed it seems that a NOX2-like gene was lost in the ancestral lineage leading to Ecdysozoa; 2) As suggested before, NOX5 and DUOX are present in all arthropod lineages; 3) NOX4-art evolved from a NOX4-like ancestor; 4) NOX4-art is functional and, although no p22-*phox* was observed in arthropods, there was no indication of neo-functionalization as this gene still produces hydrogen peroxide; 5) Although functional and present in the genomes of many species, NOX4-art was lost in a number of arthropod lineages. Public domain database's automatic blast similarity searches have already indicated that a NOX4-like gene is present in some arthropod species. Nevertheless, to our knowledge, this is the first work to show that NOX4-art is an arthropod specific new gene that originates from a NOX4-like ancestor in a phylogenetic framework. We show here the evolutionary origin of this new gene and how pervasive it is in arthropod phylogeny although it has also been lost in many species.

Evolution of NOX genes

A previous study that analyzed the NOX family in a larger context found that this ROS generating family is monophyletic and clusters together with other ferric reductase genes. It was proposed that the shift in function from metalloreductase to ROS production took place in the ancestral gene of early eukaryotes [7]. In agreement with other phylogenetic studies, it was also suggested that since NOX2-like is present in choanoflagellates and that sea anemones have both NOX2-like and NOX4, the split between these two clades happened in the ancestral metazoan. We have not found NOX4 in the sponge genome, indicating that the NOX4 gene might have actually emerged later at the time of the cnidarian-bilaterian divergence or may have been lost in this sponge species. These two clades arose from EF-hand containing genes (Ca^{2+}-dependent NOXes) and the ancestral NOX2-like gene later lost these domains [7]. This would suggest that both NOX5 and DUOX were later lost in choanoflagellates and cnidarians. Gene loss seems to be common in the NOX family evolution [6]. NOX5, for example, was also lost in the phylum Nematoda [6, 32] and in the mammalian order Rodentia [6]. Our results also corroborate the loss of NOX2-like in the Ecdysozoa lineage [6, 32]. In fact, Ecdysozoans seem to have suffered extensive gene losses in many gene families throughout their evolution [42].

Gene duplication followed by subfunctionalization or neofunctionalization has been proposed to facilitate the evolution of new gene functions [43–45]. Nevertheless, the evolutionary impact of lineage-specific gene losses has never gained much attention. If no paralogs are present, a gene function that is exclusively associated with a certain gene may disappear if that gene is lost. This outcome has been thought of as detrimental to the species, rendering it less adaptable to the changing environment. However, recently, with new methodological and technological advances (great number of genomes sequenced), more evidence of the pervasiveness of gene loss has been gathered [46]. It is now contemplated that gene loss can be neutral or even adaptive and thus relevant for species evolution (for a review see [42]). Castro et al. 2014, for example, have found that gene loss, related to the gastric function gene kit, correlates with the evolution of different stomach types in vertebrates, which might be associated with their diet [47]. Also, the absence of some urea cycle and essential amino acid synthesis enzymes in the hematophagous *Rhodnius prolixus* was speculated to be due to relaxation of these pathways in an amino acid rich diet [48].

The evidence for gene loss is negative and can pass unnoticed or not be considered due to uncertainties in the completion or assembly of sequenced genomes. Therefore, the impact of gene loss in the evolution and function of surviving paralogs is not well investigated. It is easier to recognize gene duplication and the appearance of a new gene function as adaptive. In the evolution of the NOX family, gene duplication followed by neofunctionalization seems to have happened very early since both the ability to produce superoxide and hydrogen peroxide were present in the ancestral calcium binding enzymes (NOX5 and DUOX, respectively) [7]. In addition, further evidence of subfunctionalization has recently been gathered in vertebrates. Vertebrate NOX2 seems to be expressed mostly in phagocytes whilst NOX1, NOX3 and NOX4 have other specific functions and patterns of subcellular localization and tissue distribution [2, 19, 49]. Still, the major characteristic of this family of enzymes seems to be evolution by gene loss.

Gene loss and NOX4-art evolution

Our more extensive search shows that, as with NOX2-like, the NOX4/NOX4-art gene is present in more animal groups than previously thought [32] but was also lost in many lineages throughout metazoan evolution. This gene seems to have been lost in Nematoda and Annelida and, of the four molluscan genomes searched, it was found in only one. In these phyla, however, only a small number of genomes were searched and no transcriptomes. In arthropods, where a larger sampling scheme was performed, we show here that NOX4-art was lost many times during evolution (Fig. 4). Although present in Chelicerates, Crustaceans and Hexapods, it was not found in the two Myriapoda genomes searched (Fig. 4).

In Hexapods, sometimes the gene seems to have been lost in the ancestral lineages of whole orders (in Hymenoptera, for example) but this has happened mostly in different species within different taxonomic groups (clades). It is important to differentiate species where only transcriptomes were analyzed from those where complete genome assemblies are present (Fig. 4) since the absence from a transcriptome may only mean the gene was not being transcribed at that particular time. Nevertheless, even if we take into consideration that the gene could not be found in the genomes of a few species due to assembly errors, we would still see that it is absent from a variety of Hexapod species. In Diptera, for example, NOX4-art is found in mosquitoes but not in the extensively studied genomes of the different *Drosophila* species. This pattern of evolution certainly fits the "patchy ortholog" design that is commonly found in ancient gene families [42].

The pervasiveness of gene loss begs the question of how many genes are actually indispensable in any given

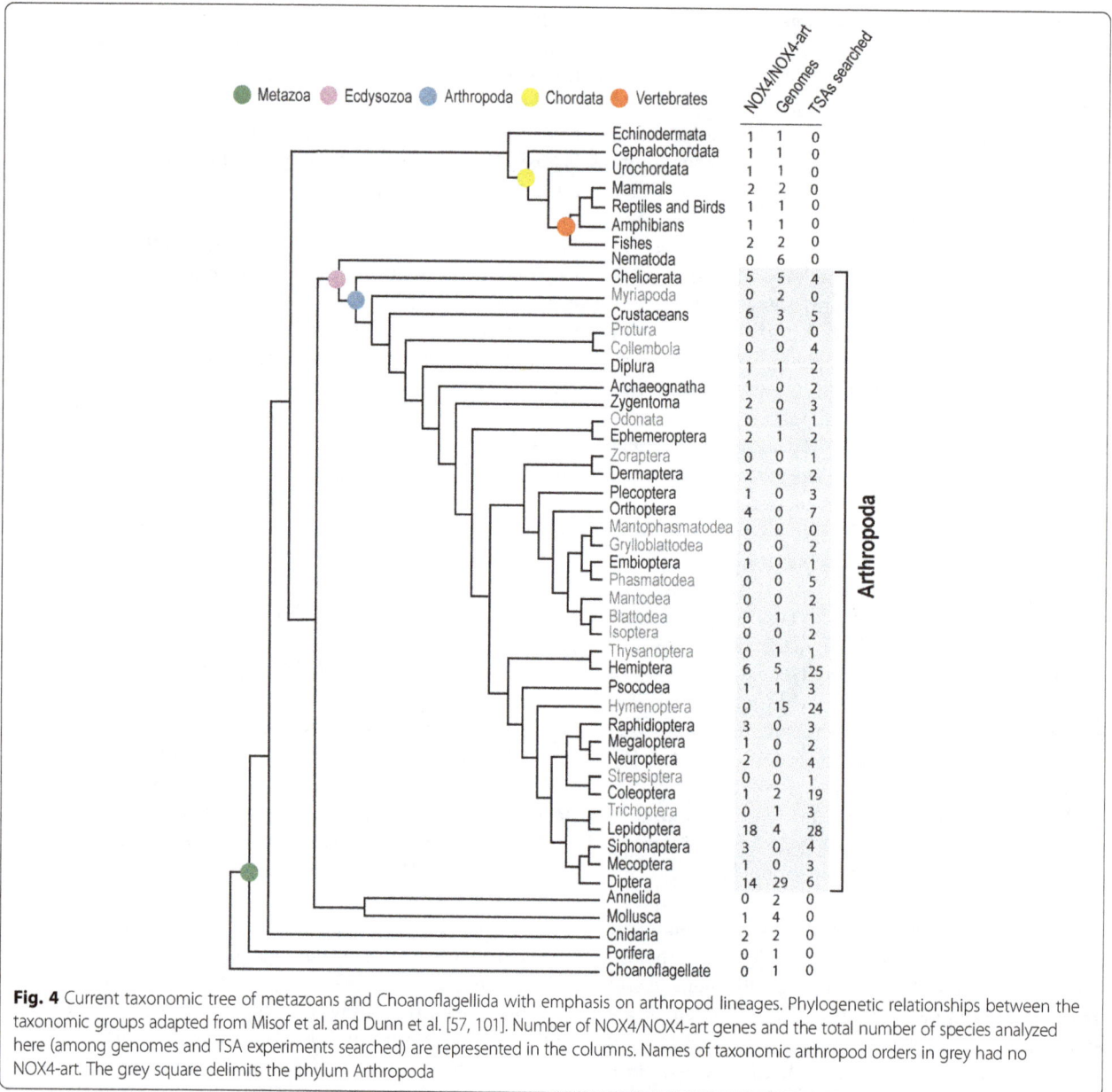

Fig. 4 Current taxonomic tree of metazoans and Choanoflagellida with emphasis on arthropod lineages. Phylogenetic relationships between the taxonomic groups adapted from Misof et al. and Dunn et al. [57, 101]. Number of NOX4/NOX4-art genes and the total number of species analyzed here (among genomes and TSA experiments searched) are represented in the columns. Names of taxonomic arthropod orders in grey had no NOX4-art. The grey square delimits the phylum Arthropoda

genome. In *Drosophila melanogaster*, it has been suggested that around 85% of its genes are dispensable [50]. Reductive evolution in parasitic or symbiotic species is not a new concept but for other species gene loss has usually been linked to reduction in fitness or has been associated to adaptiveness to specific environments such as the loss of genes related to vision in cave dwelling animals, for example [51–53]. Another possibility is the presence of other genes such as paralogs, analogs or even whole different pathways that serve the same or very similar functions and therefore the loss of a specific gene does not mean loss of function (mutation robustness). Since NOX4-art produces the same byproduct (hydrogen peroxide) as DUOX enzymes, it is plausible

that this redundancy might be the reason why it was lost in some lineages. However, this does not explain why in other lineages the gene is present and functional. If DUOX and NOX4-art roles and expression patterns are exactly the same, one questions if a dosage balance problem could not have arisen.

Comparative analyses of gene losses according to gene ontology (GO) categories have shown that the differences in dispensability observed between different genes might not be stochastic. GOs related to signal transduction, one of the main functions attributed to NOX4 enzymes [54], and other functions that are more sensitive to dosage imbalance are more prone to be lost [42]. Other GO categories of more ancient biochemical processes such as protein modification and immune response, among others, were also deemed more prone to gene loss in different organisms. All of these are functions in which the production of ROS by NOX enzymes seems to be involved in. This family of enzymes is certainly a success story that has arisen early in the evolution of multicellular life and thus their participation in ancient biochemical processes and patterns of gene gain and loss are expected [2, 55, 56]. However, it is also possible that the loss of a gene function altogether might not have a detrimental or adaptive effect on a species and, in fact, can be neutral [46]. Neutral or nearly neutral gene losses can be fixed in a species through genetic drift and, in ancient gene families such as NOXes, the effect on a phylogenetic framework can be seen as the patchy distribution of orthologs observed [42, 46].

Although it would have been desirable to be sure whether incomplete genome coverage and/or assembly or true gene loss was the cause behind the apparent absence of NOX4/NOX4-art in many species, a synteny analysis to examine the genomic *loci* was simply not possible for most groups. A strong degree of synteny was only possible to be seen in closely related species probably due to the great divergence among the many species studied (diverse metazoan groups). Even though hexapods are a more closely related group, the lack of synteny found among the different orders may be due to their ancient radiation (479 million years ago [57]) and to the great amount of genome rearrangements that might have occurred. Indeed, much faster rates of chromosomal rearrangements have been found for *Drosophila* species when compared to other eukaryotes [58, 59]. In addition, for this genus, high rearrangement rates were also found for co-expressed genes within short intergene sequences [60]. Extensive genome re-arrangements among mosquito (*Anopheles gambiae*) and fly lineages have also been found [61]. Therefore, it is not surprising then that in our study we have not found synteny between the three mosquitoes analyzed and the *Drosophila melanogaster* genome. Furthermore, the lepidopteran orthologous groups belong to completely different ones than those of the mosquitoes, indicating that there is no special region on the genome for the NOX4-art gene. It seems that it has changed places in the genomes as much as any other old gene and does not seem to belong to any specific expression or otherwise constrained synteny block [58, 59].

NOX4-art function and activation

The arthropod NOX4-art clade includes the mosquito gene previously described as NOXm [32], indicating that this gene is more widely distributed within the phylum and is not related to hematophagy as previously thought.

Arthropod NOX4-art formed a well-supported distinct group closely related to NOX4 (Fig. 1). Indeed, our functional analysis indicates that this gene still produces hydrogen peroxide as other NOX4 genes.

Although NOX4-art is closely related to NOX4, its expression and stability might be different. No p22-*phox* was found in any arthropod indicating that a different protein or mode of action might be responsible for NOX4-art activation. When the expression of p22-*phox* is inhibited, it greatly diminishes the production of ROS by NOX4. Nevertheless, neither truncated nor mutated forms of p22-*phox* that disrupted the PRR had any effect on ROS production on vertebrate NOX4 genes [21]. Since this region is the most characteristic of p22-*phox* in most metazoans, it might be possible that another protein that we could not find with our searches using these proteins as queries could be acting as a NOX4-art activator. It has been proposed that NOX4/p22-*phox* complex is structurally different from NOX1-3 with loop D being the most important feature for the formation of the complex. NOX4 chimeras with NOX2 loop D sequences did not translocate to the cell surface nor produced ROS [38]. Nevertheless, a specific motif or residue has not been attributed to the formation of this complex and a better understanding of the interaction of these two proteins is still needed.

Motifs and residues in loop B and DH domain are essential for ROS production since their interaction approximates the heme, FAD and NADPH binding sites, facilitating the transfer of electrons. Although the characteristic mammalian RRXRR loop B motif is not present in NOX4-art, it is also absent in other metazoan NOX4. This motif has been determined as important for ROS production in mammalian NOX4 but mostly because of the first arginine residue [37]. Arginine residues are present in most NOX4-art loop B sequences although not in that specific motif. It seems plausible that although different from the vertebrate motif other positively charged motifs might be responsible for the interaction of the DH domain and loop B and, therefore, NOX4-art activity. The C-terminal region of NOX4 is also important for catalytic activity. Substitution of the amino acids histidine and glutamate or changes in the size of this region substantially decrease ROS production in vertebrates. These changes alter the distance between FAD and NADPH binding regions and hinder the proper folding of the DH domain [37]. The glutamate residue seems to be present in most NOX4-art sequences and the number of residues in the C-terminal region does not vary much among arthropod groups being well within the range found for vertebrates (Table 1). The position of the histidine residue varies in NOX4-art; which might not be a problem since in loop B the motifs are different from vertebrates and the relative position of these two regions, that have to interact for proper ROS production, may be different for this gene when compared to vertebrate NOX4. Indeed, although there are many differences in important functional residue positions and motifs between vertebrate NOX4, other metazoan NOX4 and NOX4-art, the silencing of *Aedes aegypti* NOX4-art shows us that this gene is functional and probably produces hydrogen peroxide as expected for a NOX4 gene.

When VectorBase (www.vectorbase.org) expression maps are searched for *Anopheles gambiae* and *Aedes aegypti* NOX4-art genes (AGAP003772 and AAEL010179, respectively), we find differential expression during embryonic development [62, 63] and between tissues [64, 65] and increased expression after a blood meal [66, 67]. In *Aedes aegypti*, infection with *Micrococcus luteus* or *Wolbachia* (wMel strain) also increases NOX4-art expression [68, 69]. Thus, it seems that NOX4-art, in mosquitoes at least, might be linked to a number of physiological functions.

Conclusions

The increasing number of genomes available can significantly contribute to the study of gene family evolution. The NOXes are an intriguing gene family that is responsible for important cellular processes such as cell signaling, transcriptional regulation, stress response and gut homeostasis and defense. Genes from this family are present in all eukaryotic groups analyzed until now suggesting that the common ancestor of NOX genes emerged at an early stage in the evolution of eukaryotes.

We have found that the NOXm (now renamed NOX4-art) gene described for mosquitoes is actually widespread within arthropods. This gene is potentially functional where it is present and probably produces ROS in the form of hydrogen peroxide. Important insect specific functions were described for NOXes such as cuticular hardening [70, 71], wound healing [72, 73], smooth muscle contraction [74] and gut immunity [34, 75–77]. NOX4-art could have an arthropod specific function as well and therefore be an interesting target for new vector and pest control strategies. In addition, non- mammalian experimental systems are advantageous due to the smaller number of genes present which helps to elucidate questions regarding their expression in different developmental stages and/or tissues.

The arthropod NOX4-art gene is also an interesting example of evolution by gene loss. Within the phylum Arthropoda it has a 'patchy' distribution with some species possessing the gene while others do not. It is difficult to determine if the loss of the NOX4-art gene in different lineages is due to the presence of another enzyme with similar function (DUOX) or simply a neutral pattern of evolution. Dosage imbalance does not

seem to be a problem where NOX4-art is present, although subfunctionalization has not been discarded. One thing is certain though; the study of gene loss in gene family evolution should receive more attention, as it could be an important source of information about evolutionary patterns and processes.

Methods
Gene search

To search for NOX genes in the 95 genomes analyzed, the ferric reductase transmembrane component domain (HMM profile PF01794) [78] was used as query in HMMsearch [79] using the FAT pipeline [80]. All proteins with significant E-value (<0.001) were retrieved and used as queries on BLASTp searches [81] against the manually curated Uniprot/SwissProt protein database [82] also using FAT. To make sure that all genes belonging to the NOX gene family were found in arthropods and other genomes, tBLASTn and Exonerate (protein2-genome mode) [83] searches against the scaffolds of the whole genome databases were also performed. This makes it possible to find genes that might not have been predicted before and therefore could not be retrieved with the HMMsearch step. These searches were performed with the already predicted NOX genes of other closely related species as queries. Redundancy was eliminated with the software CD-Hit [84]. When new peptides were found they were predicted with GeneWise [85], using the closest BLASTp search homolog to predict a full-length protein sequence, when possible. The results were visualized and further edited, when needed, in the genome browser Artemis [86].

We also performed tBLASTn searches against the Transcriptome Shotgun Assembly (TSA) sequence database [35] at the NCBI website using *Anopheles gambiae* (AGAP003772), *Bombyx mori* (XP_012553051), *Catajapyx aquilonaris* (CAQU006423) and *Daphnia pulex* (EFX76139) NOX4-art sequences. These sequences were the ones that seemed better annotated and spanned a wide taxonomic diversity from our previous search. The mRNA best hits were saved in FASTA format and redundancy was removed with CD-Hit [84] with a cut off of 100%. Artemis [86] was used to extract the coding sequence and the respective peptide sequence for each hit. Conserved domain composition was confirmed by searches against the Protein Family Database (Pfam) and Conserved Domain Database (CDD) [78, 87].

Sequences encoding putative p22-*phox* accessory protein were also searched in the genomes analyzed using the same approach described for the NOX genes above. The cytochrome b558 alpha-subunit domain (HMM profile PF05038) was used as query for HMMsearch. Proteins with significant E-values (<0.001) were retrieved and used as queries on BLASTp searches against Uniprot/SwissProt. Again, whenever no hits were found tBLASTn and Exonerate searches against the scaffolds of the whole genome databases were also performed. This search ensured that we could find genes that were not automatically predicted.

The sequences of all peptides used in the phylogenetic analysis and their accession numbers are available for download in FASTA format as supplementary material [see Additional file 7].

Phylogenetic analysis

Amino acid sequences of the proteins retrieved by our searches of genomes and the TSA database were aligned locally with PASTA [88] using the JTT + G20 model and other default parameters. The alignments were visualized and converted to Phylip format using the software SeaView [89]. The same program was used to trim the sequences leaving only the region containing the ferric reductase and the dehydrogenase (FAD1-2 and NADPH1-4) domains that are common to all NOX/DUOX genes. This way, the peroxidase domain found only in DUOX genes and the calcium binding domains present in both DUOX and NOX5 genes were eliminated from the alignment. This trimmed version of the alignment was then used to construct a phylogenetic tree using the maximum likelihood method with RAxML [90] on CIPRES web server [91]. The amino acid Jones Taylor Thornton scoring matrix was used [92] and bootstrap analysis with 1000 replicates was performed to infer branch support.

Synteny and protein domain and motif analysis

For all genes CD-Search Batch (CDD v3.14 database) [93] and BLAST searches on Pfam 28.0 [78] were used to infer conserved sites and to confirm the protein domain structure. TMHMM [94] available on the web was also used to infer the presence and position of the transmembrane helixes. To investigate NOX4-art neighborhood microsynteny we used genomes of the species *Aedes aegypti*, *Culex quinquefasciatus*, *Anopheles gambiae*, *Bombyx mori*, *Manduca sexta*, *Catajapyx aquilonaris*, *Centruroides sculpturatus*, *Daphnia pulex*, *Parasteatoda tepidariorum* and *Tetranychus urticae*. We extracted all translated coding sequences present from each side of NOX4-art genes in their scaffolds/chromosomes. OrthoMCL [95] was used to group putative orthologues with an E-value lower than 10E-5. For the motif analysis, the amino acid sequences retrieved by the FAT pipeline were aligned with ClustalW [96] within BioEdit software [97]. Vertebrate NOX4 was used as a guide for NOX4-art analysis. Based on Kawahara et al. 2007 loop sizes, canonical regions and conserved amino acid residues required for ROS

production were searched in the alignment and were identified manually [32].

NOX4-art silencing

Aedes aegypti Aag-2 cells were cultivated in 25 cm^2 plastic flasks in Schneider's *Drosophila* medium supplemented with 10% fetal bovine serum until 100% confluence and then maintained at 28 °C. Cells were transfected using the cell line Nucleofector kit V according to the manufacturer's instructions (Amaxa Biosystems, Köln, Germany). Briefly, cells (1×10^6) were centrifuged and carefully resuspended in 3 µg of dsRNA (dsMAL and dsNOX4-art) and 100 µL of transfection reagent (82 µL cell line plus 18 µL supplement). The unrelated dsRNA, dsMal, specific of *Escherichia coli* MalE gene (Gene ID: 948538), was used as a control for the off-target effects of dsRNA. The cell/dsRNA suspensions were transferred into a Lonza certified cuvette and transfected in the Nucleofector I Device (Lonza, USA) with the program G-030. Immediately after transfection, 2.4 mL of Schneider's supplemented media was gently added to the cuvette and the cells were seeded into a 96 well plate at a density of 4×10^4 cells per well. After that, cells were incubated at 28 °C in an air incubator until analysis.

RNA extraction and qPCR

Total RNA was extracted from Aag-2 cells (4×10^4) using TRIzol (Invitrogen) according to the manufacturer's protocol. One microgram of RNA was treated with RNasefree DNase I (Fermentas International Inc., Burlington, Canada). The treated RNA was used to synthesize the cDNA with the High Capacity cDNA reverse transcription kit (Applied Biosystems, Foster City, CA). qPCR was performed on a StepOnePlus qPCR system (Applied Biosystems) using the Power SYBR Green PCR master mix (Applied Biosystems). The comparative *Ct* method [98] was used to compare gene expression levels. The *Aedes aegypti* ribosomal protein 49 gene (Rp49) was used as an endogenous control, based on previous data [99]. The primer pairs used for the amplification of cDNA fragments for both conventional and qPCR were: *NOX4-art*: forward 5-TTG TGT TCG CAC ATC CAA CT-3 and reverse 5-GGT CCA ACG AAA AAT ATC CAA A-3; *Rp49*: forward, 5-TGT CGG TGT AAC TGG CAT GT-3 and reverse, 5-TCG GCC AAC AAA AGT ACA CA-3. Statistical analysis was performed with Graphpad Prism software with Student's t-test.

ROS measurement

Hydrogen peroxide production by Aag-2 cells was measured by monitoring resorufin fluorescence due to the oxidation of Amplex Red (Invitrogen, USA). Cells (4×10^4 cells/well) were incubated at 28 °C for 4 days after transfection with dsRNA and then, assayed in 50 µM Amplex Red (Invitrogen, USA), 40 units of horseradish peroxidase (Sigma, USA) and 25 units of superoxide dismutase (Sigma, USA) in Schneider's *Drosophila* medium, for a final volume of 200 µL. Immediately after reagent mixture, the endpoints of Amplex Red oxidation were recorded at room temperature using a Spectra Max spectrofluorimeter (Varian, USA), operating at excitation and emission wavelengths of 530 nm and 590 nm. Statistical analysis was performed with Graphpad Prism software with One-way ANOVA, Sidak's test.

Additional files

Additional file 1: Table with a list of the 95 species with complete genomes analyzed, number of proteins found for each NOX gene, their genome source and version. (XLSX 18 kb)

Additional file 2: Table with a list of the species where NOX4-art genes could be found on NCBI's TSA database. Their TSA identification number and amino acid length. (XLSX 12 kb)

Additional file 3: Table with the list of species where p22-*phox* was found, their gene identification, amino acid length and composition of the polybasic and proline rich region. (XLSX 13 kb)

Additional file 4: Maximum Likelihood phylogeny of aligned NOX and DUOX proteins identified in this study. Numbers on branches are bootstrap support values from 1000 replicates; only numbers above 60% are shown. The image was created using iTOL [100]. (TIFF 7473 kb)

Additional file 5: Micro-synteny around NOX4-art gene (orange arrow) in Diptera and Lepidoptera. Orthologous genes are represented by grey arrows, paralogs by blue arrows and genes where no orthologous or paralogous relationship could be determined within the genomes are depicted in grey-wired arrows. A) Complete chromosomes 2 and 3 of *Drosophila melanogaster* and scaffold/chromosome regions, where NOX4-art was found, in *Anopheles gambiae*, *Culex quinquefasciatus* and *Aedes aegypti*. B) Scaffold regions, where NOX4-art was found, in *Bombyx mori* and *Manduca sexta*. Estimated divergence between *Anoheles gambiae* and other diptera species and the two lepidopteran species are given in million years (MY) [57]. (TIFF 1688 kb)

Additional file 6: Alignment of NOX4/NOX4-art genes organized by taxonomic group. Important regions and residues are highlighted in different colors: green – transmembrane regions; black – heme binding histidines; grey – loops between transmembranes; cyan – different important residues; magenta – FAD binding regions; orange – NADPH binding regions; yellow – C-terminal region. (PDF 3860 kb)

Additional file 7: FASTA file with the peptides used in the phylogenetic analysis and their NCBI accession numbers. (TXT 287 kb)

Acknowledgements
We thank all the members of the Laboratory of Biochemistry of Hematophagous Arthropods especially MSc. OA Talyuli, for maintenance of Aag-2 cell culture, C Cosme, SR Cássia, SRD Hob and F Nur for technical assistance. Thanks to RM Albano for critically reading earlier versions of the manuscript. The authors would also like to thank the two anonymous reviewers for valuable comments that greatly improved the work.

Funding
This work was supported by grants from Fundação Carlos Chagas Filho de Amparo à Pesquisa do Estado do Rio de Janeiro (FAPERJ), Coordenação de Aperfeiçoamento de Pessoal de Nível Superior (CAPES), the Brazilian National Research Committee (CNPq) and Fiocruz.

Authors' contributions
Planned and designed the study: RS, PLO, ACPG, AT; Performed the experiments analyzed and interpreted the bioinformatics data: RS, PLO, ACPG and AT; Performed the experiments analyzed and interpreted the molecular lab data: PLO, ACPG and ACB; wrote the paper: RS, PLO, ACPG, AT and ACB. All authors read and approved the final manuscript.

Competing interests
The authors declare that they have no competing interests.

Author details
[1]Instituto de Bioquímica Médica Leopoldo de Meis, Universidade Federal do Rio de Janeiro, Rio de Janeiro, Brazil. [2]Laboratório de Biologia Computacional e Sistemas, Instituto Oswaldo Cruz, Fiocruz, Rio de Janeiro, Brazil. [3]Instituto de Biofísica, Universidade Federal do Rio de Janeiro, Rio de Janeiro, Brazil. [4]Instituto Nacional de Ciência e Tecnologia em Entomologia Molecular – INCT-EM, Rio de Janeiro, Brazil.

References
1. Aguirre J, Lambeth JD. Nox enzymes from fungus to fly to fish and what they tell us about Nox function in mammals. Free Radical Bio Med. 2010;49(9):1342–53.
2. Lambeth JD, Neish AS. Nox enzymes and new thinking on reactive oxygen: a double-edged sword revisited. Annu Rev Pathol. 2014;9:119–45.
3. Bonini MG, Consolaro MEL, Hart PC, Mao M, de Abreu ALP, Master AM. Redox control of enzymatic functions: the electronics of life's circuitry. IUBMB Life. 2014;66(3):167–81.
4. Ha E-M, Oh C-T, Bae YS, Lee W-J. A direct role for dual oxidase in *Drosophila* gut immunity. Science. 2005;310(5749):847–50.
5. Lambeth JD. NOX enzymes and the biology of reactive oxygen. Nat Rev Immunol. 2004;4(3):181–9.
6. Sumimoto H. Structure, regulation and evolution of Nox-Family NADPH oxidases that produce reactive oxygen species. FEBS J. 2008;275(13):3249–77.
7. Zhang X, Krause KH, Xenarios I, Soldati T, Boeckmann B. Evolution of the ferric reductase domain (FRD) superfamily: modularity, functional diversification, and signature motifs. PLoS One. 2013;8(3):e58126.
8. Banfi B, Molnar G, Maturana A, Steger K, Hegedus B, Demaurex N, et al. A Ca^{2+}-activated NADPH oxidase in testis, spleen, and lymph nodes. J Biol Chem. 2001;276(40):37594–601.
9. Ameziane-El-Hassani R, Morand S, Boucher JL, Frapart YM, Apostolou D, Agnandji D, et al. Dual oxidase-2 has an intrinsic Ca^{2+}-dependent H_2O_2-generating activity. J Biol Chem. 2005;280(34):30046–54.
10. Banfi B, Tirone F, Durussel I, Knisz J, Moskwa P, Molnar GZ, et al. Mechanism of Ca^{2+} activation of the NADPH oxidase 5 (NOX5). J Biol Chem. 2004;279(18):18583–91.
11. Tirone F, Cox JA. NADPH oxidase 5 (NOX5) interacts with and is regulated by calmodulin. FEBS Lett. 2007;581(6):1202–8.
12. Kawahara T, Lambeth JD. Molecular evolution of Phox-related regulatory subunits for NADPH oxidase enzymes. BMC Evol Biol. 2007;7:178.
13. Bedard K, Jaquet V, Krause K-H. NOX5: from basic biology to signaling and disease. Free Radical Bio Med. 2012;52(4):725–34.
14. Sumimoto H, Miyano K, Takeya R. Molecular composition and regulation of the Nox Family NAD(P)H oxidases. Biochem Bioph Res Co. 2005;338(1):677–86.
15. Lambeth JD, Kawahara T, Diebold B. Regulation of Nox and Duox enzymatic activity and expression. Free Radical Bio Med. 2007;43(3):319–31.
16. Roos D, Md B, Kuribayashi F, Meischl C, S.Weening R, Segal AW, et al. Mutations in the X-linked and Autosomal recessive forms of chronic Granulomatous disease. J Am Soc Hemato. 1996;87(5):1663–81.
17. Rossi F. The O_2-forming NADPH oxidase of the phagocytes nature, mechanisms of activation and function. Biochim Biophys Acta. 1986;853:65–89.
18. Royer-Pokora B, Kunkel LM, Monaco AP, Goff SC, Newburger PE, Baehner RL, et al. Cloning the gene for an inherited human disorder - chronic granulomatous disease - on the basis of its chromosomal location. Nature. 1986;322(3):32–8.
19. Bokoch GM, Knaus UG. NADPH oxidases: not just for leukocytes anymore! Trends Biochem Sci. 2003;28(9):502–8.

20. Yu L, Zhen L, Dinauer MC. Biosynthesis of the phagocyte NADPH Oxidase Cytochrome b_{558}. Role of Heme incorporation and Heterodimer formation in maturation and stability of gp91[phox] and p22[phox] subunits. J Biol Chem. 1997;272(43):27288–94.

21. Kawahara T, Ritsick D, Cheng G, Lambeth JD. Point mutations in the proline-rich region of p22[phox] are dominant inhibitors of Nox1- and Nox2-dependent reactive oxygen generation. J Biol Chem. 2005;280(31):31859–69.

22. Iyer SS, Pearson DW, Nauseefl WM, Clark RA. Evidence for a readily dissociable complex of p47phox and p67phox in Cytosol of Unstimulated human Neutrophils. J Biol Chem. 1994;269(35):22405–11.

23. Cheng G, Diebold BA, Hughes Y, Lambeth JD. Nox1-dependent reactive oxygen generation is regulated by Rac1. J Biol Chem. 2006;281(26):17718–26.

24. Miyano K, Sumimoto H. Role of the small GTPase Rac in p22[phox]-dependent NADPH oxidases. Biochimie. 2007;89(9):1133–44.

25. Takeya R, Ueno N, Kami K, Taura M, Kohjima M, Izaki T, et al. Novel human homologues of p47[phox] and p67[phox] participate in activation of superoxide-producing NADPH oxidases. J Biol Chem. 2003;278(27):25234–46.

26. Martyn KD, Frederick LM, von Loehneysen K, Dinauer MC, Knaus UG. Functional analysis of Nox4 reveals unique characteristics compared to other NADPH oxidases. Cell Signal. 2006;18(1):69–82.

27. Nisimoto Y, Jackson HM, Ogawa H, Kawahara T, Lambeth JD. Constitutive NADPH-dependent electron transferase activity of the Nox4 dehydrogenase domain. Biochemistry. 2010;49(11):2433–42.

28. Ha E-M, Lee K-A, Park SH, Kim S-H, Nam H-J, Lee H-Y, et al. Regulation of DUOX by the Gαq-Phospholipase Cβ-Ca^{2+} pathway in Drosophila gut immunity. Dev Cell. 2009;16(3):386–97.

29. Morand S, Ueyama T, Tsujibe S, Saito N, Korzeniowska A, Leto TL. Duox maturation factors form cell surface complexes with Duox affecting the specificity of reactive oxygen species generation. FASEB J. 2008;23(4):1205–18.

30. Rigutto S, Hoste C, Grasberger H, Milenkovic M, Communi D, Dumont JE, et al. Activation of dual Oxidases Duox1 and Duox2: differential regulation mediated by cAMP-dependent protein Kinase and protein Kinase C-dependent Phosphorilation. J Biol Chem. 2009;284(11):6725–34.

31. Chen F, Wang Y, Barman S, Fulton DJR. Enzymatic regulation and functional relevance of NOX5. Curr Pharm Des. 2015;21(41):5999–6008.

32. Kawahara T, Quinn MT, Lambeth JD. Molecular evolution of the reactive oxygen-generating NADPH oxidase (Nox/Duox) family of enzymes. BMC Evol Biol. 2007;7:109.

33. Pan X, Zhou G, Wu J, Bian G, Lu P, Raikhel AS, et al. Wolbachia induces reactive oxygen species (ROS)-dependent activation of the toll pathway to control dengue virus in the mosquito Aedes aegypti. P Natl Acad Sci USA. 2012;109(1):E23–31.

34. Park Y, Stanley DW, Kim Y. Eicosanoids up-regulate production of reactive oxygen species by NADPH-dependent oxidase in Spodoptera exigua phagocytic hemocytes. J Insect Physiol. 2015;79:63–72.

35. Benson DA, Karsch-Mizrachi I, Clark K, Lipman DJ, Ostell J, Sayers EW. GenBank. Nucleic Acids Res. 2012;40(Database issue):D48–53.

36. Inoue Y, Ogasawara M, Moroi T, Satake M, Azumi K, Moritomo T, et al. Characteristics of NADPH oxidase genes (Nox2, p22, p47, and p67) and Nox4 gene expressed in blood cells of juvenile Ciona intestinalis. Immunogenetics. 2005;57(7):520–34.

37. Von Lohneysen K, Noack D, Hayes P, Friedman JS, Knaus UG. Constitutive NADPH oxidase 4 activity resides in the composition of the B-loop and the penultimate C terminus. J Biol Chem. 2012;287(12):8737–45.

38. Von Lohneysen K, Noack D, Wood MR, Friedman JS, Knaus UG. Structural insights into Nox4 and Nox2: motifs involved in function and cellular localization. Mol Cell Biol. 2010;30(4):961–75.

39. Takac I, Schroder K, Zhang L, Lardy B, Anilkumar N, Lambeth JD, et al. The E-loop is involved in hydrogen peroxide formation by the NADPH oxidase Nox4. J Biol Chem. 2011;286(15):13304–13.

40. Jackson HM, Kawahara T, Nisimoto Y, Smith SM, Lambeth JD. Nox4 B-loop creates an interface between the transmembrane and dehydrogenase domains. J Biol Chem. 2010;285(14):10281–90.

41. Zhu Y, Marchal CC, Casbon AJ, Stull N, von Lohneysen K, Knaus UG, et al. Deletion mutagenesis of p22[phox] subunit of flavocytochrome b_{558}: identification of regions critical for gp91[phox] maturation and NADPH oxidase activity. J Biol Chem. 2006;281(41):30336–46.

42. Albalat R, Cañestro C. Evolution by gene loss. Nat Rev Genet. 2016;17(7):379–91.

43. Magadum S, Banerjee U, Murugan P, Gangapur D, Ravikesavan R. Gene duplication as a major force in evolution. J Genet. 2013;92:155–61.

44. Zhang J. Evolution by gene duplication: an update. Trends Ecol Evol. 2003;18(6):292–8.

45. Dittmar K, Liberles D. Evolution after gene duplication. Wiley-Blackwell: Hoboken; 2010.

46. Wyder S, Kriventseva EV, Schröder R, Kadowaki T, Zdobnov EM. Quantification of ortholog losses in insects and vertebrates. Genome Biol. 2007;8(11):R242.

47. Castro LF, Goncalves O, Mazan S, Tay BH, Venkatesh B, Wilson JM. Recurrent gene loss correlates with the evolution of stomach phenotypes in gnathostome history. P Roy Soc B - Biol Sci. 2014;281(1775):20132669.

48. Mesquita RD, Vionette-Amaral RJ, Lowenberger C, Rivera-Pomar R, Monteiro FA, Minx P, et al. Genome of Rhodnius prolixus, an insect vector of Chagas disease, reveals unique adaptations to hematophagy and parasite infection. P Natl Acad Sci USA. 2015;112(48):14936–41.

49. Manea SA, Constantin A, Manda G, Sasson S, Manea A. Regulation of Nox enzymes expression in vascular pathophysiology: focusing on transcription factors and epigenetic mechanisms. Redox Biol. 2015;5:358–66.

50. Dietzl G, Chen D, Schnorrer F, Su KC, Barinova Y, Fellner M, et al. A genome-wide transgenic RNAi library for conditional gene inactivation in Drosophila. Nature. 2007;448(7150):151–6.

51. Protas ME, Hersey C, Kochanek D, Zhou Y, Wilkens H, Jeffery WR, et al. Genetic analysis of cavefish reveals molecular convergence in the evolution of albinism. Nat Genet. 2006;38(1):107–11.

52. Protas ME, Trontelj P, Patel NH. Genetic basis of eye and pigment loss in the cave crustacean, Asellus aquaticus. P Natl Acad Sci USA. 2011;108(14):5702–7.

53. Leys R, Cooper SJ, Strecker U, Wilkens H. Regressive evolution of an eye pigment gene in independently evolved eyeless subterranean diving beetles. Biol Lett. 2005;1(4):496–9.

54. Brown DI, Griendling KK. Nox proteins in signal transduction. Free Radical Bio Med. 2009;47(9):1239–53.

55. Bedard K, Lardy B, Krause KH. NOX Family NADPH oxidases: not just in mammals. Biochimie. 2007;89(9):1107–12.

56. Lalucque H, Silar P. NADPH oxidase: an enzyme for multicellularity? Trends Microbiol. 2003;11(1):9–12.

57. Misof B, Liu S, Meusemann K, Peters RS, Donath A, Mayer C, et al. Phylogenomics resolves the timing and pattern of insect evolution. Science. 2014;346(6210):763–7.

58. Bhutkar A, Schaeffer SW, Russo SM, Xu M, Smith TF, Gelbart WM. Chromosomal rearrangement inferred from comparisons of 12 Drosophila genomes. Genetics. 2008;179(3):1657–80.

59. Ranz JM, Casals F, Ruiz A. How malleable is the eukaryotic genome? Extreme rate of chromosomal rearrangement in the genus Drosophila. Genome Res. 2001;11:230–9.

60. Weber CC, Hurst LD. Support for multiple classes of local expression clusters in Drosophila melanogaster, but no evidence for gene order conservation. Genome Biol. 2011;12:R23.

61. Holt RA, Subramanian GM, Halpern A, Sutton GG, Charlab R, Nusskern DR, et al. The genome sequence of the malaria mosquito Anopheles gambiae. Science. 2002;298(4):129–49.

62. Goltsev Y, Rezende GL, Vranizan K, Lanzaro G, Valle D, Levine M. Developmental and evolutionary basis for drought tolerance of the Anopheles gambiae embryo. Dev Biol. 2009;330(2):462–70.

63. Harker BW, Behura SK, BS dB, Lovin DD, Mori A, Romero-Severson J, et al. Stage-specific transcription during development of Aedes aegypti. BMC Dev Biol. 2013;13(29):1–12.

64. Neira-Oviedo M, VanEkeris L, Corena-Mcleod MDP, Linser PJ. A microarray-based analysis of transcriptional compartmentalization in the alimentary canal of Anopheles gambiae larvae. Insect Mol Biol. 2008;17(1):61–72.

65. Baker DA, Nolan T, Fischer B, Pinder A, Crisanti A, Russell S. A comprehensive gene expression atlas of sex- and tissue-specificity in the malaria vector, Anopheles gambiae. BMC genomics. 2011;12:296.

66. Marinotti O, Calvo E, Nguyen QK, Dissanayake S, Ribeiro JMC, James AA. Genome-wide analysis of gene expression in adult Anopheles gambiae. Insect Mol Biol. 2006;15(1):1–12.

67. Dissanayake SN, Ribeiro JM, Wang MH, Dunn WA, Yan G, James AA, et al. aeGEPUCI: a database of gene expression in the dengue vector mosquito, Aedes aegypti. BMC Res Notes. 2010;3:248.

68. Choi YJ, Fuchs JF, Mayhew GF, Yu HE, Christensen BM. Tissue-enriched expression profiles in Aedes aegypti identify hemocyte-specific transcriptome responses to infection. Insect Biochem Mol Biol. 2012;42(10):729–38.

69. Rancès E, Ye YH, Woolfit M, McGraw EA, O'Neill SL. The relative importance of innate immune priming in *Wolbachia*-mediated dengue interference. PLoS Pathog. 2012;8(2):e1002548.

70. Anh NTT, Nishitani M, Harada S, Yamaguchi M, Kamei K. A *Drosophila* model for the screening of bioavailable NADPH oxidase inhibitors and antioxidants. Mol Cell Biochem. 2011;352(1-2):91–8.

71. Dias FA, Gandara ACP, Queiroz-Barros FG, Oliveira RLL, Sorgine MHF, Braz GRC, et al. Ovarian dual Oxidase (Duox) activity is essential for insect eggshell hardening and waterproofing. J Biol Chem. 2013;288(49):35058–67.

72. Juarez MT, Patterson RA, Sandoval-Guillen E, McGinnis W. *Duox*, *Flotillin-2*, and *Src42A* are required to activate or delimit the spread of the transcriptional response to epidermal wounds in *Drosophila*. PLoS Genet. 2011;7(12):e1002424.

73. Razzell W, Evans Iwan R, Martin P, Wood W. Calcium flashes orchestrate the wound inflammatory response through DUOX activation and hydrogen peroxide release. Curr Biol. 2013;23(5):424–9.

74. Ritsick DR, Edens WA, Finnerty V, Lambeth JD. Nox regulation of smooth muscle contraction. Free Radical Bio Med. 2007;43(1):31–8.

75. Oliveira GA, Lieberman J, Barillas-Mury C. Epithelial nitration by a Peroxidase/NOX5 system mediates mosquito Antiplasmodial immunity. Science. 2012;335(6070):856–9.

76. Oliveira JHM, Gonçalves RLS, Lara FA, Dias FA, Gandara ACP, Menna-Barreto RFS, et al. Blood meal-derived Heme decreases ROS levels in the Midgut of *Aedes aegypti* and allows proliferation of intestinal Microbiota. PLoS Pathog. 2011;7(3):e1001320.

77. Kumar S, Molina-Cruz A, Gupta L, Rodrigues J, Barillas-Mury C. A Peroxidase/dual Oxidase system modulates Midgut epithelial immunity in *Anopheles gambiae*. Science. 2010;327(5973):1644–8.

78. Finn RD, Bateman A, Clements J, Coggill P, Eberhardt RY, Eddy SR, et al. Pfam: the protein families database. Nucleic Acids Res. 2014;42(Database issue):D222–30.

79. Eddy SR. Accelerated profile HMM searches. PLoS Comput Biol. 2011;7(10):e1002195.

80. Seabra-Junior ES, Souza EM, Mesquita RD. In: Istituto Nacional de Propriedade Industrial, editor. FAT - functional analysis tool. Brazil: IFRJ; 2011.

81. Altschul SF, Gish W, Miller W, Myers EW, Lipman DJ. Basic local alignment search tool. J Mol Biol. 1990;215:403–10.

82. The Uniprot Consortium. UniProt: a hub for protein information. Nucleic Acids Res. 2014;43(D1):D204–12.

83. Slater GS, Birney E. Automated generation of heuristics for biological sequence comparison. BMC Bioinformatics. 2005;6:31.

84. Fu L, Niu B, Zhu Z, Wu S, Li W. CD-HIT: accelerated for clustering the next-generation sequencing data. Bioinformatics. 2012;28(23):3150–2.

85. Birney E, Clamp M, Durbin R. GeneWise and Genomewise. Genome Res. 2004;14:988–95.

86. Rutherford K, Parkhill J, Crook J, Horsnell T, Rice P, Rajandream MA, et al. Artemis: sequence visualization and annotation. Bioinformatics. 2000;16(10):944–5.

87. Marchler-Bauer A, Derbyshire MK, Gonzales NR, Lu S, Chitsaz F, Geer LY, et al. CDD: NCBI's conserved domain database. Nucleic Acids Res. 2015;43(Database issue):D222–6.

88. Mirarab S, Nguyen N, Warnow T. PASTA: ultra-large multiple sequence alignment. In Research in Computational Molecular Biology-18th Annual International Conference, RECOMB 2014, Proceedings. 2014;8394:177–91.

89. Gouy M, Guindon S, Gascuel O. SeaView version 4: a multiplatform graphical user interface for sequence alignment and phylogenetic tree building. Mol Biol Evol. 2010;27(2):221–4.

90. Stamatakis A. RAxML version 8: a tool for Phylogenetic analysis and post-analysis of large phylogenies. Bioinformatics. 2014;30(9):1312–3.

91. Miller MA, Pfeiffer W, Schwartz T. Creating the CIPRES Science Gateway for inference of large phylogenetic trees in Proceedings of the Gateway Computing Environments Workshop (GCE), 14 Nov. 2010. New Orleans. 2010. p. 1-8.

92. Jones DT, Taylor WR, Thornton JM. The rapid generation of mutation data matrices from protein sequences. Comput Appl Biosci. 1992;8:275–82.

93. Marchler-Bauer A, Anderson JB, Chitsaz F, Derbyshire MK, DeWeese-Scott C, Fong JH, et al. CDD: specific functional annotation with the conserved domain database. Nucleic Acids Res. 2009;37:D205–10.

94. Krogh A, Larsson B, Heijne GV, Sonnhammer ELL. Predicting transmembrane protein topology with a hidden Markov model: application to complete genomes. J Mol Biol. 2001;305(3):567–80.

95. Li L, Stoeckert Jr CJ, Roos DS. OrthoMCL: identification of Ortholog groups for eukaryotic genomes. Genome Res. 2003;13:2178–89.

96. Larkin MA, Blackshields G, Brown NP, Chenna R, McGettigan PA, McWilliam H, et al. Clustal W and Clustal X version 2.0. Bioinformatics. 2007;23(21):2947–8.

97. Hall TA. BioEdit: a user-friendly biological sequence alignment editor and analysis program for windows 95/98/NT. Nucleic Acids Symp. 1999;41:95–8.

98. Schmittgen TD, Livak KJ. Analyzing real-time PCR data by the comparative CT method. Nat Protoc. 2008;3:1101–8.

99. Gentile C, Lima JBP, Peixoto AA. Isolation of a fragment homologous to the *rp49* constitutive gene of *Drosophila* in the Neotropical malaria vector *Anopheles aquasalis* (Diptera: Culicidae). Mem I Oswaldo Cruz. 2005;100(6):545–7.

100. Letunic I, Bork P. Interactive tree of life v2: online annotation and display of phylogenetic trees made easy. Nucleic Acids Res. 2011;39(Web Server issue):W475–8.

101. Dunn CW, Giribet G, Edgecombe GD, Hejnol A. Animal phylogeny and its evolutionary implications*. Annu Rev Ecol Evol S. 2014;45(1):371–95.

6

Reconstruction of the evolution of microbial defense systems

Pere Puigbò[1,2], Kira S. Makarova[1], David M. Kristensen[1,3], Yuri I. Wolf[1] and Eugene V. Koonin[1*]

Abstract

Background: Evolution of bacterial and archaeal genomes is a highly dynamic process that involves intensive loss of genes as well as gene gain via horizontal transfer, with a lesser contribution from gene duplication. The rates of these processes can be estimated by comparing genomes that are linked by an evolutionary tree. These estimated rates of genome dynamics events substantially differ for different functional classes of genes. The genes involved in defense against viruses and other invading DNA are among those that are gained and lost at the highest rates.

Results: We employed a stochastic birth-and-death model to obtain maximum likelihood estimates of the rates of gain and loss of defense genes in 35 groups of closely related bacterial genomes and one group of archaeal genomes. We find that on average, the defense genes experience 1.4 fold higher flux than the rest of microbial genes. This excessive flux of defense genes over the genomic mean is consistent across diverse microbial groups. The few exceptions include intracellular parasites with small, degraded genomes that possess few defense systems which are more stable than in other microbes. Generally, defense genes follow the previously established pattern of genome dynamics, with gene family loss being about 3 times more common than gain and an order of magnitude more common than expansion or contraction of gene families. Case by case analysis of the evolutionary dynamics of defense genes indicates frequent multiple events in the same locus and widespread involvement of mobile elements in the gain and loss of defense genes.

Conclusions: Evolution of microbial defense systems is highly dynamic but, notwithstanding the host-parasite arms race, generally follows the same trends that have been established for the rest of the genes. Apart from the paucity and the low flux of defense genes in parasitic bacteria with deteriorating genomes, there is no clear connection between the evolutionary regime of defense systems and microbial life style.

Background

Horizontal gene transfer (HGT) that can result in genome expansion and acquisition of new functions and gene loss leading to genome reduction are recognized as key processes in the evolution of bacterial and archaeal genomes [1–3]. It has been quantitatively demonstrated that gene gain via HGT rather than intragenomic gene duplication is the principal factor of innovation in the evolution of prokaryotes [4]. Reconstructions of gene gain and loss in groups of closely related bacteria and archaea reveal genomes that are in constant flux, with high rates of gain and loss [5–7]. In many cases, several events of gene gain and loss have been estimated to occur over the time that it takes for a single nucleotide substitution to be fixed in an average gene [7]. According to these reconstructions, gene loss appears to be the most common process in microbial evolution that occurs in a roughly clock-like manner [8]. Gene gain appears to be a more episodic process that occurs at rates that on average are two to three-fold lower than the gene loss rate which might be compensated for by bursts of gene gain [8].

Nearly all genes can be transferred and lost during the evolution of bacteria and archaea but the rates of these processes markedly differ between functional classes of genes [7]. The genes involved in key processes of information transmission, in particular translation, comprise the most stable group whereas genes that encode components of the mobilome, such as transposons or prophages, predictably are lost and gained more often. On

* Correspondence: koonin@ncbi.nlm.nih.gov
[1]National Center for Biotechnology Information, National Library of Medicine, National Institutes of Health, Bethesda, MD 20894, USA
Full list of author information is available at the end of the article

the genome dynamics scale, the mobilome is closely followed by the genes that encode various defense systems. Bacteria and archaea exist under constant pressure from parasites, such as viruses, transposons and plasmids, and have evolved elaborate, diverse, multilayer defense strategies [9–12]. The defense systems include those that provide innate immunity, such as restriction-modification (RM), toxin-antitoxin (TA), and abortive infection (AI) systems, and the CRISPR-Cas (Clustered Regularly Interspaced Repeats and CRISPR-associated proteins) systems of adaptive immunity. The incessant arms race between the parasites and the defense systems results in rapid evolution on both sides which involves sequence change as well as intense gene flux [7, 13–15].

Over the last several decades, microbial defense mechanisms have been intensely exploited as source of tools for genome editing, engineering and regulation. In the early days of genetic engineering, the key tools were RM enzymes, and the need in endonucleases with different specificities has greatly stimulated the study of the diversity of RM systems [16–18]. Later, the new era of genome editing has been ushered by the characterization of the programmable, RNA-guided endonuclease activity of CRISPR-Cas systems, particular those of Class 2 that possess a single-protein effector modules and therefore can be introduced into mammalian cells in a straightforward manner [19–21]. Notably, systematic screening of bacterial and archaeal genomes for new Class 2 CRISPR-Cas systems resulted in the discovery of several novel variants that have been immediately harnessed for applications [22–24]. Even more recently, bacterial and archaeal Argonaute proteins have been shown to represent a distinct, RNA- or DNA-guided innate immunity machinery in bacteria and archaea [25, 26] that eventually might lead to yet another generation of genome editing tools notwithstanding the controversy around the early attempts on application [27, 28]. The remarkable utility

of microbial defense systems as tools for genome manipulation stems from their naturally evolved ability to recognize and cleave specific DNA or RNA sequence, which is a pre-requisite for self vs non-self discrimination. These features of defense systems undoubtedly enhance the incentive for detailed exploration of their diversity and evolution.

We undertook a detailed analysis of the gain and loss dynamics of defense systems in 36 clusters of closely related prokaryotic genome cluster (35 bacteria and 1 archaea) from an updated version of the ATGC (Alignable Tight Genome Clusters), a database of orthologous genes from closely related prokaryotic genomes and a research platform for microevolution of prokaryotes [7, 29, 30] and compared the identified trends with those for the overall genome dynamics. The results support the previous observations on the highly dynamic character of the evolution of defense systems but also show that, despite the evolution in the arms race regime, the relative contributions of different types of evolutionary events are roughly the same for defense systems as they are for the rest of microbial genes.

Results and discussion
Distribution of defense systems among the ATGCs
In extension of the previous observations [31], the total number of defense systems (DS) genes strongly correlates with the total number of COGs in an ATGC such that about 75% of the variation is explained by the characteristic genome size (Fig. 1a). In contrast, the fraction of defense-related genes does not significantly depend on the genome size (Fig. 1b) although considerable variation is observed across the ATGCs, from <1% in Chlamydia-Chlamydophila to >4% in Sulfolobus (the only archaeon in the dataset). Not surprisingly, ATGCs with the lowest representation of defense systems include

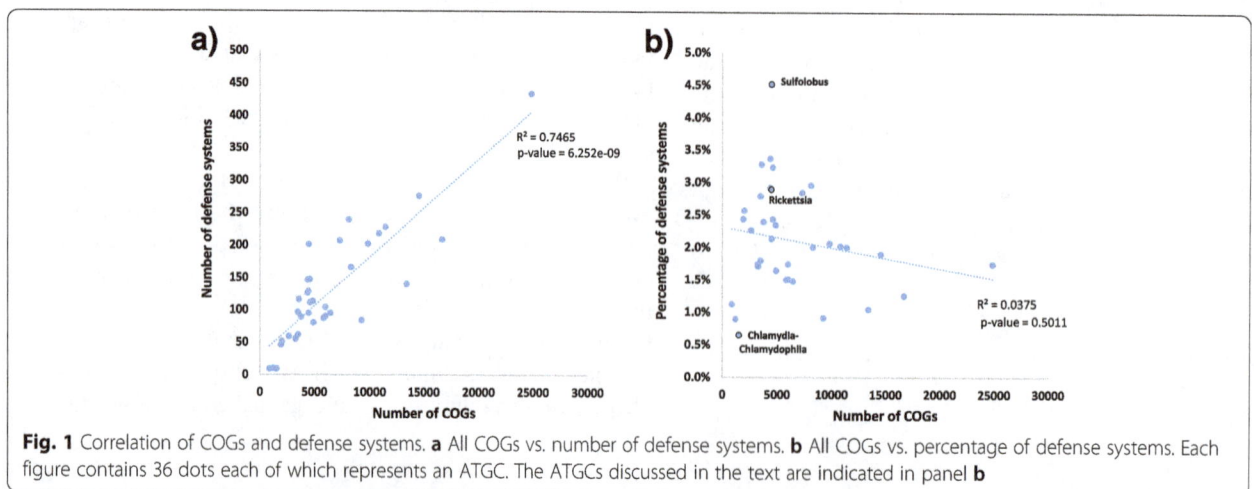

Fig. 1 Correlation of COGs and defense systems. a All COGs vs. number of defense systems. b All COGs vs. percentage of defense systems. Each figure contains 36 dots each of which represents an ATGC. The ATGCs discussed in the text are indicated in panel b

intracellular parasites *Chlamydia-Chlamydophila* (ATGC022), followed by another group of *Chlamydia* (ATGC021), and *Mycoplasma* (ATGC032), facultative, host-associated bacteria with heavily reduced genomes (Additional file 1: Table S1). *Rickettsia*, intracellular parasites with average-size genomes, also encompass an average number of defense systems (Additional file 1: Table S1). Thus, apparently, the number of defense systems is determined primarily by the size of the genome and is largely independent of the life style of the microbes.

The distribution of the different types of defense systems is largely consistent across the ATGCs (Fig. 2 and Additional file 1: Table S1). Statistically, the fractions of each class of defense systems did not differ from the respective means for any of the ATGCs (Additional file 1: Table S2).

Overall, TA comprise the most abundant class of defense systems (on average, ~30% out of the total defense systems in an ATGC), followed by RM (~20%) and CRISPR (~15%) (here we analyzed only the type 2 TA which are more directly implicated in defense, considerably more abundant and more readily recognizable than other types of TA [32]). There is also, on average, about 30% of uncharacterized (predicted) defense systems, i.e. those that contain genes associated with defense but could not be classified into one of the defined categories. As shown in Fig. 2b (see also Additional file 1: Table S1), in 20 of the 36 ATGCs, TA are the most abundant defense systems. The RM systems are extremely abundant in *Helicobacter* (ATGC050), representing 54% of the DS in this ATGC, and also more abundant than other DS in *Listeria* (ATGC108), *Francisella* (ATGC138) and *Propionibacterium* (ATGC159). The CRISPR-Cas systems are the most abundant DS in *Sulfolobus* (ATGC093), followed by *Corynebacterium* (ATGC067) and *Mycoplasma* (ATGC032).

Rates of different types of genome dynamics events in defense systems

We reconstructed the evolution of the DS in large ATGCs that include at least 10 and up to 109 genomes (Additional file 1: Table S1). In each ATGC, the gene families involved in DS were identified, and the rates of Gene Dynamics Events (GDE), including gene family gain, loss, expansion and reduction (henceforth, gain and loss refers to appearance and disappearance, respectively, of a new gene family in the given genome; expansion and reduction refer to gain and loss of a new member of a family of paralogs, respectively), were analyzed using COUNT [7, 33].

COUNT employs a phylogenetic birth-and-death model to infer three parameters: κ (rate of gene family gain), λ (individual gene duplication rate) and μ (individual gene loss rate) These parameters are used to estimate the posterior probabilities of the four types of transitions for each gene family and on all edges of the species tree, namely gain of a gene family (absence - > presence transition), gene family expansion (family of k - > k + 1, k > 0), gene family reduction (family of k - > k-1, k > 1) and loss of a gene family (presence - > absence, i.e. same as reduction but for k = 1). The posterior probabilities can be used to calculate the actual rates (effective number of events, normalized per gene) that can differ from the internal rates in the model. We chose to investigate the evolution of genomes in terms of such effective rates of the four classes of events rather than in terms of the underlying model rates because the former provide a more realistic account of the actual phyletic patterns observed by genome comparison.

The results were compared with the overall rates of the respective events within each ATGC (See Methods and Additional file 1: Figure S1 for the details on the analysis pipeline). The relative rates of gene family gain,

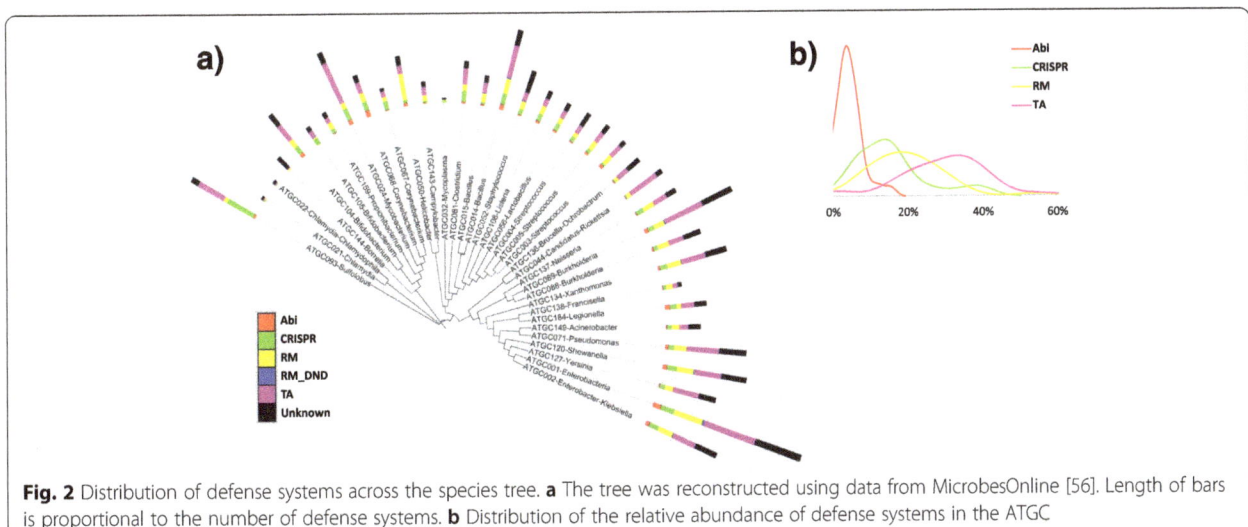

Fig. 2 Distribution of defense systems across the species tree. **a** The tree was reconstructed using data from MicrobesOnline [56]. Length of bars is proportional to the number of defense systems. **b** Distribution of the relative abundance of defense systems in the ATGC

loss, expansion and reduction among the DS are similar to those previously estimated for all genes in the ATGC [7]. The rates of gain, loss, expansion and reduction were divided by the total number of GDE and the resulting normalized relative rates were compared for DS genes vs all genes using the Chi-square test. This test yielded a *p*-value of 0.24 indicating that the DS genes do not significantly deviate from the overall pattern.

As in the case of the overall genome dynamics, gene loss is the dominant mode of evolution of defense systems, the loss rate being approximately 3-fold higher than gene gain rate and an order of magnitude higher than the gene family expansion and reduction rates (Fig. 3). Under the assumption that the genomes are at equilibrium, long term, it appears likely that the substantial excess of losses over gains is offset by sporadic gain of multiple genes [7, 8]. The proportional differences in the rates of genome dynamics events are similar for all classes of defense systems (Abi, CRISPR-Cas, RM and TA) (Fig. 3).

As expected, the number of genome dynamics events depends on the size of the ATGC. Thus, ATGC001 that consists of *Enterobacteria* (>100 genomes) encompasses the majority of the inferred events (Table 1). In order to compare genome dynamics within and across ATGCs, we normalized the estimated number of events by the number of COGs and the number of genomes in an ATGC (Additional file 1: Table S2).

The number of events per COG per genome ranged between approximately 0.1 and 0.4 (Additional file 1: Table S3). The median value is 0.16, and only 4 ATGCs show values greater than 0.3 (ATGC068-*Corynebacterim* [CRISPR and TA], ATGC120-*Shewanella* [Abi], ATGC149-*Acinetobacter* [TA] and ATGC184 - *Legionella* [Abi]). There were no significant differences in genome dynamics rates between different microbial life styles (free-living vs facultative host-associated vs parasites) or across the major bacterial taxa (Additional file 1: Tables S4 and S5).

High flux vs low flux in defense systems

Rates of gain, loss, expansion and reduction per COG in defense systems strongly correlate with the rates among all genes (Fig. 4). To identify genomes with high and low relative rates of genome dynamics in defense systems, the results were normalized by the overall rates of genome dynamics events in all genes. There are 9 ATGCs with high flux (top quartile), 9 ATGCs with low flux (bottom quartile) and 18 ATGCs with average flux of defense systems (Table 1 and Additional file 1: Table S6). *Helicobacter* and *Listeria* have the highest gene flux, whereas *Chlamydia* and *Brucella* show the lowest flux (Additional file 1: Table S6). The majority of the ATGCs with a high DS gene flux are compressing, i.e. appear to be shedding defense systems (*Enterobacteria*, in ATGC002 as well as Streptococcus, *Listeria*, *Xanthomonas* and *Campylobacter*) although two ATGCs (*Helicobacter* and *Legionella*) show high rates of both gene gain and loss. Moreover, *Helicobacter* presents the highest rates of gain, loss, expansion and reduction (Additional file 1: Table S3). In contrast, ATGCs with low flux rate tend to maintain a balanced gain/loss rate in 6 ATGCs or to slowly expand their defense systems (ATGC003-*Streptococcus*, ATGC088-*Burkolderia* and *Propionibacterium*). The same trend is observed in ATGCs with average flux: 14 ATGCs have balanced gain-loss rates, two tend to gain DS genes (*Shewanella* and *Acinetobacter*) and two tend to lose DS genes (*Clostridium* and *Francisella*).

On average, the rates of DS gene dynamics per COG are about 1.4 times higher than the rates for all genes (the ratios are, approximately: 1.5 for gene loss, 1.3 for gene gain, 1.5 for gene family reduction, and 1.04 for gene family expansion) (Fig. 5a and Table 1). The

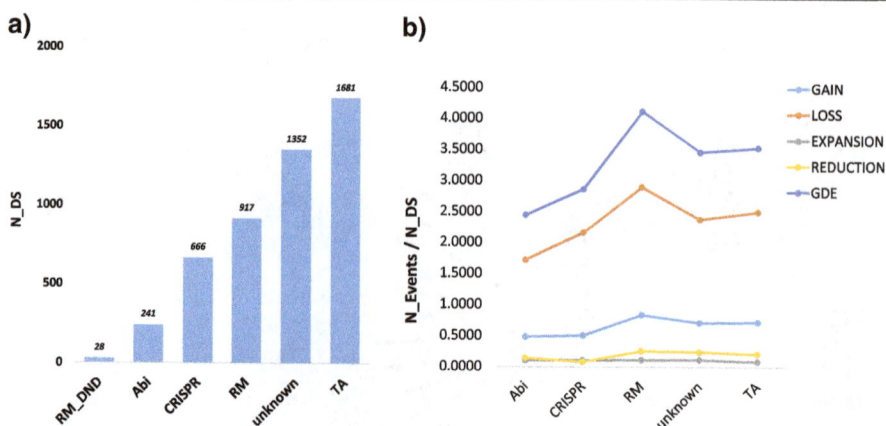

Fig. 3 Number of genes assigned to defense systems (N_DS) and genome dynamics by type of defense system. **a** Number of defense genes (COGs) distributed by type of defense systems. **b** Number of genome dynamic events (GDE), including gain, loss, expansion and reduction, by type of defense systems

Table 1 Genome dynamics in defense systems relative to the dynamics in all genes

ATGC	Genera	(DS/N_DS) / (ALL/N_ALL)				
		Gain	Loss	Expansion	Reduction	All GDE
ATGC149	Acinetobacter	1.73	1.31	1.09	1.89	1.39
ATGC014	Bacillus	1.06	1.18	0.71	1.16	1.13
ATGC015	Bacillus	1.17	1.68	0.62	0.73	1.47
ATGC104	Bifidobacterium	1.22	1.31	0.59	0.43	1.22
ATGC105	Bifidobacterium	1.90	1.73	0.55	1.11	1.76
ATGC144	Borrelia	0.78	1.63	1.12	1.71	1.51
ATGC136	Brucella-Ochrobactrum	1.20	0.45	0.02	0.19	0.70
ATGC088	Burkholderia	1.76	0.76	1.18	0.78	0.91
ATGC089	Burkholderia	1.34	1.24	0.24	0.51	1.20
ATGC143	Campylobacter	1.16	1.92	1.37	1.99	1.77
ATGC044	Candidatus-Rickettsia	1.01	1.44	1.22	1.85	1.34
ATGC021	Chlamydia	0.20	0.68	0.01	0.82	0.65
ATGC022	Chlamydia-Chlamydophila	0.01	0.79	1.01	0.26	0.54
ATGC081	Clostridium	1.70	1.36	1.04	2.59	1.40
ATGC067	Corynebacterium	1.46	1.17	0.70	1.08	1.19
ATGC068	Corynebacterium	1.18	1.89	1.44	1.33	1.79
ATGC002	Enterobacter-Klebsiella	1.40	1.85	1.10	3.14	1.63
ATGC001	Enterobacteria	1.18	1.98	1.61	3.51	1.88
ATGC138	Francisella	1.46	1.63	0.59	2.86	1.59
ATGC050	Helicobacter	1.22	2.90	3.15	4.97	2.63
ATGC056	Lactobacillus	1.58	1.34	0.75	1.17	1.36
ATGC184	Legionella	2.13	1.75	1.19	1.92	1.78
ATGC108	Listeria	1.23	2.69	0.42	2.29	2.29
ATGC024	Mycobacterium	0.84	1.38	0.88	0.54	1.16
ATGC032	Mycoplasma	0.34	1.51	1.29	0.19	1.32
ATGC137	Neisseria	1.35	1.18	1.43	0.76	1.22
ATGC159	Propionibacterium	1.86	0.72	1.05	0.17	0.93
ATGC071	Pseudomonas	1.54	1.45	0.69	0.96	1.42
ATGC120	Shewanella	2.87	0.99	1.31	2.52	1.34
ATGC052	Staphylococcus	1.06	1.22	0.73	1.65	1.21
ATGC003	Streptococcus	1.11	1.02	2.03	0.68	1.05
ATGC004	Streptococcus	1.29	1.45	1.48	1.57	1.42
ATGC005	Streptococcus	1.36	1.64	0.69	1.54	1.49
ATGC093	Sulfolobus	0.96	1.66	1.55	1.62	1.55
ATGC134	Xanthomonas	1.11	2.19	1.43	3.01	1.71
ATGC127	Yersinia	1.29	1.38	1.20	1.21	1.34

difference between the rates for DS and the means over all functional categories of genes are statistically significant for the total of all GDE, loss and gain (p-values of 0.0056, 0.0071 and 0.0284, respectively) as calculated from the differences in the log-likelihoods (see Methods for details). These ratios vary within a broad range, from nearly absent gene gain in *Chlamydia-Chlamydophila* to a near threefold excess of the loss rate over the genomic average in *Helicobacter*. Nevertheless, the trend of higher than average genome dynamics rates for DS is consistent: it holds for 30 of the 36 ATGCs for gene gain and 31 of the 36 ATGCs for loss (Table 1). The observed differences cannot be explained by taxonomy (Fig. 5b) or life style (Fig. 5c).

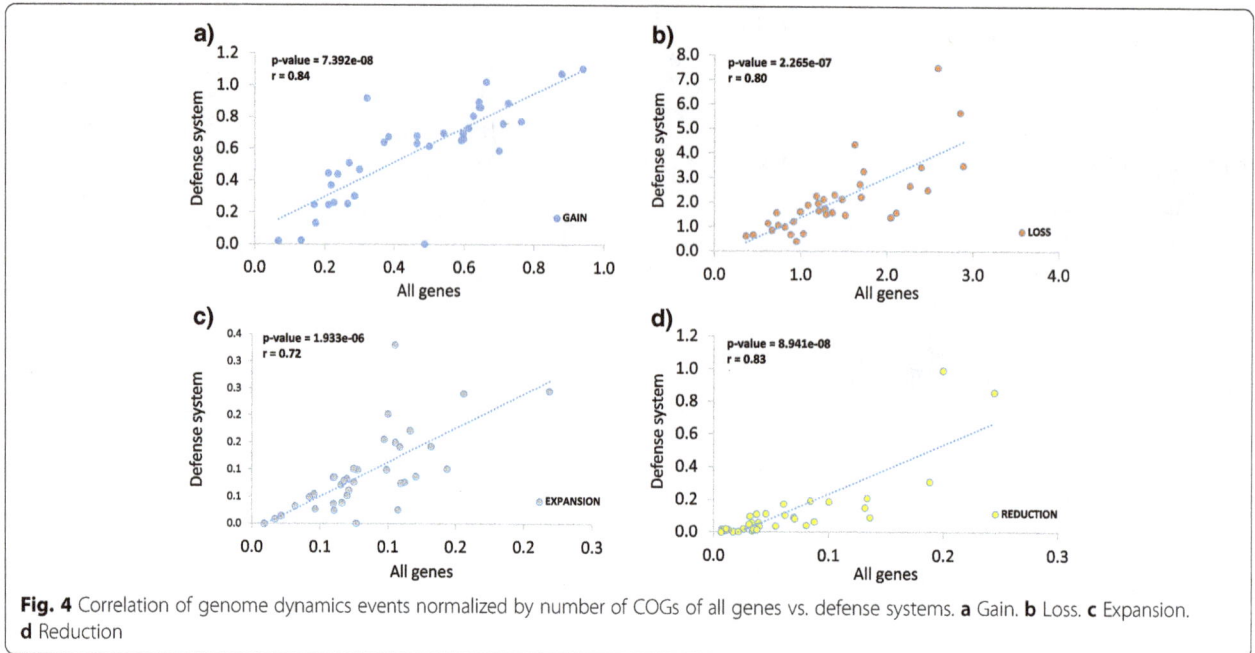

Fig. 4 Correlation of genome dynamics events normalized by number of COGs of all genes vs. defense systems. **a** Gain. **b** Loss. **c** Expansion. **d** Reduction

In the principal component analysis (PCA), the first principal component (PC1) explains 54% of the variance and separates ATGCs with high and low gene flux in DS genes. The second principal component (PC2) explains 25% of the variance and separates the gainers of DS genes from the rest of the ATGCs (Fig. 6a, b). Predictably, the evolution of most of the ATGCs is dominated by PC1 but several ATGCs that are active gainers are primarily characterized by PC2 (Fig. 6c). *Helicobacter* shows the highest flux among the ATGCs whereas *Shewanella* presents the highest gene gain rate.

Case by case analysis of the evolution of defense systems
We examined in detail several specific cases of gain or loss of defense-related genes in different ATGCs, in an attempt to illustrate various evolutionary trends. The ATGCs for this case by case, namely ATGC068: *Corynebacterium* (Additional file 1: Figure S1), ATGC081: *Clostridium* (Additional file 1: Figure S2) and ATGC159:*Propionibacterium* (Additional file 1: Figure S3), were selected based by the following criteria: 1) approximately 10 genomes in an ATGC; 2) high genome synteny allowing for whole genome alignments and 3)

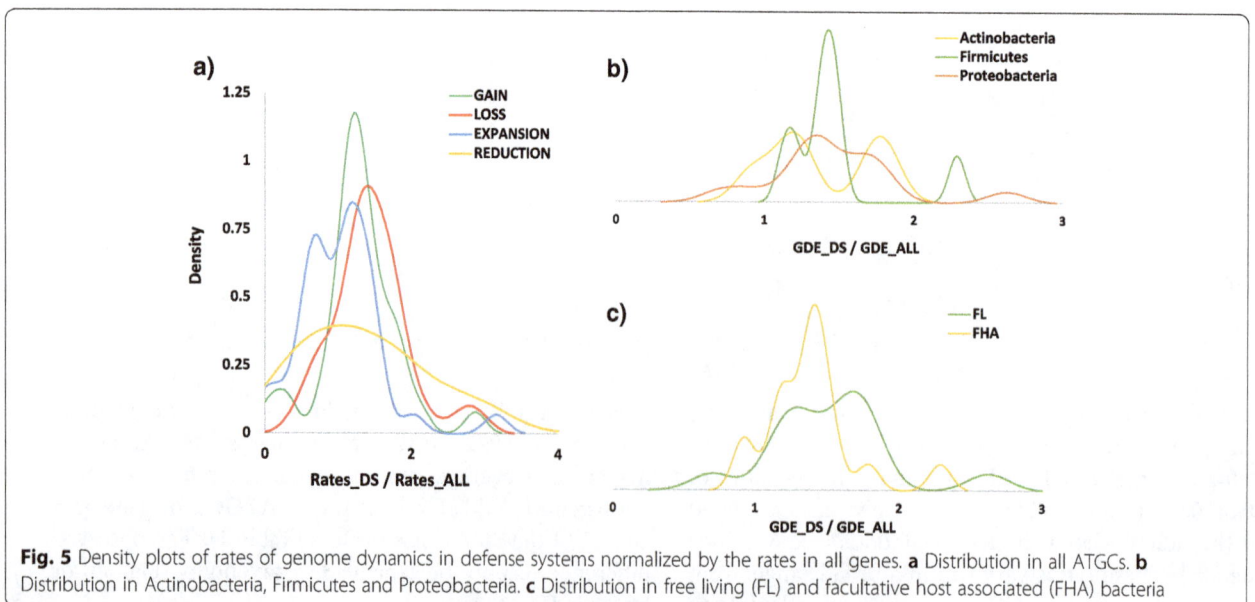

Fig. 5 Density plots of rates of genome dynamics in defense systems normalized by the rates in all genes. **a** Distribution in all ATGCs. **b** Distribution in Actinobacteria, Firmicutes and Proteobacteria. **c** Distribution in free living (FL) and facultative host associated (FHA) bacteria

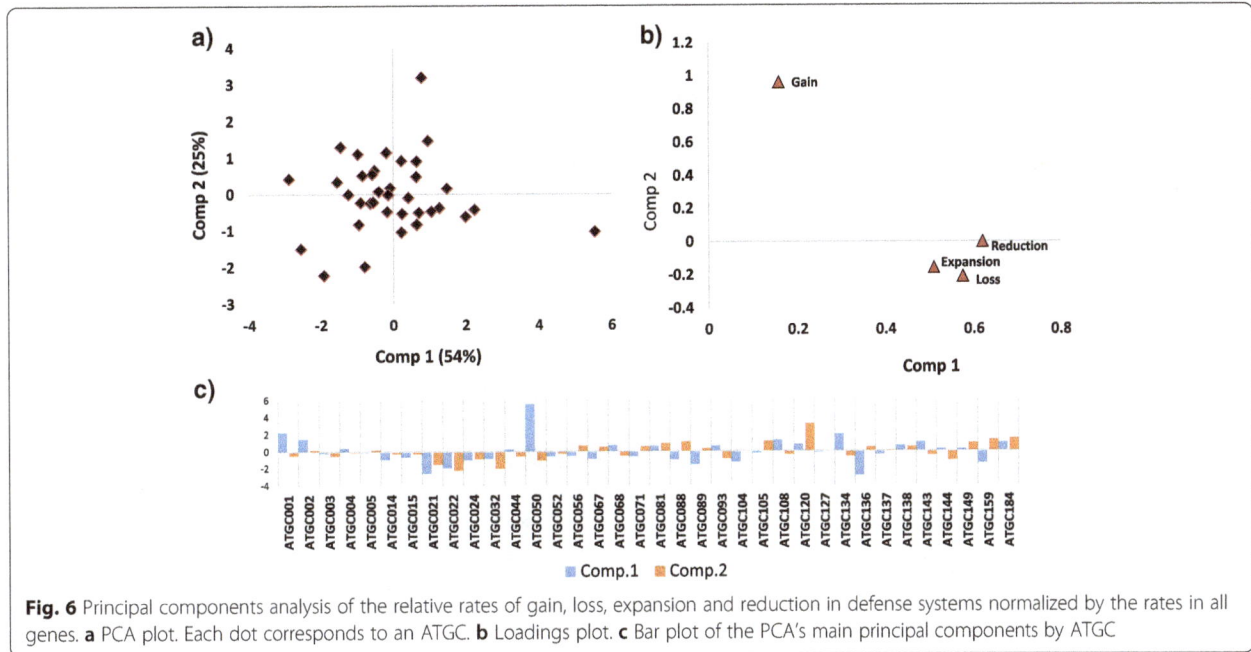

Fig. 6 Principal components analysis of the relative rates of gain, loss, expansion and reduction in defense systems normalized by the rates in all genes. **a** PCA plot. Each dot corresponds to an ATGC. **b** Loadings plot. **c** Bar plot of the PCA's main principal components by ATGC

moderate number of GDE allowing for comprehensive analysis.

ATGC068: Corynebacterium

The ATGC068 consists of 13 closely related *Corynebacterium diphtheriae* strains. The COUNT reconstruction indicates a strong DS gene loss trend in all branches of the tree (Additional file 1: Figure S4). Below we describe several examples of loss and gain of defense genes in this ATGC. The COUNT reconstruction indicates several losses of *cas* genes during the evolution of *Corynebacterium diphtheriae* (Additional file 1: Table S6). All of these genes apparently have been lost from the genomic region between the genes for hydroxymethylpyrimidine/phosphomethylpyrimidine kinase *thiD* and serine/threonine kinase *pknB* genes (Additional file 1: Figure S5). A CRISPR-Cas system is present in this locus in each genome but interestingly, these belong to two different subtypes [34], namely II-C (in 8 genomes) and I-E (in the remaining 5 genomes). The organisms that possess the I-E system are not monophyletic in the respective species tree (Additional file 1: Figure S5). In the phylogenetic tree of Cas1 [34], the II-C and I-E Cas1 proteins from this locus form two clades within the II-C and I-E branches, respectively. Thus, the most parsimonious scenario is that type II-C is ancestral in this locus but had been replaced with I-E systems from sources within the same bacterial lineage on several independent occasions. The reversal of such a replacement, i.e. restoration of type II-C system in this locus might have occurred as well as suggested by the mapping of the cas9 gene tree onto the species tree (Additional file 1: Figure S5) which

shows that Cas9 from *C. diphtheriae* VA01 is more closely related to that of *C. diphtheriae* 31A than to the Cas9 proteins from related strains on the species tree (Additional file 1: Figure S5). Transposases that are present in these loci might have facilitated recombination. This example shows that *in situ* replacements of CRISPR-Cas systems by different types might be common among related microbial strains.

Two RM genes, a EcoRII-like restriction endonuclease and a diverged predicted helicase containing a Z1 domain previously found to be associated with RM systems [35, 36], were inferred to have been gained on the terminal branch corresponding to *C. diphtheriae* BH8 (Fig. 7a). These two genes belong to an island of 6 singletons, i.e. genes restricted to a single genome within the analyzed data set, (CDBH8_0988-CDBH8_0994) inserted between galactokinase *galK* and RNA helicase *srmB* genes. In addition to the two RM genes, the island encodes a Dcm-like cytosine-C5 specific DNA methylase fused to a Xre family HTH (helix-turn-helix) domain, a DUF4420 (pfam14390) and three unclassified proteins (Fig. 7a). The best BLASTP matches for these proteins in the NR database are to homologs from other *Corynebacterium* species (although none from this ATGC) but the similarity is comparatively low, around 40% identity (data not shown), suggesting acquisition of this region via HGT from a relatively distant bacterium. This island replaces two genes that are present in all other genomes from this ATGC, namely a Superfamily II DNA helicase fused to a phospholipase D family nuclease domain (pfam11907) and a MutT-like

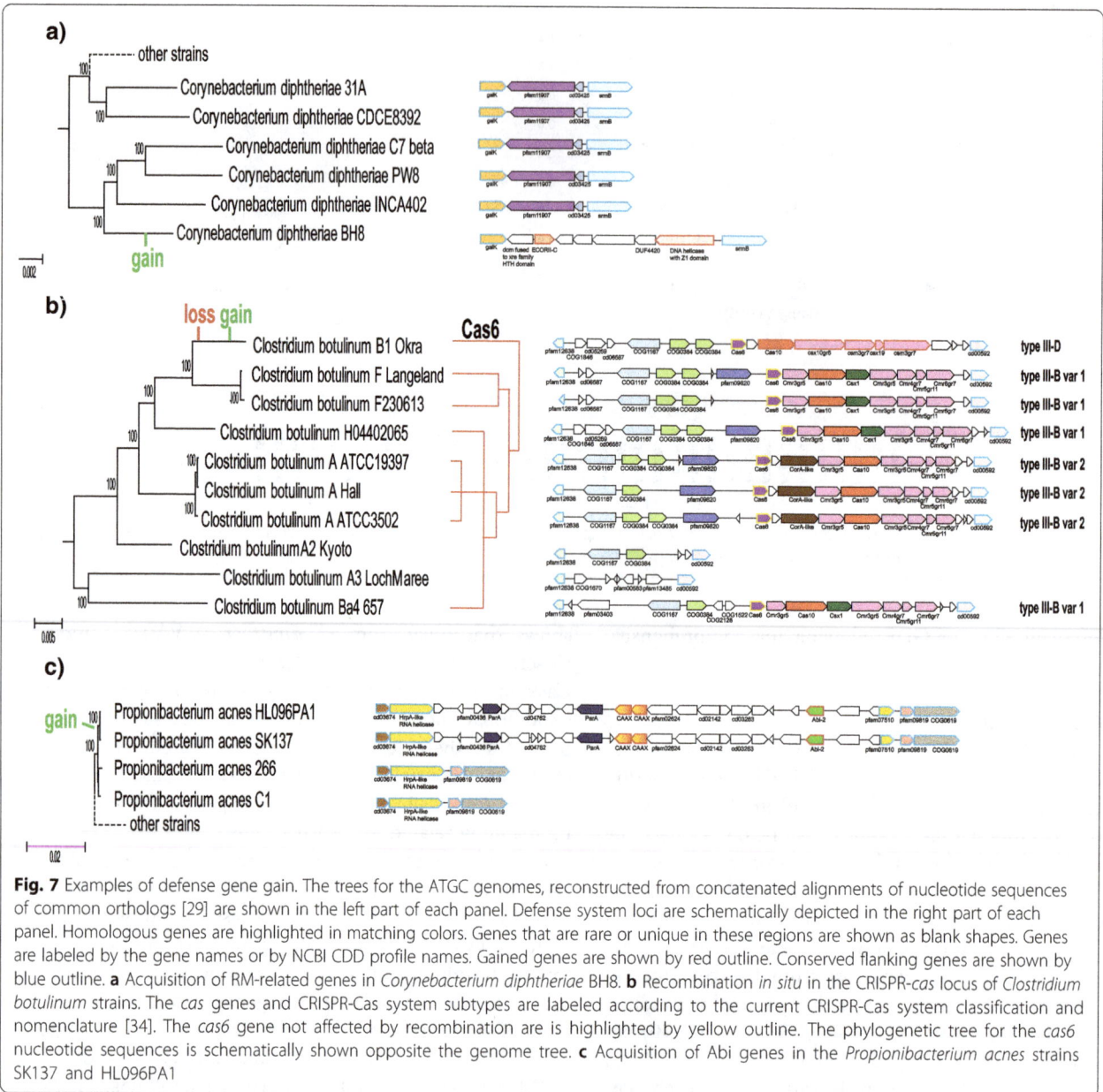

Fig. 7 Examples of defense gene gain. The trees for the ATGC genomes, reconstructed from concatenated alignments of nucleotide sequences of common orthologs [29] are shown in the left part of each panel. Defense system loci are schematically depicted in the right part of each panel. Homologous genes are highlighted in matching colors. Genes that are rare or unique in these regions are shown as blank shapes. Genes are labeled by the gene names or by NCBI CDD profile names. Gained genes are shown by red outline. Conserved flanking genes are shown by blue outline. **a** Acquisition of RM-related genes in *Corynebacterium diphtheriae* BH8. **b** Recombination *in situ* in the CRISPR-*cas* locus of *Clostridium botulinum* strains. The *cas* genes and CRISPR-Cas system subtypes are labeled according to the current CRISPR-Cas system classification and nomenclature [34]. The *cas6* gene not affected by recombination are is highlighted by yellow outline. The phylogenetic tree for the *cas6* nucleotide sequences is schematically shown opposite the genome tree. **c** Acquisition of Abi genes in the *Propionibacterium acnes* strains SK137 and HL096PA1

pyrophosphohydrolase (cd03425). The latter pair of genes is missing in the genome of *C. diphtheriae* BH8 suggesting that it was lost in the process of the island integration. Despite the presence of two genes that have been previously identified in the context of known RM systems, no full gene complement for any characterized type of RM systems could be identified in this region. Thus, either these proteins are functionally unrelated and are encoded together in a defense island by chance or some subset of the genes in the island comprises a novel, multicomponent defense system.

The COUNT reconstruction predicts a gain of an Abi2 (HEPN domain-containing protein, predicted ribonuclease) gene in *C. diphtheriae* VA01. This gene is absent from all other genomes of this ATGC and shows the highest similarity to homologs from *Corynebacterium pseudotuberculosis* suggesting acquisition of this gene from a more distant species of the same genus. The region upstream of the superoxide dismutase gene, where the Abi2 gene apparently was inserted, shows variable gene content. In *C. diphtheriae* VA01, it contains several uncharacterized genes some of which encode predicted DS proteins such as three subunits of a Type I RM system, a PD-(D/E)xK nuclease and Fic/DOC, and HTH domain-containing protein shared with several other species from this ATGC (Additional file 1: Figure S6). Three genes in this region (CDVA01_2131-

CDVA01_2133) including one for a PD-(D/E)xK nuclease (CDVA01_2132, often annotated as "P51 protein") are found next to each other in Gram-positive bacterial prophages of the phi-APSE family although no other phage-associated genes were identified in this region. Nevertheless, the presence in this neighborhood of a gene encoding a ParA family ATPase known to be involved in plasmid partitioning [37] implies that this region is associated with a mobile element.

In a case of apparent loss of a TA gene pair (a toxin containing a truncated PIN domain, apparently an inactived ribonuclease, and an antitoxin containing a Xre family HTH domain) in *Corynebacterium diphtheria* VA01, the two genes apparently were excised together with a mobile element (as inferred by the presence of an integrase gene in this region), which had integrated next to the TA module in the ancestor of the three related *C. diphtheriae* strains (Additional file 1: Figure S7). Thus, this case is an example of mobilization of chromosomal genes. In the corresponding region in several other genomes that lack the mobile element, the toxin had been independently truncated, whereas the antitoxin remained intact, suggesting that toxin inactivation could be a general trend in evolution of toxin-antitoxin modules (Additional file 1: Figure S7).

ATGC081: *Clostridium botulinum*
The ATGC081 contains 10 closely related *Clostridium botulinum* strains. This ATGC also shows an overall tendency to lose DS genes according to the COUNT prediction, from 49 at the root to 30–35 at the tips (Additional file 1: Figure S8).

Four DS genes predicted to have been gained in *Clostridium botulinum* A2 str Kyoto belong to a genomic island (CLM_2285- CLM_2304) between Radical SAM superfamily protein and UDP-N-acetylglucosamine 4-epimerase genes (Additional file 1: Figure S9). These four DS genes encode an AIPR family Abi gene and three subunits of a type I restriction modification system (Additional file 1: Figure S9). In addition, the island encodes two proteins that are common in phages, namely a holin (CLM_2290) and an N-acetylmuramoyl-L-alanine amidase (CLM_2289), suggesting that this region is of a phage origin. This region shows considerable variability in other genomes from this ATGC (Additional file 1: Figure S9).

A replacement of a type III-B CRISPR-*cas* locus with type III-D is observed in *Clostridium botulinum* B1 Okra (Fig. 7b). The replacement apparently occurred *in situ*, without affecting the *cas6* gene, which is present in this region in all Clostridium botulinum species from this ATGC. These *cas6* sequences are highly similar on the nucleotide level, whereas even on the amino acid sequence level, there is only weak similarity between the Cas6 proteins of subtypes III-D and III-B. The Cas6 protein is an endoribonuclease that is required for pre-crRNA transcript processing and is often encoded in the type III loci, even those that lack a CRISPR array [34]. This particular III-D locus is not found in any of the *C. botulinum* species and shows the closest similarity to the III-D locus of *Clostridium lundense*, suggesting horizontal transfer from a relatively distant species. Furthermore, the phylogeny of the Cas10 protein [34] suggests that at least one additional *in situ* gene replacement occurred in this clade because Cas10 and, as we show here, the respective gene set of type III-B systems are distinct in the strains of *C. botulinum* strains ATCC 19397, ATCC 3502 and Hall, and in the branch that encompasses strains H04402 065, F str. 230,613 and F str. Langeland. In contrast, *cas6* genes can be confidently aligned at the nucleotide level and thus apparently were unaffected by these events (Fig. 7b).

ATGC159: *Propionibacterium*
The ATGC159 consists of 10 closely related *Propionibacterium acnes* strains and one strain of *Propionibacterium avidum*. Based on the COUNT estimates, the *P. acnes* strains have lost DS genes (from 20 at the root to 11–18 at the tips) whereas *P. avidum* gained 9 additional DS genes (Additional file 1: Figs. S3 and S10).

The gain of several Abi genes (one Abi2 family genes and two CAAX family protease genes) most likely occurred due to the insertion of a large DNA segment containing 23 genes (PAGK_0160- PAGK_0182 locus in *P. acnes* HL096PA1 (for details, see Additional file 1: Figure S11) into the genome of the common ancestor of *P. acnes* SK137 and *P. acnes* HL096PA1 (Fig. 7c). This DNA fragment was inserted into the HrpA-like RNA helicase gene which was disrupted in the process. There are two ParA ATPase genes encoded in this locus (Fig. 7c). The DNA-binding ATPase ParA, typically together with another DNA-binding protein, ParB, is essential for plasmid partitioning [38]. Although the *parB* gene is missing in *P. acnes,* the presence of the *parA* genes suggests that this genomic segment derives from a plasmid.

Another example of DS gene gain also involves acquisition of a large genomic island (PAC1_03830- PAC1_03980) in *P. acnes* C1 (Additional file 1: Figure S12). The island includes a toxin gene of the Fic/Doc family (PAC1_03955) that is located next to a gene encoding an antitoxin of the VbhA family [39]. Recently, an antitoxin of this family has been shown to repress growth arrest mediated by the FIC domain-containing toxin VbhT in the mammalian pathogen *Bartonella schoenbuchensis* [39]. In *P. acnes,* the island is inserted into an alpha-beta family hydrolase gene and is flanked by transposases. It includes a number of genes unrelated to defense but implicated in stress response, particularly metal resistance (Additional file 1: Figure S12). We were unable to identify an identical or substantially similar island in any other complete genome present in our data set but the presence of a ParA family ATPase and the transposases again suggests that the island could be a mobile element.

Distribution of defense systems and genome dynamic events on bacterial chromosomes

Given the previous observations on defense islands in microbial genomes [36], we examined possible clustering of defense genes in the ATGC genomes. Defense genes show a significant trend of co-localization in 57% of the genomes (Additional file 1: Figure S13a), for all classes of defense systems (Additional file 1: Figure S13b). However, when the analysis was repeated with "directons", i.e. groups of co-directed genes that comprise putative operons [40], random distribution of DS was observed in 83% of the genomes (Additional file 1: Figure S13a (Additional file 1: Figure S13c). Additionally, we analyzed in detail the distribution of inferred DS gene gains and losses in the chromosomes of ATGC068 (*Corynebacterium*), ATGC081 (*Clostridium*) and ATGC159 (*Propionibacterium*). In 10 of the 18 analyzed genomes (14 genomes for directons), gains and losses were found to be randomly distributed across the chromosome (Additional file 1: Figure S14).

Conclusions

The results of the present reconstruction of the evolution of defense systems in prokaryotes perhaps can be considered "disappointing" in that no distinct evolutionary regime was discovered for the defense genes. The evolution of the DS is consistently and significantly more dynamic than the genomic mean for the respective groups of microbes. However, the difference in the GDE rates is moderate (less than 1.5 fold), perhaps surprisingly, given the common view of the evolution of defense systems in the regime of incessant arms race with parasites [41–43]. The evolution of defense systems is shaped by several factors that seem to exert different, in some cases opposite effects on the gain and loss rates of defense genes. The arms race is only one of such factors. The others include the fitness cost of defense systems stemming from energetic burden, autoimmunity and barriers to horizontal gene transfer and potentially enhancing loss over gain [44, 45]; the selfish behavior of defense system, particularly RM and TA, which often become addictive to the host cells, such that loss is inhibited, and are frequently transferred on plasmids, enhancing gain [46, 47]; and additional, non-defense functions of some of these systems, e.g. CRISPR-Cas [48], which favor retention of the respective genes over loss. It appears that the net outcome of these distinct

effects constrains the mobility of defense systems, resulting in the moderate excess of the GDE rates over the genomic averages.

The relative contributions of gene family loss, gain, expansion and contraction are, on average, the same as they are for the rest of microbial genes. Furthermore, with the exception of some parasitic bacteria that encode few DS and show low flux of defense genes, there is no obvious connection between the dynamics of the DS and the life style of the respective organisms. Thus, apparently, the evolutionary dynamics of the DS follows the general trends of microbial genome evolution. These trends seem to stem from the inherent deletion bias of genome evolution combined with the neutral, largely clock-like mode of gene loss which contrasts the less uniform and partially adaptive mode of gene gain [7, 8, 49, 50]. Although the quantitative trends in the evolution of the DS seem to recapitulate the overall tendencies of microbial genome evolution, case by case analysis points to notable phenomena, such as frequent involvement of mobile elements and *in situ* replacement of the DS, particularly for different types of CRISPR-Cas systems. These features appear to reflect the frequent localization of the DS in defense islands that are enriched also in mobile elements and are likely to be responsible for the enhanced dynamics of DS evolution.

Methods
ATGC dataset

Genomic data were obtained from an updated version of the ATGC database [30] that contains alignable (>85% conserved synteny) and tight (synonymous substitution rate < 1.5) genomic clusters. We selected only ATGC clusters that contain at least 10 genomes. Complete genome alignments of three ATGCs (ATGC068, ATGC081 and ATGC159) were performed with the program MAUVE [51].

Defense systems

A list of COGs, Pfam and CDD domains involved in defense systems was constructed from the results of previous studies [11, 32], including restriction-modification (RM), toxin-antitoxin (TA), and abortive infection (AI) systems, and the CRISPR-Cas systems of adaptive immunity. All defense genes were mapped onto the ATGCs using a custom Perl script and the overlapping data set was used to quantify rates of gene dynamics. In order to harness sufficient analytical power, only defense systems with more than 10 genes and more than 10 events (including gain, loss, expansion or reduction) were analyzed.

Species trees

The program COUNT [33], which was employed for evolutionary reconstruction (see description bellow), requires rooted phylogenetic trees as an input. Thus, a species tree of each ATGC was reconstructed from a concatenated alignment of all universal genes with conserved synteny among species. Protein sequences were aligned with MUSCLE [52] and back translated to respective nucleotide sequences using an in-house script. All alignments of genes were concatenated in a single alignment and used to reconstruct a species tree for each ATGC using the program FastTree [53] under the General Time Reversible (GTR), Among Site Rate Variation (gamma) nucleotide substitution model [54]. Accordingly, all species trees were rooted using the least-squares modification of the mid-point method [55].

Phylogenetic birth-and-death analysis

We estimated the rates of gain, loss, expansion and reduction based on a phylogenetic birth-and-death model implemented in the program COUNT [33]. The program estimates the rates of gene gain (κ), individual gene duplication (λ) and individual gene loss (μ). Thus, a gene family of size n decreases at a rate $n\mu$ and increases at a rate ($\kappa + n\lambda$). The parameters (κ,λ,μ) are different for each gene family and across edges of the species tree. The parameters were optimized iteratively, as recommended [33], as previously described [7]. Thus, the parameters were optimized in each ATGC individually (using all gene families) through 10 rounds of increasing complexity, from uniform rates of gain, loss and duplication to up to 4 discrete categories for the gamma distribution. The parameter values obtained in the final round were used to estimate the numbers of gains, losses, expansions and reductions at different branches of the species tree for gene families involved in defense systems. The final number of events for each ATGC was estimated as the sum over all branches and across all families. The p-values for the differences in the rates of gene dynamics events between DS and the rest of the genes were calculated from the differences in log-likelihoods using the Welch Two Sample t-test implemented in R.

Principal component analysis

The input variables for the PCA were the relative rates of gain, loss, expansion and reduction in defense systems in each ATGC (Table 1). These values were transformed into the logarithmic scale prior to the analysis. The PCA was performed using the function *princomp* from the R statistical package.

Distribution of genes and dynamic events in the chromosome

We analyzed all genomes from each ATGC individually to assess whether DS genes are randomly distributed in the chromosome (Additional file 1: Figure S15). First, we calculated the median distance of each DS gene to the closest DS gene. Then, we performed a randomization test, randomly sampling the same number of genes on each chromosome as there are defense genes and calculating the median distance between sampled genes. Repeating this procedure 10,000 times allowed us to test the null hypothesis that DS genes are not closer together than chance expectations. Alternatively, directons (i.e. strings of consecutive co-directed genes [40]) were marked on each chromosome and labeled as defense directons if at least one of the genes in a directon belongs to DS. Then the median distance between the defense directons was compared to that between randomly sampled directons as described above.

Similarly, the distribution of gains and losses of DS genes on the terminal branches in ATGC068, ATGC081 and ATGC159 was analyzed by mapping these events on the corresponding chromosomes and calculated the median distance between these locations. For comparison, the same number of genes were randomly sampled from all defense genes on these chromosomes and the median distance calculated; the procedure was repeated 10,000 times. The test was performed with both individual genes and at the directon level.

Software availability

All custom scripts used for this analysis are available at ftp://ftp.ncbi.nlm.nih.gov/pub/wolf/_suppl/ATGCdefense.

Additional file

Additional file 1: Figure S1. Gains and losses in ATGC068-*Corynebacterium*. Figure S2. Gains and losses in ATGC081-*Clostridium*. Figure S3. Gains and losses in ATGC159-*Propionibacterium*. Figure S4. Count output for ATGC068-*Corynebacterium*. Figure S5. Example of multiple independent substitutions of CRISPR-Cas system type II-C to type I-E. Figure S6. Example of predicted gain by COUNT of Abi2 gene in a large loci with multiple gains and losses. Figure S7. Example of TA gene loss. Figure S8. Count output for ATGC081-*Clostridium*. Figure S9. Example of gain of four DS genes in *Clostridium botulinum* A2 str Kyoto. Figure S10. Count output for ATGC159-*Propionibacterium*. Figure S11. Locus details for Fig. 7c. Figure S12. Example of TA gene gain (within a large locus) in *Propionibacterium acnes* C1. Figure S13. Density distribution of p-values from the randomization test. Figure S14. Distribution of defense systems and dynamic events in 18 genomes. Figure S15. Scheme of the methodology used to test randomness in the distribution defense genes and dynamic events in the chromosome. Table S1. Distribution of defense systems COGs in ATGCs. Table S2. Number of the defense systems normalized by the total number of genes (COGs). Table S3. Genome dynamics in defense systems, including gain, loss, expansion and reduction: (a) total number of events; (b) events relative to the number of COGs and (c) events relative to the number of COGs and genomes. Defense systems with less than 10 genes or less than 10 events are left empty. Table S4. Comparison of the genome dynamics in defense systems (relative to the dynamics in all genes) between life styles using the Welch Two Sample t-test implemented in R. Table S5. Comparison of the genome dynamics in defense systems (relative to the dynamics in all genes) between taxa using the Welch Two Sample t-test implemented in R. Table S6. Relative fluxes in defense systems. Table S7. Description of genes in Additional file 1: Figures. S1, S2 and S3. Table S8. Locus description of Additional file 1: Figure S9. Table S9. Locus description of Additional file 1: Figure S12. Table S10. P-values of Additional file 1: Figure S14. (DOCX 2297 kb)

Abbreviations

AI: Abortive infection; ATGC: Alignable Tight Genome Clusters; CRISPR-Cas: Clustered Regularly Interspaced Repeats and CRISPR-associated (proteins); DS: Defense systems; HGT: Horizontal gene transfer; RM: Restriction-modification; TA: Toxin-antitoxin

Acknowledgements

The authors thank Koonin group members for helpful discussions.

Funding

This work was supported by intramural funds of the US Department of Health and Human Services (to the National Library of Medicine).

Authors' contributions

Conception and design of the study: EVK, YIW. Data collection: DMK. Data analysis: PP, KSM. Manuscript drafting: PP, KSM. Manuscript revision for critical intellectual content: EVK. All authors read and approved the final manuscript.

Competing interests

The authors declare that they have no competing interests.

Author details

[1]National Center for Biotechnology Information, National Library of Medicine, National Institutes of Health, Bethesda, MD 20894, USA. [2]Present address: Division of Genetics and Physiology, Department of Biology, University of Turku, Turku, Finland. [3]Present address: Department of Biomedical Engineering, University of Iowa, Iowa City, IA, USA.

References

1. Koonin EV, Wolf YI. Genomics of bacteria and archaea: the emerging dynamic view of the prokaryotic world. Nucleic Acids Res. 2008;36(21): 6688–719.
2. Doolittle WF. Phylogenetic classification and the universal tree. Science. 1999;284(5423):2124–9.
3. Doolittle WF. Lateral genomics. Trends Cell Biol. 1999;9(12):M5–8.
4. Treangen TJ, Rocha EP. Horizontal transfer, not duplication, drives the expansion of protein families in prokaryotes. PLoS Genet. 2011;7(1): e1001284.
5. Snel B, Bork P, Huynen MA. Genomes in flux: the evolution of archaeal and proteobacterial gene content. Genome Res. 2002;12(1):17–25.
6. Mirkin BG, Fenner TI, Galperin MY, Koonin EV. Algorithms for computing parsimonious evolutionary scenarios for genome evolution, the last universal common ancestor and dominance of horizontal gene transfer in the evolution of prokaryotes. BMC Evol Biol. 2003;3(1):2.
7. Puigbo P, Lobkovsky AE, Kristensen DM, Wolf YI, Koonin EV. Genomes in turmoil: quantification of genome dynamics in prokaryote supergenomes. BMC Biol. 2014;12:66.
8. Wolf YI, Koonin EV. Genome reduction as the dominant mode of evolution. BioEssays. 2013;35(9):829–37.
9. Comeau AM, Krisch HM. War is peace–dispatches from the bacterial and phage killing fields. Curr Opin Microbiol. 2005;8(4):488–94.
10. Stern A, Sorek R. The phage-host arms race: shaping the evolution of microbes. BioEssays. 2011;33(1):43–51.
11. Makarova KS, Wolf YI, Koonin EV. Comparative genomics of defense systems in archaea and bacteria. Nucleic Acids Res. 2013;41(8):4360–77.
12. Koskella B, Brockhurst MA. Bacteria-phage coevolution as a driver of ecological and evolutionary processes in microbial communities. FEMS Microbiol Rev. 2014;38(5):916–31.
13. Lopez-Pascua L, Buckling A. Increasing productivity accelerates host-parasite coevolution. J Evol Biol. 2008;21(3):853–60.
14. Buckling A, Brockhurst M. Bacteria-virus coevolution. Adv Exp Med Biol. 2012;751:347–70.
15. Takeuchi N, Wolf YI, Makarova KS, Koonin EV. Nature and intensity of selection pressure on CRISPR-associated genes. J Bacteriol. 2012;194(5): 1216–25.
16. Roberts RJ, Belfort M, Bestor T, Bhagwat AS, Bickle TA, Bitinaite J, Blumenthal RM, Degtyarev S, Dryden DT, Dybvig K, et al. A nomenclature for restriction enzymes, DNA methyltransferases, homing endonucleases and their genes. Nucleic Acids Res. 2003;31(7):1805–12.
17. Roberts RJ, Vincze T, Posfai J, Macelis D. REBASE–a database for DNA restriction and modification: enzymes, genes and genomes. Nucleic Acids Res. 2015;43(Database issue):D298–9.
18. Pingoud A, Wilson GG, Wende W. Type II restriction endonucleases - a historical perspective and more. Nucleic Acids Res. 2016;42(12):7489–527.
19. Doudna JA, Charpentier E. Genome editing. The new frontier of genome engineering with CRISPR-Cas9. Science. 2014;346(6213):1258096.
20. Ran FA. Adaptation of CRISPR nucleases for eukaryotic applications. Anal Biochem. 2016;S0003-2697(16):30354–2.
21. Komor AC, Badran AH, Liu DR. CRISPR-Based Technologies for the Manipulation of Eukaryotic Genomes. Cell. 2016, http://dx.doi.org/10.1016/j.cell.2016.10.044.
22. Mohanraju P, Makarova KS, Zetsche B, Zhang F, Koonin EV, van der Oost J. Diverse evolutionary roots and mechanistic variations of the CRISPR-Cas systems. Science. 2016;353(6299):aad5147.
23. Shmakov S, Abudayyeh OO, Makarova KS, Wolf YI, Gootenberg JS, Semenova E, Minakhin L, Joung J, Konermann S, Severinov K, et al. Discovery and Functional Characterization of Diverse Class 2 CRISPR-Cas Systems. Mol Cell. 2015;60(3):385–97.
24. Shmakov S, Smargon A, Scott D, Cox D, Pyzocha N, Yan W, Abudayyeh OO, Gootenberg JS, Makarova KS, Wolf YI et al: Diversity and evolution of class 2 CRISPR-Cas systems. Nat Rev Microbiol. 2017;15(3):169–82.
25. Hur JK, Olovnikov I, Aravin AA. Prokaryotic Argonautes defend genomes against invasive DNA. Trends Biochem Sci. 2014;39(6):257–9.
26. Swarts DC, Makarova K, Wang Y, Nakanishi K, Ketting RF, Koonin EV, Patel DJ, van der Oost J. The evolutionary journey of Argonaute proteins. Nat Struct Mol Biol. 2014;21(9):743–53.
27. Gao F, Shen XZ, Jiang F, Wu Y, Han C. DNA-guided genome editing using the Natronobacterium gregoryi Argonaute. Nat Biotechnol. 2016;34(7):768–73.
28. Cyranoski D. Replications, ridicule and a recluse: the controversy over NgAgo gene-editing intensifies. Nature. 2016;536(7615):136–7.
29. Novichkov PS, Ratnere I, Wolf YI, Koonin EV, Dubchak I. ATGC: a database of orthologous genes from closely related prokaryotic genomes and a research platform for microevolution of prokaryotes. Nucleic Acids Res. 2009;37(Database issue):D448–54.
30. Kristensen DM, Wolf YI, Koonin EV. ATGC database and ATGC-COGs: an updated resource for micro- and macro-evolutionary studies of prokaryotic genomes and protein family annotation. Nucleic Acids Res. 2016, in press.
31. Makarova KS, Wolf YI, Koonin EV. Comparative genomics of defense systems in archaea and bacteria. Nucleic Acids Res. 2013;41(8):4360–77.
32. Makarova KS, Wolf YI, Koonin EV. Comprehensive comparative-genomic analysis of type 2 toxin-antitoxin systems and related mobile stress response systems in prokaryotes. Biol Direct. 2009;4:19.
33. Csuros M. COUNT: evolutionary analysis of phylogenetic profiles with parsimony and likelihood. Bioinformatics. 2010;26(15):1910–2.
34. Makarova KS, Wolf YI, Alkhnbashi OS, Costa F, Shah SA, Saunders SJ, Barrangou R, Brouns SJ, Charpentier E, Haft DH, et al. An updated evolutionary classification of CRISPR-Cas systems. Nat Rev Microbiol. 2015;13(11):722–36.
35. Iyer LM, Abhiman S, Aravind L. MutL homologs in restriction-modification systems and the origin of eukaryotic MORC ATPases. Biol Direct. 2008;3:8.
36. Makarova KS, Wolf YI, Snir S, Koonin EV. Defense islands in bacterial and archaeal genomes and prediction of novel defense systems. J Bacteriol. 2011;193(21):6039–56.
37. Bignell C, Thomas CM. The bacterial ParA-ParB partitioning proteins. J Biotechnol. 2001;91(1):1–34.
38. Surtees JA, Funnell BE. Plasmid and chromosome traffic control: how ParA and ParB drive partition. Curr Top Dev Biol. 2003;56:145–80.
39. Engel P, Goepfert A, Stanger FV, Harms A, Schmidt A, Schirmer T, Dehio C. Adenylylation control by intra- or intermolecular active-site obstruction in Fic proteins. Nature. 2012;482(7383):107–10.
40. Wu H, Mao F, Olman V, Xu Y. On application of directons to functional classification of genes in prokaryotes. Comput Biol Chem. 2008;32(3):176–84.
41. Forterre P, Prangishvili D. The great billion-year war between ribosome- and capsid-encoding organisms (cells and viruses) as the major source of evolutionary novelties. Ann N Y Acad Sci. 2009;1178:65–77.
42. Forterre P, Prangishvili D. The major role of viruses in cellular evolution: facts and hypotheses. Curr Opin Virol. 2013;3(5):558–65.

43. Koonin EV, Dolja VV. A virocentric perspective on the evolution of life. Curr Opin Virol. 2013;3(5):546–57.

44. Koonin EV, Wolf YI. Evolution of the CRISPR-Cas adaptive immunity systems in prokaryotes: models and observations on virus-host coevolution. Mol BioSyst. 2015;11(1):20–7.

45. Vale PF, Lafforgue G, Gatchitch F, Gardan R, Moineau S, Gandon S. Costs of CRISPR-Cas-mediated resistance in Streptococcus thermophilus. Proc Biol Sci. 2015;282(1812):20151270.

46. Kobayashi I. Behavior of restriction-modification systems as selfish mobile elements and their impact on genome evolution. Nucleic Acids Res. 2001; 29(18):3742–56.

47. Mruk I, Kobayashi I. To be or not to be: regulation of restriction-modification systems and other toxin-antitoxin systems. Nucleic Acids Res. 2014;42(1):70–86.

48. Westra ER, Buckling A, Fineran PC. CRISPR-Cas systems: beyond adaptive immunity. Nat Rev Microbiol. 2014;12(5):317–26.

49. Wolf YI, Makarova KS, Yutin N, Koonin EV. Updated clusters of orthologous genes for Archaea: a complex ancestor of the Archaea and the byways of horizontal gene transfer. Biol Direct. 2012;7:46.

50. Sela I, Wolf YI, Koonin EV. Theory of prokaryotic genome evolution. Proc Natl Acad Sci U S A. 2016;113(41):11399–407.

51. Darling AE, Tritt A, Eisen JA, Facciotti MT. Mauve assembly metrics. Bioinformatics. 2011;27(19):2756–7.

52. Edgar RC. MUSCLE: multiple sequence alignment with high accuracy and high throughput. Nucleic Acids Res. 2004;32(5):1792–7.

53. Price MN, Dehal PS, Arkin AP. FastTree 2–approximately maximum-likelihood trees for large alignments. PLoS One. 2010;5(3):e9490.

54. Rogers JS. Maximum likelihood estimation of phylogenetic trees is consistent when substitution rates vary according to the invariable sites plus gamma distribution. Syst Biol. 2001;50(5):713–22.

55. Wolf YI, Aravind L, Grishin NV, Koonin EV. Evolution of aminoacyl-tRNA synthetases–analysis of unique domain architectures and phylogenetic trees reveals a complex history of horizontal gene transfer events. Genome Res. 1999;9(8):689–710.

56. Dehal PS, Joachimiak MP, Price MN, Bates JT, Baumohl JK, Chivian D, Friedland GD, Huang KH, Keller K, Novichkov PS, et al. MicrobesOnline: an integrated portal for comparative and functional genomics. Nucleic Acids Res. 2010;38(Database issue):D396–400.

7

Natural selection drove metabolic specialization of the chromatophore in *Paulinella chromatophora*

Cecilio Valadez-Cano[1], Roberto Olivares-Hernández[2], Osbaldo Resendis-Antonio[3,4], Alexander DeLuna[5] and Luis Delaye[1*] (iD)

Abstract

Background: Genome degradation of host-restricted mutualistic endosymbionts has been attributed to inactivating mutations and genetic drift while genes coding for host-relevant functions are conserved by purifying selection. Unlike their free-living relatives, the metabolism of mutualistic endosymbionts and endosymbiont-originated organelles is specialized in the production of metabolites which are released to the host. This specialization suggests that natural selection crafted these metabolic adaptations. In this work, we analyzed the evolution of the metabolism of the chromatophore of *Paulinella chromatophora* by in silico modeling. We asked whether genome reduction is driven by metabolic engineering strategies resulted from the interaction with the host. As its widely known, the loss of enzyme coding genes leads to metabolic network restructuring sometimes improving the production rates. In this case, the production rate of reduced-carbon in the metabolism of the chromatophore.

Results: We reconstructed the metabolic networks of the chromatophore of *P. chromatophora* CCAC 0185 and a close free-living relative, the cyanobacterium *Synechococcus sp.* WH 5701. We found that the evolution of free-living to host-restricted lifestyle rendered a fragile metabolic network where >80% of genes in the chromatophore are essential for metabolic functionality. Despite the lack of experimental information, the metabolic reconstruction of the chromatophore suggests that the host provides several metabolites to the endosymbiont. By using these metabolites as intracellular conditions, in silico simulations of genome evolution by gene lose recover with 77% accuracy the actual metabolic gene content of the chromatophore. Also, the metabolic model of the chromatophore allowed us to predict by flux balance analysis a maximum rate of reduced-carbon released by the endosymbiont to the host. By inspecting the central metabolism of the chromatophore and the free-living cyanobacteria we found that by improvements in the gluconeogenic pathway the metabolism of the endosymbiont uses more efficiently the carbon source for reduced-carbon production. In addition, our in silico simulations of the evolutionary process leading to the reduced metabolic network of the chromatophore showed that the predicted rate of released reduced-carbon is obtained in less than 5% of the times under a process guided by random gene deletion and genetic drift. We interpret previous findings as evidence that natural selection at holobiont level shaped the rate at which reduced-carbon is exported to the host. Finally, our model also predicts that the ABC phosphate transporter (pstSACB) which is conserved in the genome of the chromatophore of *P. chromatophora* strain CCAC 0185 is a necessary component to release reduced-carbon molecules to the host.

(Continued on next page)

* Correspondence: luis.delaye@cinvestav.mx
[1]Departamento de Ingeniería Genética, Centro de Investigación y de Estudios Avanzados del Instituto Politécnico Nacional, Unidad Irapuato, Km. 9.6 Libramiento Norte Carr. Irapuato-León, 36821 Guanajuato, Irapuato, Mexico
Full list of author information is available at the end of the article

(Continued from previous page)

Conclusion: Our evolutionary analysis suggests that in the case of *Paulinella chromatophora* natural selection at the holobiont level played a prominent role in shaping the metabolic specialization of the chromatophore. We propose that natural selection acted as a "metabolic engineer" by favoring metabolic restructurings that led to an increased release of reduced-carbon to the host.

Keywords: Endosymbiont, Metabolic evolution, Adaptation, Metabolic integration

Background

Paulinella chromatophora is an amoeba dispensed with phototrophic nutrition that contains blue-green photosynthetic organelles of cyanobacterial origin termed chromatophores [1, 2]. These novel organelles have a monophyletic origin in different strains of photosynthetic *Paulinella* that have been described [3] and were acquired through a primary endosymbiotic event about ~90 to 140 Mya [2–6].

Chromatophore genome sequencing from two strains of *P. chromatophora* (FK 01 [7] and CCAC 0185 [5]), revealed a size of 0.977 and 1.02 Mbp, respectively. This represents about 1/3 of the genome size of *Synechococcus sp.* WH 5701, the closest free-living relative cyanobacterium with a sequenced genome. *Synechococcus sp.* WH 5701 has a genome of ~3 Mbp and 3346 protein-coding genes [5]. It indicates that the chromatophore evolved by genome reduction. However, genome reduction in *P. chromatophora* is not as extreme as in plastids which rarely exceed 200 Kbp [2].

Chromatophores are genetically integrated with their host. More than 30 nuclear encoded genes of chromatophore origin have been identified [7, 8]. And some of the protein products coded by these genes are imported back into the chromatophore and participate in the photosynthetic apparatus [9]. Accordingly, chromatophores have been described as plastids in the making.

P. chromatophora nutrition relies on the reduced-carbon photosynthetically assimilated by the chromatophore [10]. This endosymbiotic-nutrient dependency has been observed in other organisms such as aphids and tsetse flies housing prokaryotic endosymbionts [11]. Particularly for aphids, host essential amino acids are provided by an endosymbiotic bacterium called *Buchnera aphidicola* [12]. Sequencing of the genome of *B. aphidicola* revealed a high degree of genetic degradation, while genes necessary for the syntrophic relationship with its host have been retained [12].

Prokaryotic endosymbionts evolve small genomes mainly by the combined action of genetic drift and negative selection [13–16]. In host-restricted conditions, the endosymbiont experiences a lack of recombination and horizontal gene transfer, as well as recurrent population bottlenecks lowering its effective population size (N_e) and a concomitant relaxation of natural selection [15–17]. The combined action of these factors allows the accumulation of slightly deleterious mutations through a process called Muller's ratchet [14, 17]. As a consequence, many genes become pseudogenes and are subsequently lost. In addition, selection at holobiont level by mechanisms like "partner fidelity feedback" have been proposed to promote the evolution of mutualistic interactions [18].

Something that should be considered is that, differing from free-living relatives, the metabolism of mutualistic endosymbionts is specialized in the production of metabolites that are released to their host as nutrients [19, 20]. This metabolic specialization is the consequence of metabolic restructuring caused by gene loss and genome reduction [20]. Resulting reduced genomes code for fewer genes, however, they are more integrated to the host. The extreme cases are organelles of endosymbiotic origin such as chloroplasts [21]. Therefore, if mutualistic endosymbionts show metabolic adaptations to provide nutrients to their hosts [19, 20], natural selection must have participated in the evolution of these systems.

During early stages of organellogenesis, the cyanobacteria that evolved into the chromatophore, had access to metabolites provided by the host. It is likely that the availability these metabolites render of some metabolic routes dispensable in the endosymbiont. The loss of these biosynthetic pathways in the endosymbiont led to restructurings and changes in the remaining metabolic fluxes. Taking into consideration all these modifications experienced by the chromatophore and the nutrient dependency of the holobiont for the photosynthetic function of the chromatophore, we made the analogy of natural selection acting as a "metabolic engineer" directing the strategies for the metabolic specialization of the chromatophore. In general, the objective of metabolic engineering is the directed improvement of metabolic capabilities through the deletion of metabolic genes or the introduction of new ones [22]. By using these strategies, microorganisms have been engineered for the improvement of the yield and the production and consumption rates of desired metabolites. For instance, for the of 1-butanol production in cyanobacteria [23], many more examples can be found elsewhere [24, 25].

In this work, we reconstructed the genome based metabolic models of the chromatophore of *Paulinella chromatophora* and the cyanobacteria *Synechococcus sp.* WH 5701. We inquired into the metabolic capabilities of the chromatophore; the possible metabolic interaction of the chromatophore with its host; and in silico simulate the process of metabolic evolution experienced by the chromatophore in host-restricted conditions.

Results

Differential gene retention of functional categories in the chromatophore genome

Our first objective was to determine to what extent genetic loss affected functional metabolic categories in the chromatophore (i.e. which functional gene categories were preferentially preserved) when compared to the genome of *Synechococcus sp.* WH 5701. We compared against *Synechococcus sp.* WH 5701 because is the closest free-living cyanobacterium with a sequenced genome and it is likely to be similar in gene content to the ancestor of the chromatophore. To assess the statistical significance we used a hypergeometric distribution.

As is shown in Fig. 1, genes belonging to 13 functional categories have been less affected by genome erosion. In particular, photosynthesis and fatty acid biosynthesis categories are less affected. Retention of these 13 functional categories in the chromatophore can be attributed to a host-level selection protecting from gene loss. Conserved genes very likely play an adaptive role in the holobiont.

In silico metabolic reconstruction of the chromatophore of *P. chromatophora* and *Synechococcus sp.* WH 5701

To better understand the role in the symbiosis played by remaining genes in the chromatophore, we reconstructed two metabolic models. One for the chromatophore of *P. chromatophora* CCAC 0185 [5] and the other for *Synechococcus sp.* WH 5701, the closest free-living cyanobacterium with a sequenced genome. The rationale behind this is to use *Synechococcus sp.* WH 5701 as a proxy of the ancestral cyanobacterium that evolved into the chromatophore.

Metabolic model reconstruction of the free-living cyanobacterium *Synechococcus sp.* WH 5701 was done by identifying orthologs to those protein-coding genes reported in the metabolic model of *Synechocystis sp.* PCC 6803 (*i*JN678) [26]. The resulting metabolic model of the free-living organism (*i*CV498) comprised 743 metabolic reactions with 698 metabolites and 498 protein-coding genes. Metabolic model reconstruction of the chromatophore was done by identifying those genes in the genome of the chromatophore of *P. chromatophora* CCAC 0185 that are orthologous to the free-living metabolic model (*i*CV498). The metabolic model of the chromatophore (*i*CV265) comprised 627 reactions, 615 metabolites and 265 protein-coding genes. Because *Synechococcus sp.* WH 5701 is a close

free-living relative of the chromatophore, it could be considered that 158 reactions were lost along genome reduction in the chromatophore (Table 1).

By using the biomass equation of the cyanobacterium *Synechocystis sp.* PCC 6803 [26], we tested the functionality of the *i*CV498 and the *i*CV265 metabolic models with Flux Balance Analysis (FBA). Biomass production was set as objective function. In silico growth was simulated under autotrophic conditions. CO_2 and photons uptake were set to 3.7 mmol $\times$ gDW^{-1} $\times$ h^{-1} and 100 mmol $\times$ gDW^{-1} $\times$ h^{-1} respectively and set as constraining metabolites as in [26].

In model *i*CV498, almost every metabolic pathway for biomass production is complete. The exceptions were 9 reactions for which no orthologous exist in *Synechococcus sp.* WH 5701 when compared to *i*JN678 (see model *i*CV498 in Additional file 1). These reactions had to be added to the *i*CV498 model in order to produce all the components necessary for the biomass equation. In this way, *i*CV498 showed an in silico growth rate of 0.0884 h^{-1} which is identical to the in silico growth reported for *Synechocystis sp.* PCC 6803 metabolic model under autotrophic conditions [26].

Under these conditions, the metabolic model of the chromatophore (*i*CV265) did not show in silico growth. This was obviously due to the reduced metabolic capabilities caused by the genomic reduction process experienced by the photosynthetic endosymbiont. Genome reduction has affected the metabolic capabilities of the chromatophore in two ways: a) some biosynthetic pathways were completely lost; while b) some other were partially lost.

For example, in *Synechocystis sp.* PCC 6803 riboflavin is synthesized by four genes that perform six reactions by using Guanosine 5′-triphosphate (G5P) and D-Ribulose 5-phosphate (R5P) as precursors metabolites [26]. All these genes for riboflavin biosynthesis were lost in the chromatophore. In this case, we assumed that the host provides riboflavin to the chromatophore. The possible explanation for this loss is that riboflavin is the main precursor for flavin mononucleotide (riboflavin 5′-monophosphate, FMN) and flavin adenine dinucleotide, two main compounds that work as coenzymes for many of the enzymes such as oxidoreductases including NADH dehydrogenase as well as in biological blue-light photo receptors. This observation is concomitant with the loss in some functional gene categories; as in oxidative phosphorylation (Fig. 1). As the hypothesis is that the metabolic network must preserve its functionality, whenever we found a similar situation, exchange reactions were added to the metabolic model to simulate the incorporation of riboflavin and other metabolites as additional nutrients from the host. These metabolites included amino acids, cofactors, vitamins and other

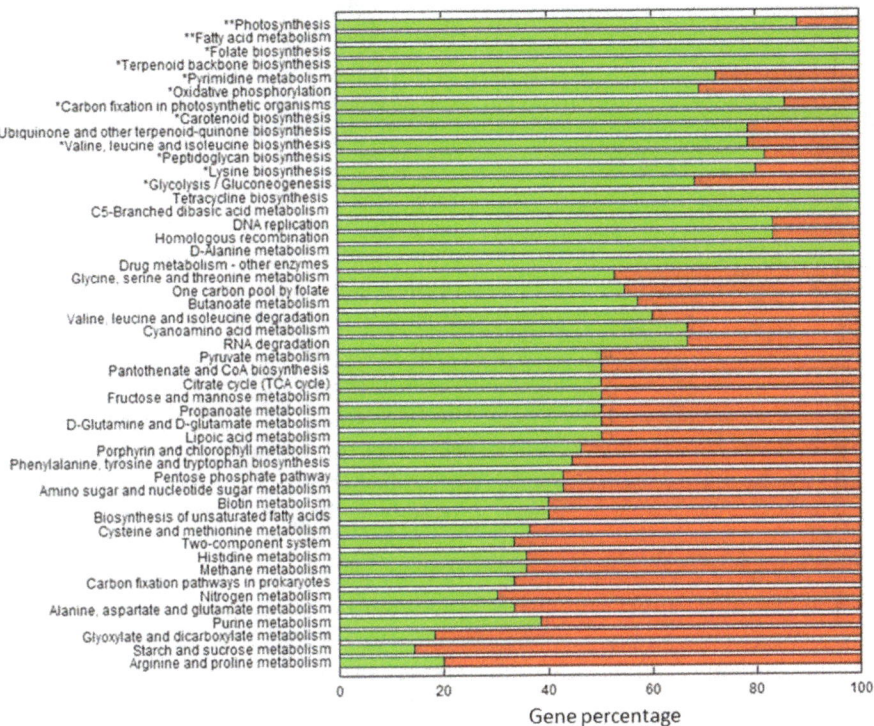

Fig. 1 Conservation of functional gene categories in the chromatophore when compared to *Synechococcus sp.* WH 5701. For each functional category we show in green and red the percentage of gene conservation and lost in the chromatophore, respectively. For instance, if a gene category is completely green, it indicates that all orthologs in *Synechococcus sp.* WH 5701 are conserved in the chromatophore. As shown, gene loss affects differentially each one of the functional categories in the chromatophore. Functional categories particularly well conserved are indicated with asterisks (*p-value* < 0.05* or <0.05**, Bonferroni corrected). Statistical significance calculated by using a hypergeometric distribution [63]. Genes were classified following KEGG database (http://www.kegg.jp)

molecules which are essential for the biomass equation but cannot be produced by the chromatophore (Fig. 2).

Some other biosynthetic pathways are truncated in the chromatophore because single gene coding enzymes were lost. For example, in the biosynthetic pathway of leucine, most gene coding enzymes are present in the chromatophore except for the gene coding for

Table 1 Characteristics of metabolic models of *Synechococcus sp.* WH 5701 (*i*CV498) and the chromatophore (*i*CV265)

	Metabolic model	
	*i*CV498	*i*CV265
Genes	498	265
Metabolites	698	615
Intracellular metabolites	661	578
Extracellular metabolites	37	37
Reactions	743	627
Enzymatic reactions	624	478
Transport reactions	82	70
Exchange reactions	37	37

3-isopropylmalate dehydrogenase. In this case, we assumed that host encoded enzymes complement the pathway in the endosymbiont. Either by importing host encoded enzymes to the chromatophore or by exchanging intermediated metabolites between the symbionts. Similar situations have been proposed for other host-endosymbiont systems [12]. For this reason, we assumed that the production of these metabolites is shared between the host and the endosymbiont (see model *i*CV265 in Additional file 2).

In addition, some reactions in the chromatophore model *i*CV265 for which no orthologous genes exist with the free-living model *i*CV498 but are essential for in silico growth were assumed to be present (see model *i*CV265 in Additional file 2).

Finally, chromatophores lost the ability to store photosynthates as well as the capacity to synthesize sucrose [5]. Because of that, glycogen was removed from the biomass equation in *i*CV265. Under these conditions, in silico growth of the *i*CV265 model was 0.1568 h^{-1}. This is an unrealistic rate because growth of the chromatophores is restricted to host division which is much lower than growth rate reported for

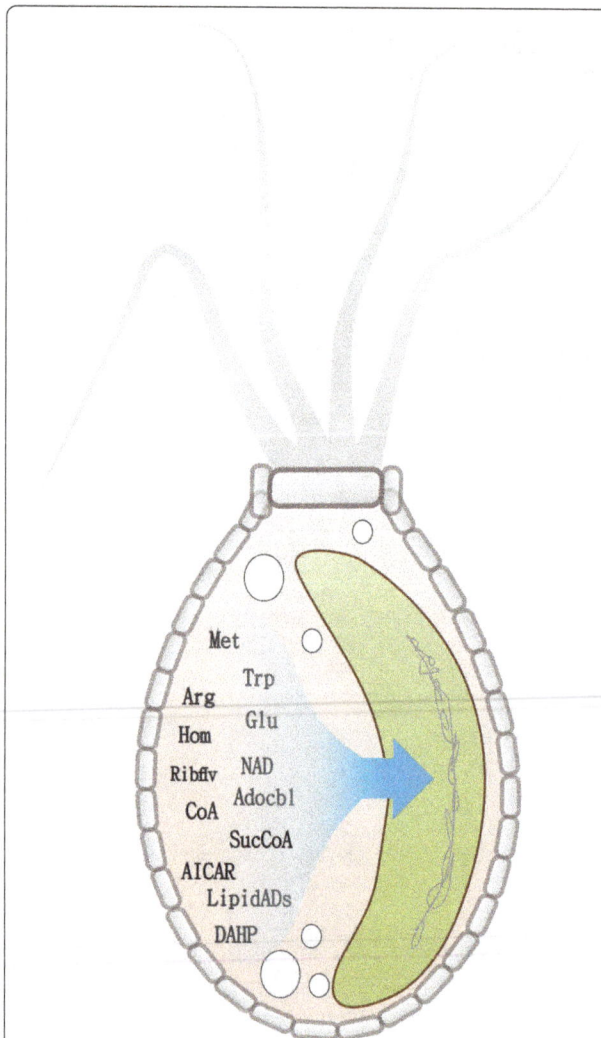

Fig. 2 Nutrients uptake simulation in the chromatophore model (*iCV265*). Metabolites that cannot be produced by the chromatophore (with respect to the free-living model, *iCV498*) include: amino acids (Met = L-Methionine, Trp = L-Tryptophan, Arg = L-Arginine, Glu = L-Glutamate, Hom = L-Homoserine), cofactors (NAD = Nicotinamide adenine dinucleotide, Adocbl = Adenosylcobalamin, CoA = Coenzyme A), vitamins (Ribflv = Riboflavin) and others (AICAR = 1-(5'-Phosphoribosyl)-5-amino-4-imidazolecarboxamide, SucCoA = Succinyl-CoA, LipidADs = Lipid A Disaccharide, DAHP = 2-Dehydro-3-deoxy-D-arabino-heptonate 7-phosphate)

the biomass production over 99% (Fig. 3). This result shows that *iCV498* is less robust than the metabolic model of *Synechocystis sp.* PCC 6803 where 51.6% of the genes are essential under these same conditions [26]. In addition, there is a decreasing robustness in the model of the chromatophore where 222 of the 265 genes (83.77%) are essential (Fig. 3). This indicates that the genomic reduction experimented by the chromatophore rendered its metabolic network fragile. The same result has been observed for other metabolic networks of endosymbionts [20, 28, 29].

Interestingly, we found that there are 3 non-essential genes in the metabolic model *iCV498* whose single deletion decreases in silico growth rate. These include genes encoding the enzymes acetyl-CoA synthetase, malic enzyme (NAD) and fumarase. Of these three, the last enzyme is the only one decreasing the in silico growth rate in *iJN678* when it is deleted (data not shown). In the *iCV265* model, all these 3 genes were lost. In addition, the non-essential gene in *iCV498* coding for an enzyme with arginase activity is the only one whose deletion decreases the growth rate in the *iCV265* chromatophore model. This suggests that genome reduction leading to *iCV265* caused metabolic restructuring because deletion of this enzyme with arginase activity in *iCV498* has no effect.

In silico simulation of metabolic-gene loses in the chromatophore of *P. chromatophora*

Based on the metabolic network of the free living *Synechococcus sp.* WH 5701, we simulated in silico the gene loss. We evaluated the impact of intracellular conditions (metabolite availability) on the evolution of the chromatophore. In particular, we asked whether the set of metabolites predicted to be provided by the host in the *iCV265* model (Fig. 2) determined actual gen content of the chromatophore after genome reduction.

Two in silico intracellular conditions were evaluated. In the first one, we simulated genetic reduction under in silico intracellular conditions where available nutrients were those predicted in the *iCV265* model (we refer to them as Proposed Nutrients) (Fig. 2). In the second one, we randomly selected metabolites from the *iCV498* model (the same number as in the first condition) and assigned as available nutrients under intracellular conditions (we refer to them as Randomized Nutrients; see Additional file 3: Table S1). The algorithm to simulate genome reduction is explained in detail in the methods section.

This algorithm allowed us to obtain in silico evolved chromatophores whose metabolic capabilities regarding the biomass production are equivalent to those of *iCV265*; but differing in their in silico evolutionary history and gene content.

free-living cyanobacteria and even other photosynthetic eukaryotes [27].

Robustness analysis of metabolic models

We assessed the robustness of the *iCV498* and the *iCV265* models to single gene deletions. Genetic robustness was defined as the capacity of the models to maintain its metabolic capabilities (in silico biomass production) after a genetic deletion. Under phototrophic conditions, model *iCV498* showed 333 genes (66.86%) to be essential because its deletion decreases

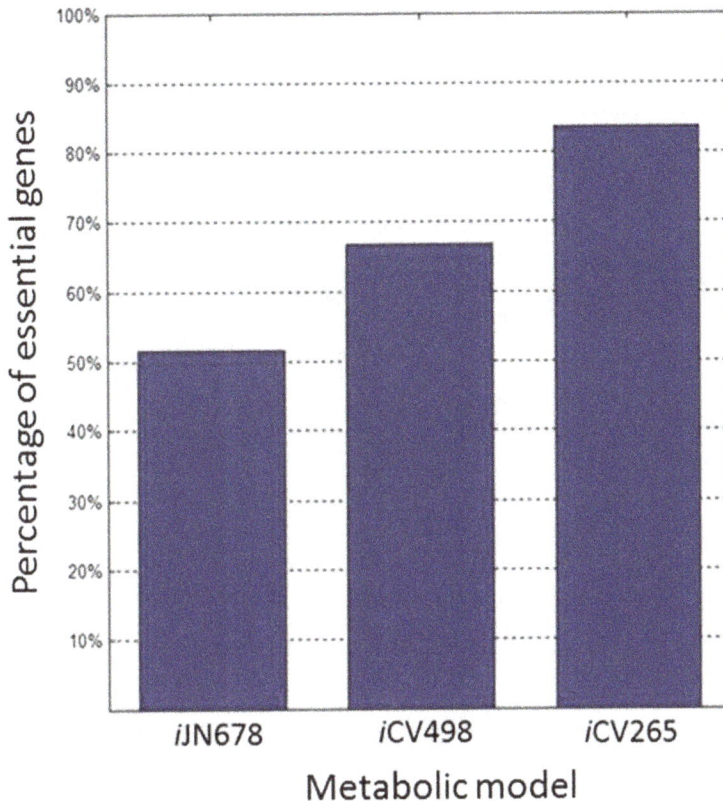

Fig. 3 Genetic robustness analysis of metabolic models of *Synechocystis sp*. PCC 6803 (*iJN678*) [26], *Synechococcus sp*. WH 5701 (*iCV498*) and the chromatophore of *Paulinella chromatophora* (*iCV265*). Percentage of essential genes (Y axis) for each metabolic model (X axis)

Simulations under the Proposed Nutrients conditions resulted in reduced metabolic networks with 295.1 ($\pm$ 2.63) genes on average. In these reduced networks, of the 498 genes present in the free-living ancestor (model *iCV498*), 52.2% are strictly conserved in the 500 simulations and 26.7% are dispensable in all of them. In Randomized Nutrients simulations, reduced networks have an average size of 326.8 ($\pm$ 5.26) genes and 54% and 16.2% are conserved and dispensable in the 500 simulations, respectively.

As is shown in Fig. 4, the proportion of: i) essential genes (predicted to be essential in 500 simulations); ii) variable genes (predicted to be conserved in 1 to 499 simulations); and iii) dispensable genes (predicted to be lost in 500 simulations), varies between metabolic pathways. These proportions also vary between treatments (i.e., Proposed Nutrients or Randomized Nutrients). Surprisingly, the most extreme case is that of the genes participating in photosynthetic activity. In Proposed Nutrients 77.6%, 18.3% and 4.1% are predicted to be essential, variable and dispensable, respectively. While for Randomized Nutrients none was predicted to be essential nor dispensable because 100% of them were variable.

Genetic concordance was evaluated between these simulated minimal networks and the *real* chromatophore model (*iCV265*). This was done by measuring sensitivity and specificity as in [30]. In Fig. 5, we show the fraction of true-positives and false-positives for every cutoff (1 to 500). True-positive and false-positive for every cutoff (1 to 500) form a curve whose area under the curve represents the probability that a gene conserved in *iCV265* is present in more simulations than a gene which has been lost.

The area under the curve shows the contribution of the nutrients available in intracellular conditions explaining the evolutionary history experimented by the chromatophore. Accordingly, the accuracy obtained under the Proposed Nutrients condition was 77.4%, while that of the Randomized Nutrients was 59.8%. The difference between the areas under the curve from both conditions is statistically significant (p-value < 0.001, Chi-square test of homogeneity).

Modeling selection and drift to explain metabolic evolution of the chromatophore

We are interested in understanding the role played by natural selection during the evolution of the metabolic capabilities of the chromatophore. Chromatophores provide the host with reduced-carbon, probably a hexose. This in analogy to the origin of plastids. It has been proposed that during the early stages of plastid evolution,

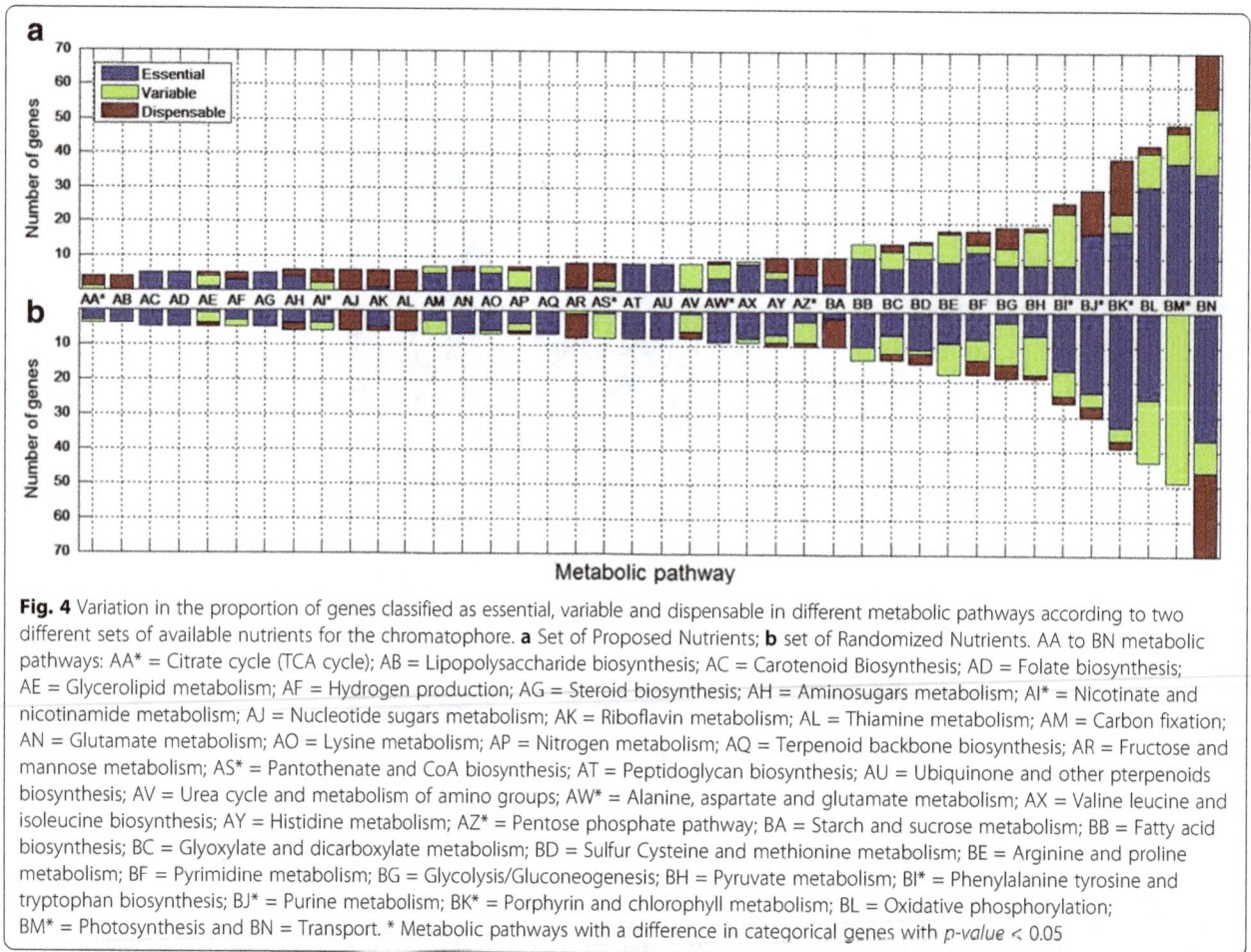

Fig. 4 Variation in the proportion of genes classified as essential, variable and dispensable in different metabolic pathways according to two different sets of available nutrients for the chromatophore. **a** Set of Proposed Nutrients; **b** set of Randomized Nutrients. AA to BN metabolic pathways: AA* = Citrate cycle (TCA cycle); AB = Lipopolysaccharide biosynthesis; AC = Carotenoid Biosynthesis; AD = Folate biosynthesis; AE = Glycerolipid metabolism; AF = Hydrogen production; AG = Steroid biosynthesis; AH = Aminosugars metabolism; AI* = Nicotinate and nicotinamide metabolism; AJ = Nucleotide sugars metabolism; AK = Riboflavin metabolism; AL = Thiamine metabolism; AM = Carbon fixation; AN = Glutamate metabolism; AO = Lysine metabolism; AP = Nitrogen metabolism; AQ = Terpenoid backbone biosynthesis; AR = Fructose and mannose metabolism; AS* = Pantothenate and CoA biosynthesis; AT = Peptidoglycan biosynthesis; AU = Ubiquinone and other pterpenoids biosynthesis; AV = Urea cycle and metabolism of amino groups; AW* = Alanine, aspartate and glutamate metabolism; AX = Valine leucine and isoleucine biosynthesis; AY = Histidine metabolism; AZ* = Pentose phosphate pathway; BA = Starch and sucrose metabolism; BB = Fatty acid biosynthesis; BC = Glyoxylate and dicarboxylate metabolism; BD = Sulfur Cysteine and methionine metabolism; BE = Arginine and proline metabolism; BF = Pyrimidine metabolism; BG = Glycolysis/Gluconeogenesis; BH = Pyruvate metabolism; BI* = Phenylalanine tyrosine and tryptophan biosynthesis; BJ* = Purine metabolism; BK* = Porphyrin and chlorophyll metabolism; BL = Oxidative phosphorylation; BM* = Photosynthesis and BN = Transport. * Metabolic pathways with a difference in categorical genes with *p-value* < 0.05

the photosynthetic endosymbiont exported reduced-carbon in the form of an hexose-phosphate through an hexose phosphate transporter of bacterial origin (non-cyanobacterial) [31, 32].

To study how the potential rate of carbon exportation evolved, a hexose export reaction was added to the metabolic models. This reaction was defined as objective function. To ensure biomass components production, the biomass reaction was fixed to 0.0884 h^{-1} which, as stated previously, is the growth rate of *Synechocystis sp.* PCC 6803 metabolic model under autotrophic growth conditions [26].

Under these conditions, there is no exportation of reduced-carbon in the *i*CV498 model. However, in the *i*CV265 chromatophore model, the potential rate of hexose exported without affecting the in silico growth rate was 0.2689 mmol × gDW^{-1} × h^{-1}. In Fig. 6, we show the fluxes calculated with FBA of the central metabolism of the models of the chromatophore and the free-living cyanobacteria in conditions previously mentioned.

Fluxes calculated for production of metabolites precursors used to produce biomass components are produced in less quantity in the chromatophore's model (Fig. 6).

This is a consequence of the loss of metabolic capabilities in the metabolism of the chromatophore which allow the redirection of carbon through the gluconeogenic pathway for the production of hexoses as metabolic objective, instead of being used in the production of biomass components.

To analyze the efficiency of the metabolic networks in terms of hexose production at overall metabolism, we calculated the yields. The yields are parameters that measure the efficiency of the metabolic network and allow the comparison across different microorganisms. For instance, the yield of the ethanol production is higher in *Saccharomyces cerevisiae* compared to *Zymomonas mobilis*, this was the result of the specialization of the microorganism to produce specific metabolites [33]. Therefore, we calculated the yields of carbon, energy and reducing equivalents (extracellular CO_2, ATP and NADPH) required to produce hexose (Table 2). These results show that the model of the chromatophore is more efficient for producing hexose from the external carbon than the free-living cyanobacteria. It means that metabolic restructurings experienced by the chromatophore

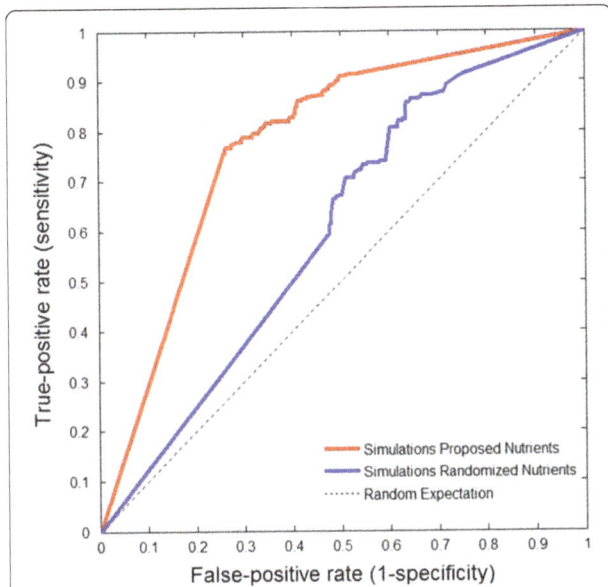

Fig. 5 Genetic concordance between in silico evolved chromatophores and the chromatophore model (iCV265). The area under the curve indicates the predictive accuracy for evolved chromatophores (i.e. simulations) obtained under Proposed and Randomized Nutrients conditions. Proposed Nutrients: simulation with metabolites proposed as nutrients in the iCV265 model (Fig. 2) and Randomized Nutrients: simulations with randomized metabolites assigned as nutrients (Additional file 3: Table S1). Area under the curve: Proposed Nutrients =0.7742 (*p-value* = 2.853E-26); Randomized Nutrients =0.5987 (*p-value* = 1.42E-4)

rendered its metabolism more efficient to produce hexose which can be provided to the host.

The yields suggest that the loss of some metabolic capabilities in the ancestral cyanobacterium caused a redirection of fixed CO_2 causing changes in metabolic fluxes and consequently increasing the rate of reduced-carbon exported to the extracellular compartment.

We then inquired about the evolutionary forces that determined genetic conservation and metabolic functionality in chromatophores. Specifically, we wanted to infer if these metabolic capabilities of the iCV265 model (the potential rate of hexose exportation of 0.2689 mmol $\times$ gDW^{-1} $\times$ h^{-1}) could have been possible under a random model of evolution or were the consequence of natural selection for metabolic specialization of the chromatophore and its positive impact at the holobiont. To tests this, we simulated the metabolic reduction process with hexose export and biomass production as evolutionary restrictions in a random model where hexose exported must be greater than zero. It means that every gene affecting the in silico growth rate of 0.0884 h^{-1} and impairing hexose export was considered as essential, while the hexose export rate could always vary while being greater than zero (purifying selection restriction for hexose export).

Minimal networks obtained in silico under these conditions were variable in size and gene content. Of 500 simulations, only 175 (66.03%) of the genes conserved in model iCV265 are conserved in all the simulations. Conversely, there are 45 genes predicted to be essential in all these 500 simulations which are not conserved in model iCV265. The metabolic networks from these 500 simulations are different in gene content and show different hexose export rates however they are equivalent in biomass production (Additional file 3: Figure S1).

As shown in Fig. 7, after in silico metabolic reduction, hexose export rate in minimal networks obtained under these conditions tend to be minimal and close to zero (hexose export rate could not reach zero because of the restriction we imposed). On the other hand, only 2.6% of simulations have a potential rate of hexose exported equal or higher than the metabolic model of the chromatophore (0.2689 mmol $\times$ gDW^{-1} $\times$ h^{-1}). This suggests that the probability of obtaining a potential rate of hexose exported similar to that of iCV265 under a random model is less than 5%. We got a similar result by varying the growth rate constraint of 0.0884 h^{-1} under plausible biological values (see Additional file 3: Figure S2).

Although variable, our simulations evolved metabolic networks that have approximately the same number of reactions than iCV265. The average number of reactions with non-zero fluxes in the reduced metabolic models of the 500 simulations is 416.15 ± 3.91. This is slightly less than the number of reactions with non-zero fluxes in the iCV265 model (442 reactions). This shows that the small percentage of simulations (2.6%) showing a potential rate of hexose exported equal or greater than that of the chromatophore (0.2689 mmol $\times$ gDW^{-1} $\times$ h^{-1}) is not due simply to smaller size of the simulated metabolic networks.

These in silico experiments suggest that the potential rate of hexose exported in model iCV265 is unlikely to be the outcome of only genetic drift and purifying selection (i.e., less than 5% of the simulated networks export hexose at a rate comparable to that of iCV265). This suggests that the potential rate of hexose exported was the result of a process of functional specialization in which the increasing rate of hexose exportation was favored by natural selection due to the positive impact at the holobiont level.

Interestingly all these 2.6% of in silico evolved chromatophores have in common the conservation of a phosphate transporter via ABC system which is also present in the chromatophore model (iCV265). Conservation of this phosphate transporter allows the simulated network to get the phosphate necessary to be able to export fixed carbon. Without this transporter most of fixed carbon is oxidized in the pentose phosphate pathway releasing only a small amount to the extracellular compartment (data not shown).

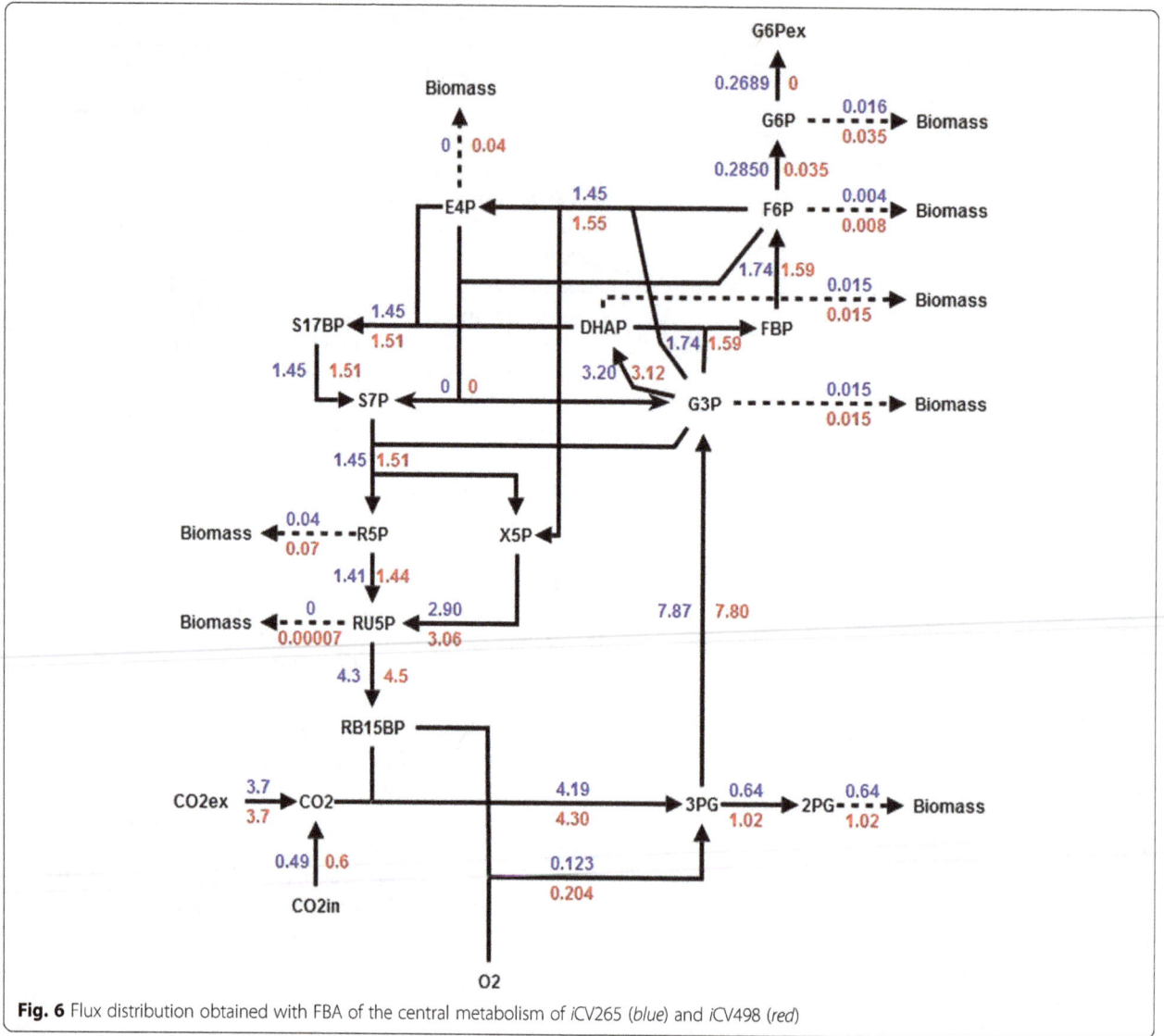

Fig. 6 Flux distribution obtained with FBA of the central metabolism of *i*CV265 (*blue*) and *i*CV498 (*red*)

Metabolic integration of the chromatophore to its host

In our previous simulations, we assumed that nutrients (Fig. 2) were available simultaneously for the chromatophore since the beginning of the evolutionary process at the onset of the endosymbiosis. However, it is likely that this has not been the case and transporters for these nutrients were gained (or lost) sequentially. For instance, metabolic transport activity in the chromatophore is

reduced due to loss of most transporters in comparison to free-living cyanobacteria [5]. And it has been reported that a large percentage of solute transporters in plastids from Plantae have host and bacterial (non-cyanobacterial) origin [31, 34].

Therefore, we simulated the evolutionary acquisition of transporters and its consequences in gene loss and the capability of the chromatophore to export fixed carbon to its host.

For every simulation, we used *i*CV498 as a free-living ancestor of the chromatophore under nutrient-rich conditions (Fig. 2). However, in this experiment, the model *i*CV498 did not have access to all nutrients since the beginning of the simulation. Instead, we randomly assigned a transport allowing the uptake of the respective nutrient. We then randomly deleted one gene at a time from *i*CV498. If the deleted gene affected the growth rate (0.0884 h^{-1}) or impaired hexose

Table 2 Yields ($Y_{P/S}$) analysis of extracellular CO_2, ATP and NADPH consumed in hexose production for both models

	Metabolic model	
	*i*CV265	*i*CV498
	mmol Hexose/ mmol	mmol Hexose/ mmol
CO_2	0.069	0.009
ATP	0.023	0.002
NADPH	0.036	0.004

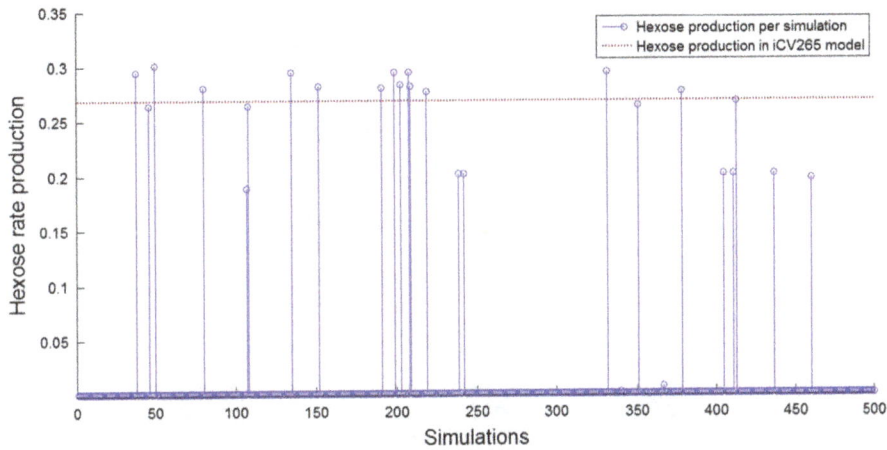

Fig. 7 Hexose exportation rate of the chromatophore model (*i*CV265) is achieved in only 2.6% of the simulations under a random model of evolution. Hexose exported rate (Y axis) for 500 independent simulations (X axis). Red-dotted line indicates the hexose export rate in the chromatophore model (*i*CV265)

exportation, we considered this gene as essential and we restored it to the model. In this way we analyzed the selective impact caused by gene loss due to the addition of a single transporter and the concomitant relaxation of natural selection for retention of specific biosynthetic pathways. Once we analyzed every gene in the model, we randomly assigned a second transport and then we repeated the gene loss simulation mentioned above. Simulation stops when in silico chromatophore has access to the 13 nutrients (Fig. 2) and all genes have been evaluated for their essentiality.

As shown in Fig. 8, after the incorporation of the 13 transporters, the probability of getting a potential rate of hexose exported equal or higher than the metabolic model of the chromatophore (0.2689 mmol $\times$ gDW^{-1} $\times$ h^{-1}) is less than 5% in 500 simulations, in agreement with our previous result (Fig. 8).

During the process of metabolic integration, it is noted that the maximum rate of hexose exportation becomes greater with every metabolite obtained as nutrient. However, by inspecting the frequency distribution of simulations with different potential rates of hexose exported, it is obvious that as metabolic integration advances, the probability of getting the maximum rate of hexose exportation decreases (i.e. the frequency of networks with large export rate becomes smaller).

These changes in frequency distributions during the process of metabolic integration can be interpreted in terms of the functional specialization of the chromatophore. As metabolic integration proceeds (with the addition of more transporters), the chromatophore increased its capacity to provide fixed carbon to its host. However, continued gene loss led to a simplified metabolic network and a smaller fraction of in silico evolved chromatophores can export as much fixed carbon as

*i*CV265. The evolutionary landscape becomes smaller as evolution proceeds.

The metabolism of the chromatophore is specially adapted to produce carbon for its host

As shown above, the potential rate of hexose exported in the chromatophore model (*i*CV265) is highly dependent on phosphate consumption. In addition, the growth rate of the chromatophore is coupled to the host's growth rate. As shown above, the potential rate of hexose exported is unlikely to be the outcome of a random evolution.

To test the impact of these two restrictions, we analyzed the metabolic properties of models *i*CV265 and *i*CV498 in potential rate of hexose exported under growth rate and phosphate uptake restrictions. As shown in Fig. 9, the potential rate of hexose exported in the *i*CV498 model is robust with respect to growth rate and phosphate uptake (i.e. a given growth rate can sustain the hexose rate exportation with different rates of phosphate consumption). This contrasts with the chromatophore model (*i*CV265) where, for a given growth rate, only a specific consumption of phosphate is necessary to sustain hexose release. In addition, the capacity of hexose export in the *i*CV265 model for a determined growth rate restriction is greater than the *i*CV498 model.

Discussion

In this work, we show that the metabolic network of the chromatophore of *P. chromatophora* is different to the metabolic network of its free-living relative *Synechococcus sp.* WH 5701. We suggest that these differences evolved by natural selection. Gene-loss and carbon flux redirection guided by natural selection led to metabolic

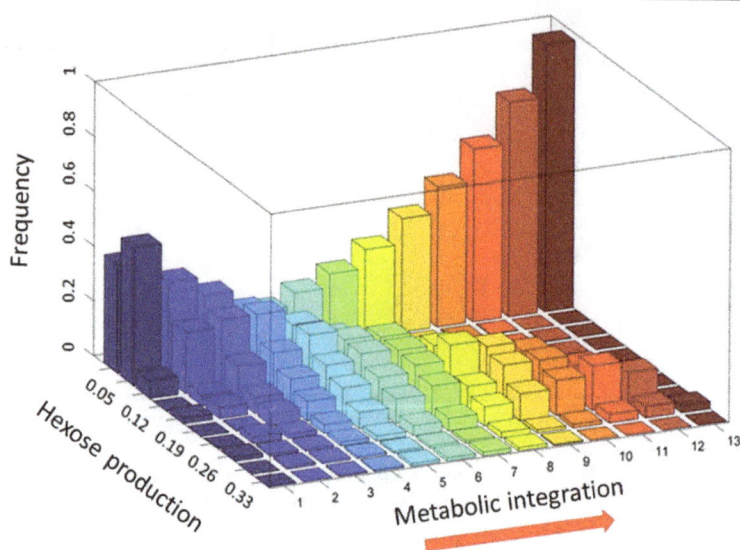

Fig. 8 Metabolic integration with the host determines the rate of hexose exported by in silico chromatophores as well as the frequency of simulations that provide higher rates of reduced-carbon to the host. X axis, hexose export rate; Y axis, frequency of simulations; Z axis, transporters added to the model (metabolic integration)

specialization of the chromatophore as a reduced-carbon provider.

Purifying selection and the maintenance of the symbiosis

Our analysis showed that some metabolic pathways have been preferentially conserved in the chromatophore (Fig. 1). These preserved metabolic pathways (i.e. photosynthesis, carbon fixation, and gluconeogenesis) very likely play a prominent role in the symbiosis. This pattern is analogous to the one observed in many other endosymbionts e.g. *Buchnera aphidicola* [35]. In this later case, biosynthetic pathways producing essential amino acids for the host [12, 35] are preserved by host-level natural selection.

As mentioned above, differential conservation of gene category functions suggests that purifying selection is preserving relevant symbiotic functions. Accordingly, estimation of the rate of nucleotide substitution in 681 DNA alignments of protein-coding genes orthologous between chromatophores of two different strains of *P. chromatophora* (CCAC 0185 [5] and FK 01 [7]) showed that most of them have signals of purifying selection [7].

It has been suggested that host-level selection prevents the fixation of deleterious mutations in endosymbionts thus lowering the chances of a mutational meltdown resulting in extinction [36, 37]. And, of course, this prevents the consequent replacement of non-functional endosymbionts [38]. In addition, selective pressure to maintain functional proteins increases with the time of host-endosymbiont interaction [36] and combined with very strong bottlenecks may help to reduce the

accumulation of deleterious mutations. This has been proposed to explain mitochondrial genome evolution [39].

Metabolic integration of the chromatophore to its host

Comparison of the metabolic models of the chromatophore and the cyanobacterium *Synechococcus sp.* WH 5701 allowed us to inquire into the evolution of the metabolic interaction of the chromatophores with its host. For example, several metabolic pathways in the chromatophore are incomplete. It is likely that the host supplies these metabolites as nutrients to the chromatophore. Metabolic pathway sharing is a hallmark of endosymbiotic organisms. For example *Wolbachia*, which are endosymbionts of many animal species, show a degraded genome [40, 41] whose limited metabolic capabilities are complemented by its host [42]. In turn, the endosymbiont provides the host with nutrients such as riboflavin, positively impacting host fitness [42]. Equally remarkable is the likely coupled production of some metabolites between the chromatophore and its host. As mentioned above, this collaboration in metabolite biosynthesis has been observed in other symbiotic systems [43–46].

Fragility of a reduced metabolic network

To study the metabolic capabilities of the chromatophores we used FBA. This stoichiometric approach can predict cellular phenotypes in specific environmental conditions. Generally, biomass production is fixed as objective function. In absence of biomass composition, the use of a biomass equation from a related organism is a valid starting point for metabolic analysis [47–49]. In

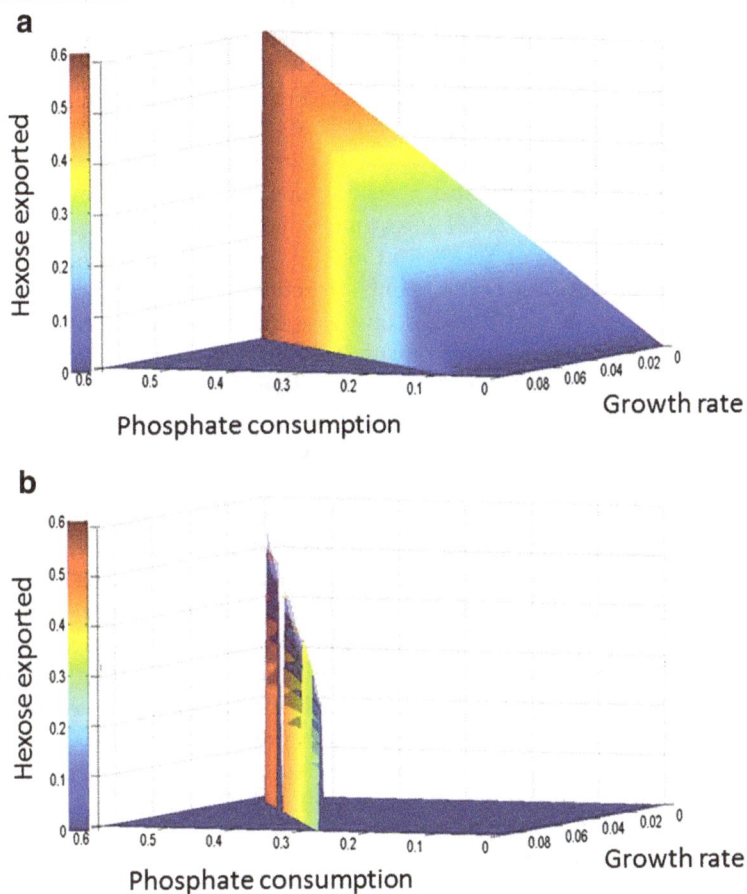

Fig. 9 The metabolic model of the chromatophore (*iCV265*) is specially adapted to produce higher rates of reduced-carbon under phosphate and growth restrictions in comparison with *Synechococcus sp.* WH 5701 (*iCV498*). **a** Hexose export rate in the free-living model *iCV498*. **b** Hexose export rate in the chromatophore model *iCV265*. Y axis, growth rate restriction. X axis, phosphate uptake restriction. Z axis, potential rate of hexose exported. Colors in axis Z are only for drawing purposes and do not represent an extra variable

this way, FBA has been used to infer the metabolic capabilities of different organisms whose cultivation and experimental management is challenging or not yet possible, as in the case of endosymbionts. For example, biomass composition and the metabolic model of *Escherichia coli* were used for metabolic analysis of *Buchnera aphidicola* [20, 30], *Sodalis glossinidius* [29], and *Blattabacterium cuenoti* [28]. In the same way, we used the biomass composition and stoichiometric model of *Synechocystis sp.* PCC 6803 as a starting point to model the metabolism of the chromatophore and *Synechococcus sp.* WH 5701 [26].

We found that the metabolism of the chromatophore is highly fragile to gene deletions. Approximately 84% of the genes in the model are essential when singly deleted in comparison with ~67% of the genes in *Synechococcus sp.* WH 5701. A similar difference in metabolic fragility was found by [20] when comparing the models of *B. aphidicola* and its free-living relative *Escherichia coli* where 84% and 19% of genes were essential, respectively.

In the same way, the metabolic network of two strains of *Blattabacterium cuenoti* (Bge and Pam), the obligated primary endosymbiont of cockroaches, were shown to be highly fragile to single gene deletion. It was found that 76.1% and 79.6% were essential genes, respectively [28]. Finally, in *Sodalis glossinidius* (the secondary non-obligated endosymbiont in early stages of tsetse flies), 44.54% metabolic genes were found to be essential, compared with its ancestral network where only 25.48% are predicted to be essential [29].

Our robustness analyses of the *iCV265* and the *iCV498* models agree with the generalization that metabolic networks of endosymbionts are more fragile than their free-living counterparts. This metabolic fragility of endosymbionts contrasts with theoretical estimations that suggest that, in general, metabolic systems are robust and complex [50]. However, the metabolic systems of endosymbionts are considered more robust [28] than minimalist metabolic networks [51]. The difference in metabolic fragility of the chromatophore when compared to *Synechococcus sp.* WH 5701 reflects the transition from a free-living style to a more stable condition inside *Paulinella chromatophora*.

Metabolic environment as a determinant of gene content

It has been shown that retention of metabolic genes in endosymbionts is determined by the metabolic requirements and molecular environment of the host [52, 53]. With the use of FBA and the metabolic model of *Synechococcus sp.* WH 5701 as a proxy of the ancestor of the chromatophore, we evaluated the impact of the host-metabolic environment in the reduction of the metabolic system of the endosymbiont. The proposed host-metabolic environment (Proposed Nutrients) predicted with 77.42% of accuracy the actual gene content of the chromatophore. This is in contrast with the 59.8% of accuracy obtained when using a randomly set of host-provided metabolites (Randomized Nutrients). This emphasizes the contribution of the intracellular metabolic environment to the evolution of the metabolism in the chromatophore.

Similar reductive simulations have been used to predict the set of essential genes of pathogens located in certain environmental niches (like the bloodstream) within the human body [52]. In the same way, reductive evolution simulations using *E. coli* as free-living ancestor predicts with 80% of accuracy the metabolic gene content of *B. aphidicola* and *Wigglesworthia glossindia* [30].

Inspection of the proportion of dispensable, variable, and essential genes by in silico reductive simulations (i.e. Proposed Nutrients and Randomized Nutrients) predicts differential gene retention patterns between different metabolic pathways. For example, in Randomized Nutrients simulations, photosynthesis pathway (which is the *raison d'être* of the symbiosis) 100% of genes are predicted as "variable" (none of the genes are predicted to be retained in the 500 simulations) while in Proposed Nutrients ~78% are essential. This means that under Randomized Nutrients, photosynthesis function could be useful but not essential and could have been lost in the chromatophore by chance. Clearly, the set of metabolites comprising Randomized Nutrients cannot account for the metabolic gene content of extant chromatophores.

Maximization of biomass production is regularly used as objective function in FBA analysis. It allows predicting the distribution of fluxes through a metabolic network [54]. The maximization of biomass function is used as a proxy of evolutionary fitness. However, many other objective functions can be used [54, 55]. For instance, it was estimated that *Chlorella* (the photosynthetic endosymbiont of *Paramecium bursaria*), releases 57% of its photosynthates to its host [56]. This means that most carbon photosynthetically assimilated is destined to symbiotic interaction instead of biomass production of *Chlorella* itself. In the same way, *P. chromatophora* has phototrophic nutrition. It depends on carbon assimilates which derive from the endosymbiotic cyanobacterium whose inorganic carbon rate assimilation is the same as a free-living cyanobacteria [10]. But unlike its free-living relatives, its growth rate is restricted by *P. chromatophora*. Considering the above metabolic analysis of the chromatophore, which predict an in silico growth rate of $0.1568\ h^{-1}$, it is difficult to consider the biomass as the only objective function in chromatophores. Taking into consideration that chromatophores provide the host with reduced-carbon, a reaction simulating hexose export to extracellular compartment was added. This reaction was defined as objective function. And to ensure biomass components production, biomass reaction was fixed to $0.0884\ h^{-1}$ which is the growth rate of a free-living relative cyanobacterium. Interestingly, under these conditions the metabolic model of the chromatophore predicts a potential rate of hexose exportation of $0.2689\ mmol \times gDW^{-1} \times h^{-1}$. As far as we know, this is the first metabolic reductive evolutionary analysis where metabolic functionality (i.e. hexose export) of the endosymbiont is explored as objective function, differing from previous analyses where biomass is set as objective function of mutual endosymbionts as *B. aphidicola* [20, 30], *S. glossinidius* [29] and *B. cuenoti* [28].

ABC phosphate transporter is an essential component of the chromatophore

All simulations showing a hexose exportation rate equivalent to that of the chromatophore model (*i*CV265) share the ABC phosphate transporter. This P_i-dependency in the chromatophore agrees with that observed in isolated spinach chloroplasts [57]. It has been shown that photosynthesis declines dramatically (less than 10% of the maximum rate) in chloroplast in the absence of P_i in the reaction medium. Also, carbon export from the chloroplast is inhibited [58], with up to 60% of ^{14}C fixed being retained in the chloroplast [57]. As mentioned above, this observation agrees with the more than 95% of simulations which predict that lack of ABC phosphate transport favors carbon retention in the chromatophore instead of being released to the host. Therefore, we predict that lack of ABC transporter in the genome of the chromatophore of *Paulinella* FK01 is compensated by a phosphate transporter coded in the host [7].

The role of natural selection on the evolution of the metabolism of the chromatophore

Inspection of FBA calculated central metabolic fluxes in the chromatophore and in the free-living cyanobacteria showed that the endosymbiont is better at producing hexose. This is likely a host related adaptation. To investigate whether this and other characteristics of the metabolic model of the chromatophore evolved by natural selection, we simulated in silico reductive evolution with a null model not including positive selection. As a proxy of genome reduction by purifying selection and random genetic drift, we submitted the metabolic model of *Synechococcus sp.* WH 5701 with the following algorithm: a) first, we simulated host-level purifying selection by requiring that the rate of hexose exportation of the model must be always greater than 0 and biomass is produced at 0.0884 mmol $\times$ gDW^{-1} $\times$ h^{-1}; b) next, we performed rounds of single gene deletion until no more genes could be deleted; c) finally, we repeated this process 500 times. By this, we obtained a population of 500 reduced metabolic networks all of them capable of producing 0.0884 mmol $\times$ gDW^{-1} $\times$ h^{-1} of biomass, but differing in hexose rate exportation. Differences in rates of hexose exportation were due to contingency-dependent loss of alternative pathways [30]. With this experiment, we could determine if the potential rate of hexose exported in *i*CV265 (0.2689 mmol $\times$ gDW^{-1} $\times$ h^{-1}) is easily obtained by host-level purifying selection (hexose exportation >0) and contingency-dependent evolution on random gene deletion. Our evolutionary reductive analyses showed that <5% of simulations were predicted to export hexose at a similar rate as the model *i*CV265. This suggests that metabolic functionality of *i*CV265 is unlikely to be determined by genetic drift alone. Therefore, we conclude that natural selection at holobiont level may have contributed to shape metabolic functionality of the chromatophore.

Natural selection as metabolic engineer

According to the above mentioned, we consider suitable to make the analogy of natural selection as metabolic engineer. Metabolic engineering can be defined as "the directed improvement of product formation or cellular properties through the modification of specific biochemical reactions or introduction of new ones" [22]. One of the objectives of metabolic engineers is to redirect the flux of mass through the metabolism of organisms towards a desired metabolic product. Some genetic strategies to redirect metabolic flux toward production of a desired metabolite include: increasing the precursor supply; altering the regulation (overexpressing) genes; increasing the efficiency of bottleneck enzymes; reducing flux toward unwanted byproducts; or eliminating competing pathways by gene-deletion [59]. It has been proposed that cellular metabolism of free-living microorganisms is primed, through natural selection, for the maximum responsiveness to the history of selective pressures rather than for the overproduction of specific chemical compound [60]. In host-restricted conditions this responsiveness to free-living selective pressures are

no longer needed. Instead, new biological objectives are defined now related to holobiont survival.

For instance, it was proposed that the chloroplast metabolic network has improved photosynthetic properties in comparison to free-living cyanobacteria [21]. For example, the metabolic network in chloroplast has: i) a longer average path length; ii) a larger diameter; iii) is Calvin Cycle-centered; iv) and presents better modular organization when compared with the network of free-living cyanobacteria [21]. In a similar way, the metabolism of the chromatophore (*i*CV265) seems to be tailored for the exportation of reduced-carbon; that is, when comparing the export of reduced-carbon between the *i*CV265 and the *i*CV498 models (with phosphate as restrictive nutrient) we found that *i*CV265 shows higher rates of hexose exported than the free-living *i*CV498 model at the cost of increased consumption of phosphate (Fig. 9).

The evolutionary mechanism outlined above applies when the host benefits from the endosymbiont. In particular, mechanisms such as "partner fidelity feedback" (PFF) promote cooperation between symbionts. PFF requires individuals to be "associated for an extended series of exchanges that last long enough that a feedback operates" [18]. Similar mechanisms likely operated in other symbiotic systems. For example, *Buchnera* [61] and *Blochmannia* [62] overproduce essential amino acids (EAAs) to its host. This overproduction of EAAs was consequence of metabolic restructurings due to metabolic-gene losses. For example, the truncation of the purine biosynthesis pathway which allows the endosymbiont to produce histidine at higher rates than free-living relatives [20]. Reductive evolutionary simulations carried out by [20] showed that this truncation is an improbable evolutionary event under conditions tested.

Conclusion

Our main objective was to better understand the metabolic changes experienced by the free-living cyanobacteria to become a chromatophore. In addition, we assessed the evolutionary forces driving organellogenesis. We found evidence that certain metabolic pathways are preferentially conserved in the chromatophore. We also found that the pattern of metabolic gene loss strongly depends on the availability of nutrients from its host. The high fragility of the chromatophore network reflects the transition to a more stable environment and, consequently, its simplification. The chromatophore is specialized in producing reduced-carbon which could be released to the host. This specialization was consequence of metabolic restructurings which could not be possible in free-living conditions. We interpret this specialization as consequence of natural selection acting as a metabolic engineer which modifies intrinsic metabolic properties of the endosymbiont impacting positively at the holobiont level. Our in silico simulations allowed us to determine that metabolic specialization of the chromatophore is an unlikely result of purifying host-level selection and genetic drift alone. In this way, computational analysis of biological systems allows to obtain new insights on the evolutionary forces shaping metabolic evolution of mutualistic endosymbionts.

Methods

Differential gene retention of functional categories in the chromatophore genome

To identify metabolic pathways preferentially conserved in the chromatophore we carried out a statistical analysis using the program GeneMerge [63]. First, we classified each of the genes in both genomes (the chromatophore of *Paulinella chromatophora* CCAC 0185 [5] and *Synechococcus sp.* WH 5701) according to the functional categories of KEGG orthology (http://www.genome.jp/kegg/ko.html). Then we carried out the statistical analysis with GeneMerge. GeneMerge is a program written in Perl which allows the identification of overrepresented functions or categories in a sample by using a hypergeometric distribution [63].

Metabolic reconstruction of the *i*CV498 and the *i*CV265 models

A draft metabolic model was initially reconstructed by identifying orthologous genes between *Synechococcus sp.* WH 5701 and the metabolic model of *Synechocystis sp.* PCC 6803 (*i*JN678) [26]. Because this draft metabolic network had many inconsistencies we performed a manual refinement. This consisted in reviewing literature and databases to fill gaps in the model. We followed recommendations of [64].

The metabolic network of the endosymbiont was reconstructed by identifying orthologs between the chromatophore and *Synechococcus sp.* WH 5701. *Synechococcus sp.* WH 5701 is the closest free-living relative of the chromatophore with a sequenced genome [5].

The metabolic capabilities of both organisms were tested with Flux Balance Analysis [65]. FBA is an optimization algorithm based on lineal programming provided in the Matlab COBRA toolbox [66]. FBA determines the flux distribution of all reactions in the model by maximizing an objective function [30].

The functionality of metabolic models is evaluated by their capacity to produce every metabolite that is necessary for in silico growth. For this, the biomass equation of *Synechocystis sp.* PCC 6803 was assigned as objective function in both models. In silico growth was simulated under autotrophic conditions with CO_2 and

photons uptake set to 3.7 mmol $\times$ gDW^{-1} $\times$ h^{-1} and 100 mmol $\times$ gDW^{-1} $\times$ h^{-1}, respectively. These were restrictive metabolites in the systems. Nutrient assignment for metabolic functionality of the chromatophore was based on the literature [5] and metabolite requirements predicted by the model for in silico biomass production.

Network robustness analysis

In both models, robustness to gene deletions was analyzed by using the function singleGeneDeletion of the COBRA toolbox. If deletion of a single gene decreases the biomass production over 99%, compared with wild type, this gene was consider as essential for biomass production.

Simulation of metabolic reductive evolution in the chromatophore

To simulate genome reduction, we used the metabolic model of *Synechococcus sp.* WH 5701 (*i*CV498) as a proxy of the free-living ancestor of the chromatophore (Fig. 2). Genetic loss was simulated under Proposed Nutrients and Randomized Nutrients intracellular conditions. All nutrients were available simultaneously since the beginning of the simulations. The algorithm starts by randomly deleting a gene from the *i*CV498 model (i.e., setting its flux to zero) and then evaluating the impact of this deletion in the metabolic functionality by using FBA. If in silico growth rate in this network (lacking a gene) was equal to or above the growth rate of a free-living cyanobacteria ($\geq$ 0.0884 h^{-1}), then this gene was considered as non-essential and permanently removed. In contrast, if the growth rate was below 0.0884 h^{-1} then this gene was considered as essential and retained in the model. This process was repeated until each of the genes in the model was evaluated. The whole process is initiated 500 times which results in a population of 500 reduced metabolic networks.

Genetic concordance between the 500 reduced metabolic networks and chromatophore model (*i*CV265) was analyzed as in [30]. In each of the 500 simulations, a binary variable was assigned for each gene in *i*CV498 depending on whether the gene is predicted to be conserved or not among the 500 simulations. This allowed us to determine the number of occurrences that a gene is predicted as essential in the 500 simulated reduced networks.

Measures of sensitivity and specificity were obtained calculating the fraction of true-positives (fraction of genes predicted to be conserved by the simulations and present in *i*CV265) and false-positives (fraction of genes predicted by the simulations and not present in *i*CV265) for every cutoff (minimal fraction of simulated genomes in which a gene must be present to be predicted as conserved in iCV265). Figure 5 plots true-positive and false-positive (1-specificity) predictions for every cutoff (1 to 500) to form a ROC curve. The area under the curve represents how well the simulations recover gene content in iCV265. The area under the curve was empirically calculated as in [67].

Simulation of metabolic integration of the chromatophore with its host

We performed this analysis by using the same algorithm used in the simulation of reductive evolution. However, this analysis was performed only in Proposed Nutrient conditions (Fig. 2). In addition, a reaction simulating hexose export from the chromatophore to the host was defined as objective function and the growth rate equation (biomass equation) was fixed to 0.0884 h^{-1}. Also, in this simulation, a non-essential gene was defined as one whose deletion does not affect the growth rate (0.0884 h^{-1}) and the hexose export. Specifically, the rate of hexose export could vary while being always greater than zero. Otherwise, the gene was defined as essential.

In this analysis the model does not have access to all 13 nutrients at the same time from the beginning of the simulation. Instead, we randomly allow the model to have access to one of the 13 Proposed Nutrients (Fig. 2) and subsequently applied our algorithm of reductive evolution. Once we evaluated the impact of singly deleting each one of the genes, we randomly allowed the model to have access to a second nutrient and newly applied our algorithm of reductive evolution. The analysis stops when iCV498 has access to all 13 nutrients and all genes have been tested for essentiality.

Additional files

Additional file 1: Metabolic model of *Synechococcus sp.* WH 5701 (*i*CV498). (XLSX 187 kb)

Additional file 2: Metabolic model of the chromatophore (*i*CV265). (XLSX 121 kb)

Additional file 3: Figure S1 and S2 and Tables S1. (DOCX 149 kb)

Acknowledgements
Not applicable.

Funding
CV thanks to CONACYT for the doctoral fellowship granted.

Authors' contributions
CV performed the model construction, evolutionary simulations and wrote the manuscript. RO and OR provided support in mathematical modelling of metabolic reconstructions. AL and LD contributed to interpretation and discussion of the data. CV and LD designed the study. All authors approved the final version of the manuscript.

Competing interests
The authors declare that they have no competing interests.

Author details
[1]Departamento de Ingeniería Genética, Centro de Investigación y de Estudios Avanzados del Instituto Politécnico Nacional, Unidad Irapuato, Km. 9.6 Libramiento Norte Carr. Irapuato-León, 36821 Guanajuato, Irapuato, Mexico. [2]Departamento de Procesos y Tecnología, Universidad Autónoma Metropolitana-Cuajimalpa, Av. Vasco de Quiroga 4871, Santa Fe, Del. Cuajimalpa, C.P. 05348 Ciudad de Mexico, México, Mexico. [3]Human Systems Biology Laboratory, Coordinación de la Investigación Científica-Red de Apoyo a la Investigación (RAI), UNAM, México City, Mexico. [4]Instituto Nacional de Medicina Genómica (INMEGEN), 14610 México City, Mexico. [5]Unidad de Genómica Avanzada (Langebio), Centro de Investigación y de Estudios Avanzados del IPN, Guanajuato, Irapuato, Mexico.

References
1. Nakayama T, Archibald JM. Evolving a photosynthetic organelle. BMC Biol. 2012;10:35–7.
2. Nowack ECM. Paulinella chromatophora - Rethinking the transition from endosymbiont to organelle. Acta Soc Bot Pol. 2014;83:387–97.
3. Yoon HS, Nakayama T, Reyes-Prieto A, Andersen RA, Boo SM, Ishida K-I, et al. A single origin of the photosynthetic organelle in different Paulinella lineages. BMC Evol Biol. 2009;9:98.
4. Marin B, Nowack ECM, Melkonian M. A plastid in the making: Evidence for a second primary endosymbiosis. Protist. 2005;156:425–32.
5. Nowack ECM, Melkonian M, Glöckner G. Chromatophore Genome Sequence of Paulinella Sheds Light on Acquisition of Photosynthesis by Eukaryotes. Curr Biol. 2008;18:410–8.
6. Delaye L, Valadez-Cano C, Pérez-Zamorano B. How Really Ancient Is Paulinella Chromatophora?. PLoS Curr Tree Life. 2016:1–12. Edition 1. doi:10.1371/currents.tol.e68a099364bb1a1e129a17b4e06b0c6b.
7. Reyes-Prieto A, Yoon HS, Moustafa A, Yang EC, Andersen RA, Boo SM, et al. Differential gene retention in plastids of common recent origin. Mol Biol Evol. 2010;27:1530–7.
8. Nowack ECM, Vogel H, Groth M, Grossman AR, Melkonian M, Glöckner G. Endosymbiotic gene transfer and transcriptional regulation of transferred genes in Paulinella chromatophora. Mol Biol Evol. 2011;28:407–22.
9. Nowack ECM, Grossman AR. Trafficking of protein into the recently established photosynthetic organelles of Paulinella chromatophora. Proc Natl Acad Sci U S A [Internet]. 2012;109:5340–5. Available from: http://www.pnas.org/content/109/14/5340.
10. Kies L, Kremer BP. Function of cyanelles in the thecamoeba Paulinella chromatophora. Naturwissenschaften. 1979;66:578–9.
11. Moya A, Peretó J, Gil R, Latorre A. Learning how to live together: genomic insights into prokaryote-animal symbioses. Nat Rev Genet [Internet]. 2008;9:218–29. Available from: http://www.ncbi.nlm.nih.gov/pubmed/18268509.
12. Shigenobu S, Watanabe H, Hattori M, Sakaki Y, Ishikawa H. Genome sequence of the endocellular bacterial symbiont of aphids Buchnera sp. APS. Nature. 2000;407:81–6.
13. Lee MC, Marx CJ. Repeated, selection-driven genome reduction of accessory genes in experimental populations. PLoS Genet. 2012;8:2–9.
14. McCutcheon JP, Moran N. a. Extreme genome reduction in symbiotic bacteria. Nat. Rev. Microbiol. [Internet]. Nat Publ Group. 2011;10:13–26. Available from: http://dx.doi.org/10.1038/nrmicro2670.
15. Marais GAB, Calteau A, Tenaillon O. Mutation rate and genome reduction in endosymbiotic and free-living bacteria. Genetica. 2008;134:205–10.
16. Kuo C, Moran N. a, Ochman H. The consequences of genetic drift for bacterial genome complexity The consequences of genetic drift for bacterial genome complexity. Genome Res. 2009:1450–4.
17. Moran NA. Accelerated evolution and Muller's ratchet in endosymbiotic bacteria. Proc Natl Acad Sci U S A. 1996;93:2873–8.
18. Shou W. Acknowledging selection at sub- organismal levels resolves controversy on pro-cooperation mechanisms. elife. 2015;4:1–19.

19. Bennett GM, Moran NA. Heritable symbiosis: The advantages and perils of an evolutionary rabbit hole. Proc Natl Acad Sci [Internet]. 2015;112:10169–76. Available from: http://www.ncbi.nlm.nih.gov/pubmed/25713367.

20. Thomas GH, Zucker J, Macdonald SJ, Sorokin A, Goryanin I, Douglas AE. A fragile metabolic network adapted for cooperation in the symbiotic bacterium Buchnera aphidicola. BMC Syst Biol. 2009;3:24.

21. Wang Z, Zhu X-G, Chen Y, Li Y, Hou J, Li Y, et al. Exploring photosynthesis evolution by comparative analysis of metabolic networks between chloroplasts and photosynthetic bacteria. BMC Genomics [Internet]. 2006;7: 100. Available from: http://www.biomedcentral.com/1471-2164/7/100.

22. Stephanopoulos G. Metabolic Fluxes and Metabolic Engineering. Metab Eng. 1999;1(1):1–11.

23. Lan EI, Liao JC. Metabolic engineering of cyanobacteria for 1-butanol production from carbon dioxide. Metab Eng [Internet]. 2011;13:353–63. [cited 2017 13]. Available from: http://linkinghub.elsevier.com/retrieve/pii/S1096717611000474.

24. Lee JW, Kim TY, Jang YS, Choi S, Lee SY. Systems metabolic engineering for chemicals and materials [Internet]. Trends Biotechnol Elsevier; 2011. p. 370–378. [cited 2017 Mar 13] Available from: http://www.ncbi.nlm.nih.gov/pubmed/21561673.

25. Lee SY, Mattanovich D, Villaverde A, et al. Microb Cell Fact [Internet]. 2012; 11:156. Available from: http://www.microbialcellfactories.com/content/11/1/156.

26. Nogales J, Gudmundsson S, Knight EM, Palsson BO, Thiele I. Detailing the optimality of photosynthesis in cyanobacteria through systems biology analysis. Proc. Natl. Acad. Sci. U. S. A. [Internet]. 2012;109:2678–83. Available from: http://www.pnas.org/content/109/7/2678.abstract.

27. Lürling M, Eshetu F, Faassen EJ, Kosten S, Huszar VLM. Comparison of cyanobacterial and green algal growth rates at different temperatures. Freshw Biol. 2013;58:552–9.

28. González-Domenech CM, Belda E, Patiño-Navarrete R, Moya A, Peretó J, Latorre A. Metabolic stasis in an ancient symbiosis: genome-scale metabolic networks from two Blattabacterium cuenoti strains, primary endosymbionts of cockroaches. BMC Microbiol [Internet] BioMed Central Ltd.; 2012;12(Suppl 1):S5. Available from: http://www.pubmedcentral.nih.gov/articlerender.fcgi?artid=3287516&tool=pmcentrez&rendertype=abstract.

29. Belda E, Silva FJ, Peretó J, Moya A. Metabolic networks of Sodalis glossinidius: A systems biology approach to reductive evolution. PLoS One. 2012;7(1):e30652. doi:10.1371/journal.pone.0030652.

30. Pál C, Papp B, Lercher MJ, Csermely P, Oliver SG, Hurst LD. Chance and necessity in the evolution of minimal metabolic networks. Nature. 2006;440: 667–70.

31. Karkar S, Facchinelli F, Price DC, Weber APM, Bhattacharya D. Metabolic connectivity as a driver of host and endosymbiont integration. Proc Natl Acad Sci U S A [Internet]. 2015;201421375 Available from: https://www.ncbi.nlm.nih.gov/pmc/articles/PMC4547263/.

32. Facchinelli F, Colleoni C, Ball SG, Weber APM. Chlamydia, cyanobiont, or host: Who was on top in the ménage à trois? Trends Plant Sci. 2013;18:673–9.

33. Villadsen J, Nielsen J, Lidén G. Bioreaction Engineering Principles. New York: Kluwer Academic/Plenum Publishers; 2003. Available from: http://www.springerlink.com/index/10.1007/978-1-4419-9688-6.

34. Tyra HM, Linka M, Weber APM, Bhattacharya D. Host origin of plastid solute transporters in the first photosynthetic eukaryotes. Genome Biol [Internet]. 2007;8:R212. Available from: http://genomebiology.com/2007/8/10/R212.

35. Canbäck B, Tamas I, Andersson SGE. A phylogenomic study of endosymbiotic bacteria. Mol Biol Evol. 2004;21:1110–22.

36. Allen JM, Light JE, Perotti MA, Braig HR, Reed DL. Mutational meltdown in primary endosymbionts: Selection limits Muller's Ratchet. PLoS One. 2009; 4(3):e4969. doi:10.1371/journal.pone.0004969.

37. Lynch M, Gabriel W. Mutation Load and the Survival of Small Populations. Evolution (N Y). 1990;44:1725–37.

38. Andersson SG. e, Kurland CG. Reductive evolution of resident genomes. Trends Microbiol. 1998;6:263–8.

39. Bergstrom CT, Pritchard J. Germline bottlenecks and the evolutionary maintenance of mitochondrial genomes. Genetics. 1998;149:2135–46.

40. Foster J, Ganatra M, Kamal I, Ware J, Makarova K, Ivanova N, et al. The Wolbachia genome of Brugia malayi: Endosymbiont evolution within a human pathogenic nematode. PLoS Biol. 2005;3:0599–614.

41. Wu M, Sun LV, Vamathevan J, Riegler M, Deboy R, Brownlie JC, et al. Phylogenomics of the reproductive parasite Wolbachia pipientis wMel: A streamlined genome overrun by mobile genetic elements. PLoS Biol. 2004;2:327–41.

42. Moriyama M, Nikoh N, Hosokawa T, Fukatsu T. Riboflavin Provisioning Underlies Wolbachia's Fitness Contribution to Its Insect Host. MBio. 2015;6:1–8.

43. Wilson ACC, Ashton PD, Calevro F, Charles H, Colella S, Febvay G, et al. Genomic insight into the amino acid relations of the pea aphid, Acyrthosiphon pisum, with its symbiotic bacterium Buchnera aphidicola. Insect Mol Biol. 2010;19:249–58.

44. Husnik F, Nikoh N, Koga R, Ross L, Duncan RP, Fujie M, et al. Horizontal gene transfer from diverse bacteria to an insect genome enables a tripartite nested mealybug symbiosis. Cell [Internet]. Elsevier. 2013;153:1567–78. Available from: http://dx.doi.org/10.1016/j.cell.2013.05.040.

45. Sloan DB, Nakabachi A, Richards S, Qu J, Murali SC, Gibbs RA, et al. Parallel histories of horizontal gene transfer facilitated extreme reduction of endosymbiont genomes in sap-feeding insects. Mol Biol Evol. 2014;31:857–71.

46. Hansen AK, Moran NA. The impact of microbial symbionts on host plant utilization by herbivorous insects. Mol Ecol. 2014;23:1473–96.

47. Oberhardt MA, Puchałka J, Fryer KE, VAP MDS, Papin JA. Genome-scale metabolic network analysis of the opportunistic pathogen Pseudomonas aeruginosa PAO1. J Bacteriol. 2008;190:2790–803.

48. Ates O, Oner ET, Arga KY. Genome-scale reconstruction of metabolic network for a halophilic extremophile, Chromohalobacter salexigens DSM 3043. BMC Syst Biol [Internet] BioMed Central Ltd. 2011;5:12. Available from: http://www.biomedcentral.com/1752-0509/5/12.

49. Puchałka J, Oberhardt MA, Godinho M, Bielecka A, Regenhardt D, Timmis KN, Papin JA, Martins dos Santos VA. Genome-scale reconstruction and analysis of the Pseudomonas putida KT2440 metabolic network facilitates applications in biotechnology. PLoS Comput Biol. 2008;4(10):e1000210. doi: 10.1371/journal.pcbi.1000210.

50. Wagner A. Robustness and evolvability: A paradox resolved. Proc R Soc Biol Sci. 2008;275:91–100.

51. Gabaldón T, Peretó J, Montero F, Gil R, Latorre A, Moya A. Structural analyses of a hypothetical minimal metabolism. Philos Trans R Soc Lond B Biol Sci [Internet]. 2007;362:1751–62. Available from: http://www.pubmedcentral.nih.gov/articlerender.fcgi?artid=2442391&tool=pmcentrez&rendertype=abstract.

52. Ding T, Case KA, Omolo MA, Reiland HA, Metz ZP, Diao X, et al. Predicting Essential Metabolic Genome Content of Niche-Specific Enterobacterial Human Pathogens during Simulation of Host Environments. PLoS One [Internet]. 2016;11:e0149423. Available from: http://dx.plos.org/10.1371/journal.pone.0149423.

53. Mendonça AG, Alves RJ, Pereira-Leal JB. Loss of genetic redundancy in reductive genome evolution. PLoS Comput Biol. 2011;7(2):e1001082. doi: 10.1371/journal.pcbi.1001082.

54. Khannapho C, Zhao H, Bonde BK, Kierzek AM, Avignone-Rossa CA, Bushell ME. Selection of objective function in genome scale flux balance analysis for process feed development in antibiotic production. Metab Eng. 2008;10:227–33.

55. Patil KR, Rocha I, Forster J, Nielsen J. Evolutionary programming as a platform for in silico metabolic engineering. BMC Bioinformatics [Internet]. 2005;6:308. Available from: http://eutils.ncbi.nlm.nih.gov/entrez/eutils/elink.fcgi?dbfrom=pubmed&id=16375763&retmode=ref&cmd=prlinks.

56. Johnson MD. The acquisition of phototrophy: Adaptive strategies of hosting endosymbionts and organelles. Photosynth Res. 2011;107:117–32.

57. Giersch C, Robinson SP. Regulation of photosynthetic carbon metabolism during phosphate limitation of photosynthesis in isolate spinach chloroplast. Photosynth Res. 1987;14:211–27.

58. Heldt HW, Chon CJ, Maronde D. Role of orthophosphate and other factors in the regulation of starch formation in leaves and isolated chloroplasts. Plant Physiol. 1977;59:1146–55.

59. Pickens LB, Tang Y, Chooi Y-H. Metabolic Engineering for the Production of Natural Products. Annu Rev Chem Biomol Eng. 2011;2:211.

60. Burgard AP, Pharkya P, Maranas CD. OptKnock: A Bilevel Programming Framework for Identifying Gene Knockout Strategies for Microbial Strain Optimization. Biotechnol Bioeng. 2003;84:647–57.

61. Russell CW, Poliakov A, Haribal M, Jander G, van Wijk KJ, Douglas AE. Matching the supply of bacterial nutrients to the nutritional demand of the animal host. Proc R Soc B Biol Sci [Internet]. 2014;281:20141163 Available from: http://rspb.

royalsocietypublishing.org/content/281/1791/20141163.short?rss=1

62. Zientz E, Dandekar T, Gross R. Metabolic Interdependence of Obligate Intracellular Bacteria and Their Insect Hosts. Microbiol Mol Biol Rev. 2004; 68:745–70.

63. Castillo-Davis CI, Hartl DL. GeneMerge - Post-genomic analysis, data mining, and hypothesis testing. Bioinformatics. 2003;19:891–2.

64. Thiele I, Palsson B. A protocol for generating a high-quality genome-scale metabolic reconstruction. Nat Protoc. 2010;5:93–121.

65. Orth JD, Thiele I, Palsson BO. What is flux balance analysis? Nat Biotech [Internet]. Nat Publ Group. 2010;28:245–8. Available from: http://dx.doi.org/10.1038/nbt.1614%5Cnhttp://www.nature.com/nbt/journal/v28/n3/abs/nbt.1614.html.

66. Schellenberger J, Thiele I, Orth JD. Quantitative prediction of cellular metabolism with constraint- based models: the COBRA Toolbox v2.0. Nat Protoc. 2012;6:1290–307.

67. Hanley JA, BJ MN. The Meaning and Use of the Area under a Receiver Operating (ROC) Curvel Characteristic. Radiology [Internet]. 1982;143:29–36. Available from: http://www.ncbi.nlm.nih.gov/pubmed/7063747.

8

168 million years old "marine lice" and the evolution of parasitism within isopods

Christina Nagler[1][*] (ID), Matúš Hyžný[2,3] and Joachim T. Haug[1,4]

Abstract

Background: Isopods (woodlice, slaters and their relatives) are common crustaceans and abundant in numerous habitats. They employ a variety of lifestyles including free-living scavengers and predators but also obligate parasites. This modern-day variability of lifestyles is not reflected in isopod fossils so far, mostly as the life habits of many fossil isopods are still unclear. A rather common group of fossil isopods is *Urda* (190-100 million years). Although some of the specimens of different species of *Urda* are considered well preserved, crucial characters for the interpretation of their lifestyle (and also of their phylogenetic position), have so far not been accessible.

Results: Using up-to-date imaging methods, we here present morphological details of the mouthparts and the thoracopods of 168 million years old specimens of *Urda rostrata*. Mouthparts are of a sucking-piercing-type morphology, similar to the mouthparts of representatives of ectoparasitic isopods in groups such as Aegidae or Cymothoidae. The thoracopods bear strong, curved dactyli most likely for attaching to a host. Therefore, mouthpart and thoracopod morphology indicate a parasitic lifestyle of *Urda rostrata*. Based on morphological details, *Urda* seems deeply nested within the parasitic isopods of the group Cymothoida.

Conclusions: Similarities to Aegidae and Cymothoidae are interpreted as ancestral characters; *Urda* is more closely related to Gnathiidae, which is therefore also interpreted as an ingroup of Cymothoida. With this position *Urda* provides crucial information for our understanding of the evolution of parasitism within isopods. Finally, the specimens reported herein represent the oldest parasitic isopods known to date.

Keywords: Cymothoida, Isopoda, *Urda*, Fossil life habits, Evolution, Fossil parasitism

Background

Parasitism is a widespread strategy among animals (Metazoa), if not the most widespread one. Most, if not all parasites originated from free-living relatives. Still our understanding of how the evolution of a parasitic lifestyle evolved is not fully understood. It has been suggested that there are morphological, physiological, or ecological pre-adaptations to parasitism [1–4].

For improving our understanding of the evolution of parasitism, insects have been considered to be an especially interesting group. It seems that in various insect lineages clear pre-adaptations, such as elongated mouthparts, can be identified [5]. One model example for studying evolution of parasitism and co-evolution

---
* Correspondence: christina.nagler@palaeo-evo-devo.info
[1]Functional morphology group, Department of Biology II, Ludwig-Maximilians-University, Großhaderner Strasse 2, 82152 Planegg-Martinsried, Germany
Full list of author information is available at the end of the article

between the parasite and host are lice, possibly due to human medical health and livestock health interest. Chewing lice ('Mallophaga'), specialized for a parasitic lifestyle on birds [6], have been proposed to have evolved from a free-living relative [7].

Comparable to lice, an evolutionary origin from free-living relatives, has been reconstructed for other parasitic groups, for example several worms, such as parasitic nematode worms [2, 8], parasitic flatworms [9], acanthocephalan worms [10], but also other groups closer related to mallophagan lice, such as mites [11], or parasitic isopod crustaceans [12].

Isopod crustaceans – woodlice, slaters, pill bugs and their relatives – are very diverse and successful malacostracan crustaceans (the group containing e.g. crabs, lobsters, shrimps, krill and crayfish). Isopods inhabit various habitats, including marine, freshwater and terrestrial environments [12–18]. They have developed various kinds of lifestyles, among them free-living [19], scavenging [20–22] or predatory [23], but also parasitic forms of varying degrees of specialization [24–27]. This is nicely exemplified by the isopod ingroup Cymothoida sensu Wägele [12]. Within this group numerous lifestyles have evolved, some quite soon after the appearance of the group [28]. Also, as isopods have potential to be preserved as fossils this group allows a degree of estimation of the appearance of such strategies within Earth history:

- A scavenging lifestyle is known from representatives of Cirolanidae. Fossil representatives of this group, indirectly suggesting a similar lifestyle, have been reported from the Jurassic [21] and Cretaceous [22]. Representatives of Corallanidae and Aegidae have a lifestyle reminiscent of that of a mosquito; one may interpret this as quasi-predatory behavior, yet more precisely it is a temporary parasitic lifestyle; they attach briefly to a host, a fish, only during feeding. An aegiid fossil [29] has been reported from the Late Miocene, indicating a similar lifestyle at this time. Phylogenetic inference would suggest an older origin of a "marine mosquito" strategy.
- Representatives of Cymothoidae feed similarly to aegiid isopods when they are juveniles. Yet, as adults they attach to a host fish permanently. The oldest fossil indicating such a type of parasitism in Cymothoidae has been reported from the Jurassic [30].
- During a specific larval phase, representatives of Gnathiidae feed in a comparable way to representatives of adult Aegidae and juvenile Cymothidae [20, 31]. Yet, as adults gnathiid isopods are not parasitic. An ingroup position of gnathiids within Cymothoida is equivocal ([32] vs. [33]). So far no fossils of this lineage have been reported.
- A host change respective to their ontogenetic phase can be observed also in representatives of Epicaridea. Larval epicaridids parasitize small crustaceans, e.g. copepods. Adult epicaridids infest mainly larger crustaceans, some are even quasi-endoparasitic. Based on malformations on the host [24, 27] or by comparing the life habits of modern relative groups [34], this lifestyle must have been present since the Jurassic.

These examples illustrate not only the diversity of life styles within Cymothoida. They also illustrate different ways of inferring a specific lifestyle in fossils [5]: 1) The most direct case is finding a parasite directly associated with a host [30]. 2) A more indirect way is finding isolated specimens with specific morphologies [34]. More indirect cases are (3) findings of developmental stages with a different lifestyle [35] and (4) teratological changes in the morphology of a host [24].

For 2) functional morphology and comparison to extant relatives can support interpretations of different lifestyles.

Hook-like claws at the end of thoracopods for attachment in an isopod give a clear hint to a parasitic lifestyle in contrast to small, straight and pointed tips that could be used for walking locomotion.

Similarly, also for phylogenetic interpretations of fossils morphological characters, such as details of appendages on the head and thorax are crucial [12, 28, 32]. Currently most fossil isopods are mainly interpreted based on dorsal characters, as ventral morphological characters of most fossil isopods are not accessible [29, 36–38].

Yet, under certain preservation conditions more or less complete fossil isopods can be recovered, preserving ventral details, such as appendages and even appendage sub-structures, such as spines and setae. Numerous such well-preserved fossil isopods have been reported from the Mesozoic, especially from Jurassic Konservat Lagerstätten with exceptional preservation [22, 36, 39–44].

One group of isopods that is regularly found in the Jurassic is *Urda*. This genus currently includes eight species (see Table 1). So far it has neither been possible to reliably interpret the phylogenetic position of *Urda* nor its lifestyle as descriptions concentrated on dorsal characters. Yet, some authors have suggested a closer relationship of *Urda* to parasitic isopod groups, such as Aegidae, Cymothoidae or Gnathiidae (see discussion).

Here we present two specimens of *Urda rostrata* from the Bathonian (168 mya) of Bethel-Bielefeld (Germany). Specimens were documented with the aid of micro CT and reveal crucial characters indicating that these fossils represent the oldest fossil parasitic isopod known to date. With this they contribute novel information to the evolution of parasitism within Cymothoida.

Methods
Material
We investigated two fossil isopod specimens, both interpreted as representatives of *Urda rostrata*. Both specimens are preserved in an ironstone-geode and come from Bethel-Bielefeld (Germany). They are therefore interpreted as being of Bathonian age, Middle Jurassic, about 168 million years old. Both specimens were found by K. Lenzer in July 1970 and first reported by Büchner [54];

Specimen 1 (BSPG 2011I50, Figs. 1a-b, d, 3a-c and 6a-e) is 32 mm long from the anterior end of the functional head to the posterior end of the telson and 8 mm wide. Specimen 2 (BPSG 2011I51, Figs. 1c and 3d-f) is 20 mm long from the anterior end of the cephalothorax to seventh thorax segment. Pleotelson is missing.

Additionally two extant specimens of parasitic isopods (Cymothoidae) were used for comparison. For comparisons

Table 1 Summary of *Urda* spp. occurrences in literature and their preservation

Current taxonomic assignment	Original taxonomic assignment	Age (stage and range in mya) after [92]	Country	Reference(s), original reference indicated by	Preservation group (1 = entire animal, 2 = anterior portion only, 3 = posterior portion only)
U. cretacea	*U. cretacea* [45]	Berriasian, 140-145	Germany	[45]	3 (tIII-pIV)
U. cretacea	*U. cretacea* [45]	Berriasian, 140-145	Germany	[45]	2 (ct-tIV)
U. cretacea	*U. cretacea* [45]	Berriasian, 140-145	Germany	[45, 46]	1
U. cf. cretacea	*U. cretacea* [47]	Aptian, 113-125	Antarctica	[47]	3 (pII-pt)
U. liasica	*U. liasica* [48]	Toarcian, 174-182	Germany	[48]	3 (tVI-pt)
U. mccoyi	*Palaega mccoyi* [49]	Oxfordian, 157-163	Scotland	[37]	1
U. moravica	*U. moravica* [50]	Bathonian, 166-168	Bohemia	[50, 46]	3 (tVI-pt)
U. punctata	*U. punctata* [51]	Tithonian, 145-152	Germany	[51, 45, 46, 52, 53]	1
U. rhodanica	*U. rhodanica* [46]	Callovian, 163-166	France	[46]	3 (tV-pt)
U. rostrata.	*Urda* sp.[54, 55]	Bathonian, 166-168	Germany	this study	1
U. rostrata.	*Urda* sp.[54, 55]	Bathonian, 166-168	Germany	this study	2 (ct-tIII)
U. rostrata	*U. rostrata* [51]	Tithonian, 145-152	Germany	[51, 45, 46, 52, 53, 56, 57]	2 (ct-tVIII)
U. rostrata	*U. "cincta"* [57]	Tithonian, 145-152	Germany	[52]	1
U. rostrata	*U. "cincta"* [51, 57]	Tithonian, 145-152	Germany	[52]	1
U. rostrata	*U. "elongata"* [51], *U. "cincta"* [57]	Tithonian, 145-152	Germany	[52]	1
U. rostrata	*U. "decorata"* [51], *U. "cincta"* [57]	Tithonian, 145-152	Germany	[52]	1
U. zelandica	*U. zelandica* [56]	Tithonian, 145-152	New Zealand	[56]	3 (tVI-pt)
Urda sp.	*Urda* sp. [48]	Pliensbachian, 183-191	Germany	[48]	2 (ct-tIII)
Urda sp.	*Urda* sp. [58]	Aalenian, 170-174	Switzerland	[58]	3 (tVI-pt)

of the mouthparts a female of *Nerocila acuminata* (ZSMA 20159001, Figs. 1f and 4d-e) was used. It originates from Cross Bay, Rovinj, Croatia (45°7.06′N 13°3.99′E), and is still attached to the caudal fin of a representative of Mugilidae, identified and found by R. Melzer in 2014. Preparation, documentation and methodological proceedings have been described in Nagler and Haug [59]. For comparison of the thoracopods a female of *Anilocra physodes* (ZSMA 04con034, Fig. 1h) was used. It was collected in the Atlantic (21°19.5′N, 17°13.1′W) by L. Tiefenbacher in 1975.

Documentation methods

Specimens were investigated with macro-photography and x-ray micro-CT scanning.

Macro-photography, combined with composite imaging (stacks of images of several adjacent image details) was performed following e.g. [60–62] under cross-polarized light. We used a Canon EOS Rebel T3i camera, either with a Canon EFS (18-55 mm) lens (for overview images) or a Canon MP-E (65 mm) macro lens (for close-up images). Illumination was provided by a Canon Macro Twin Lite MT-24EX flash from the two opposing sides to provide even illumination.

Fluorescence microscopy of the sixth thoracopods of *A. physodes* was performed on an inverse fluorescence microscope BZ-9000 (BIOREVO, Keyence) with a DAPI filter (λ = 358-461 nm) recording auto fluorescence and 10x objective resulting in about 100x magnification. Several focus layers (stacks of images) were recorded.

Stacks of images were processed with the freeware packages CombineZP (Alan Hadley), ImageAnalyzer (Meesoft) and ImageJ (Wayne Rasband). Assembling of stereo images and final processing (levels, sharpness, and saturation) was performed in Adobe Photoshop CS4.

Micro-CT scanning was performed on a Nanotom m Phoenix (GE Sensing & Inspection Technologies GmbH). An overview scan of specimen 1 ran 60 min with 140 kV and 60 μA, resulting in a calculated voxel size of 15.8 μm^3. For specimen 2, the scan took 53 min with 140 kV and 60 mA, resulting in a calculated voxel size of 16.6 μm^3. Scans were reconstructed to tiff stacks with the built-in software. Tiff stacks were further processed with ImageJ and Osirix 5.8.2 (Antoine Rosset). Surface models and volume renderings of both specimens, of thoracopods of specimen 1 and of mouthparts of specimens 1 and 2 were created ("segmented" or by thresholds) in Osirix. The surface models were further modified and rendered with

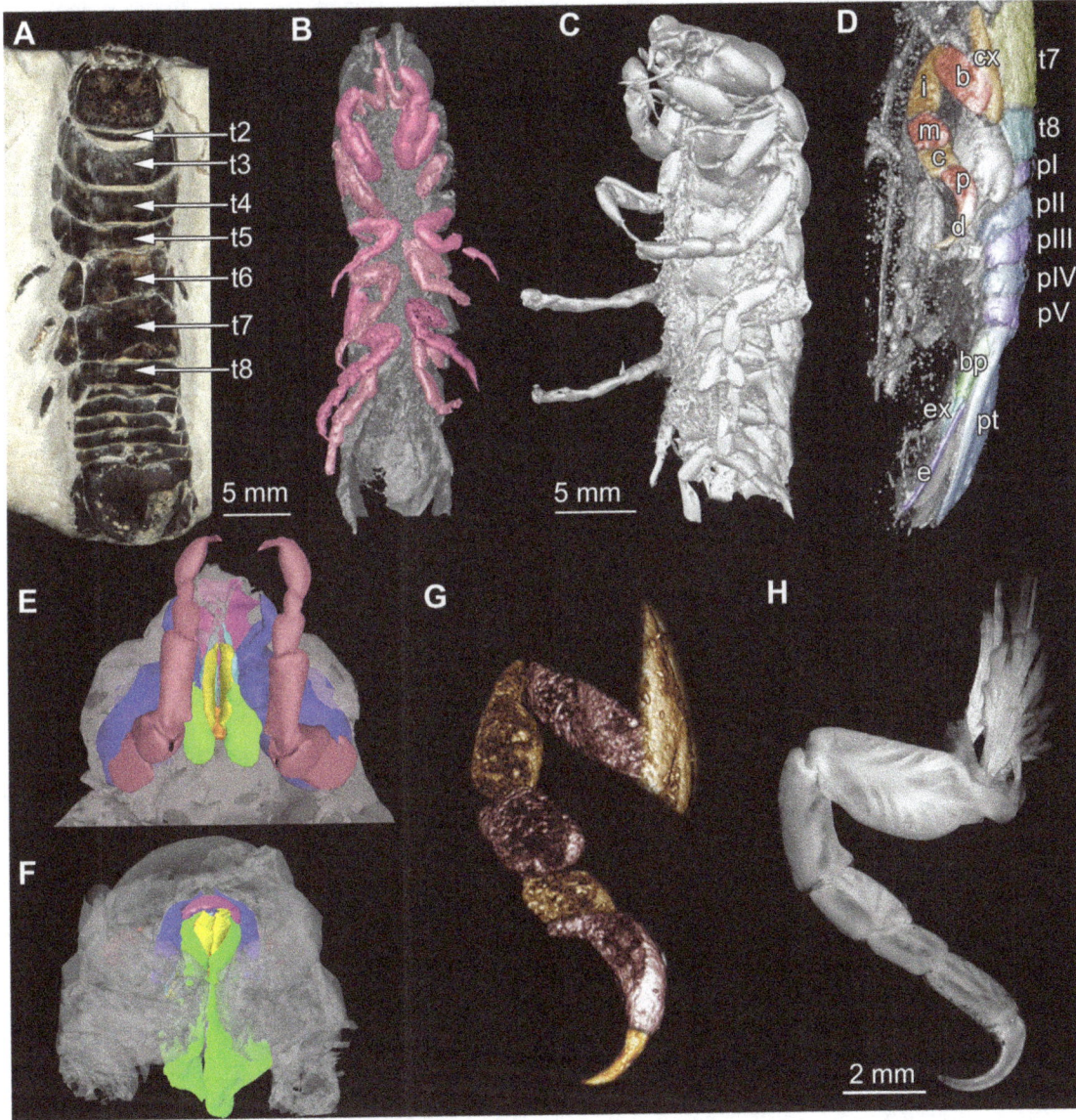

Fig. 1 Fossil specimens and modern counterparts of isopod crustaceans. *Urda rostrata:* specimen 1 (BPSG 2011I50; **a-b**, **d**, **g**); specimen 2 (BPSG 2011I51; **c**, **e**); *Nerocila bivitatta* (**f**); *Anilocra physodes* (**h**). Color-marks: labrum = purple, mandibles = blue, paragnaths = orange, maxillulae = cyan, maxillae = yellow, maxillipeds = green, thoracopods = pink, thoracopod elements = red and orange, pleon segments = blue and light blue. **a** Macro photograph with indicated free thorax segments (t2-t8), dorsal view. **b-g** Reconstructed surface models. **b** Specimen 1 with colored thoracopods, ventral view. **c** Specimen 2, ventro-lateral view. **d** Specimen 1 with colored elements of sixth free thoracopod and segments of thorax, pleon and pleotelson. **e** Specimen 1, ventral view. **f** Functional head. **g** Sixth thoracopod of specimen 1. **h** Fluorescence microscopic photography of sixth thoracopod

Blender 2.49 (Blender Foundation). Stereo volume renderings and surface models of specimen 1 and specimen 2 were created ("segmented" or by thresholds) in Osirix. Surface models were further modified and rendered with Blender 2.49 (Blender Foundation).

Presentation method

The description is focused on preserved structures that give information about the lifestyle of these isopods, i.e. functional morphology of the mouthparts and thoracopods, as these appendages are in direct contact with the

host. For a better recognition we present colour-marked images of the important appendages.

Terminology

Due to the necessity for a uniform terminology among arthropod workers [63], we choose expressions that allow an unambiguous connection of term and structure. Therefore, we use thoracopod II-VIII instead of 'pereiopod 1-7' (or 'peraeopod 1-7'). We also avoid terminology implying serial homology of structures that have independent evolutionary origins;

hence we use maxillula and maxilla (instead of maxilla one and two).

Results

Description of the two fossils (specimen 1 BSPG 2011I50; specimen 2 BSPG 2011I51) is focused on preserved structures that give information about the lifestyle of the two isopod crustaceans. Therefore especially the morphology of mouthparts and thoracopods is the focus of the description, as these appendages are in direct contact with the host.

Specimen 1 is more or less complete (Figs. 1a-b, d, g, 3a-c, 4a$_6$, 5a-b, d-e and 6a-e); specimen 2 is only preserved anteriorly including the first pleon segment; further posterior structures are missing (Figs. 1c, e, 2 and 3d-f, 4a$_{1-5, 7}$, b-c and 5c).

Body organization

Both specimens three-dimensionally preserved (Figs. 1b-c, 3a-f and 6a-e). In visible sclerotised body areas cuticle appears tuberculate. Entire body elongate slightly dorsoventrally flattened, with more or less constant width, but tapering anteriorly and posteriorly.

Body organized into functional head (cephalothorax; seven segments), posterior thorax (pereon; seven free segments) and pleon (five free pleon segments and pleotelson, conjoined structure of pleon segment six and telson).

Functional head consists of a eucrustacean head (ocular segment plus five appendage-bearing segments) and first original thorax segment (Fig. 1e). Head segments form dorsally a single capsulate shield. Head shield sub-rectangular in dorsal view (Fig. 5d). Head segments bear mouthparts ventrally. Each free thorax segment, dorsally forms a sclerotisation, tergite and bears a pair of appendages ventrally (Fig. 5). Tergite of first free thorax segment (thorax segment II) much smaller than those of following segments; not extending as far laterally as functional head or other tergites; very short in anterior-posterior dimensions. Tergite of second free thorax segment (thorax segment III) shows slight anterior indentation medially to match up with smaller anterior tergite. Five pleon segments, each of them roughly as long as one third of the most posterior free thorax segment. Pleotelson shape partly unclear, with a rounded posterior edge in dorsal view; slightly wider than pleon segments (Fig. 6).

Structures of the anterior body

Description of structures of functional head is largely based on morphology of specimen 2.

Large lateral compound eyes (Figs. 1a, c and 5a-e) situated antero-laterally on functional head; reniform outline. Appendages of post-ocular segments 1 and 2 (antennula and antenna) are not preserved.

Mouthparts (appendages of ocular segment and posterior head appendages) together forming truncated cone (Figs. 1e, 4c and 5a-e); labrum confines the cone from anterior, maxillipeds and thoracopods II seal cone from the posterior. Oral opening is Y-shaped (Fig. 5e).

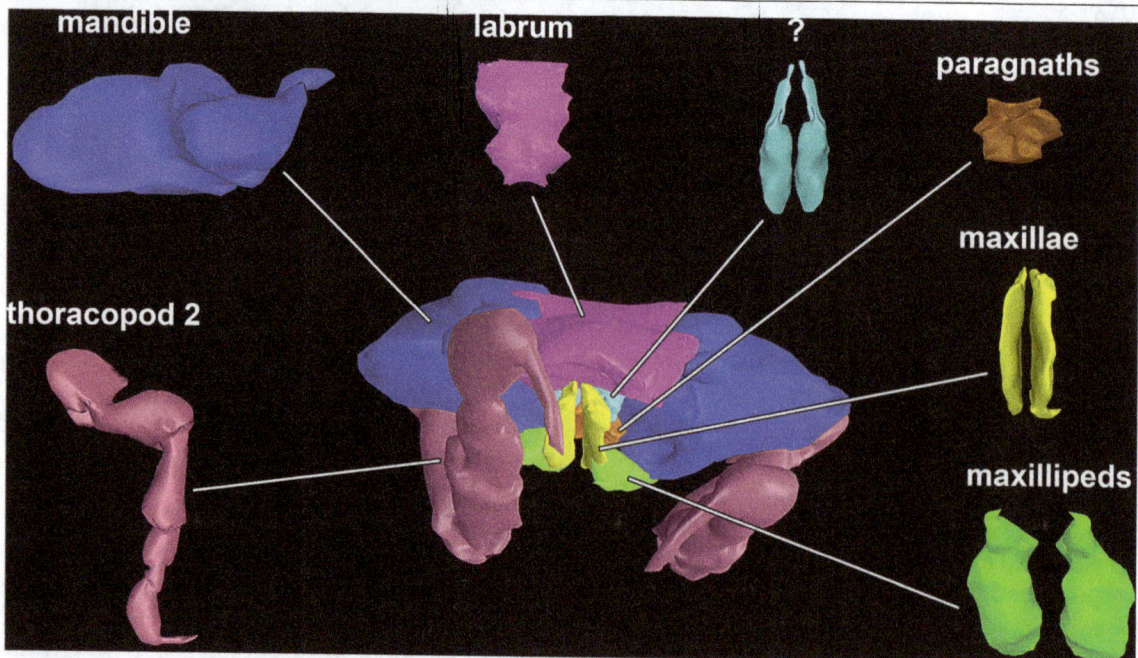

Fig. 2 Reconstructed surface model of the mouthparts of the fossil isopod *Urda rostrata* (BSPG 2011I51). Mouthparts together in anterior view, individual mouthparts in dorsal view. Mouthpart of unclear identity (?) may either represent the maxillula or the distal region of the paragnaths. Not to scale

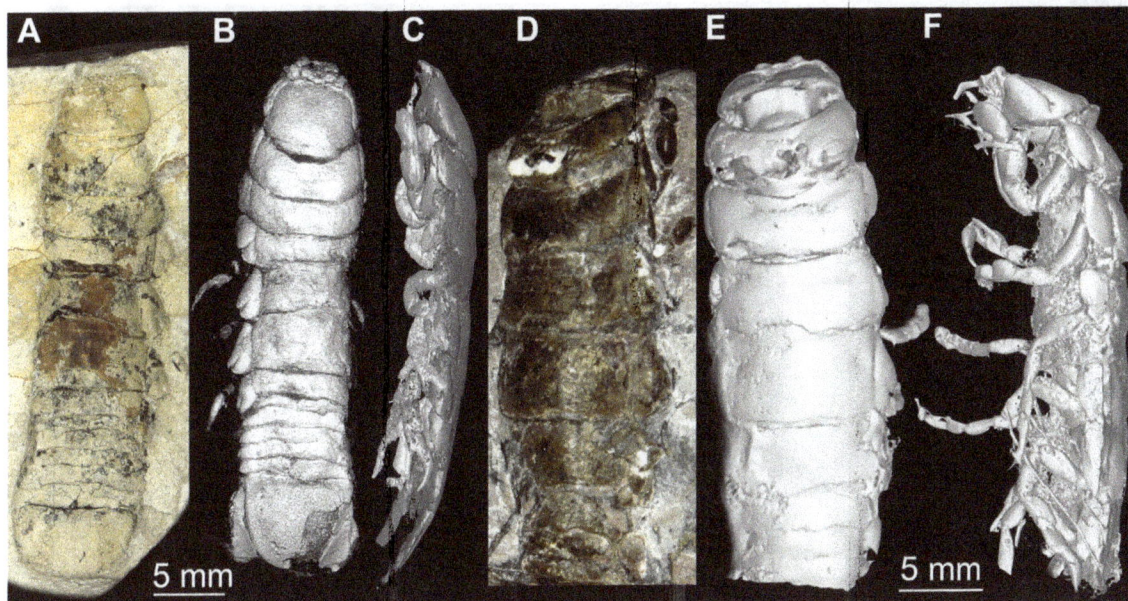

Fig. 3 Macro photographs (**a, d**), volume renderings (**b, e**) and reconstructed surface models (**c, f**) of fossil isopod *Urda rostrata*. **a-c)** *U. rostrata* specimen 1 (BPSG 2011I50). **a** Counterpart, dorsal view. **b** Dorsal view. **c** Lateral view. **d-f** *U. rostrata* specimen 2 (BPSG 2011I51). **d** Latero-dorsal view. **e** Dorsal view. **f** Lateral view

Labrum (upper lip, medially fused appendages of ocular segment) elongated; projecting forward from the functional head (Figs. 2, 4a, b_7, c and 5b-c, e); crescent-shaped in distal view (doming anteriorly; functionally dorsally), trapezoid in dorsal view; partly covering remaining mouthparts from anterior.

Appendages of post-ocular segment 3 (mandible) with prominent and stout proximal part (coxa; Figs 2, $4a_{1-6}$, b_5, c and 5b-c, e). Coxa rounded rectangular in anterior view. Disto-medially coxa is drawn out into process with massive sharp tip (incisor) at distal end. Distal part (mandibular palp) not preserved (unclear if original condition).

Paired protrusions of the mandibular sternum, paragnaths appear partly preserved (Figs. 2 and $4a_{1-5}$, b_6, c); medially conjoined. More or less square-shaped in ventral view, with pointed projections in the middle. Medially forming a funnel, due to a depression (Figs. 2 and $4b_6$).

Paired mouthpart of unclear identity slender and elongated with two possible elements or region (large tube-shaped proximal element and thinner flattened distal element); each ending in pointed tip (Figs. 2 and $4a_{1-4}$, b_4, c). Mouthpart is protruding further distally than any other mouthpart. Mouthpart may represent the distal part of the paragnaths or the appendages of post-ocular segment 4 (maxillula).

Appendages of post-ocular segment 5 (maxilla) elongated, chisel-like, (3.5 times as long as broad), flattened in anterior-posterior axis (Figs. 2, $4a_{1-3}$, b_3, c and 5b-d), distal end with three spines.

Appendages of post-ocular segment 6 (maxilliped, thoracopod I), broad and stout with two elements (Figs. 1e, 2 and $4a_{1-2}$, b_2, c); distal part bears prominent hook-like spine with a sharp, pointed, recurved tip.

Appendages of post-ocular segment 7 (thoracopod II) with seven elements along main axis. Element 1 (coxa), prominent, stout. Element 2 (basipod) longer, widening distally. Element 3 (ischium) as long as element 2, also widening distally. Element 4 (merus) shorter (50%), tube-shaped. Element 5 (carpus) shorter (50%). Element 6 (propodus) swollen, as long as two third of basipod. Element 7 (dactylus) recurved, hook-like (Figs. 2, $4a_1$, b_1, c_{1-2} and 5a-e).

Appendages on posterior thorax and pleon

Description of appendages of post-ocular segment 8-9 (thoracopods II-III) mainly based on morphology of specimen 2 (Figs. 1e, 2, $4a_1$, b_1, c_{1-2} and 5c), description of appendages of post-ocular segment 10-13 (thoracopods IV-VIII) and post-ocular segment 19 (uropods) based on morphology of specimen 1 (Figs. 1d and 6a-e).

Appendages of post-ocular segment 8-13 (thoracopods II-VIII) each consist of seven elements (similar to thoracopod I); all about twice the size of thoracopod I, largely similar in organization to thoracopod I. All elements 7 (dactyli) at least as long as elements 6 (propodus) (Figs. 1g and 6a-d); elongated, strongly curved, hook-like (Figs. 1g, $4b_1$, 5a-e and 6a-d).

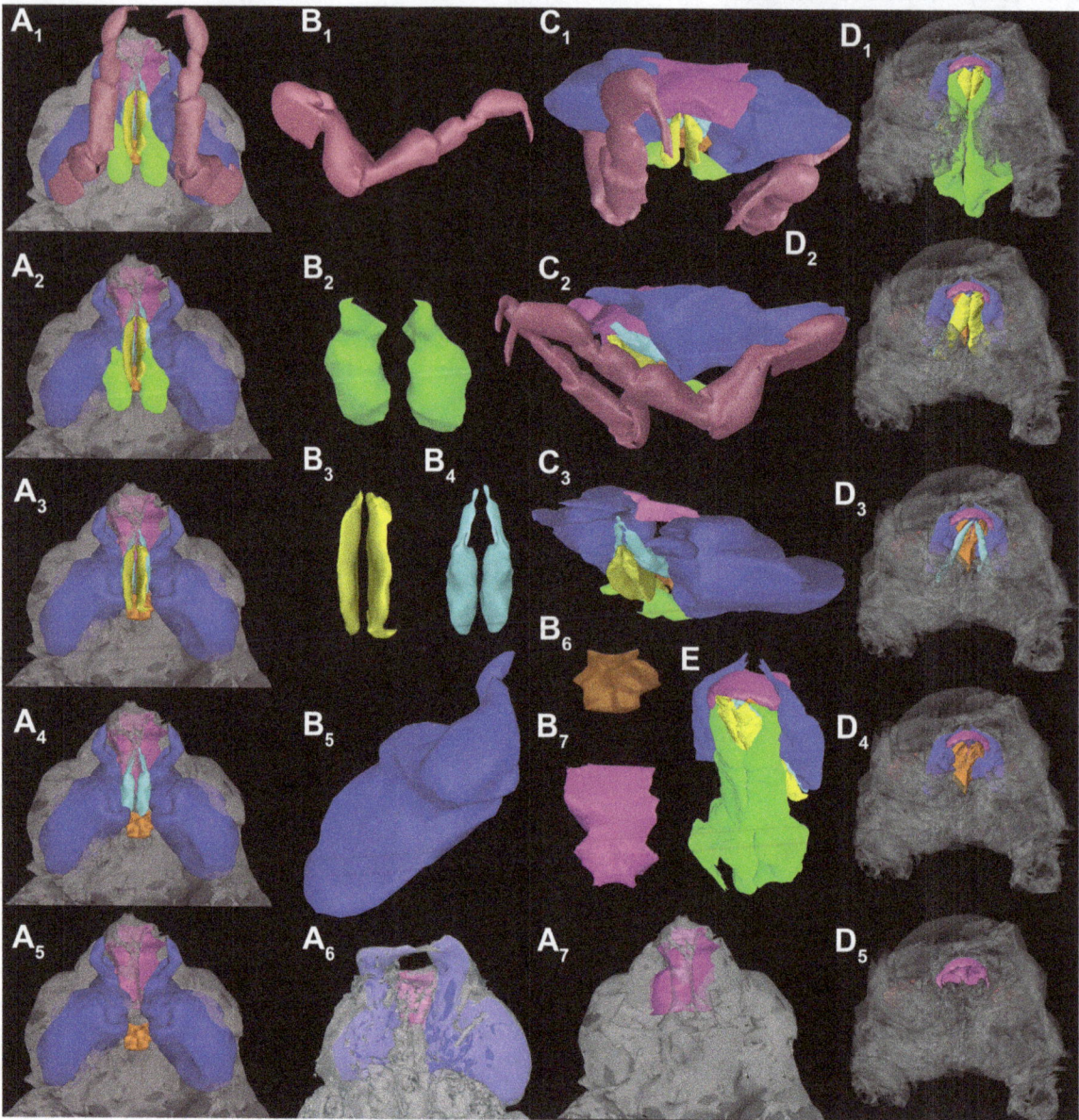

Fig. 4 Reconstructed surface models of the fossil isopod *Urda rostrata* and modern isopod *Nerocila acuminata*. Abbreviations: tp = first thoracopods, mxp = maxillipeds, mxa = maxillae, uc = unclear mouth part, md = mandibles, pg = paragnaths, lb = labrum. $a_{1-5, 7}$ Functional head of *U. rostrata* specimen 2 (BPSG 2011I51) with the first free thorax segment, successively one appendage removed from posterior to anterior, ventral view. a_6 Functional head of *U. rostrata* specimen 1 (BPSG 2011I50). b_{1-7} Mouthparts of *U. rostrata* specimen 2. b_1) Tp. b_2) Mxp. b_3) Mxa. b_4) Uc. b_5) Md. b_6) Pg. b_7) Lb. c_{1-3} Mouthparts of *U. rostrata* specimen 2, from different angles. d_{1-5}) Functional head of *N. acuminata*, successively one appendage removed from posterior to anterior, ventral view. Not to scale

Thoracopods II-IV rotated roughly 30° degrees forward, resulting in elements 4-6 (merus, carpus and propodus) being directed diagonally towards the anterior outer edge of body. Thoracopods V-VIII rotated roughly 30° degrees backward resulting in elements 4-6 (merus, carpus, propodus) being directed diagonally towards the posterior outer edge of body. Dactyli of thoracopods III-IV curved in a ventro-median direction (Figs. 1b, 5c and 6a-d), dactyli of thoracopods V-VIII more inclined backwards (Figs. 1b and 6a-d).

Appendages of post-ocular segment 14-18 (pleopods I-V) not preserved (Figs. 1b and 3a-f). Appendages of post-ocular segment 14-19 (uropods) with basipod carrying two distal rami, endopod and exopod; both similar sized. Uropod forming tail fan with pleotelson (Figs. 1d and 6c-e).

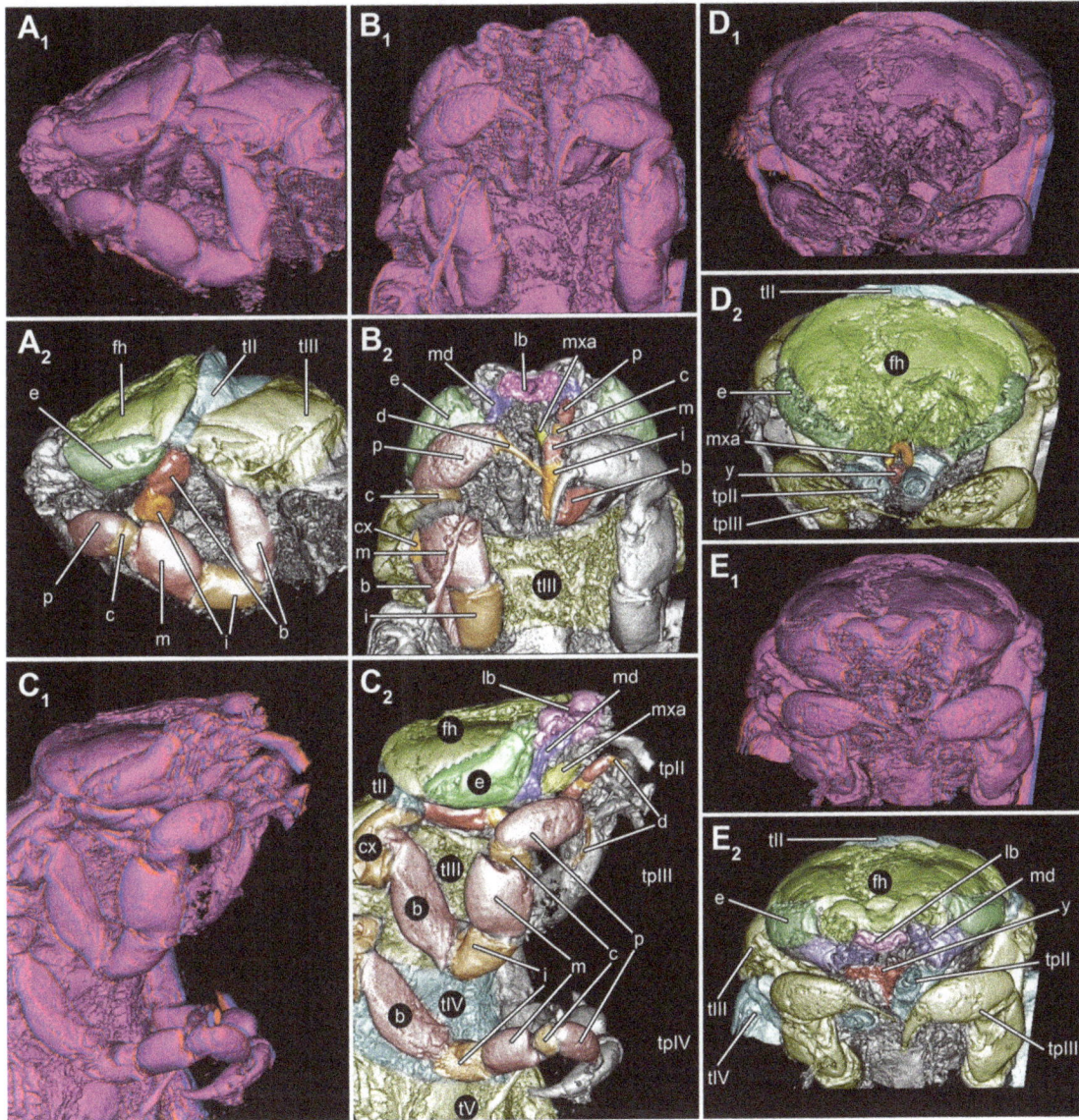

Fig. 5 Stereo images and respectively colour marked versions of volume rendering of the functional head and the anterior region of the thorax of fossil isopod *Urda rostrata*. Abbreviations: fh = functional head, e = eye, lb = labrum, md = mandible, mxa = maxillae, y = y-shaped mouth opening, cx = coxal plate, b = basis, i = ischium, m = merus, c = carpus, p = propodus, d = dactylus, tII-V = thorax segments II-V, tpII-IV = thoracopods II-IV. **a-d** Specimen 1 (BPSG 2011I50). **a**1-2 Lateral view. **b**1-2 Ventral view. **c**1-2 Specimen 2 (BPSG 2011I51), lateral view. **d-e** Specimen 1 (BPSG 2011I50). **d**1-2 Anterior view. **e**1-2 Anterior-ventral view. Not to scale

Discussion

Inferring the lifestyle

The lifestyle of representatives of *Urda* has so far largely been discussed in an anecdotal way. This led, for example, to interpretations of these isopods as scavengers due to a proposed position within Cirolanidae [58] or a swimming lifestyle [64]. Yet, alternative interpretations have also been forward, for example a supposed closer relationship to Gnathiidae [46], Aegidae [39] or Cymothoidae [52, 65]. Such phylogenetic interpretations would indicate an at least partly parasitic lifestyle for representatives of *Urda*. Support for such

interpretations has been largely lacking, as the critical morphological characters, such as the mouthparts and thoracopods, were not visible or only partly preserved [45–47, 58]. With our finding we can now contribute to this aspect.

Mouthparts

Modern parasitic isopods (especially within Cymothoida) in general have some of their mouthparts elongated, rotated on their axis distally (45° from the horizontal axis of the animal), and together forming a more or less tight sucking and/or piercing mouth cone, (Fig. 1f) [59, 66].

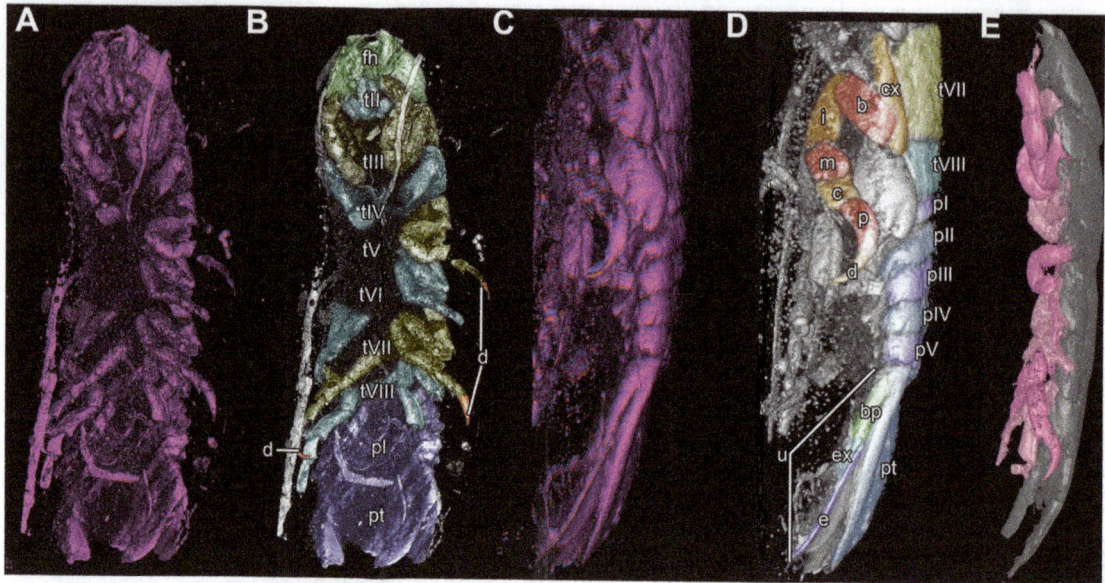

Fig. 6 Stereo images, respectively colour marked versions of volume rendering of the thoracopods of fossil isopod *Urda rostrata* specimen 1 (BPSG 2011I50). Abbreviations: fh = functional head, tII-VIII = thoraxsegments II-VIII, cx = coxal plate, b = basis, i = ischium, m = merus, c = carpus, p = propodus, d = dactylus, pI-V = pleon segments I-V, pt = pleotelson, bp = basipod, ex = exopod, en = endopod, u = uropod. **a-b** Ventral view. **c-e** Lateral view. Not to scale

Such an arrangement also appears to be present in the fossil specimens studied here (Figs. 1e, 2, 4c-d and 5a-e).

The forward projecting labrum of the here presented fossils (Figs. 2 and $4a_{6-7}$, b_7, c) is similar in shape and position to that of representatives of modern parasitic isopods of the groups Aegidae and Cymothoidae (Fig. 5d-e). In representatives of these groups this type of labrum prohibits loss of fluids when feeding on the host by sealing the mouth cone from the anterior [59, 67].

The maxilla and the second mouthpart of unclear identity in the fossil are very elongated and rotated off axis (Fig. 2). Such a type of maxilla is known in representatives of Aegidae and Cymothoidae and act as piercing structures (Fig. $4d_3$).

The second mouthpart in the fossil is more difficult to interpret as, based on its position, it may represent either the maxillula or parts of the paragnaths. In representatives of Aegidae and Cymothoidae the maxillula is also elongated, as it is in larval, parasitic forms of Gnathiidae [12, 68–70]. The structure seen in the fossil could thus be interpreted as the maxillula. Yet, in adult representatives of Gnathiidae the maxillula is absent and the distal parts of the paragnaths are comparably elongated. This is therefore also a possible interpretation for the fossils.

While the arrangement of the fossils' mouthparts clearly shows that these possess a mouth cone, it differs from that of representatives of Aegidae and Cymothoidae (Figs. 1f and 4d-e). The mouthparts of the fossils appear to form a more loose type of mouth cone (Figs. 1e, 2, 4c and 5a-e). This more loose appearance is caused by 1) the absence of

a mandibular palp (Figs. 2 and $4b_5$) that in representatives of Cymothoidae "grasps" around the labrum further sealing it [59], and 2) the relatively smaller maxillipeds (Fig. 4). The arrangement is therefore more comparable to that in larval representatives of Gnathiidae, where the mouthparts also only form a very loose type of cone [12, 65]. In these larvae the labrum and maxillipeds leave even more areas open than in the fossil.

The mandibles of representatives of Cymothoidae have a triangular blade-like incisor region, to cut pieces of tissues off the host [59, 71, 72]. This is different to the fossils, where the pointed, hooked mandibles were most likely used for piercing movements.

A further important observation is that the first free thorax segment (second thorax segment) is partly incorporated into the functional head (Figs. 1e, 2, $4a_1$, c and 5a-e). This condition is indicated by the small size of the tergite as well as the far anterior position of the appendage, as well as its size. This appendage may have been used to grasp into the host to provide pressure when inserting the mouthparts into the host. This distantly resembles the condition in Gnathiidae, yet in representatives of this group the segment of the second thoracopod is fully integrated into the functional head [65].

We can conclude that the mouthparts of the fossils investigated here strongly resemble those of modern parasitic forms, such as Aegidae, Cymothoidae and Gnathiidae in many aspects. This makes a parasitic lifestyle, probably on a fish host, for the fossil isopods likely.

Thoracopods

The dactyli of all thoracopods of the fossils are strongly curved, i.e. roughly modified into a hook (Figs. 1b, g, 3 and 6). This resembles dactyli of modern parasitic isopods (Fig. 1h). In modern forms, such as representatives of Cymothoidae, such hook-like dactyli are used for attaching to the host [25, 59, 73]. Additionally in modern forms the arrangement of the dactyli is adapted for prohibiting removal from the host [59]. A very similar pattern is seen in the fossil specimens, although the first free thoracopod, tII, is partly incorporated into the head. Still, the dactyli of thoracopods II-V grasp into the host at a 90° angle to the isopod's body, whereas the dactyli of the remaining thoracopods VI-VIII are more inclined backwards, and in this way strongly resemble modern forms [59].

To summarize, not only the morphology of the mouthparts (Figs. 1e, 2, 4 and 5), but also that of the thoracopods (Figs. 1d, g, 3 and 6) exhibit strong resemblance to similar structures in modern parasites. This similarity indicates a parasitic lifestyle of the fossil specimens studied herein.

Additional, although weaker, hints include: 1) The palaeo-environmental setting of the Bethel-Bielefeld limestone. It was interpreted as a tropical to subtropical back-reef lagoon [74]. Representatives of the obligate parasitic group Cymothoidae are nowadays most diverse in such ecosystems [32, 75]. 2) The dorso-ventrally flattened body of the fossils (Figs. 1a-d and 3a-f). This body shape is in contrast to, e.g., free-living cirolanids [76]. This can be understood as an additional adaptation for parasitism, reducing water resistance for the host. Similar adaptations are known in modern parasitic forms [77]. 3) The size of the fossils. With at least 30 mm the specimens are relatively large. Isopods parasitizing fishes have been reported to be larger than most free-living species [78] (except for deep sea forms, such as representatives of *Bathynomus*).

The eyes of the here described specimens appear well-developed. With this they give no additional indication for a reduced visual capability, as reported for permanently attached parasitic isopods [67, 73]. Together with the appendages that are clearly modified for parasitism we suggest a lifestyle similar to modern representatives of juvenile Gnathiidae. The animal would have attached to a host for a longer time (more or less permanently), but feed on the host only for a short time. With this the overall behaviour of the fossils could be compared to a mallophagan louse. Hence, the fossils are a kind of "marine mallophagan".

Evolution of parasitism within Cymothoida

Historically, *Urda* has been interpreted as closely related to Cirolanidae [39, 79], to Gnathiidae [46], to both of these two groups [47, 53, 80–82], or as a subgroup of Cymothoidae [52, 65].

One major challenge to resolving this issue, besides the lack of knowledge of mouthparts and thoracopods, was a dispute on the body organization of representatives of *Urda*. Available descriptions vary between five [45, 82], six [54–56, 58, 80] and seven free thorax segments [37, 46, 51, 52]. Our specimens clearly show seven free thorax segments for *Urda rostrata* (Fig. 1a). Part of the former confusion might have been caused by the rather small tergite of the first free thorax segment.

With our newly observed features we can therefore provide a phylogenetic interpretation of *Urda*. Additionally, we can provide a new reconstruction of character evolution for parasitic isopods within Cymothoida.

Phylogenetic interpretation

Cymothoida is a large group within Isopoda, including most (if not all, see below) of the parasitic isopods. There was most likely an evolutionary switch to parasitism that was followed by a large adaptive radiation and thus diversification of the different parasitic isopod groups.

Among the parasitic cymothoidans, it seems well established that Corallanidae is the sister group to all remaining parasitic forms [12, 28, 32, 83]. Corallanidae is united with all remaining forms by the specialization of a hook-like dactylus on thorax appendage II. Such a specialization is absent in free-living closer relatives such as representatives of Cirolanidae, and therefore appears to be an autapomorphy of the group including all the parasites.

The sister group to Corallanidae is a group including Aegidae, Cymothoidae and Epicaridea. This group has been largely accepted as monophyletic. The group is characterized (autapomorphy) by not one, but three hook-like dactyli on thorax appendages II, III and IV. Additionally they share a specialization of the mouthparts, which form a mouth cone allowing piercing and sucking (partly further specialized and reduced in different epicarids).

Epicaridea and Cymothoidae both have more than three hook-like dactyli, indicating a closer relationship between the two, and together being the sister group to Aegidae. Yet, we have not mentioned Gnathiidae. This group has been "shifted around the tree" in numerous studies [12, 28, 32, 33, 73, 83]. In this sense the fossils described here are interesting as they share certain aspects of their morphology with representatives of Gnathiidae and others with representatives of Cymothoidae and Aegidae. Representatives of Gnathiidae share certain morphological aspects with Epicaridea and Cymothoidae: all have the posterior six thoracopods modified for attaching (in contrast to Aegidae, where there are only three). We therefore suggest that *Urda* and

Gnathiidae are nested within the other parasitic isopods of Cymothoida.

Yet, six hook-shaped dactyli are restricted to a specific (larval) life stage in Gnathiidae. Also the exact attachment appears modified in Gnathiidae as the propodus appears to be a functional part of the "hook". Still the non-parasitic lifestyle of other life stages and the different attachment structure can be understood as secondary modifications. Gnathiidae and Epicaridea share the absence of the maxillula [73]. We therefore interpret these two as more closely related to each other than either of these two to Cymothoidae.

Urda could thus either represent the sister group of (Epicaridea + Gnathiidae) or of Gnathiidae alone. Similarities of *Urda* with Aegidae and Cymothoidae would represent plesiomorphies; a less tightly packed mouth cone and possible absence of the maxillula (if the

unclear mouthpart represents the paragnaths) would unite *Urda*, Gnathiidae and Epicaridea. The partial incorporation of the thorax segment II into the head could represent a synapomorphy of *Urda* and Gnathiidae.

Character evolution
The proposed phylogeny would lead to a character evolution as follows (Fig. 7):

1) The ground pattern (= reconstructed morphology of the stem species) of Cymothoida (character state transition 1) includes mouthparts indicating a carnivorous mode of feeding [12, 32, 73], thoracopods are of a swimming-type [84]. The stem species was most likely a scavenger or predator on fish.

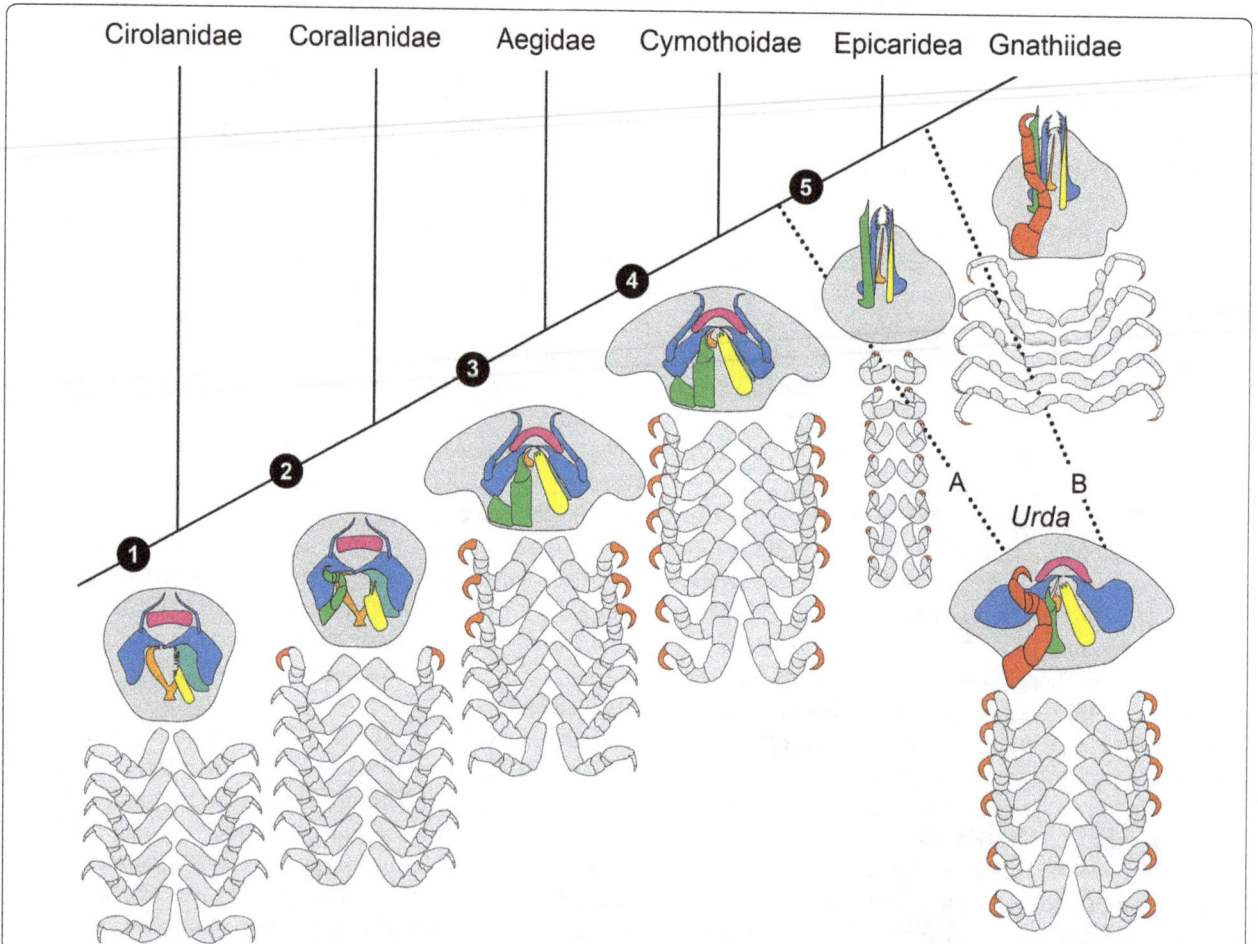

Fig. 7 Reconstructed relationship of some groups within the isopod group Cymothoida with a schematic character evolution of the mouthparts and thoracopods. Color-marks: labrum = purple, mandibles = blue, paragnaths = orange, maxillula = cyan, unclear mouthpart = white, maxilla = yellow, maxilliped = green, first free thoracopod = red, claw-like dactyli on thoracopods = dark orange. Important steps are: 1) cirolanid-like ancestor with carnivorous mouthparts, swimming thoracopods; 2) modified mouthparts for piercing, maxilliped and first free thoracopod modified for attachment; 3) mouthparts forming a sucking mouth cone, three thoracopods modified for attachment; 4) sucking mouth cone, all seven thoracopods modified for attachment; 5) mouthparts elongated still forming a mouth cone for piercing, reduced mandibular palp and maxillula, reduced seventh thoracopod. Position of *Urda* either in position A or B (see discussion)

2) The ground pattern of the unnamed sister group to Cirolanidae (including Corallanidae, Aegidae, Cymothoidae, *Urda*, Gnathiidae, Epicaridea; character state transition 2) is characterized by slightly longer, thinner and more pointed mouthparts that facilitate piercing; but still a slender maxilliped. At least the second pair of thoracopods appears to have a hook-like dactylus with which they attach to fish for temporary parasitism.

3) The ground pattern of the unnamed sister group to Corallanidae (including Aegidae, Cymothoidae, *Urda*, Gnathiidae, Epicaridea; character state transition 3) is characterized by a further specialization of the mouthparts; these form a sucking mouth cone (sealed by the labrum, paragnath, maxilla, maxilliped, and maxillula and mandible are used for piercing and cutting pieces of the host) [66]. Thoracopods two to five (three pairs of thoracopods) have strong hook-like dactyli [67, 73]. Adults have retained well-developed eyes [85]. The morphology of the mouthparts and thoracopods is strongly modified for a temporary parasitism on fish, comparable to a "marine mosquito".

4) The ground pattern of the unnamed sister group to Aegidae (including Cymothoidae, *Urda*, Gnathiidae, Epicaridea, character transition state 4) still includes a mouth cone [59, 66, 73]. Seven thoracopods (II-VIII) are now modified for better attachment to the host with hook-like dactyli [67, 73].

5) The ground pattern of the unnamed sister group of Cymothoidae (Gnathiidae, *Urda* Epicaridea, character transition state 5) is characterized by the lack of the mandibular palp and the maxillula [73]. Also the labrum is smaller and not covering the other mouthparts, resulting in a sucking and piercing mouth cone that is not as tight as in representatives of Cymothoidae of Aegidae. Thoracopods retain the hook-like dactyli.

In summary, the specialized and diverse morphology and parasitic lifestyle of representatives of Epicaridea, Gnathiidae and *Urda* originated from the scavenging life style in representatives of Cirolanidae. It occurred stepwise via a lifestyle as seen in representatives of Corallanidae, attaching to a fish and feeding on it, the still temporary parasitic lifestyle with already sucking-feeding in representatives of Aegidae and the permanent parasitic lifestyle of for example Cymothoidae. A similar evolutionary reconstruction has been proposed for other parasitic arthropods, such as lice [6]. Modern book lice, supposedly the sister group to true lice, are known to live in nests and pelage of mammals and birds and feed

on remains of these larger animals [7, 86, 87]. Modern true lice, e.g. chewing and sucking lice (mallophagan and anopluran respectively) are obligatory ectoparasites on birds and mammals [88]. The ancestors of true lice had simple chewing mouthparts and were free living in the nests of vertebrates (similar to book lice). Later in their evolutionary history, they adapted from associates to parasites by feeding directly from their hosts; hence, representatives of true lice evolved more specialized mouthparts for specific hosts and consequently a large diversity of forms [5, 89]. Due to the similarity between the evolutionary reconstruction of lice and representatives of Cymothoida and the similar lifestyle of chewing lice and the fossil specimens studied herein, we refer to these fossils and similar behaving isopods as "marine mallophagans".

Conclusion

We provide here indirect evidence for a case of palaeo-parasitology by a 168 million years old isopod. This represents the oldest possible fossil parasitic isopod to date. Furthermore, the fossils contribute important data towards the origin and diversification of parasitism within the isopod group Cymothoida. Parasitism appears to have arisen only once, further diversifying within the group. As these fossils appear deeply nested within the parasitic Cymothoida, the origin of the group and with this of a parasitic lifestyle within isopods must be even older than 168 million years.

Abbreviations
BSPG: Bavarian state collection for palaeontology and geology; ZSM: Bavarian state collection of zoology

Acknowledgements
We thank Martin Nose and Alexander Nützel, Bayerische Staatssammlung für Paläontologie und Geologie (BSPG, Bavarian State Collection for Palaeontology and Geology, Munich) for providing the material of the fossil specimens. We thank Jason Dunlop, Museum für Naturkunde Berlin (Museum of Natural History Berlin), for proofreading the manuscript. Special thanks to Carolin Haug, LMU Munich, for several discussions about the proposed phylogeny of Cymothoida. We want to thank Roland Melzer, Enrico Schwabe and Stefan Friedrich, Zoologische Staatssammlung München (ZSM, Zoological State Collection, Munich) for providing this material of extant specimens. We thank J. Matthias Starck, LMU Munich, Christine Dunkel, Antoinette v. Sigriz-Pesch, LMU Munich and Roland Melzer, ZSM Munich, for their support. We thank all people involved in providing free and low-cost software, such as OpenOffice, CombineZM, CombineZP, and Image Analyzer.

Funding
CN is gratefully funded by Studienstiftung des deutschen Volkes with a PhD-fellowship. MH was supported by VEGA 02/0136/15 and the Slovak Research and Development Agency under APVV-0644-10 and APVV-0436-12. The German Research Foundation (DFG) under Ha 6300/3-1 kindly funded JTH.

Authors' contributions

Conceptualization, CN and JTH; Methodology, CN and JTH; Formal Analysis, CN and JTH; Investigation, CN, MH and JTH; Resources, BSPG (MN and AN), ZSM (RM, ES and SF), MH; Data Curation, CN; Writing – Original Draft, CN; Writing – Review & Editing, MH and JTH; Visualization, CN; Supervision, MH and JTH; Funding acquisition, CN, MH and JTH. All authors read and approved the final manuscript.

Competing interests

The authors declare that they have no competing interests.

Author details

[1]Functional morphology group, Department of Biology II, Ludwig-Maximilians-University, Großhaderner Strasse 2, 82152 Planegg-Martinsried, Germany. [2]Department of Geology and Palaeontology, Faculty of Natural Sciences, Comenius University, Mlynská dolina, Ilkovičova 6, 84215 Bratislava, Slovakia. [3]Geological-Paleontological Department, Natural History Museum Vienna, Burgring 7, A-1010 Vienna, Austria. [4]GeoBio-Center, Richard-Wagner Strasse 10, 80333 Munich, Germany.

References

1. Osche G. Die präadaptation freilebender nematoden an den parasitismus. Zool Anz. 1956;19:391–6.
2. Osche G. Beiträge zur Morphologie, Ökologie und Phylogenie der Ascaridoidea (Nematoda). Z Parasitenkd. 1958;18:479–572.
3. Piekarski G. Neue Ergebnisse parasitologischer Forschung. Naturwissenschaften. 1973;60:139–44.
4. Poulin R. Evolutionary ecology of parasites. Princeton: Princeton University Press; 2011.
5. Nagler C, Haug JT. From fossil parasitoids to vectors: insects as parasites and hosts. Adv Parasit. 2015;90:137–200.
6. Johnson KP, Clayton DH. The biology, ecology and evolution of chewing lice. Syst Biol. 2003;53:449–76.
7. Yoshizawa K, Lienhard C. In search of the sister group of true lice: a systematic review of booklice and their relatives, with an updated checklist of Liposcelididae (Insecta: Psocodea). Athropod Syst Phylog. 2010;68:181–95.
8. Blaxter ML, De Ley P, Garey JR, Liu LX, Scheldeman P, Vierstraete A, Vanfleteren JR, Mackey LY, Dorris M, Frisse LM, Vida JI, Thomas K. A molecular evolutionary framework for the phylum Nematoda. Nature. 1998;392:71–5.
9. Littlewood DTJ, Rohde K, Bray RA, Herniou EA. Phylogeny of the Platyhelminthes and the evolution of parasitism. Biol J Linn Soc. 1999;68:257–87.
10. Near TJ. Acanthocephalan phylogeny and the evolution of parasitism. Integr Comp Biol. 2002;42:668–77.
11. Mironov SV, Bochkov AV, Fain A. Phylogeny and evolution of parasitism in feather mites of the families Epidermoptidae and Dermationidae (Acari: Analgoidea). Zool Anz. 2015;243:155–79.
12. Wägele JW. Evolution und phylogenetisches System der Isopoda. Zoologica. 1989;140:1–262.
13. Brusca R, Coelho VR, Taiti S. Isopoda. In: Carlton JT, editor. The Light and Smith manual: intertidal invertebrates from central California to Oregon. Berkley: University of California Press; 2007. p. 503–42.
14. Hornung E. Evolutionary adaptation of oniscidean isopods to terrestrial life: Structure, physiology and behavior. Terr Arthropod Rev. 2011;4:95–130.
15. Kensley B, Schotte M. Guide to the marine isopod crustaceans of the Caribbean. Washington: Smithsonian Institute Press; 1989.
16. Kussakin O. Marine and brackish-water Crustacea (Isopoda) of cold and temperate waters of the Northern Hemisphere. Nat Acad Sci Zoo. 1979;122:1–470.
17. Poore GC, Bruce NL. Global diversity of marine isopods (except Asellota and crustacean symbionts). PLoS One. 2012;7:e43529.
18. Sfenthourakis S, Taiti S. Patterns of taxonomic diversity among terrestrial isopods. Zookeys. 2015;2015:13.
19. Kensley B. Estimates of species diversity of free-living marine isopod crustaceans on coral reefs. Coral Reefs. 1998;17:83–8.
20. Lowry JK, Dempsey K. The giant deep-sea scavenger genus *Bathynomus* (Crustacea, Isopoda, Cirolanidae) in the Indo-West Pacific. In: Richer DFB, Justine J-L, editors. Tropical Deep-Sea Benthos. Paris: Mémoires du Muséum National d'Histoire Naturelle, Paris; 2006. p. 163–92.

21. Polz H. Asselansammlung auf einer Wasserwanze aus den Solnhofener Plattenkalken. Archaeopteryx. 2004;22:51–60.
22. Wilson GD, Paterson JR, Kear BP. Fossil isopods associated with a fish skeleton from the Lower Cretaceous of Queensland, Australia–direct evidence of a scavenging lifestyle in Mesozoic Cymothoida. Palaeontology. 2011;54:1053–68.
23. Wallerstein BR, Brusca RC. Fish predation: a preliminary study of its role in the zoogeography and evolution of shallow water idoteid isopods (Crustacea: Isopoda: Idoteidae). J Biogeogr. 1982;1982:135–50.
24. Klompmaker AA, Artal P, van Bakel BW, Fraaije RH, Jagt JW. Parasites in the fossil record: a Cretaceous fauna with isopod-infested decapod crustaceans, infestation patterns through time, and a new ichnotaxon. PLoS One. 2014;9;e92551.
25. Smit NJ, Bruce NL, Hadfield KA. Global diversity of fish parasitic isopod crustaceans of the family Cymothoidae. Int J Parasitol Parasites Wildl. 2014; 3:188–97.
26. Trilles J-P, Hipeau-Jacquotte R. Symbiosis and parasitism in the Crustacea. In: Scram F, Vauple Klein C, editors. Treatise on Zoology-Anatomy, Taxonomy, Biology. Leiden: Brill; 2012. p. 239–317.
27. Klompmaker AA, Boxshall GA. Fossil Crustaceans as Parasites and Hosts. In: DeBaets K, Littlewood T, editors. Advances in Parasitology. London: Elsevier; 2015. p. 233–89.
28. Dreyer H, Wägele JW. Parasites of crustaceans (Isopoda: Bopyridae) evolved from fish parasites: molecular and morphological evidence. Zoology. 2001; 103:157–78.
29. Hansen T, Hansen J. First fossils of the isopod genus *Aega* Leach, 1815. J Paleontol. 2010;84:141–7.
30. Nagler C, Haug C, Resch U, Kriwet J, Haug JT. 150 million years old isopods on fishes: a possible case of palaeo-parasitism. Bull Geosci. 2016;91:1–12.
31. Hispano C, Bultó P, Blanch AR. Life cycle of the fish parasite *Gnathia maxillaris* (Crustacea: Isopoda: Gnathiidae). Folia Parasitol. 2014;61:277.
32. Brandt A, Poore GC. Higher classification of the flabelliferan and related Isopoda based on a reappraisal of relationships. Invertebr Syst. 2004;17:893–923.
33. Wilson GD. The phylogenetic position of the Isopoda in the Peracarida (Crustacea: Malacostraca). Arthropod Syst Phylogeny. 2009;67:159–98.
34. Serrano-Sánchez ML, Nagler C, Haug C, Haug JT, Centeno-García E, Vega FJ. The first fossil record of larval stages of parasitic isopods: cryptoniscus larvae preserved in Miocene amber. Neues Jahrb Geol Palaontol Abh. 2016;279:97–106.
35. Radwańska U, Poirot E. Copepod-Infested Bathonian (Middle Jurassic) echinoids from Northern France. Acta Geol Pol. 2010;60:549–55.
36. Basso D, Tintori A. New Triassic isopod crustaceans from northern Italy. Palaeontology. 1995;37:801–10.
37. Feldmann RM, Wieder RW, Rolfe WI. *Urda mccoyi* (Carter 1889), an isopod crustacean from the Jurassic of Skye. Scott J Geol. 1994;30:87–9.
38. Polz H, Schweigert G, Maisch M. Two new species of *Palaega* (Isopoda: Cymothoida: Cirolanidae) from the Upper Jurassic of the. Palaeodiversity. 2006;362:1–17.
39. Brandt A, Crame J, Polz H, Thomson M. Late Jurassic tethyan ancestry of recent southern high-latitude marine isopods (Crustacea, Malacostraca). Palaeontology. 1999;42:663–75.
40. Etter W. A well-preserved isopod from the Middle Jurassic of southern Germany and implications for the isopod fossil record. Palaeontology. 2014;57:931–49.
41. Feldmann RM. A new cirolanid isopod (Crustacea) from the Cretaceous of Lebanon: dermoliths document the pre-molt condition. J Crust Biol. 2009;29:373–8.
42. Feldmann RM, Charbonnier S. *Ibacus cottreaui* Roger, 1946, reassigned to the isopod genus *Cirolana* (Cymothoida: Cirolanidae). J Crust Biol. 2011;31:317–9.
43. Gaillard C, Hantzpergue P, Vannier J, Margerard AL, Mazin JM. Isopod trackways from the Crayssac Lagerstätte, Upper Jurassic, France. Palaeontology. 2005;48:947–62.
44. Jones WT, Feldmann RM, Garassino A. Three new isopod species and a new occurrence of the tanaidacean *Niveotanais brunnensis* Polz, 2005 from the Jurassic Plattenkalk beds of Monte Fallano, Italy. J Crust Biol. 2014;34:739–53.
45. Stolley E. Über zwie neue Isopoden aus norddeutschem Mesozoikum. Jber niedersächs Geol Ver. 1910;6:191–216.
46. Straelen VE. Contribution à l'étude des isopodes méso-et cénozïques. Mem Acad r Belg. 1928;9:1–66.
47. Taylor BJ. An urdidid isopod from the Lower Cretaceous of south-east Alexander Island. Brit Antarct Surv. 1972;27:97–103.

48. Frentzen K. Paläontologische Skizzen aus den Badischen Landessammlungen für Naturkunde, Karlsruhe i. Br. II. *Mecochirus eckerti* nov. spec. aus dem Lias Epsilon (Posidonienschiefer) von Langenbrücken. Carolinea. 1937;2:103–5.

49. Carter J. On fossil isopods, with a description of a new species. Geol Mag. 1889;6:193–6.

50. Remeš M. *Urda moravica* n. sp. z doggeru Chřibů. Acta Mus Moraviae Sci biol. 1912;12:173–7.

51. Münster G. Ueber einige Isopoden in den Kalkschiefern von Bayern. Beitr Petrefactenkunde. 1840;3:19–23.

52. Kunth A. Über wenig bekannte Crustaceen von Solnhofen. Ger J Geol. 1870; 22:771–802.

53. von Ammon JG. Ein Beitrag zur Kenntniss der vorweltlichen Asseln. Abh Math-Phys Kl, K Bayer Akad Wiss. 1882;12:507–50.

54. Büchner M. Eine fossile Meeresassel (Isopoda, Malacostraca) aus den Parkinsonienschichten (Mittlerer Jura) von Bethel, Kreis Bielefeld. Ber Nat wiss Ver Belef. 1971;20:27–35.

55. Werner W. *Urda* sp. - Zwei Meeresasseln aus dem Mitteljura von Bielefeld. Mitt. Bayer. Staatssaml. Paläont hist Geol. 2012;40:39–42.

56. Grant-Mackie J, Buckeridge J, Johns P. Two new Upper Jurassic arthropods from New Zealand. Alcheringa. 1996;20:31–9.

57. Oppel A. Über jurassische Crustaceen. Mitt Bayer Staatssaml Paläont hist Geol. 1862;1:1–120.

58. Etter W. Isopoden und Tanaidaceen (Crustacea, Malacostraca) aus dem unteren Opalinuston der Nordschweiz. Eclogae Geol Helv. 1988;81:857–77.

59. Nagler C, Haug JT. Functional morphology of parasitic isopods: understanding morphological adaptations of attachment and feeding structures in *Nerocila* as a pre-requisite for reconstructing the evolution of Cymothoidae. PeerJ. 2016;4:e2188.

60. Haug C, Kutschera V, Ahyong ST, Vega FJ, Maas A, Waloszek D, Haug JT. Re-evaluation of the Mesozoic mantis shrimp *Ursquilla yehoachi* based on new material and the virtual peel technique. Palaeontol Electron. 2013;16:16.2.5T.

61. Haug C, Mayer G, Kutschera V, Waloszek D, Maas A, Haug JT. Imaging and documenting gammarideans. Int J Zoo. 2011;38:380829.

62. Haug C, Van Roy P, Leipner A, Funch P, Rudkin DM, Schöllmann L, Haug JT. A holomorph approach to xiphosuran evolution—a case study on the ontogeny of *Euproops*. Dev Genes Evol. 2012;222:253–68.

63. Simonetta AM, Delle CL. An essay in the comparative and evolutionary morphology of Palaeozoic arthropods. Accad Naz Lincei Rome. 1981;49: 389–439.

64. Bunkley-Williams L, Williams EH. Isopods associated with fishes: a synopsis and corrections. J Parasitol. 1998;84:893–6.

65. Monod T. Les Gnathiidae: essai monographique. Mém Soc Sci Nat Maroc. 1926;13:1–668.

66. Günther K. Bau und Funktion der Mundwerkzeuge bei Crustaceen aus der Familie der Cymothoidae (Isopoda). Zoomorphology. 1931;23:1–79.

67. Brusca RC. A monograph on the Isopoda Cymothoidae (Crustacea) of the eastern Pacific. Zool J Linn Soc. 1981;73:117–99.

68. Coetzee ML, Smit NJ, Grutter AS, Davies AJ. Gnathia trimaculata n. sp. (Crustacea: Isopoda: Gnathiidae), an ectoparasite found parasitising requiem sharks from off Lizard Island, Great Barrier Reef, Australia. Syst Parasitol. 2009;72:97–112.

69. Davies AJ. A scanning electron microscope study of the praniza larva of *Gnathia maxillaris* Montagu (Crustacea, Isopoda, Gnathiidae), with special reference to the mouthparts. J Nat Hist. 1981;15:545–54.

70. Smit NJ, Basson L. Gnathia pantherina sp. n. (Crustacea: Isopoda: Gnathiidae), a temporary ectoparasite of some elasmobranch species from southern Africa. Folia Parasitol. 2002;49:137–51.

71. Jithendran K, Natarajan M, Azad I. Crustacean parasites and their management in brackishwater finfish culture. Aquac Mag. 2008;5:47–50.

72. Thatcher V. Mouthpart morphology of six freshwater species of Cymothoidae (Isopoda) from Amazonian fish compared to that of three marine forms, with the proposal of Artystonenae subfam. nov. Amazonia. 1997;14:311–22.

73. Brusca RC, Wilson GD. A phylogenetic analysis of the Isopoda with some classificatory recommendations. Mem Queensl Mus. 1991;31:143–204.

74. Barthel K, Swinburne, NHM & Conway Morris, S. Solnhofen. A Study in Mesozoic Palaeontology. Cambridge: Cambridge University Press; 1990.

75. Lester R. Isopoda. In: Rohde K, editor. Marine Parasitology. Collingwood: Csiro Publishing; 2005. p. 138–44.

76. Hansen HJ. Cirolanidæ et familiæ nonnullæ propinquæ musei Hauniensis: et bidrag til kundskaben om nogle familier af isopode krebsdyr. Naturvidenskabelig og Mathematisk Afdelning. 1890;5:237–426.

77. Rand TG. The histopathology of infestation of *Paranthias furdfer* (L.) (Osteichthyes: Serranidae) by *Nerocila acuminata* (Schioedte and Meinert)(Crustacea: Isopoda: Cymothoidae). J Fish Dis. 1986;9:143-6.

78. Poulin R. Evolutionary influences on body size in free-living and parasitic isopods. Biol J Linn Soc. 1995;54:231–44.

79. Hyžný M, Bruce NL, Schloegl J. An appraisal of the fossil record for the Cirolanidae (Malacostraca: Peracarida: Isopoda: Cymothoida), with a description of a new cirolanid isopod crustacean from the Early Miocene of the Vienna Basin (Western Carpathians). Palaeontology. 2013;56:615–30.

80. Hessler RR. Peracarida. In: Morre RC, editor. Part R: Arthropoda, vol. 4. Kansas: Geological Society of America and University of Kansas Press; 1969. p. 360–93.

81. Menzies R. The zoogeography, ecology and systematics of the Chilean isopods. Acta Univ Lund. 1961;57:1–162.

82. Wittler F. Bemerkungen zu *"Palaega"*. Arbeitskreis Paläontologie Hannover. 2001;29:19–23.

83. Wetzer R, Perez-Losada M, Bruce NL. Phylogenetic relationships of the family Sphaeromatidae Latreille, 1825 (Crustacea: Peracarida: Isopoda) within Sphaeromatidea based on 18S-rDNA molecular data. Zootaxa. 2013;3599:161–77.

84. Kensley B. Guide to the marine isopods of southern Africa. Cape Town: South Africa Museum; 1978.

85. Brusca RC. A monograph on the isopod family Aegidae in the tropical eastern Pacific. Los Angeles: Allan Hancock Foundation; 1983.

86. Mockford EL. Some Psocoptera from plumage of birds. Proc Entomol Soc Washington. 1967;69:307–9.

87. Mockford EL. Psocoptera from sleeping nests of the dusky-footed wood rat in southern California. Pan-Pac Entomol. 1971;47:127–40.

88. Yoshizawa K, Johnson KP. How stable is the "polyphyly of lice" hypothesis (Insecta: Psocodea)?: A comparison of phylogenetic signal in multiple genes. Mol Phylogenet Evol. 2010;55:939–51.

89. Light JE, Smith VS, Allen JM, Durden LA, Reed DL. Evolutionary history of mammalian sucking lice (Phthiraptera: Anoplura). BMC Evol Biol. 2010;10:292.

90. Nagler C. (2017a). C_Nagler_20170221-M-130.1. www.morphdbase.de/?C_ Nagler_20170221-M-130.1.

91. Nagler, C. (2017b). C_Nagler_20170221-M-131.1. www.morphdbase.de/?C_ Nagler_20170221-M-131.1.

92. Cohen KM, Finney SC, Gibbard PL, Fan JX. The ICS international chronostratigraphic chart. Episodes. 2013;36:199-204.

9

Identification of constraints influencing the bacterial genomes evolution in the PVC super-phylum

Sandrine Pinos[1,2], Pierre Pontarotti[1], Didier Raoult[2] and Vicky Merhej[2*] (iD)

Abstract

Background: Horizontal transfer plays an important role in the evolution of bacterial genomes, yet it obeys several constraints, including the ecological opportunity to meet other organisms, the presence of transfer systems, and the fitness of the transferred genes. Bacteria from the *Planctomyctetes, Verrumicrobia, Chlamydiae* (PVC) super-phylum have a compartmentalized cell plan delimited by an intracytoplasmic membrane that might constitute an additional constraint with particular impact on bacterial evolution. In this investigation, we studied the evolution of 33 genomes from PVC species and focused on the rate and the nature of horizontally transferred sequences in relation to their habitat and their cell plan.

Results: Using a comparative phylogenomic approach, we showed that habitat influences the evolution of the bacterial genome's content and the flux of horizontal transfer of DNA (HT). Thus bacteria from soil, from insects and ubiquitous bacteria presented the highest average of horizontal transfer compared to bacteria living in water, extracellular bacteria in vertebrates, bacteria from amoeba and intracellular bacteria in vertebrates (with a mean of 379 versus 110 events per species, respectively and 7.6% of each genomes due to HT against 4.8%). The partners of these transfers were mainly bacterial organisms (94.9%); they allowed us to differentiate environmental bacteria, which exchanged more with *Proteobacteria*, and bacteria from vertebrates, which exchanged more with *Firmicutes*. The functional analysis of the horizontal transfers revealed a convergent evolution, with an over-representation of genes encoding for membrane biogenesis and lipid metabolism, among compartmentalized bacteria in the different habitats.

Conclusions: The presence of an intracytoplasmic membrane in PVC species seems to affect the genome's evolution through the selection of transferred DNA, according to their encoded functions.

Keywords: Horizontal transfer, Bacteria, Environments, Lifestyle, Genomes, Functions

Background

The extensive amount of genomic data acquired over the last 20 years has provided insights into the evolutionary processes that drive bacterial evolution. The horizontal transfer of DNA (HT) appears to be major driving force of innovation [1, 2] as it provides additional functions, allowing adaptation to specific conditions and environmental changes. The HT process in bacteria depends on several conditions [3]: i. the possibility of exchanges, meaning the presence of different microorganisms in a single place; ii. the possibility of foreign sequences to enter into recipient bacteria, mediated by conjugation, transformation or transduction; iii. the ability to integrate into the recipient genome; iv. the genes expressed and the genes used v. those conserved, in relation to the benefits for recipient bacteria. This process could be regulated by intrinsic and extrinsic constraints. Two extrinsic constraints influencing the possibility of exchanges include the environment or the "ecological niches" and the lifestyle, which together constitute the habitat of bacteria [4–6]. Thus the proportions and origins of HT were more similar among bacteria from the same habitat than among bacteria from a given phylum [7]. Changing environmental conditions are also well known constraints for HT regulation; UV irradiation or starvation and other stress conditions, were shown to

* Correspondence: vicky_merhej@hotmail.com
[2]Aix Marseille Univ, CNRS, IRD, INSERM, AP-HM URMITE, IHU -Méditerranée Infection, 19-21 Boulevard Jean Moulin, Marseille 13005, France
Full list of author information is available at the end of the article

affect the mobility of transposons and insertion sequences [6, 8–10]. The habitat also seems to play an important role in the selection and conservation of transferred sequences encoding for specific functions that are involved in host's colonization, and the development of pathogenesis. Indeed, many examples in the literature indicate that genes encoding for metabolic functions [11–13] and for antibiotic resistance [14, 15] and virulence [16–18] represent commonly transferred sequences. The intrinsic constraints that influence the entrance and integration of foreign DNA into a recipient genome include the exclusion surface that limits the entrance of specific sequences in some bacteria [3], the presence of CRISPR that decreases the quantity of transferred sequences insertion in recipient genomes [19, 20] and the presence of some endonucleases that can destroy foreign DNA [3, 21].

Many studies have been conducted to explore the impact of the different extrinsic and intrinsic constraints on horizontal transfers. However, these studies involved one or a few species, or bacteria presenting only one to two habitats or lifestyles [22–25], or undergoing relatively few intrinsic constraints [26, 27]. The study of only few characteristics may one lead to miss the cumulative or overlapping effects of the different constraints. Therefore, we used a phylogenomic approach to mine a large set of bacteria with different habitats in order to decipher the impact of different constraints on genome composition, especially regarding HT. The PVC super-phylum seems to be a good model to study, as it includes seven bacterial phyla (*Planctomycetes, Verrucomicrobiae, Chlamydiae, Lentisphaera, Poribacteria, OP3, WWE2*) [28–31] with diverse habitats, three different lifestyles (intracellular allopatric, intracellular sympatric, extracellular sympatric) and numerous environments (water, soils, water and soils, metazoa, amoeba, ubiquitous...), thus varying the external constraints. Moreover, a specific cell plan is also present in all the *Planctomycetes* [32–34], in some of *Verrucomicrobiae* [35] in one *Lentisphaera* and in one *Poribacteria* [36]. The cytoplasm of these bacteria is separated into two compartments by an intracytoplasmic membrane (ICM), the pirellulosome inside (with DNA [37]) and the paryphoplasm outside. This membrane is a lipid bilayer in contact with proteins [32, 33, 38] presenting structural similarities with proteins from eukaryotic membranes like the clathrins [39, 40]. The function of this intracytoplasmic membrane is still unknown, but we hypothesize the possible impact of this intrinsic constraint on HT. In the present investigation, we analyzed 33 PVC bacteria together with 31 phylogenetically close species (*Bacteroidetes, Chlorobi* and *Spirochaetes)* that were considered as the control group, looking for evidence for horizontal transfer. Statistical analyses of the potential partners and functions involved in HT allowed us to estimate the real impact of habitat and cell plan on the genomes evolution.

Methods
Bacterial set selection, definition of lifestyles, environments and cell plan
The genomes of 64 bacteria have been retrieved from two different databases [41, 42]. These bacteria belong to different phyla (Additional file 1) including four phyla of the PVC super-phylum, *Planctomycetes, Verrucomicrobiae, Lentisphaerae* and *Chlamydia* and the phylogenetically closest phyla, *Bacteroidetes, Chlorobi* and *Spirochaeta* (determined thanks to a reference tree [43]). We reconstructed the species tree of PVC bacteria and *Bacteroidetes-Chlorobi-Spirochaetes* on the basis of 12 markers that are common to the 64 species (Additional file 2) using Mega5 [44]. Therefore, the protein sequences of each marker were aligned with Muscle [45] and non-conserved positions were removed manually. All alignments were concatenated, leading to an alignment of 5067 sites. We used a Maximum likelihood tree (substitution model JTT) based on this concatenated alignment to reconstruct the phylogeny of species. Bootstrap support values were obtained with 150 replicates (Additional file 3). The bacteria studied have different lifestyles (intracellular or extracellular, allopatric or sympatric [46, 47]) and live in different environments (amoeba, mammals, soils, water, insects). We use the term "environment" as the main place where the bacteria are living. For example, bacteria detected in sea, freshwater or wastewater are all annotated as bacteria from 'water'. If bacteria are present in two different environments, we indicate both of them (for example 'water-soil' bacteria), while bacteria living in more than 5 environments are considered 'ubiquitous'. The lifestyle of bacteria is characterized by two factors: the intracellular and extracellular conditions, and the ability to exchange with other microorganisms (in allopatric or sympatric lifestyles, respectively [46]). Lifestyles and living environments were defined for each bacterium based on a literature search [48–53]. Cell plans of the bacteria were determined via transmission electron microscopy images already available in the literature [33–36] and microscopic observations of the bacteria realized in our laboratory [54]. Three states are determined for the cell plan: compartmentalization, non compartmentalization and unknown. The selected set of species contains 4 bacteria from amoeba, 4 from insects (1 intracellular, 3 extracellular), 3 from soils, 4 living in soils and water, 3 ubiquitous, 26 from vertebrates (18 extracellular, 8 intracellular) and 20 from water. Among these 64 bacteria, 20 present a compartmentalized cell plan, 5 have an unknown cell plan, and 39 are not compartmentalized (Additional file 1).

Genome analysis: common genes, specific genes, ORFans
OrthoMCL [55] was used to obtain groups of orthologous proteins. Groups containing at least one representative member of each habitat were considered as the

common genes, and those that contain only proteins of bacteria from the same habitat are considered as specific to the corresponding habitat. We calculated the rate and determined the function of proteins that are specific to habitat in each species using two software, COGnitor and WGA and the Interpro database [56–59]. Genes that do not belong to any orthologous group are either acquired by "specific" HT or generated de novo. Blast against NR database allowed the identification of ORFans in the genome with no identifiable homologous (i.e. genes that do not have a Blast hit with an *e*-value < 10e-4 AND a query coverage >50%). We performed a clustering of all species according to their genes contents in order to detect, in some bacteria, a tendency to share the same gene contents in relation with their habitat.

HT detection, functions and partner identification

Generally, two main difficulties hinder the analysis of the Horizontal Transfer of DNA sequences (HT) according to habitat: the distinction between the ancestral and recent gene gains, as well as the difficulty of determining the ancestral habitat of the bacteria. In order to avoid these problems, we focused our study on recent transfers that occurred only in modern species of the super-phylum (in the leafs of the tree) and not in their ancestors (at the nodes of the tree). HT instances were identified using a comparative phylogenomic approach, phylogenetic profiling of proteins and phylogenetic analysis of gene trees in comparison with the species tree. Using the Phylopattern [60] pipeline, we identified the gene gain events in four steps: 1) based on the orthologous groups, a tree was reconstructed for each group. 2) The topologies of these trees were compared with that of the species tree in order to detect species missing in orthologous groups. 3) We obtained a pattern of presence/absence for each gene in the different species, allowing Phylopattern to reconstruct the ancestral states of genes by implementing the Sankoff parsimony algorithm [61]. Based on this reconstruction, the pattern-matching module in PhyloPattern allows us to infer, by parsimony, two types of genetic events that could have occurred during gene evolution: gains and losses. The gains could be a possible HT, de novo genes or artifacts, therefore, among gene gains detected by Phylopattern, we had to identify those due to HT. We focused on specific gene gains and performed a Blast to identify similar sequences in the NR database. If the first twenty hits of Blast result belong to species outside the super-phylum of the query, and they are orthologous as confirmed by a reciprocal best hit, the enquired gene could be considered horizontally acquired. The pattern permitting the automatic identification of HT among gain events is presented in Additional file 4. Sequences with *e*-value > 10– 5, coverage < 60% or identities < 30% were not

considered. The localization of these events in the genome allowed the identification of any horizontal transfers of DNA sequences. The directionality of the transfer could not always be identified, but as we were interested in the capacity of exchange of the bacteria, it did not matter if the bacterium was the donor or the recipient. If some transferred genes are side by side in genomes, and were exchanged with the same partners, we considered them to have been transferred by a single event.

We calculated quantities and proportions of proteins (proteins transferred/total proteins in proteome) and sequences (nucleotides transferred/total nucleotides in genomes) implicated in HT for each genomes and the size of the transferred sequences. We identified the function of transferred genes by using two software programs, COGnitor and WGA and the Interpro database [56–59]). These programs attribute one type of function to proteins, according to the COG to which proteins belong. We studied the possible significant differences in HT distribution among the different studied groups of bacteria concerning the HT partners and functions.

Statistical analyses

The occurrence of HT and the frequency of specific genes were compared among the different groups of bacteria from different habitats. We tested whether the data (proportions of functions for specific genes and proportions of partners and functions for HT) follows a normal (Gaussian) distribution using the Shapiro-Wilk test and we controlled the homogeneity of the data by the Levene test [62]. These tests were followed by a comparison of variance among the different habitats, using the Kruskal test [63] or the ANOVA test, according to whether or not the data had a normal distribution. The Nemenyi [64] or Tukey [65] tests were performed to obtain a comparison of each pair of habitats. We also realized a Principal Components Analysis (PCA), focused on HT proportions in genomes, size, functions and partner of transfer, followed by a hierarchical clustering (HCPC), to identify clustering of bacteria according to their transfer partners or their functions. All analyses were realized with R software.

We then used comparative phylogenetic methods to test the impact of phylogenetic relationships between species on the acquisition of studied characters. Analysis of variance was used in intergroup comparisons of categorical variables to determine whether there is statistical significance of the repartition of species on the basis of their habitat, compared to their classification according to phylogenetic distances (Additional file 3). Therefore, the Nemenyi [64] or Tukey [65] tests were performed to obtain a comparison of each pairs of groups based on the phylogenetic relationships. These groups were determined by the phylogenetic distance separating bacteria. We compared the results of these tests with the results of tests carried out for groups based on habitats, allowing us to determine if classes defined by the phylogenetic distances present different and more reliable results than classes based on habitat (t-test). If results were similar, it could be difficult to determine whether the differences observed were related to the habitat or to the phylogenetic relationships. We also performed two correlation tests. The first test, Pagel's correlation method [66], performed on Mesquite, is a test of the independent evolution of two binary characters (all features studied were tested thanks to a binarization of continuous values). This test compares the ratio of likelihoods of two models where the rates of change in each character are dependent or alternatively independent from phylogenetic relationships. The second test is the Spearman coefficient [67], weighted by the phylogenetic distances that studies the relationship between two variables. The detection by means of this correlation test of a significantly convergent character in bacteria from a single habitat is rather unrelated to the phylogenetic background.

Results

Pangenome analysis

The genome size varied widely among the 64 bacteria studied, ranging from 0.63 Mb for *Blattabacterium* to 9.76 Mb for *Singulisphaera acidiphila* with an average of 3.83 +/− 2.2 Mb. Bacteria from soils, insects and water-soils presented the largest genome sizes, with an average of 7.07, 6.45 and 5.69 Mb, respectively, while the smallest genome sizes were found among intracellular and extracellular bacteria of vertebrates, and bacteria from amoeba, with an average of 1.14, 2.56 and 2.91 Mb, respectively. Ubiquitous bacteria and bacteria from water with an average genome size of 4.63 Mb, formed the medium-sized genomes. Likewise, the protein sets were very different among the studied bacteria, ranging from 579 proteins for *Blattabacterium* sp to 7969 for *Gemmata obscuriglobus*, with an average of 3227 +/− 2513. When using OrthoMCL, 124,175 out of 206,508 proteins that form the pangenome of the PVC group bacteria, could be assigned to 16,918 different orthologous groups (OG). Among these, 1224 OGs were common to the eight different habitats studied, and constituted the common genes. The rest of the genes were present in bacteria from two or more habitats, and thus form the "shared genome." or they did not share sequence similarity with any other gene of the species of other habitats, and thus constituted the "specific genes" (Fig. 1). When species were clustered according to their gene contents, some bacteria sharing a same ecological niche were preferentially grouped together, forming subclusters within the clusters, determined by phylogenetic relationships or disturbing the phylogenetic unity of some groups, like the *Verrucomicrobiae*,

Spirochaetes and Planctomycetes (Additional file 5). Thus the ecological niche seems to have influenced the gene content of some bacteria from soil (*p*-value = 0.027) and water (*p*-value = 0.045 for internal cluster). Bacteria from insects that showed the highest proportion of their genes shared with the other groups and the lowest quantity of specific genes (29 genes) were scattered throughout the different clusters (Additional file 5).

The genes that are common to all habitats represented 26.2% of the content of each genome on average, and varied from 20.1% for bacteria from insects to 49.5% for the intracellular vertebrates. In order to determine the functional profile of the common genes, each protein was assigned to Cluster of Orthologous Groups of proteins (COGs) functional category. We could infer a putative function to the protein sequences of 74% of the common genes; of these, 43.8% encode for cellular processes and signaling, 36.8% for metabolic functions, and 19.5% for storage and processing information (Fig. 1). Among these, four functions were significantly over-represented compared to the other functions: wall/membrane/envelop biogenesis, signal transduction mechanisms, transcription and energy production and conversion (12.4, 11.6, 9.1 and 8.5%, respectively Chi2 test: *p*-value = 9.4*10-7) (Fig. 1).

The genes shared by some habitats and the specific genes represented 35.3 and 8.1% of the content of each genome, respectively on average. The specific genes varied from 0.5% in bacteria from insects to 18.4% in the intracellular vertebrates. Of all species analyzed, *Bacteroides xylanisolvens* and *Bacteroides vulgatus* in the group of extracellular vertebrates, had the most specific genes, with a total of 801 and 780 exclusive sequences (18.2 and 19.2% of their total genomes), respectively. The functional distribution of the specific genes was significantly different from that of the common genes in all the habitats with fewer genes implicated in cell process and signaling (37.9%) (t-test for comparison between specific genes and common genes; *p*-value = 4.3*10-4) (Fig. 1). Some functions were significantly over-represented in some habitats compared to other habitats (Fig. 2), including transcription within their specific genes (16.3 and 18.8%, Kruskal-Wallis test: *p*-value = 3.9*10-6 and Correlation test : *p*-value = 4.7*10-3) in bacteria from amoeba and from soils-water; the signal transduction mechanisms and defense mechanisms in the intracellular bacteria of vertebrates (15.7%, Kruskal-Wallis test : *p*-value = 2.2*10-5 and 8.1%, 1.8*10-6, respectively; Correlation test : *p*-value = 5.5*10-2 and 3.1*10-3, respectively); the transport and metabolism of amino acid and coenzyme in bacteria from insects (15.6%, Kruskal-Wallis test: *p*-value = 1.0*10-2 and 13.3%, ANOVA test : *p*-value = 1.1*10-7, respectively), the transport and metabolism of coenzymes in bacteria from soils (9.7%, ANOVA test: *p*-value = 1.1*10-7) (Fig. 2). Although they had high proportions of specific genes, ubiquitous bacteria,

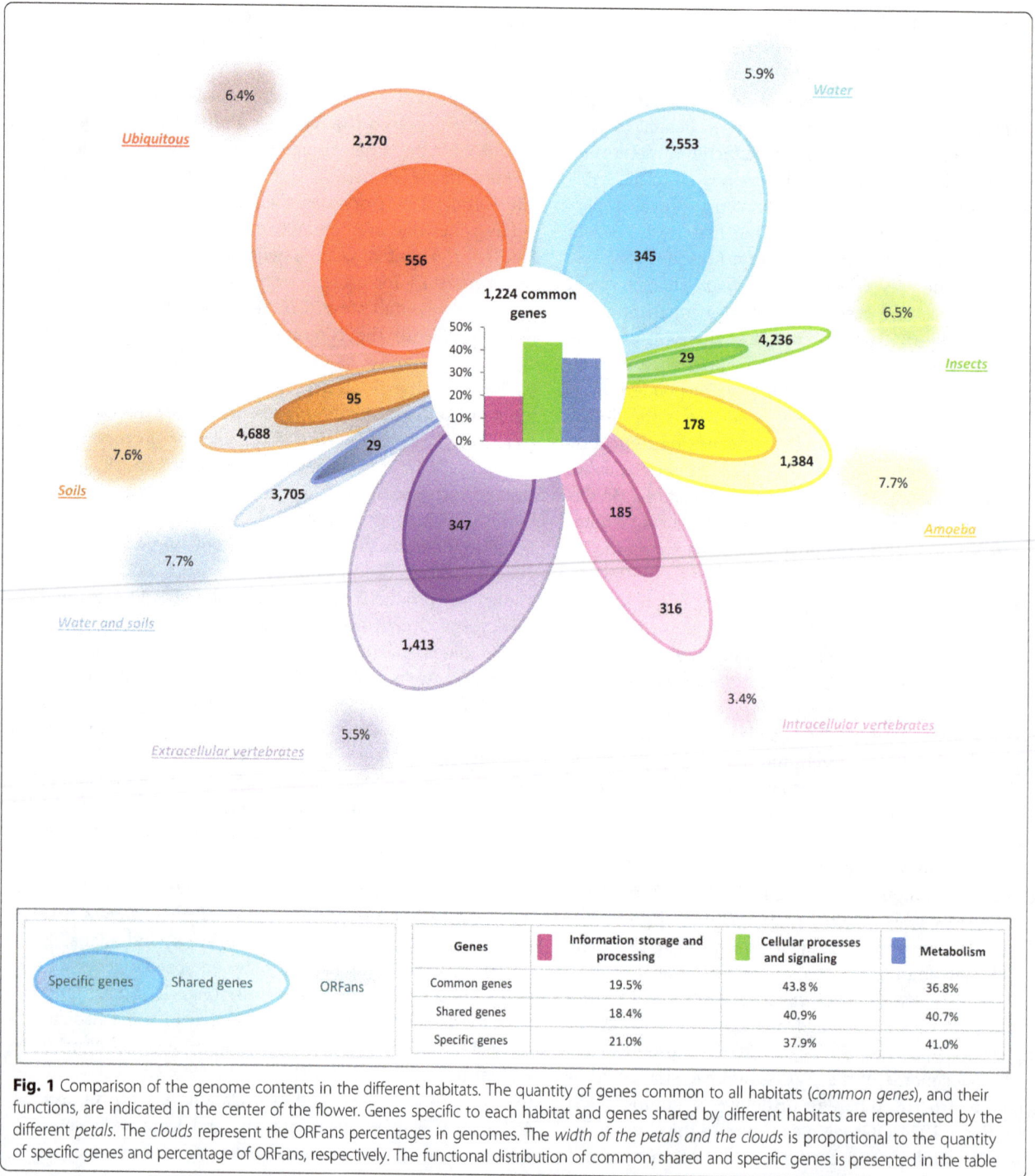

Fig. 1 Comparison of the genome contents in the different habitats. The quantity of genes common to all habitats (*common genes*), and their functions, are indicated in the center of the flower. Genes specific to each habitat and genes shared by different habitats are represented by the different *petals*. The *clouds* represent the ORFans percentages in genomes. The *width of the petals and the clouds* is proportional to the quantity of specific genes and percentage of ORFans, respectively. The functional distribution of common, shared and specific genes is presented in the table

bacteria from water and extracellular bacteria from vertebrates had no overrepresented functional category compared to the others (Fig. 2).

Horizontal transfers

Phylogenetic analyses conducted to infer the evolutionary origin of all the proteins other than the common genes indicated that 12,885 proteins (6.3% of all the

proteins) were acquired via horizontal transfer (Fig. 3). As transfer events are not necessarily confined to individual genes but they may concern a cluster of genes, we considered neighboring genes with the same horizontal transfer history to reflect only a single event of HT. We counted a total of 10,918 HT events among studied bacteria. The incidence with which the HT events occurred was as high as 170.6 +/− 116.4, yet this was highly

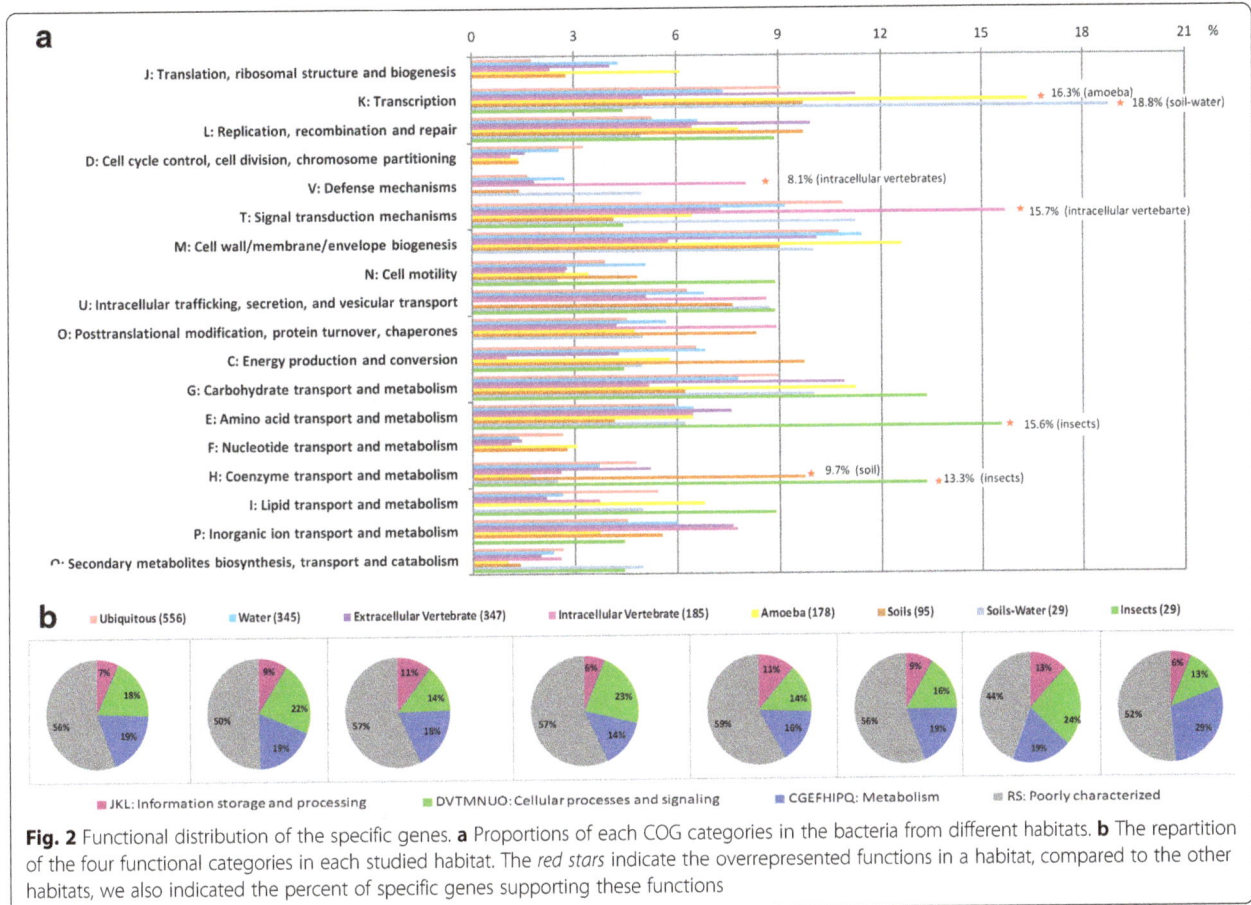

Fig. 2 Functional distribution of the specific genes. **a** Proportions of each COG categories in the bacteria from different habitats. **b** The repartition of the four functional categories in each studied habitat. The *red stars* indicate the overrepresented functions in a habitat, compared to the other habitats, we also indicated the percent of specific genes supporting these functions

variable among bacteria ranging from 0 transfers in the genome of *Blattabacterium* and *Borrelia* spp. to 803 sequences transferred in the genome of *Chthoniobacter* (Fig. 3 and Additional file 6). The count of HT events was not correlated with the genome size (Wilcoxon, p = 0.042) (Additional file 7). The transferred fragments length ranged from 859 bp for *C. tepidum* to 1492 pb for *S. acidophila*, with a mean length of 1124+/− 274 bp for all fragments. The size of the HT was 1022 +/− 383 bp on average, and was similar in the different habitats. Bacteria from soil, from insects and ubiquitous bacteria presented the highest average of HT (524.0, 343.7 and 269.3 transfer events per species, respectively), compared to bacteria living in water, extracellular bacteria from vertebrates, bacteria from amoeba and the intracellular bacteria of vertebrates (183.7, 126.7, 113.5 and 17.6 HT per species, respectively). The statistical comparison of the bacterial group from different habitats allowed us to define three classes, based on percentages of sequences due to transfer (Kruskal-Wallis test : p-value = 2.3*10-2): Bacteria of soils (8.4%) and insects (7.3%) and ubiquitous bacteria (6.6%) defined the first class. Bacteria from soil-water (5.0%), water (4.9%), amoeba (4.8%), and extracellular bacteria of vertebrates (4.6%) presented similar proportions,

and formed the second class. Intracellular bacteria of vertebrates were grouped in the third class, with 1.9% of HT events (Additional files 6).

Most of the horizontal exchanges were realized with bacteria (94.9%) and very few with *Archaea* (2.4%), *Eukaryota* (2.5%) and viruses (0.2%). Bacteria from amoeba and ubiquitous bacteria showed a significantly higher quantity of HT instances realized with eukaryotes, compared to bacteria from other habitats (9.8 and 4.2%, respectively) (Kruskal test : p-value = 3.3*10-3). The proportion of exchanges with *Archaea* was significantly higher in bacteria from water-soils and from water (3.4 and 3.4%, respectively) compared to bacteria from other habitats (Kruskal test: p-value = 3.1*10-2). The most common bacterial partners identified were *Proteobacteria* (42%), *Firmicutes* (23%) and *Cyanobacteria* or *Actinobacteria* (6%) (Fig. 4). For several transfer partners, we identified significant differences among the 64 bacteria studied, according to their habitat: bacteria from intracellular and extracellular vertebrates were both characterized by their preference for the *Firmicutes* (21.4 and 41.1%, respectively) as transfer partners (Kruskal test : p-value = 3.6*10-3), and their significantly lower proportion of transfers with *Actinobacteria* (ANOVA test : p-value = 3.8*10-2), compared to bacteria

Species	N° (Fig5)	Genome size (Mb)	Genes transferred	HT quantity	Percentages of functions
Planctomyces maris	53	7.8	322	284	
Planctomyces brasiliensis	51	6.0	146	139	
Planctomyces limnophilus	52	5.5	48	43	
Gemmata obscuriglobus	11	9.2	377	347	
Singulisphaera acidiphila	58	9.8	238	219	
Isosphaera pallida	12	5.5	245	222	
Blastopirellula marina	41	6.7	343	307	
Rhodopirellula baltica	57	7.2	207	184	
Pirellula staleyi	56	6.2	152	137	
Phycisphaera mikurensis	54	3.9	244	220	
Candidatus Kuenenia stuttgartiensis	48	4.2	542	450	
Opitutus terrae	9	6.0	351	297	
Opitutaceae bacterium	5	7.1	525	466	
Verrucomicrobiae bacterium	14	5.8	333	288	
Coraliomargarita akajimensis	42	3.8	217	186	
Methylacidiphilum infernorum	13	2.3	130	102	
Pedosphaera parvula	10	7.4	496	472	
Chthoniobacter flavus	8	7.9	1026	803	
Verrucomicrobium spinosum	7	8.2	268	244	
Akkermansia muciniphila	19	2.7	159	129	
Lentisphaera araneosa	49	6.0	358	344	
Candidatus Protochlamydia amoebophila	1	2.4	188	152	
Parachlamydia acanthamoebae	2	3.1	120	105	
Waddlia chondrophila	3	2.1	70	63	
Simkania negevensis	4	2.6	156	134	
Chlamydia trachomatis	40	1.0	34	27	
Chlamydia muridarum	36	1.1	35	34	
Chlamydophila pneumoniae	38	1.2	30	24	
Chlamydophila pecorum	37	1.1	22	20	
Chlamydophila felis	35	1.2	14	12	
Chlamydophila caviae	34	1.2	16	15	
Chlamydia psittaci	39	1.2	8	8	
Chlamydophila abortus	–	1.1	0	0	
Turneriella parva	17	4.4	338	285	
Leptospira interrogans	16	4.7	292	258	
Leptospira biflexa	15	4.0	304	265	
Brachyspira intermedia	20	3.3	416	317	
Borrelia dutonii	–	1.0	0	0	
Borrelia burgdorferi	–	1.5	0	0	
Spirochaeta africana	59	3.3	208	165	
Sphaerochaeta pleomorpha	60	3.6	506	346	
Treponema pallidum	33	1.1	24	23	
Treponema primitia	6	4.1	462	321	
Treponema brennaborense	30	3.1	211	176	
Treponema denticola	31	2.8	319	227	
Chloroherpeton thalassium	47	3.3	248	203	
Prosthecochloris aestuarii	50	2.6	67	55	
Pelodictyon phaeoclathratiforme	55	3.0	113	97	
Chlorobium phaeobacteroides	45	2.7	115	103	
Chlorobium limicola	43	2.8	77	67	
Chlorobaculum parvum	44	2.3	78	70	
Chlorobium tepidum	46	2.2	68	55	
Blattabacterium sp	–	0.6	0	0	
Capnocytophaga ochracea	23	2.6	104	94	
Riemerella anatipestifer	29	2.2	153	131	
Parabacteroides distasonis	26	4.8	188	165	
Tannerella forsythia	32	3.4	119	105	
Porphyromonas gingivalis	27	2.3	37	32	
Porphyromonas asaccharolytica	24	2.2	34	33	
Alistipes finegoldii	18	3.7	240	213	
Bacteroides xylanisolvens	22	6.0	157	133	
Bacteroides vulgatus	21	5.2	466	388	
Prevotella dentalis	25	3.4	55	51	
Prevotella intermedia	28	2.7	66	63	

Groups (right brackets): Planctomycetes, Verrucomicrobiae, Chlamydiae, Spirochaetes, Chlorobi, Bacteroidetes

Legend: ■ Information storage and processing   ■ Cellular process and signaling   ■ Metabolism   ■ Poorly characterized   ***Pedosphaera parvula*: Compartmentalized bacteria**

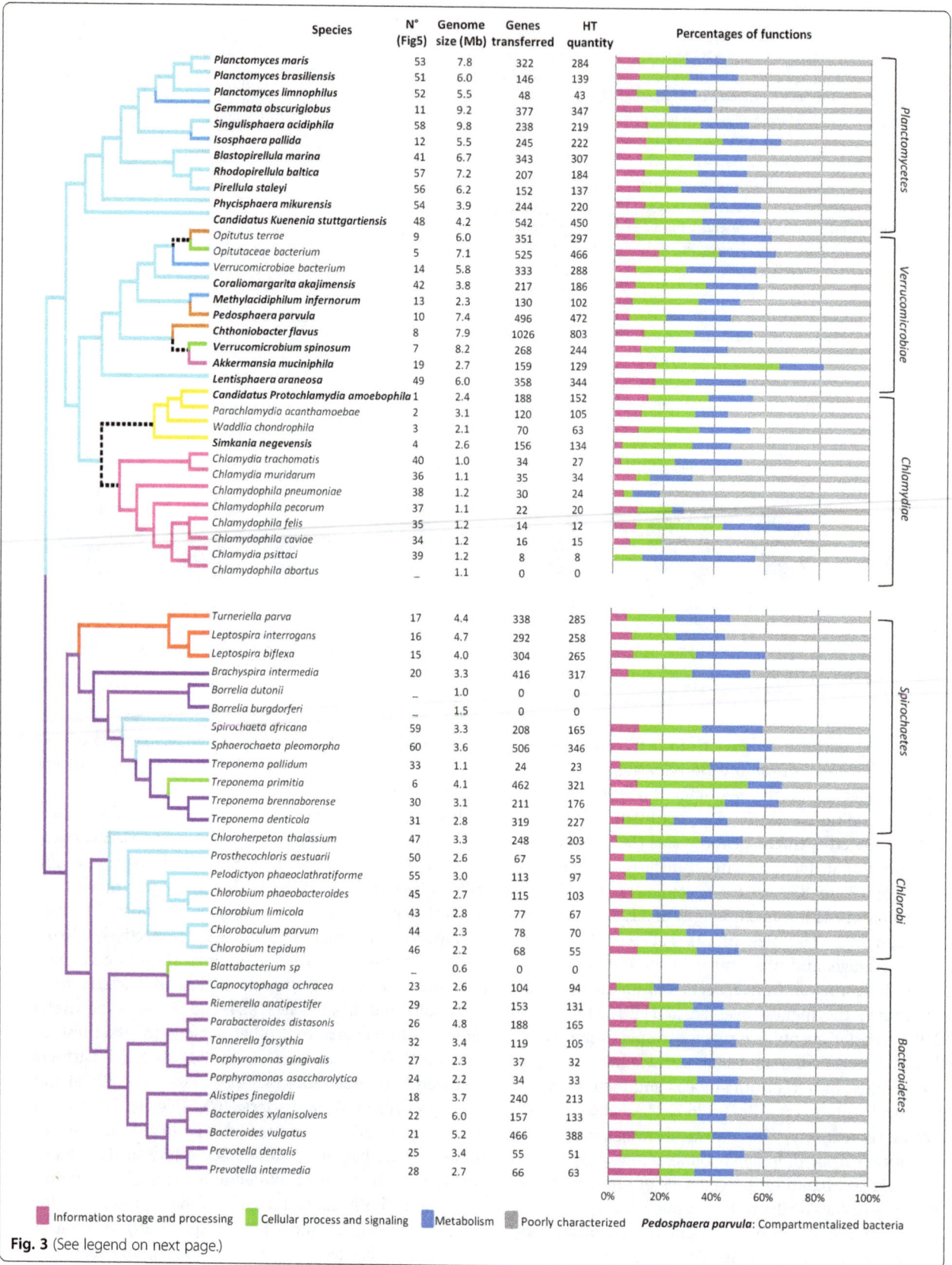

Fig. 3 (See legend on next page.)

Fig. 3 Functional description of the detected horizontally transferred genes in each studied bacteria. The habitats of species and their ancestors are indicated by the color of the *branches* (*red* : ubiquitous, *cyan* : water, *green* : insects, *yellow*: amoeba, *pink*: intracellular vertebrates, *purple*: extracellular vertebrates, *blue*: water and soils, *brown*: soils), the *black dotted branches* indicate an unknown habitat. The compartmentalized bacteria are indicated in *bold*. The first column contains the number assigned to each species in Fig. 5, the second presents the genome sizes of bacteria, the third, the quantity of genes transferred in genomes, and the fourth, the quantity of HT detected. The *bar graphic* represents the distribution of the four functional categories of COGs (Information storage and processing, Metabolism, Cellular process and signaling, Poorly characterized), which contain the 18 sub-categories studied

from other habitats. Extracellular bacteria from vertebrates also presented a significantly higher proportion of transfers with *Fusobacteria* (3.0%) compared to bacteria from other habitats (Kruskal test: *p*-value = 1.6*10-2), whereas the bacteria living in soil, or soil and water, exchanged significantly more with *Acidobacteria* (4.1 and 3.9% respectively) (Kruskal test: *p*-value = 5.3*10-4) (Fig. 4). The Principal components analysis (PCA) of data recovered for HT (HT proportions and partners) showed a relationship between bacterial habitats, the quantity of HT events, and its proportion in the genome and the partner transfers (Correlation test: *p*-value = 2.4*10-7). Hierarchical clustering analysis allowed the identification of two major clusters: the environmental bacteria (soils, water, soils-water and Ubiquitous bacteria) and bacteria from amoeba were in a first cluster, and the intracellular and extracellular bacteria of vertebrates were in the second cluster (Fig. 5). A clustering according to phylogenetic relationships among species can also be identified, but was less significant (Correlation test: *p*-value = 3.5*10-4) than clustering by habitat.

Horizontally transferred functions and phenotype

When analyzing the functions of the transferred sequences, we found that the general function distribution in the HT for the different habitats was not similar to that of the whole genomes which suggests that HT was not due to chance. Genes involved in cell processes and signaling (33 to 50%) seemed to be significantly more subject to HT, whereas genes dedicated to information storage (12 to 17%) were less subject to HT (t-test between whole genomes and transferred genes : *p*-value = 4.6*10-2 and 8.3*10-4) (Fig. 3). Moreover, there were significant differences among the habitats. Biological functions of transferred sequences were biased to three

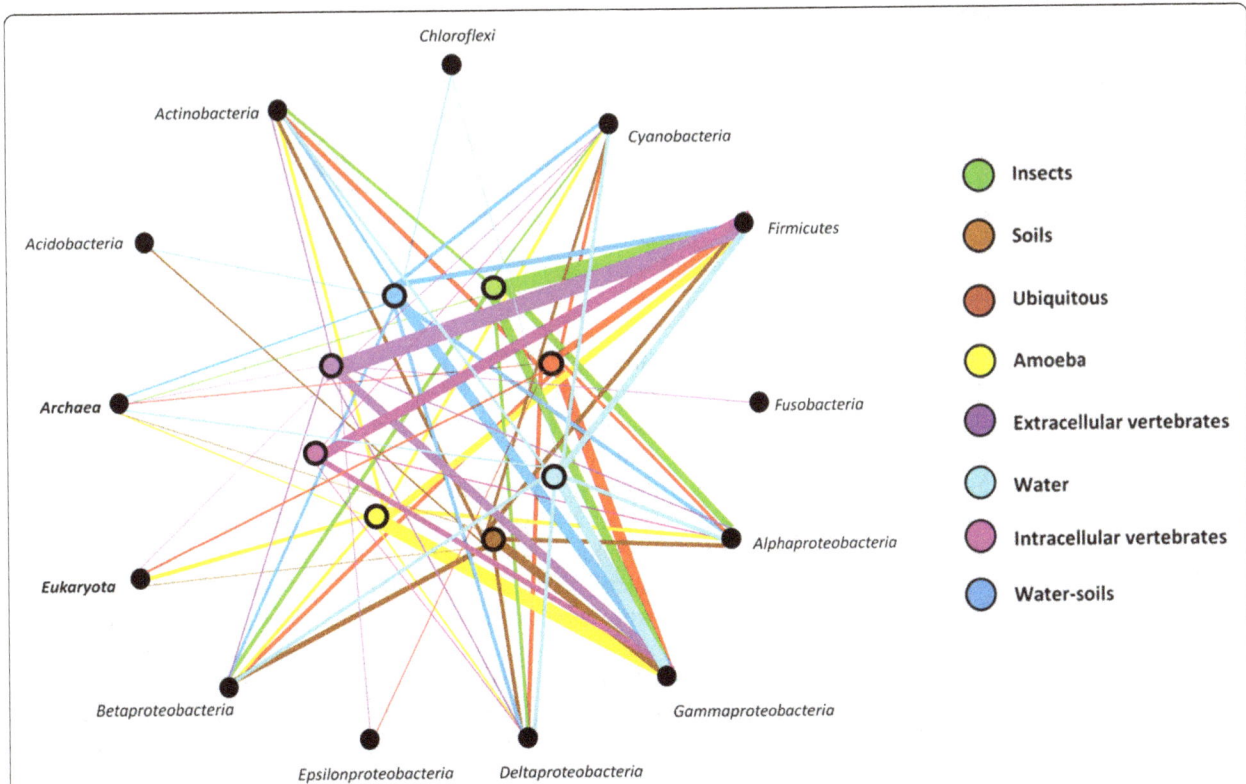

Fig. 4 Preferences for horizontal transfer partners among habitats. The *colored point* corresponds to the bacteria from the different habitats (*red* : ubiquitous, *cyan* : water, *green* : insects, *yellow*: amoeba, *pink*: intracellular vertebrates, *purple*: extracellular vertebrates, *blue*: water and soils, *brown*: soils). The traits are colored according to the habitats of studied bacteria and their thickness is proportional to the amount of genes exchanged

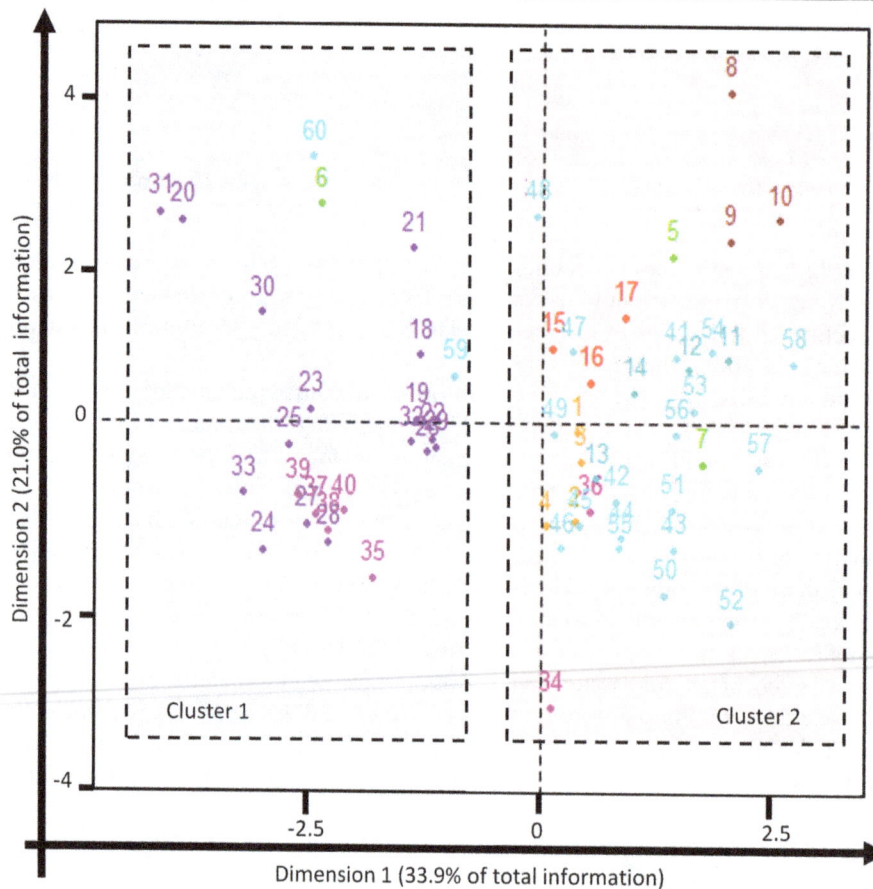

Fig. 5 Results for the principal component analysis and hierarchical clustering. They were realized with the variables: HT quantities and proportions in the studied genomes and their partners. The individuals represented are the 60 bacteria where HT were identified, the *colors of points* and numbers indicate their habitat. Two clusters were defined (*transparent color squares*) according the bacterial habitats (Chi2 test = 2.4*10-7): 91.3% of bacteria from vertebrates are gathered in the cluster 1 (*blue square*) and 92.0% of environmental bacteria belong to the cluster 2 (*red square*). Axis 1 contains 33.9% of the information and mainly represents the transfer partners (Variable represented: *Cyanobacteria*, *Firmicutes*, *Fusobacteria* and *Alphaproteobacteria* (>40%), *Actinobacteria*, *Acidobacteria* and *Epsilonbacteria* (>20%)). The axis 2 contains 21.0% of total information and represents mainly the HT quantities and proportions (Variables represented: Sequences quantities and proportions (>70%), *Actinobacteria* and *Gammaproteobacteria* (>15%)). However, 46.1% of the information is missing, concerning mainly the *Gammaproteobacteria* and *Acidobacteria*

categories according to bacterial habitat: the signal transduction mechanism function in ubiquitous bacteria and in bacteria from soils (20.2 and 17.9%, respectively; ANOVA test: *p*-value = 2.3*10-4) the transport and metabolism of amino acid in bacteria from amoeba and lipids in ubiquitous bacteria (16%, Kruskal-Wallis test and correlation test: *p*-value = 7.5*10-2 and 2.1*10-4; 10.5%, ANOVA test: *p*-value = 6.9*10-3, respectively) and the defense mechanism in bacteria from extracellular vertebrates (4.7%, Kruskal-Wallis: *p*-value = 3.5*10-2).

To test the impact of compartmentalization on HT, we compared the 11 compartmentalized with 9 non-compartmentalized bacteria in water, where the sample size allowed us to obtain statistically significant results (Fig. 3). HT proportions and preferences for partners were identical between the two groups of bacteria. However, we detected two functions that were overrepresented

among the HT events of compartmentalized bacteria, compared to non-compartmentalized, including cell wall/membrane/envelope biogenesis (5.5% of transferred genes, ANOVA test: *p*-value = 5.05*10-2) and lipid biosynthesis (2.4% of transferred genes, ANOVA test : *p*-value = 3.7*10-2) (Fig. 6). Genes encoding for these two functions were found to be horizontally transferred in compartmentalized bacteria of the different habitats, including 25 genes implicated in cell wall/membrane/envelope biogenesis (of which 12 are present in at least, one compartmentalized bacteria from each phylum) and 13 genes implicated in lipid biosynthesis (of which 3 were present in at least one compartmentalized bacteria from each phylum). Moreover, five of the genes implicated in the two functions of interest were found to be horizontally transferred in at least one compartmentalized bacteria from each habitat. These genes encoded for two different

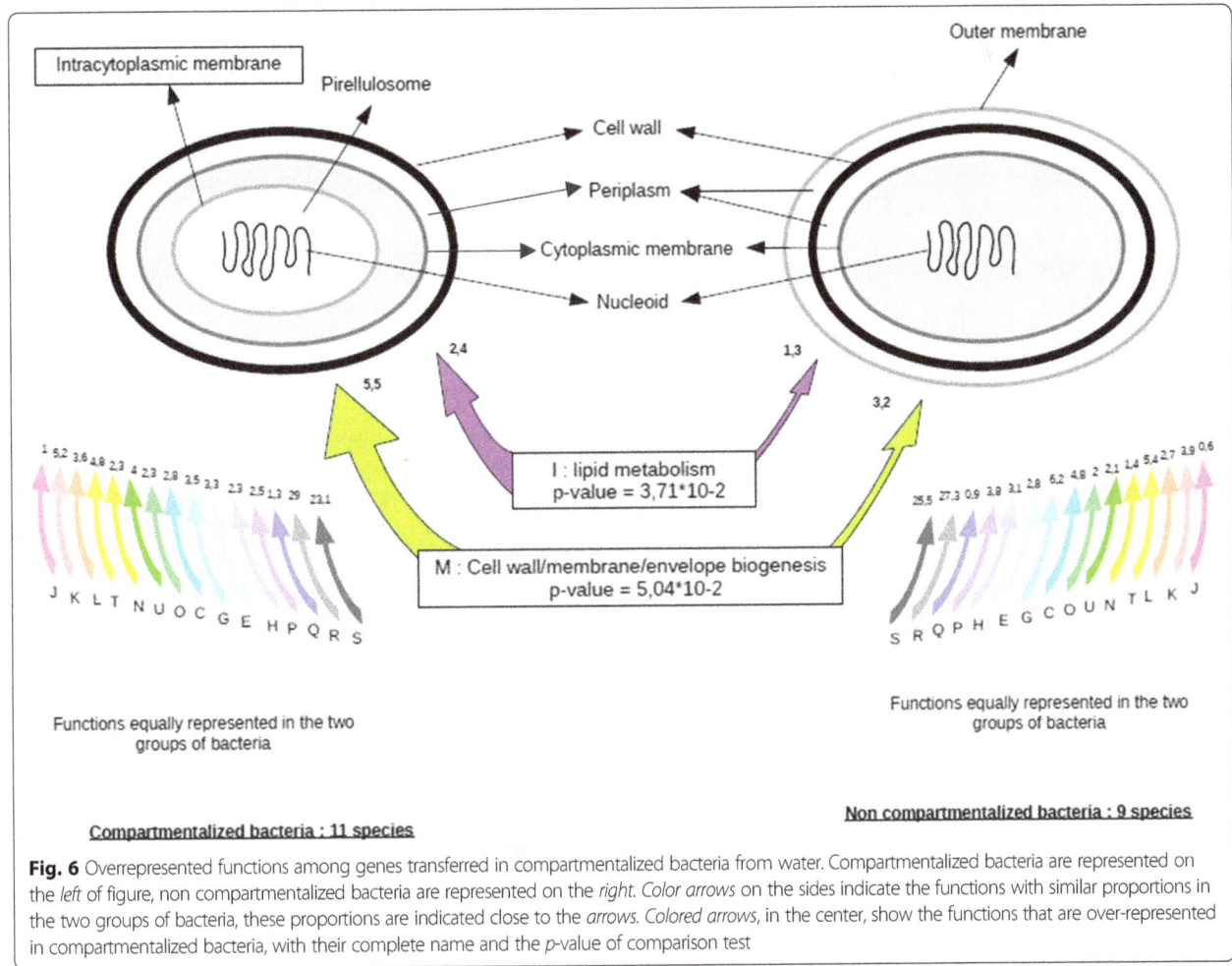

Fig. 6 Overrepresented functions among genes transferred in compartmentalized bacteria from water. Compartmentalized bacteria are represented on the *left* of figure, non compartmentalized bacteria are represented on the *right*. *Color arrows* on the sides indicate the functions with similar proportions in the two groups of bacteria, these proportions are indicated close to the *arrows*. *Colored arrows*, in the center, show the functions that are over-represented in compartmentalized bacteria, with their complete name and the *p*-value of comparison test

glycosyl transferases (Gene ID: 1791906 and 1796578), a carboxy-terminal processing protease (Gene ID: 1796308), an ACP reductase (Gene ID: 1793322), and a Resistance-Nodulation-Cell Division (RND) efflux membrane fusion protein (Gene ID: 1796098).

Discussion

The comparative analysis of 33 genomes from PVC species from four different phyla showed the influence of the living environment and compartmentalization on the genome composition of PVC bacteria. The common genes were genes encoding for transcription, signal transduction mechanisms, energy production and membrane biogenesis. Conversely, shared and specific genes encode for different functions in relation to the lifestyle of the corresponding species. Evidence for a random horizontal transfer of DNA sequences has been given using a phylogenomic approach. Genes implicated in cell wall/membrane/envelope biogenesis, and those involved in lipid metabolism, were found to be over-represented among the transferred genes of compartmentalized bacteria from different habitats, according to a convergent evolutionary selection.

Our findings replicate observations from previous studies which demonstrated the role played by shared genomes in environmental adaptation [68]. Nevertheless, our approach, by examining as many as 8 different habitat conditions, offers a large advantage over other genomic studies, and increases the reliability of our results. The low proportion of specific genes that have been detected in bacteria from insects, soils and soil-water milieu is rather due to the higher number and more distant phylogenetic relationships among the species studied [69, 70] compared to the other habitats. Intracellular bacteria from vertebrates showed a low proportion (1.9%) of horizontally transferred sequences compared to the other bacteria. This result is probably related to the physical isolation of intracellular bacteria, which prevents opportunities for HT [71, 72]. This agrees with previous studies showing that the predominant evolutionary process in intracellular bacteria is genome reduction, leading to smaller genome sizes [73, 74]. Intracellular bacteria in amoeba with 4.8% of HT are the exception [75, 76], since amoeba can phagocyte several bacteria at once, giving a particular field for potential genetic exchange and a training ground for the emergence of parasitism [75].

Likewise, results obtained for partners of transfers analysis were in agreement with previous results concerning transfers between PVC bacteria and *Proteobacteria* [43] or *Spirochaetes* and *Firmicutes* [12, 13]. Indeed, HT occurred preferentially between bacteria from the same habitats, as had already been assumed. *Firmicutes* are one of the two major phyla present in the gut microbiome [77], and this is the main partner of our bacteria from vertebrates. In the same way, *Acidobacteria* are mainly detected in soil [78] and they are overrepresented as HT partners of bacteria from soils, compared to bacteria from other habitats. The tendency of bacteria from Amoeba to exchange more with Eukaryotes, especially plants, is probably due to their ancestral habitats. Indeed, ancestral *Chlamydiae* are known to have lived in and exchanged genes with the *Archaeplastides* [79, 80]. Thus, we can support the hypothesis that part of the HT detected was acquired by the interaction between the ancestors of the *Chlamydiae* and the plants, followed by the loss in the majority of bacteria. It is worth noting that like previous studies for HT detections, it is difficult to distinguish between ancient and recent HT events; yet HT partners are the witnesses of modern and ancestral habitats of the bacteria studied, and our HT analysis helps infer the ancestral habitat of these bacteria.

Beyond the complexity hypothesis that claims that genes involved in transcription and translation are less prone to transfer than metabolic genes, our findings showed that horizontal transfers can affect any function. Thus, HT do not only concern genes encoding for metabolic mechanisms and other functions that enhance pathogenicity, like genes for virulence and antimicrobial activity [1, 15, 76]; genes involved in transcription and translation, in cell surface and DNA binding, and genes essential for defense can likely be transferred as well [46, 81–83]. Positive selection might be contributing to the overrepresentation of some functions in the category of transferable genes [84, 85]. Indeed, horizontally acquired genes that have a useful function are maintained as it follows a strategy of colonization and adaptation to the environment. Our findings confirm previous results showing that HT particularly affects the genes involved in lipid metabolism, signal transduction and membrane transport in PVC bacteria, and genes specific to outer membrane (such as O-antigen polymerase and outer membrane efflux protein) in some *Planctomycetes* [43, 86, 87]. Since, the intracytoplasmic membrane of compartmentalized bacteria is a lipid bilayer, we can assume that the over-representation of the two functions in the genes transferred could be related to the cell plan of the bacteria. These genes may be essential for the maintenance of the supplementary intracytoplasmic membrane. Knowing that the quantity of HT events was found to be similar between compartmentalized and non compartmentalized bacteria, these results revealed the possible impact of the cell plan on the transfers' positive selection. This selection that seems to be dependent on the function, and induces the recurrent maintenance of some transferred genes involved in the formation of compartments in bacteria from different habitats. It is noteworthy that genes implicated in lipid metabolism and membrane biosynthesis were not over-represented in the non-transferred part of the genome of compartmentalized bacteria, compared to the other bacteria; therefore, the selection seems to concern only the transferred genes.

One limitation of our comparative genomic approach is that the number of genomes studied leads to a small sample size in each environmental category, which hinders the realization of the statistical tests for certain categories. Moreover, our dataset was comprised of seven phyla, with only few representatives of soil bacteria, four for bacteria living in amoeba, and three extracellular bacteria from insects or ubiquitous, while some environmental categories contain only bacteria from just one phylum. Although the sample size was minimal, the results obtained were statistically usable and showed significant differences among phylogenetically close bacteria in relation with their habitat. Given the increased number of sequenced genomes, it will be interesting to characterize HT events in compartmentalized bacteria for diverse phyla, in order to elucidate the role of physical barriers in horizontal transfers.

Conclusions

The genomic study of bacteria allowed to better understand the influence of the different constraints acting on genomes evolution in bacteria, especially the impact of the habitat and the special cell plan, in PVC superphylum. The habitat influences the flux of horizontal transfer and determines the partners for genetic exchanges. The presence of an intracytoplasmic membrane in some PVC bacteria doesn't seem to limit the HT but rather, induces a selection of transferred genes, according to their functions.

Additional files

Additional file 1: Genomic features and habitats of bacteria studied. For each of the bacteria studied, we present the genomes identifiers, genomic features (proteins quantity, GC percent), the habitat (environment and lifestyle) and the cell plan. (CSV 6 kb)

Additional file 2: Proteins used for phylogenetic reconstruction. For each of the 768 sequences used for alignments and phylogenies we retrieved the accession number (column 1), the name of marker (column 2) and the species (column 3). (CSV 37 kb)

Additional file 3: Phylogeny of PVC bacteria and phylogeny of *Bacteroidetes-Chlorobi-Spirochaetes*. The phylogenies of studied bacteria were realized with Maximum likelihood method, within Mega 5 software, the bootstraps values are indicated at each node. The trees of PVC bacteria and *Bacteroidetes-Spirochaetes-Chlorobi* were rooted with *Bacteroides xylanosolvens* (*Spirochaetes*) and with *Bastopirellula marina* (*Planctomycetes*), respectively (the outgroup was removed for more clear representation). (TIF 2084 kb)

Additional file 4: The two phylogenetic patterns used for HT detection. The presence of one of these two patterns in a species tree allows us to identify an HT, these patterns are illustrated with examples specific to our data. (TIF 898 kb)

Additional file 5: Clustering of species according orthologous groups distribution. Clustering of studied species, according the pattern of presence/absence of genes in orthologous groups. The habitats of species are indicated by colors (red: ubiquitous, cyan: water, green: insects, yellow: amoeba, pink: intracellular vertebrates, purple: extracellular vertebrates, blue: water and soils, brown: soils) and phylum of species are in brackets. (TIF 1303 kb)

Additional file 6: HT distribution in habitats. The different boxplots describe the distribution of the numbers (a) and the proportions (b) of HT in each habitat. (TIF 878 kb)

Additional file 7: HT count according to genome size in studied bacteria. The trend curve (linear), its function (y) and the determination of coefficient (R^2) are indicated on the plot. (TIF 606 kb)

Acknowledgements
We thank Olivier Chabrol for assistance in computer programming during the elaboration of HT detection strategy and Manuela Royer Carenzi for her assistance in the statistical analyzes. We also thank the Xegen company for their assistance in HT detection by using of Phylopattern software. We thank TradOnline for English reviewing.

Funding
This work was supported by the Assistance Publique - Hopitaux de Marseille (Marseille Public University Hospital System). VM was supported by a Chairs of Excellence program from the Centre National de la Recherche Scientifique (CNRS). The funders had no role in study design, data collection and interpretation or the decision to submit the work.

Authors' contributions
PS carried out the design of the study, the strategy elaboration and the collection of data, performed the statistical analysis of results, and drafted the manuscript. PP participated in strategy elaboration, data interpretation and revised the manuscript. DR conceived the study, participated in its design and coordination and revised the manuscript. VM participated in the coordination of the study, strategy elaboration and the interpretation of data, and also drafted the manuscript. All authors read and approved the final manuscript.

Competing interests
The authors declare that they have no competing interests.

Author details
¹Aix Marseille Université, CNRS, Centrale Marseille, I2M UMR 7373, Evolution Biologique et Modélisation, 3 place Victor Hugo, Marseille 13331, France. ²Aix Marseille Univ, CNRS, IRD, INSERM, AP-HM URMITE, IHU -Méditerranée Infection, 19-21 Boulevard Jean Moulin, Marseille 13005, France.

References
1. Ochman H, Lawrence JG, Groisman EA. Lateral gene transfer and the nature of bacterial innovation. Nature. 2000;405(6784):299–304.
2. Le PT, Ramulu HG, Guijarro L, Paganini J, Gouret P, Chabrol O, et al. An automated approach for the identification of horizontal gene transfers from complete genomes reveals the rhizome of Rickettsiales. BMC Evol Biol. 2012; 12(1):243.
3. Thomas CM, Nielsen KM. Mechanisms of, and barriers to, horizontal gene transfer between bacteria. Nat Rev Microbiol. 2005;3(9):711–21.
4. Ghigo JM. Natural conjugative plasmids induce bacterial biofilm development. Nature. 2001;412:442–5.
5. Reisner A, Höller BM, Molin S, Zechner EL. Synergistic effects in mixed Escherichia coli biofilms: conjugative plasmid transfer drives biofilm expansion. J Bacteriol. 2006;188:3582–8.
6. Aminov RI. Horizontal gene exchange in environmental microbiota. Front Microbiol. 2011;2(158):10–3389.
7. Smillie CS, Smith MB, Friedman J, Cordero OX, David LA, Alm EJ. Ecology drives a global network of gene exchange connecting the human microbiome. Nature. 2011;480(7376):241–4.
8. Ilves H, Horak R, Kivisaar M. Involvement of sigma (S) in starvation-induced transposition of Pseudomonas putida transposon Tn4652. J Bacteriol. 2001; 183:5445–8.
9. Dorer MS, Fero J, Salama NR. DNA damage triggers genetic exchange in Helicobacter pylori. PLoS Pathog. 2010;6(7):e1001026.
10. Stanton TB, Humphrey SB, Sharma VK, Zuerner RL. Collateral effects of antibiotics: carbadox and metronidazole induce VSH-1 and facilitate gene transfer among Brachyspira hyodysenteriae strains. Appl Environ Microbiol. 2014;74:2950–6.
11. Hacker J, Carniel E. Ecological fitness, genomic islands and bacterial pathogenicity : A Darwinian view of the evolution of microbes. EMBO Rep. 2011;21:376–81.
12. Bellgard MI, Wanchanthuek P, La T, Ryan K, Moolhuijzen P, Albertyn Z. Genome sequence of the pathogenic intestinal spirochete brachyspira hyodysenteriae reveals adaptations to its lifestyle in the porcine large intestine. PLoS One. 2009;4(3):e4641.
13. Viswanathan VK. Spirochaeta and their twisted ways. Gut Microbes. 2012; 3(5):399–400.
14. Craigie R, Gellert M, Lambowitz AM. Mobile DNA. In: Craig NL, editor. American society for microbiology. 1989.
15. Baquero F, Alvarez-Ortega C, Martinez JL. Ecology and evolution of antibiotic resistance. Environ Microbiol Rep. 2009;1(6):469–76.
16. Gemski P, Lazere JR, Casey T, Wohlhieter JA. Presence of virulence-associated plasmid in Yersinia pseudotuberculosis. Infect Immun. 1980;28:1044–7.
17. Groisman EA, Ochman H. Pathogenicity islands: bacterial evolution in quantum leaps. Cell. 1996;87:791–4.
18. Hacker J, Blum-Oehler G, Muhldorfer I, Tschape H. Pathogenicity islands of virulent bacteria: structure, function and impact on microbial evolution. Mol Microbiol. 1997;23:1089–97.
19. Marraffini LA, Sontheimer EJ. CRISPR interference: RNA-directed adaptive immunity in bacteria and archaea. Nat Rev Genet. 2010;11(3):181–90.
20. Bikard D, Hatoum-Aslan A, Mucida D, Marraffini LA. CRISPR interference can prevent natural transformation and virulence acquisition during in vivo bacterial infection. Cell Host Microbe. 2012;12(2):177–86.
21. González-Candelas F, Francino MP. Barriers to Horizontal Gene Transfer: Fuzzy and Evolvable Boundaries. In: Pilar F editor. Horizontal Gene Transfer in Microorganisms. Caister: Academic Press. 2012. p. 47.
22. Besemer K, Singer G, Quince C, Bertuzzo E, Sloan W, Battin TJ. Headwaters are critical reservoirs of microbial diversity for fluvial networks. Proc R Soc Lond B Biol Sci. 2013;280(1771):20131760.
23. Maccario L, Vogel TM, Larose C. Potential drivers of microbial community structure and function in Arctic spring snow. Front Microbiol. 2014;5:413.
24. Mendes LW, Tsai SM, Navarrete AA, de Hollander M, van Veen JA, Kuramae EE. Soil-borne microbiome: linking diversity to function. Microb Ecol. 2015;70(1):1–11.
25. Caro-Quintero A, Konstantinidis KT. Inter-phylum HGT has shaped the metabolism of many mesophilic and anaerobic bacteria. ISME J. 2015;9(4):958–67.
26. Tyson GW, Chapman J, Hugenholtz P, Allen EE, Ram RJ, Richardson PM, et al. Community structure and metabolism through reconstruction of microbial genomes from the environment. Nature. 2004;428(6978):37–43.
27. Lauro FM, McDougald D, Thomas T, Williams TJ, Egan S, Rice S, et al. The genomic basis of trophic strategy in marine bacteria. Proc Natl Acad Sci. 2009;106(37):15527–33.
28. Cho JC, Vergin KL, Morris RM, Giovannoni SJ. Lentisphaera araneosa gen. nov., sp. nov, a transparent exopolymer producing marine bacterium, and the description of a novel bacterial phylum, Lentisphaerae. Environ Microbiol. 2004;6(6):611–21.

29. Wagner M, Horn M. The Planctomycetes, Verrucomicrobia, Chlamydiae and sister phyla comprise a superphylum with biotechnological and medical relevance. Curr Opin Biotechnol. 2006;17(3):241–9.

30. Siegl A, Kamke J, Hochmuth T, Piel J, Richter M, Liang C, et al. Single-cell genomics reveals the lifestyle of Poribacteria, a candidate phylum symbiotically associated with marine sponges. ISME J. 2011;5(1):61–70.

31. Gupta RS, Bhandari V, Naushad HS. Molecular signatures for the PVC clade (Planctomycetes, Verrucomicrobia, Chlamydiae, and Lentisphaerae) of bacteria provide insights into their evolutionary relationships. Front Microbiol. 2012;3:327.

32. Lindsay MR, Webb RI, Fuerst JA. Pirellulosomes: a new type of membrane-bounded cell compartment in planctomycete bacteria of the genus Pirellula. Microbiology. 1997;143(3):739–48.

33. Lindsay MR, Webb RI, Strous M, Jetten MS, Butler MK, Forde RJ, Fuerst JA. Cell compartmentalisation in planctomycetes: novel types of structural organisation for the bacterial cell. Arch Microbiol. 2001;175(6):413–29.

34. Fuerst JA, Sagulenko E. Beyond the bacterium: planctomycetes challenge our concepts of microbial structure and function. Nat Rev Microbiol. 2011;9:403–13.

35. Lee KC, Webb RI, Janssen PH, Sangwan P, Romeo T, Staley JT, Fuerst JA. Phylum Verrucomicrobia representatives share a compartmentalized cell plan with members of bacterial phylum Planctomycetes. BMC Microbiol. 2009;9(1):5.

36. Fieseler L, Horn M, Wagner M, Hentschel U. Discovery of the novel candidate phylum "Poribacteria" in marine sponges. Appl Environ Microbiol. 2004;70(6):3724–32.

37. Fuerst JA, Webb RI. Membrane-bounded nucleoid in the eubacterium Gemmata obscuriglobus. Proc Natl Acad Sci. 1991;88(18):8184–8.

38. Lage OM, Bondoso J, Lobo-da-Cunha A. Insights into the ultrastructural morphology of novel Planctomycetes. Antonie Van Leeuwenhoek. 2013;104(4):467–76.

39. Santarella-Mellwig R, Franke J, Jaedicke A, Gorjanacz M, Bauer U, Budd A, Devos DP. The compartmentalized bacteria of the planctomycetes-verrucomicrobia-chlamydiae superphylum have membrane coat-like proteins. PLoS Biol. 2010;8(1):e1000281.

40. Santarella-Mellwig R, Pruggnaller S, Roos N, Mattaj IW, Devos DP. Three-dimensional reconstruction of bacteria with a complex endomembrane system. PLoS Biol. 2013;11(5):e1001565.

41. NCBI - proteins.https://www.ncbi.nlm.nih.gov/protein. Accessed January and March 2013.

42. NCBI - genomes.http://mirrors.vbi.vt.edu/mirrors/ftp.ncbi.nih.gov/genomes/. Accessed January and March 2013.

43. Kamneva OK, Knight SJ, Liberles DA, Ward NL. Analysis of genome content evolution in PVC bacterial super-phylum: assessment of candidate genes associated with cellular organization and lifestyle. Genome Biol Evol. 2012;4(12):1375–90.

44. Tamura K, Peterson D, Peterson N, Stecher G, Nei M, Kumar S. MEGA5: molecular evolutionary genetics analysis using maximum likelihood, evolutionary distance, and maximum parsimony methods. Mol Biol Evol. 2011;28:2731–9.

45. Edgar RC. MUSCLE: multiple sequence alignment with high accuracy and high throughput. Nucleic Acids Res. 2004;32(5):1792–7.

46. Merhej V, Notredame C, Royer-Carenzi M, Pontarotti P, Raoult D. The rhizome of life: the sympatric Rickettsia felis paradigm demonstrates the random transfer of DNA sequences. Mol Biol Evol. 2011;28(11):3213–23.

47. Georgiades K. Genomics of epidemic pathogens. Clin Microbiol Infect. 2013;18(3):213–7.

48. JGI - genomes. http://genome.jgi.doe.gov/. Accessed between August and December 2014.

49. GOLD database. https://gold.jgi-psf.org/index. Accessed between August and December 2014.

50. List of Prokaryotic names with Standing in Nomenclature - Bacterio.net. http://www.bacterio.net/index.html. Accessed between August and December 2014.

51. Grigoriev IV, Nordberg H, Shabalov I, Aerts A, Cantor M, Goodstein D, et al. The genome portal of the department of energy joint genome institute. Nucleic Acids Res. 2012;D1(40):D26–32.

52. Nordberg H, Cantor M, Dusheyko S, Hua S, Poliakov A, Shabalov I, et al. The genome portal of the Department of Energy Joint Genome Institute: 2014 updates. Nucleic Acids Res. 2014;42(D1):D26–31.

53. Reddy TBK, Thomas AD, Stamatis D, Bertsch J, Isbandi M, Jansson J, et al. The Genomes OnLine Database (GOLD) v. 5: a metadata management system based on a four level (meta) genome project classification. Nucleic Acids Res. 2014;43(D1):D1099–D110.

54. Pinos S, Pontarotti P, Raoult D, Baudoin JP, Pagnier I. Compartmentalization in PVC super-phylum: evolution and impact. Biol Direct. 2016;11(1):38.

55. Fischer S, Brunk BP, Chen F, Gao X, Harb OS, Iodice JB, et al. Using OrthoMCL to assign proteins to OrthoMCL-DB groups or to cluster proteomes into New ortholog groups. Curr Protoc Bioinformatics. 2011;35(Suppl 6.12):1–19.

56. Tatusov RL, Koonin EV, Lipman DJ. A genomic perspective on protein families. Science. 1997;278(5338):631–7.

57. Wu S, Zhu Z, Fu L, Niu B, Li W. WebMGA: a customizable web server for fast metagenomic sequence analysis. BMC Genomics. 2011;12(1):444.

58. Galperin MY, Makarova KS, Wolf YI, Koonin EV. Expanded microbial genome coverage and improved protein family annotation in the COG database. Nucleic Acids Res. 2015;43(D1):D261–9.

59. Mitchell A, Chang HY, Daugherty L, Fraser M, Hunter S, Lopez R, et al. The InterPro protein families database: the classification resource after 15 years. Nucleic Acids Res. 2015;43(D1):D213–21.

60. Gouret P, Thompson JD, Pontarotti P. PhyloPattern: regular expressions to identify complex patterns in phylogenetic trees. BMC Bioinf. 2009;10(1):298.

61. Sankoff D, Morel C, Cedergren RJ. Evolution of 5S RNA and the non-randomness of base replacement. Nat New Biol. 1973;245:232–4.

62. Levene H. Robust tests for equality of variance. In: Olkin I, Ghurye SG, Hoeffeling W, Madow WG, Mann HB, editors. Contributions to Probability and Statistics: Essays in Honor of Harold Hotelling. Stanford: University Press; 1960. p. 278–292.

63. Kruskal WH, Wallis WA. Use of ranks in one-criterion variance analysis. J Am Stat Assoc. 1952;47(260):583–621.

64. Douglas CE, Michael FA. On distribution-free multiple comparisons in the one-way analysis of variance. Commun Stat Theory Methods. 1991;20(1):127–39.

65. Tukey JW. Comparing individual means in the analysis of variance. Biometrics. 1949;5:99–114.

66. Pagel M. Detecting correlated evolution on phylogenies: a general method for the comparative analysis of discrete characters. Proc R Soc London. 1994;B 255:37–45.

67. Spearman C. The proof and measurement of association between two things. Am J Psychol. 1904;15:72–101.

68. Tettelin H, Riley D, Cattuto C, Medini D. Comparative genomics: the bacterial pan-genome. Curr Opin Microbiol. 2008;11:472–7.

69. Daubin V, Gouy M, Perriere G. A phylogenomic approach to bacterial phylogeny: evidence of a core of genes sharing a common history. Genome Res. 2002;12(7):1080–90.

70. Charlebois RL, Clarke GP, Beiko RG, Jean AS. Characterization of species-specific genes using a flexible, web-based querying system. FEMS Microbiol Lett. 2003;225(2):213–20.

71. Ley RE, Peterson DA, Gordon JI. Ecological and evolutionary forces shaping microbial diversity in the human intestine. Cell. 2006;124:837–48.

72. Hirt RP, Alsmark C, Embley TM. Lateral gene transfers and the origins of the eukaryote proteome: a view from microbial parasites. Curr Opin Microbiol. 2015;23:155–62.

73. Merhej V, Royer-Carenzi M, Pontarotti P, Raoult D. Massive comparative genomic analysis reveals convergent evolution of specialized bacteria. Biol Direct. 2009;4(1):13.

74. Wolf YI, Koonin EV. Genome reduction as the dominant mode of evolution. Bioessays. 2013;35(9):829–37.

75. Molmeret M, Horn M, Wagner M, Santic M, Kwaik YA. Amoebae as training grounds for intracellular bacterial pathogens. Appl Environ Microbiol. 2005;71(1):20–8.

76. Moliner C, Fournier PE, Raoult D. Genome analysis of microorganisms living in amoebae reveals a melting pot of evolution. FEMS Microbiol Rev. 2010;34(3):281–94.

77. Qin J, Li R, Raes J, Arumugam M, Burgdorf KS, Manichanh C, et al. A human gut microbial gene catalogue established by metagenomic sequencing. Nature. 2010;464(7285):59–65.

78. Quaiser A, Ochsenreiter T, Lanz C, Schuster SC, Treusch AH, Eck J, Schleper C. Acidobacteria form a coherent but highly diverse group within the bacterial domain: evidence from environmental genomics. Mol Microbiol. 2006;50(2):563–75.

79. Subtil A, Collingro A, Horn M. Tracing the primordial Chlamydia: extinct parasites of plants? Trends Plant Sci. 2014;19(1):36–43.

80. Moustafa A, Reyes-Prieto A, Bhattacharya D. Chlamydia has contributed at

least 55 genes to Plantae with predominantly plastid functions. PLoS One. 2008;3(5):e2205.

81. Tooming-Klunderud A, Sogge H, Rounge TB, Nederbragt AJ, Lagesen K, Glöckner G, et al. From green to red: Horizontal gene transfer of the phycoerythrin gene cluster between Planktothrix strains. Appl Environ Microbiol. 2013;79(21):6803–12.

82. Nakamura Y, Itoh T, Matsuda H, Gojobori T. Biased biological functions of horizontally transferred genes in prokaryotic genomes. Nat Genet. 2004; 36(7):760–6.

83. Mongodin EF, Nelson KE, Daugherty S, DeBoy RT, Wister J, Khouri H, et al. The genome of Salinibacter ruber: Convergence and gene exchange among hyperhalophilic bacteria and archaea. PNAS. 2005;102(50):18147–52.

84. Gogarten JP, Doolittle WF, Lawrence J. G Prokaryotic evolution in light of gene transfer. Mol Biol Evol. 2002;19:2226–38.

85. Pál C, Papp B, Lercher MJ. Horizontal gene transfer depends on gene content of the host. Bioinformatics. 2005;21 Suppl 2:ii222–3.

86. Fuerst JA. The PVC superphylum: exceptions to the bacterial definition? Antonie Van Leeuwenhoek. 2013;104(4):451–66.

87. Paparoditis P, Västermark Å, Le AJ, Fuerst JA, Saier MH. Bioinformatic analyses of integral membrane transport proteins encoded within the genome of the planctomycetes species, Rhodopirellula baltica. Biochim Biophys Acta Biomembr. 2014;1838(1):193–215.

Karyotypic evolution of the *Medicago* complex: *sativa-caerulea-falcata* inferred from comparative cytogenetic analysis

Feng Yu[1,2], Haiqing Wang[1], Yanyan Zhao[1,2], Ruijuan Liu[1,2], Quanwen Dou[1*](ID), Jiangli Dong[3] and Tao Wang[3]

Abstract

Background: Polyploidy plays an important role in the adaptation and speciation of plants. The alteration of karyotype is a significant event during polyploidy formation. The *Medicago sativa* complex includes both diploid (2n = 2x = 16) and tetraploid (2n = 2x = 32) subspecies. The tetraploid *M.* ssp. *sativa* was regarded as having a simple autopolyploid origin from diploid ssp. *caerulea*, whereas the autopolyploid origin of tetraploid ssp. *falcata* from diploid form ssp. *falcata* is still in doubt. In this study, detailed comparative cytogenetic analysis between diploid to tetraploid species, as well as genomic affinity across different species in the *M. sativa* complex, were conducted based on comparative mapping of 11 repeated DNA sequences and two rDNA sequences by a fluorescence in situ hybridization (FISH) technique.

Results: FISH patterns of the repeats in diploid subspecies *caerulea* were highly similar to those in tetraploid subspecies *sativa*. Distinctly different FISH patterns were first observed in diploid ssp. *falcata*, with only centromeric hybridizations using centromeric and multiple region repeats and a few subtelomeric hybridizations using subtelomeric repeats. Tetraploid subspecies *falcata* was unexpectedly found to possess a highly variable karyotype, which agreed with neither diploid ssp. *falcata* nor ssp. *sativa*. Reconstruction of chromosome-doubling process of diploid ssp. *caerulea* showed that chromosome changes have occurred during polyploidization process.

Conclusions: The comparative cytogenetic results provide reliable evidence that diploid subspecies *caerulea* is the direct progenitor of tetraploid subspecies *sativa*. And autotetraploid ssp. *sativa* has been suggested to undergo a partial diploidization by the progressive accumulation of chromosome structural rearrangements during evolution. However, the tetraploid subspecies *falcata* is far from a simple autopolyploid from diploid subspecies *falcata* although no obvious morphological change was observed between these two subspecies.

Keywords: *Medicago sativa*, *M. sativa* ssp. *caerulea*, *M. sativa* ssp. *falcata*, Repetitive sequences, FISH, Chromosome evolution, Diploidization

Background

Polyploidy is very common in plant evolution. It plays an important role in adaptation and speciation of plants [1]. According to different chromosome set origins, polyploidy is generally classified into autopolyploid and allopolyploid [2]. The structural changes of genome including chromosome fusions, chromosome number reduction, and a variety of chromosome rearrangements were a significance event during polyploidy formation [3]. It has been illustrated in many allopolyploid species, such as *Nicotiana* [3, 4], *Tragopogon* [5], *Gossypium* [6, 7] and *Brassica* [8, 9]. In *Nicotiana*, intergenomic translocations have been detected in natural *N. tabacum* genotypes and this translocation was considered to be significant in tobacco fertility [3]. Compared with in allopolyploid, structural changes were more difficult to be discovered due to homologous genomes were duplicated in autopolyploid. However, chromosomal rearrangements were reported in induced autotetraploid *Lathyrus sativus* [10] and *Arabidopsis thaliana* [11].

The *Medicago sativa* complex includes both diploid (2n = 2x = 16) and tetraploid (2n = 2x = 32) subspecies [12]. Tetraploid subspecies *M. sativa* ssp. *sativa* L., an important world forage legume, and diploid subspecies

* Correspondence: douqw@nwipb.cas.cn
[1]Key Laboratory of Adaptation and Evolution of Plateau Biota, Northwest Institute of Plateau Biology, Chinese Academy of Sciences, Xining 810008, China
Full list of author information is available at the end of the article

M. sativa ssp. *caerulea* (Less. ex Ledeb.) Schmalh. have a similar morphology with violet flowers and coiled pods [13, 14]. Subspecies *M. sativa* ssp. *falcata* (L.) Arcang. comprises both diploid and tetraploid forms which differ morphologically from the previous two taxa by having conspicuous yellow flowers and straight to sickle-shaped pods [13, 14]. With the similar ploidy level, they intercross easily and produce viable hybrids [15]. Tetraploid ssp. *sativa* and ssp. *falcata* have been considered to be autotetraploidy due to appearance of quadrivalents at meiosis and tetrasomic inheritance [16–18]. Two diploid taxa ssp. *caerulea* and ssp. *falcata* in the complex were hypothesized to be the direct progenitor of tetraploid ssp. *sativa* and ssp. *falcata*, respectively [13, 14]. However, recent molecular evidence of chloroplast suggested *M. prostrata* may have introgression into the tetraploid ssp. *falcata* in past. Therefore, the *Medicago* complex is an interesting model for polyploidy evolutionary study especially for autopolyploid [19].

Heterochromatin distributions of diploid ssp. *falcata*, ssp. *caerulea* and tetraploid ssp. *sativa* have been analyzed by C-banding and N-banding techniques [20–24]. Comparing results showed that diploid ssp. *caerulea* had similar heterochromatin distribution with tetraploid ssp. *sativa*: constitutive heterochromatic was distributed mainly around the centromeres, telomere and interstitial region of short arms of the chromosomes and partly presented at the interstitial region of long arms of chromosomes [20–24]. On the contrary, there were few heterochromatic distributions on the telomere and interstitial region in diploid ssp. *falcata* except centromere regions [20, 21, 25]. Bauchan and Hossain's unpublished data mentioned that there were a larger number of C-bands in tetraploid ssp. *falcata* than that had been discovered in diploid ssp. *falcata* [12].

Compared with the traditional banding techniques, fluorescence in situ hybridization (FISH), a valuable molecular cytogenetic tool, can display the molecular information on the chromosome more directly, more accurately, and more stably [26, 27]. It has been widely applied to the study of plant genomic organization, chromosome identification, and species evolution by physical mapping repetitive genes or other sequences directly onto chromosomes [27–32]. In our previous study [33], 11 tandemly repetitive sequences (nine of which were novel) were isolated from a Cot-1 library in alfalfa and a FISH-based molecular cytogenetic karyotype was well developed for tetraploid ssp. *sativa*. In this study, we present an in-depth comparative molecular cytogenetic analysis between diploid and tetraploid subspecies in *Medicago sativa* complex using repetitive sequences and FISH. Chromosome changes will be described in detail in evolution process of autotetraploidy ssp. *sativa*. The relationship of tetraploid and diploid ssp. *falcata* will be discussed.

Methods
Plant materials
Four diploid ssp. *caerulea*, four diploid ssp. *falcata*, and six tetraploid ssp. *falcata* samples were used as materials in this study. Accessions beginning with 'PI' were obtained from the National Plant Germplasm System (NPGS) of the United States Department of Agriculture (USDA). Two tetraploid ssp. *falcata* accessions, XiaNH-072X-824 and Lizj0944, were acquired from the China Germplasm Bank of Wild Species. Accession 2–6 was collected from a wild population in Xinjiang, China. A list of materials with ploidy levels and origins is given in Table 1.

Chromosome preparation
Root tips with a length of 1–2 cm were harvested from germinated seeds or growing plants and pretreated in ice-cold water at 4 °C for 20–24 h. Root tips were then fixed in ethanol:glacial acetic acid (3:1, v/v) for 4 h at room temperature. Each root tip was squashed in a drop of 45% acetic acid. Finally, the slides were stored at −80 °C before use.

Probe preparation
Eleven tandemly repetitive DNA sequences developed in alfalfa by Yu et al. [33] were used in this study. Five of the sequences (*Ms*CR-1, *Ms*CR-2, *Ms*CR-3, *Ms*CR-4, and *Ms*CR-5) were centromeric or pericentromeric, three (*Ms*TR-1, clone 65, and clone 74) were subtelomeric, and three (E180, clone 68, and clone 87) produced multiple hybridization signals in alfalfa chromosomes [33]. We also used two rDNA regions, 5S and 18S–26S rDNA, as probes. The 5S rDNA sequence was amplified by polymerase chain reaction (PCR) using genomic

Table 1 Materials used in this study

Subspecies	Ploidy	Identification No.	Origin
M. sativa ssp. *caerulea*	2×	PI 464715	Turkey, Kars
	2×	PI 212798	Iran
	2×	PI 577551	Canada, Manitoba
	2×	PI 577548	Russia
M. sativa ssp. *falcata*	2×	PI 631808	Russia
	2×	PI 502447	Russia
	2×	PI 631813	Russia
	2×	PI 234815	Switzerland
M. sativa ssp. *falcata*	4×	PI 634023	Kazakhstan
	4×	PI 634118	Kazakhstan
	4×	PI 634117	Kazakhstan
	4×	XiaNH-072X-824	China
	4×	Lizj0944	China
	4×	2–6	China

DNA of alfalfa as described by Fukui et al. [34]. The plasmid pWrrn, which included fragments of wheat 18S–26S rDNA, was provided by Professor Tsujimoto (Tottori University, Japan). All purified DNA products except pWrrn were labeled by the random primer labeling method with tetramethyl-rhodamine-5-dUTP (red) or fluorescein-12-dUTP (green) (Roche Diagnostics). pWrrn was labeled with tetramethyl-rhodamine-5-dUTP (red) using the nick-translation method.

FISH and microphotometry
FISH procedure was based on Mukai's description [35] with minor modifications. Chromosome DNA denaturation was carried out in 0.2 M NaOH in 70% ethanol at room temperature for 8 min and then dehydrated with the cold ethanol series. The probe mixture (25 ng of each labeled probe DNA, 5–10 mg of sheared salmon sperm DNA, 50% formamide, 2 × SSC, and 10% dextran sulfate) was denatured for 5 min at 95 °C and cooled on ice. Then, the denatured probe mixture was applied on dehydrated chromosome slide. The slides were incubated in a humid chamber at 37 °C overnight. After hybridization, the slides were washed in 2× SSC three times for 5 min at room temperature and briefly dried. Chromosomes were counterstained with 4′, 6-diamidino-2-phenylindole in Vectashield mounting medium (Vector Laboratories, Burlingame, CA, USA). Images were acquired with a cooled charge-coupled device camera (Photometrics CoolSNAP) under a fluorescence microscope (Leica) and were processed with the MetaVue Imaging System. Finally, images were adjusted with Adobe Photoshop 6.0 for contrast and background optimization.

Results
Physical mapping of repetitive sequences on mitotic chromosomes
In Medicago sativa ssp. caerulea
Physical mapping of the 11 repetitive sequences in diploid ssp. *caerulea* accession PI 464715 was conducted by FISH (Additional file 1: Figure S1a–h). Repeat sequence *Ms*CR-1, *Ms*CR-2, *Ms*CR-3, *Ms*CR-4, and *Ms*CR-5 were physically mapped on pericentromeric regions of 14, 8, 6, 10, and 9, respectively, of the 16 chromosomes of ssp. *caerulea* (Additional file 1: Figure S1a–d). Double-target FISH further revealed that *Ms*CR-3 overlapped with *Ms*CR-2, *Ms*CR-4, and *Ms*CR-5 on four, two, and three chromosomes, respectively (Additional file 1: Figure S1b–d). All three subtelomeric sequences (*Ms*TR-1, clone 65, and clone 74) were co-localized on one end of 12–13 chromosomes (Additional file 1: Figure S1e, f). At the same time, a variation of two end of one chromosome was also detected (Additional file 1: Figure S1e). Probes E180, clone 68, and clone 87 displayed hybridization signals on 16, 15, and 14 chromosomes, respectively

(Additional file 1: Figure S1g, h). Double-target FISH revealed different FISH patterns between E180 and clones 68 or 87.

The physical mapping results revealed E180 produced the greatest number of information of hybridization signals on each chromosome. Thus, double-target FISH between each repetitive sequence and E180 were carried out. Consequently, we used E180 FISH patterns, previous double-target FISH results, and chromosome arm ratios as references to allocate each sequence to a particular chromosome (Additional file 1: Figure S2a–h). The sequences were mapped as follows (Fig. 1): 18S–26S rDNA: on chromosome 1; 5S rDNA: on chromosome 5 and 6; *Ms*CR-1: on all chromosome except 7; *Ms*CR-2: on chromosome 1, 2, 3, and 6; *Ms*CR-3: on chromosome 3, 4, and 8; *Ms*CR-4: on chromosome 1, 2, 4, 5, and one of chromosome 3; *Ms*CR-5: on chromosome 1, 2, 5, 8, and one of chromosome 3; *Ms*TR-1 (co-localized with clone 65 and clone 74): on chromosome 2, 3, 5, 6, 7, 8 and one of chromosome 1; clone 68: on all chromosomes except one of chromosome 7; and clone 87: on all chromosomes except chromosome 7.

To development a standard molecular karyotype among different ssp. *caerulea* accessions, three other accessions (PI 212798, PI 577551, and PI 577548) were also used in cytogenetic analysis. Because 18S–26S rDNA, 5S rDNA, E180, *Ms*CR-3, and *Ms*TR-1 repeats showed a strong ability to distinguish chromosomes according to the results of chromosome allocation, two FISH cocktails—one consisting of 18S–26S rDNA, 5S rDNA, and E180 (Additional file 1: Figure S3a–d, Fig. 3) and the other comprising E180, *Ms*CR-3, and *Ms*TR-1 (Additional file 1: Figure S3i–l and Figure S6, Fig. 3)—were applied to these four accessions by FISH. The detailed distributions are presented in Fig. 3. Although polymorphic FISH patterns were detected on a few chromosomes among accessions, a relatively conserved karyotype was still described in Fig. 6 (a).

In diploid Medicago sativa ssp. falcata
Physical mapping of the 11 repetitive sequences in diploid *M. sativa* ssp. *falcata* accession PI 631808 was also conducted using FISH (Additional file 1: Figure S1i–p). Repeat sequence *Ms*CR-1, *Ms*CR-2, *Ms*CR-3, *Ms*CR-4, and *Ms*CR-5 were physically mapped on pericentromeric regions of 16, 10, 0–1, 9, and 8 of the 16 total chromosomes, respectively (Additional file 1: Figure S1i–l). In addition, *Ms*CR-4 and *Ms*CR-5 showed an extra band on one and two chromosomes, respectively. The three subtelomeric probes (*Ms*TR-1, clone 65, and clone 74) were co-localized on only one end of one chromosome (Additional file 1: Figure S1m, n). Probe E180 was mostly localized at a single site (mainly around the centromere) of 10 to 11 chromosomes rather than the

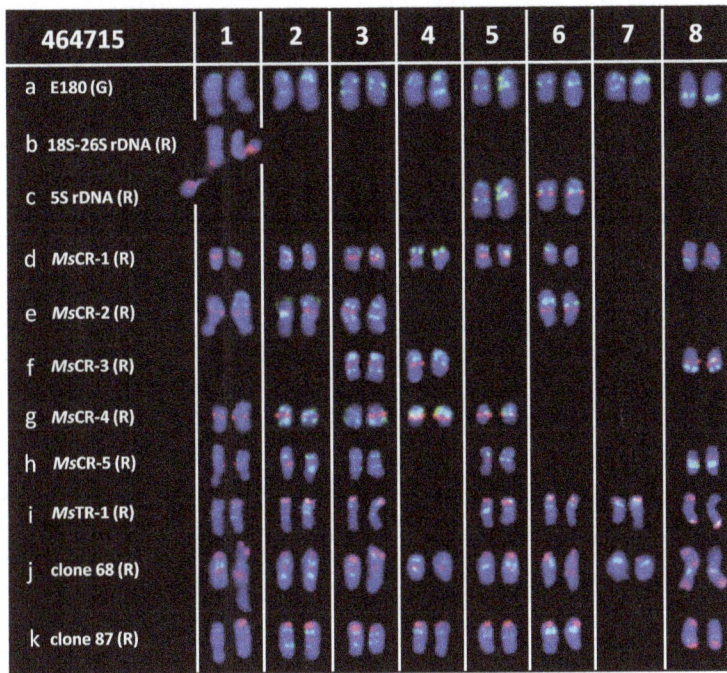

Fig. 1 Localization of repeats on *Medicago sativa* ssp. *caerulea* PI 464715 somatic chromosomes using **a** probe E180 (*green*) in combination with **b** 18S–26S rDNA, **c** 5S rDNA, **d** *Ms*CR-1, **e** *Ms*CR-2, **f** *Ms*CR-3, **g** *Ms*CR-4, **h** *Ms*CR-5, **i** *Ms*TR-1, **j** clone 68, or **k** clone 87

multiple sites observed on nearly all chromosomes in ssp. *caerulea*. Similarly, clones 68 and 87 were also mostly co-localized in centromeric regions on 16 chromosomes (Additional file 1: Figure S1o, p). Double-target FISH revealed that E180 overlapped with clones 68 and 87 on nine chromosomes.

To further characterize the chromosomes of ssp. *falcata*, hybridizations were also carried out using probe E180 and each repetitive sequence (Additional file 1: Figure S2i–p). Each sequence was allocated to a particular chromosome as follows (Fig. 2): 18S–26S rDNA: on chromosome 1; 5S rDNA: on chromosome 3 and 6; *Ms*CR-1: on all chromosome; *Ms*CR-2: on chromosome 1, 3, 4, 5, and one of chromosome 6 and 8; *Ms*CR-4: on chromosome 1, 2, 4, 5, and one of chromosome 6; *Ms*CR-5: on chromosome 2, 4, 6, and 7; *Ms*TR-1 (co-localized with clone 65 and clone 74): on one of chromosome 4; and clone 68 and clone 87: all chromosomes.

The same two FISH cocktails with ssp. *caerulea* were applied to diploid ssp. *falcata* PI 631808 and three other accessions (PI 234815, PI 502447, and PI 631813) to develop a standard molecular karyotype (Additional file 1: Figure S3e-h and m-p). The polymorphic distributions were presented in Fig. 3. A relatively conserved karyotype pattern was described in Fig. 6 (b).

In tetraploid Medicago sativa ssp. falcata

Physical mapping of the 11 repetitive sequences in tetraploid *M. sativa* ssp. *falcata* accession XiaNH-072X-824 was

also carried out by FISH (Additional file 1: Figure S4a–h). Repeat sequence *Ms*CR-1, *Ms*CR-2, *Ms*CR-3, *Ms*CR-4, and *Ms*CR-5 were physically mapped on pericentromeric regions of 30, 16, 7–8, 18, and 16 of the 32 chromosomes of tetraploid ssp. *falcata*, respectively (Additional file 1: Figure S4a–d). Double-target FISH revealed that *Ms*CR-3 overlapped with *Ms*CR-2, *Ms*CR-4, and *Ms*CR-5 on 4, 6, and 7 chromosomes, respectively. The subtelomeric sequence *Ms*TR-1 was co-localized with clone 65 on one end of 13 chromosomes and with clone 74 on one end of 17 chromosomes (Additional file 1: Figure S4e, f). Clone 65 produced more weak signals in the subtelomeric regions of two chromosomes than *Ms*TR-1 did (Additional file 1: Figure S4e), while clone 74 produced extra weak signals in the interstitial regions of two chromosomes compared with *Ms*TR-1 (Additional file 1: Figure S4f). E180, clone 68, and clone 87 were hybridized on 23–24, 29, and 32 chromosomes, respectively (Additional file 1: Figure S4g, h). Double-target FISH revealed that E180 was co-distributed with clone 68 and clone 87 on 20 and 24 chromosomes, respectively (Additional file 1: Figure S4g, h).

According to hybridization results between E180 and each repeat sequences, the chromosomal distribution of each sequence was allocated as follows (Fig. 4): 18S–26S rDNA: on chromosome 1 and 2; 5S rDNA: on chromosome 3, 5, 10, and 13; *Ms*CR-1: on all chromosomes; *Ms*CR-2: on chromosome 4, 5, 9–11, 13, 14, 16, and one of chromosome 12; *Ms*CR-3: on chromosome 7, 10, 12,

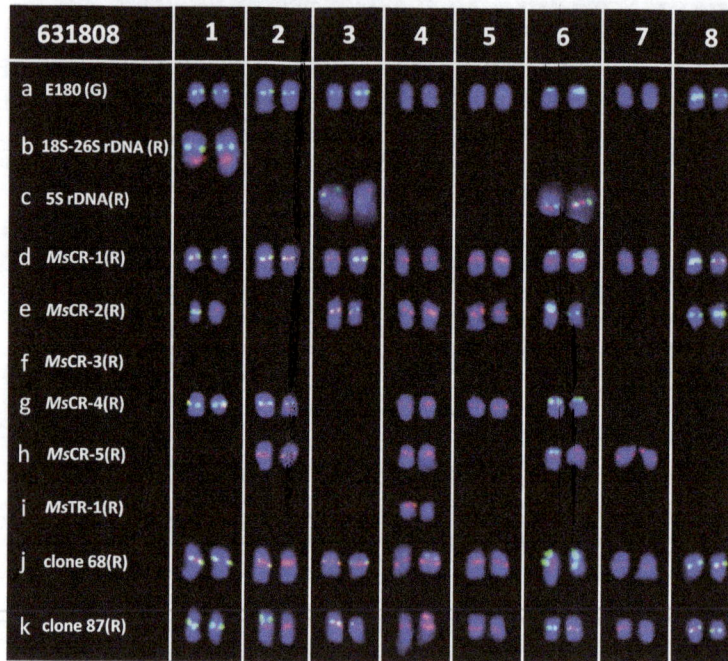

Fig. 2 Localization of repeats on diploid *Medicago sativa* ssp. *falcata* PI 631808 somatic chromosomes using **a** probe E180 (*green*) in combination with **b** 18S–26S rDNA, **c** 5S rDNA, **d** *Ms*CR-1, **e** *Ms*CR-2, **f** *Ms*CR-3, **g** *Ms*CR-4, **h** *Ms*CR-5, **i** *Ms*TR-1, **j** clone 68, or **k** clone 87

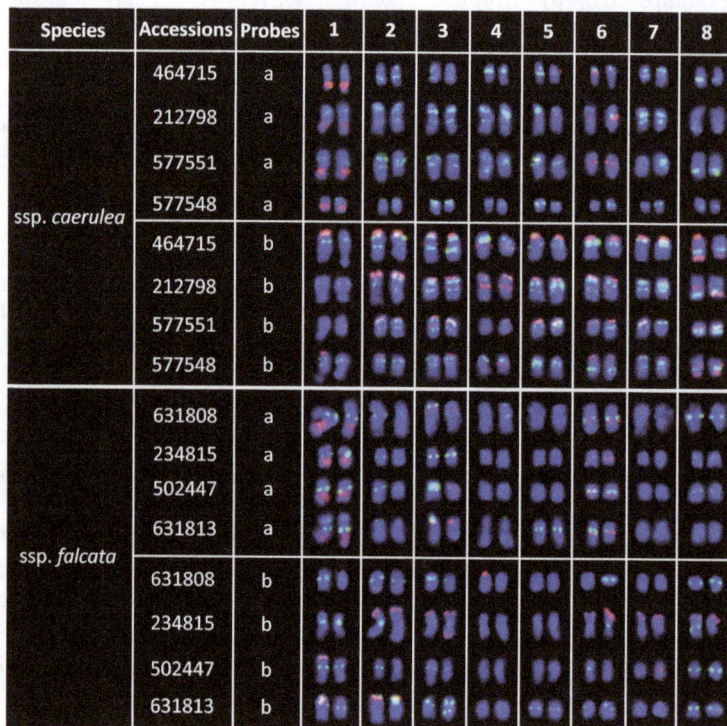

Fig. 3 Karyotypes of four *Medicago sativa* ssp. *caerulea* accessions and four diploid *M. sativa* ssp. *falcata* accessions based on two FISH combinations. **a** Probed by E180 (*green*) combined with 18S–26S rDNA (*red*) and 5S rDNA (*red*). **b** Probed by E180 (*green*) combined with *Ms*TR-1 (*red*) and *Ms*CR-3 (*red*)

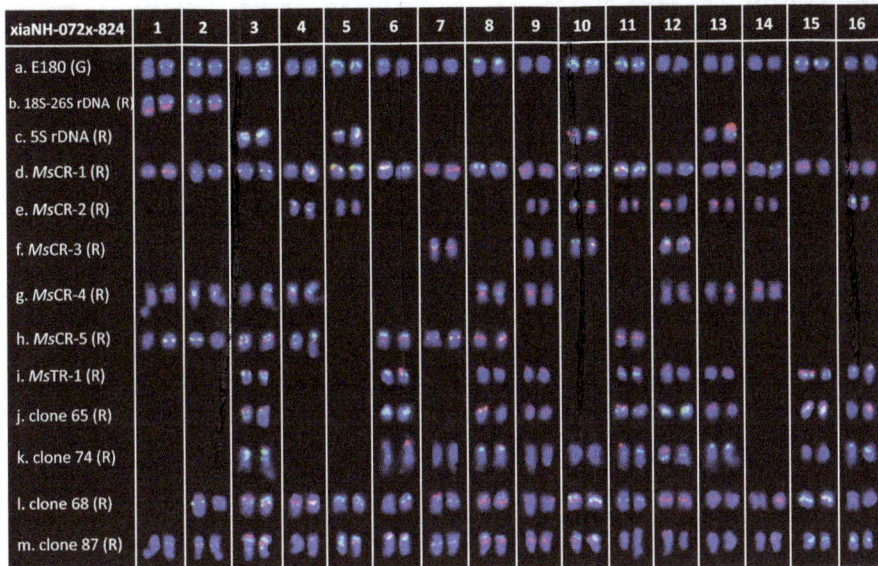

Fig. 4 Localization of repeats on tetraploid *Medicago sativa* ssp. *falcata* XiaNH-072X-824 somatic chromosomes using **a** probe E180 (*green*) in combination with **b** 18S–26S rDNA, **c** 5S rDNA, **d** *Ms*CR-1, **e** *Ms*CR-2, **f** *Ms*CR-3, **g** *Ms*CR-4, **h** *Ms*CR-5, **i** *Ms*TR-1, **j** clone 65, **k** clone 74, **l** clone 68, and **m** clone 87

and one of chromosome 9; *Ms*CR-4: on chromosome 1–4, 8, 9, 12, 13, and 14; *Ms*CR-5: on chromosome 1–3, 6–8, 11, and one of chromosome 4; *Ms*TR-1: on chromosome 3, 6, 8, 9, 11–13, 15, and one of chromosome 16; clone 65: on chromosome 3, 6, 8, 9, 11–13, 15, and one of chromosome 16; clone 74: on chromosome 3, 6, 7–9, 11–13, 15, and one of chromosome 16 and 10; clone 68: on chromosome 2–10, 12–16, and one of chromosome 11; and clone 87: all chromosomes.

The same two FISH cocktails with ssp. *caerulea* were applied to tetraploid ssp. *falcata* accession XiaNH-072X-824 and five other accessions (PI 634023, PI 634118, PI 634117, Lizj0944, and 2–6) to develop a standard molecular karyotype (Additional file 1: Figure S5). Marked variability was detected among the six tetraploid ssp. *falcata* accessions (Fig. 5). Thus only polymorphism schematic diagram was built in Fig. 6 (c).

Comparative cytogenetic analysis between diploid and tetraploid subspecies

M. sativa ssp. caerulea and ssp. sativa

The comparative results of chromosome distribution of each repeat sequence (Table 2) showed that signals of all probes had similar chromosomal locations between ssp. *caerulea* (PI 464715) and ssp. *sativa* (Zhongmu No. 1). Moreover, the signal numbers of each probe were nearly twice between ssp. *caerulea* (PI 464715) and ssp. *sativa* (Zhongmu No. 1). Cocktail FISH results revealed the chromosome distributions of repeat sequences were highly conserved in four ssp. *caerulea* accessions and four ssp. *sativa* accessions, respectively (Fig. 6(a) and

(d)). The high similar FISH patterns between ssp. *caerulea* and ssp. *sativa* facilitated the recognition of homoeologous chromosomes. Furthermore, autopolyplidization from ssp. *caerulea* to ssp. *sativa* was tentatively reconstructed. The reconstructed results showed though collinearity was well maintained in most chromosomes between diploid and tetraploid by FISH patterns, significant and stable variations were also detected in a few chromosomes (Fig. 7 (a)). Compared with chromosome 3 of ssp. *caerulea*, chromosome 6′ of ssp. *sativa* was missing the *Ms*CR-3 signal at pericentromeric region. Compared with chromosome 8 of ssp. *caerulea*, chromosome 16′ of ssp. *sativa* was missing the *Ms*CR-3 signal at pericentromeric region. Compared with chromosome 5 of ssp. *caerulea*, chromosome 9′ of ssp. *sativa* had an extra *Ms*CR-3 signal at pericentromeric region. Compared with chromosome 6 of ssp. *caerulea*, chromosome 12′ of ssp. *sativa* had 5S and *Ms*TR-1 repeat signals on long arm instead of short arm. Compared with chromosome 7 of ssp. *caerulea*, chromosome 14′ of ssp. *sativa* had one of E180 signals near the centromere of short arm instead of long arm. Compared with chromosome 8 of ssp. *caerulea*, chromosome 15′ of ssp. *sativa* had E180 signals on short arm instead of long arm. Combining all chromosome changes, chromosome deletion was speculated to occur on the long arm of chromosome 12′ of ssp. *sativa* during evolution. And pericentric inversions were speculated to occur in chromosome 14′ and chromosome 15′ of ssp. *sativa* during evolution. Putative chromosome changes were presented on Fig. 7 (b). Furthermore, significant variations

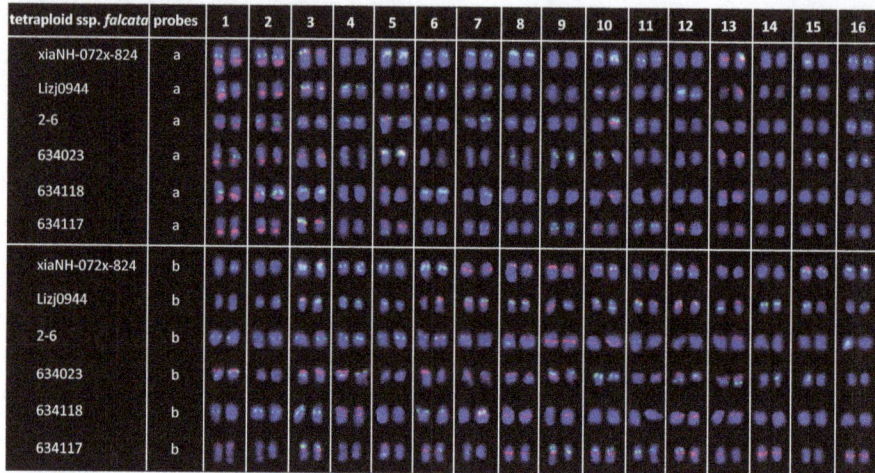

Fig. 5 Karyotypes of six tetraploid *Medicago sativa* ssp. *falcata* accessions based on different FISH combinations. **a** Probed by E180 (*green*) combined with 18S–26S rDNA (*red*) and 5S rDNA (*red*). **b** Probed by E180 (*green*) combined with *Ms*TR-1 (*red*) and *Ms*CR-3 (*red*)

between four groups of homologous chromosomes of tetraploid alfalfa were also detected (Fig. 7 (a)). Compared with homologous chromosome 8′, chromosome 7′ was missing the *Ms*TR-1 signals at subtelomeric region of short arm. Compared with homologous chromosome 11′, chromosome 12′ had 5 s signal on short arm instead of

long arm. Compared with homologous chromosome 13′, chromosome 14′ had an extra E180 signals at the pericentromeric region of short arm. Compared with homologous chromosome 15′, chromosome 16′ was missing *Ms*CR-3 signal at pericentromeric region and had signals of *Ms*TR-1and E180 repeats on long arm instead of short arm.

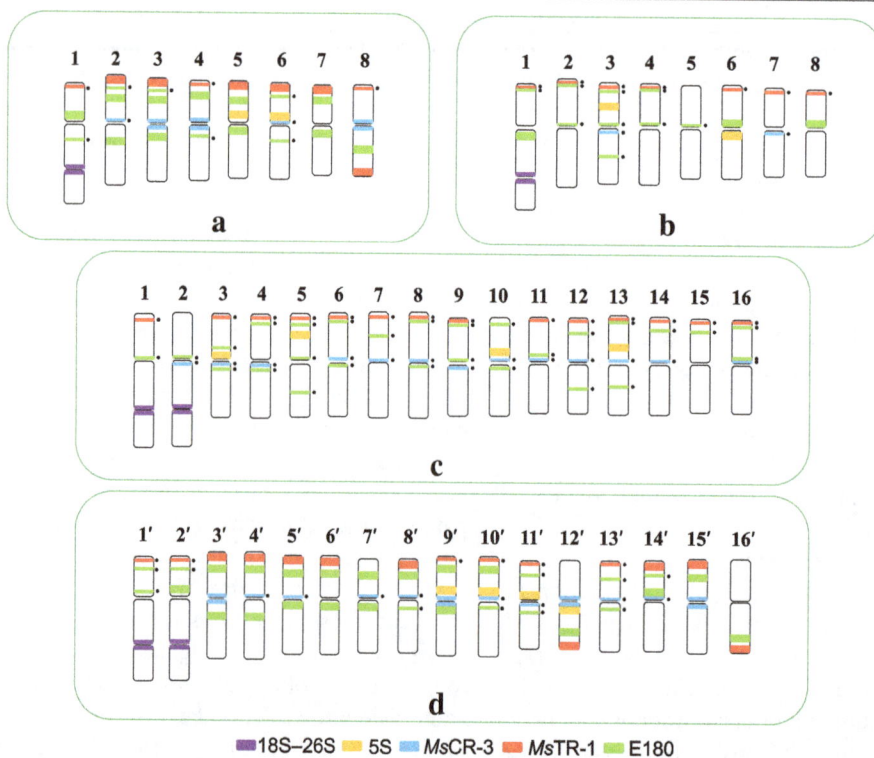

Fig. 6 Idiogram of FISH-banded chromosomes of (**a**) *Medicago sativa* ssp. *caerulea*, (**b**) diploid *M. sativa* ssp. *falcata*, (**c**) tetraploid *M. sativa* ssp. *falcata*, and (**d**) *M. sativa* ssp. *sativa*. Idiogram of FISH-banded chromosomes of *M. sativa* ssp. *sativa* was summarized from four accessions [33]. Chromosomes of tetraploid ssp. *sativa* are marked with "''" on chromosome numbers. A small black dot next to the FISH signal indicates that the signal is polymorphic across accessions

Karyotypic evolution of the Medicago complex: sativa-caerulea-falcata inferred from comparative cytogenetic...

135

Table 2 The comparison of chromosome distributions of each repeat sequence among four subspecies. Chromosome distributions of each repeat sequence in *Medicago sativa* ssp. *sativa* were summarized from Yu et al. [33]

Probes	ssp. *caerulea* (PI 464715)	ssp. *sativa* (Zhongmu No. 1)	Diploid ssp. *falcata* (PI 631808)	Tetraploid ssp. *falcata* (XiaNH-072X-824)
18S–26S	1 (secondary constriction)	2 (secondary constriction)	1 (secondary constriction)	2 (secondary constriction)
5S	2 (near centromeric)	4 (near centromeric)	2 (near centromeric and interstitial region)	4 (near centromeric and interstitial region)
*Ms*CR-1	14 (pericentromeric)	30–32 (pericentromeric)	16 (pericentromeric)	30–32 (pericentromeric)
*Ms*CR-2	8 (pericentromeric)	15 (pericentromeric)	10 (pericentromeric)	16–17 (pericentromeric)
*Ms*CR-3	6 (pericentromeric)	16 (pericentromeric)	0–1 (pericentromeric)	7–8 (pericentromeric)
*Ms*CR-4	10 (pericentromeric)	17 (pericentromeric)	9 (pericentromeric)	18 (pericentromeric)
*Ms*CR-5	9 (pericentromeric)	19 (pericentromeric)	8 (pericentromeric)	16–17 (pericentromeric)
*Ms*TR-1	12–13 (subtelomeric)	24–26 (subtelomeric)	1 (subtelomeric)	17 (subtelomeric)
clone 65	12–13 (subtelomeric)	26 (subtelomeric)	1 (subtelomeric)	13 (subtelomeric)
clone 74	12–13 (subtelomeric)	26 (subtelomeric)	1 (subtelomeric)	17 (subtelomeric)
E180	16 (multiple distribution)	30–32 (multiple distribution)	10–11 (mainly on pericentromeric)	23–24 (multiple distribution)
clone 68	15 (multiple distribution)	28 (multiple distribution)	16 (mainly on pericentromeric)	29 (multiple distribution)
clone 87	14 (multiple distribution)	32 (multiple distribution)	16 (mainly on pericentromeric)	32 (multiple distribution)

Diploid ssp. *falcata* and tetraploid ssp. *falcata*

The comparative results of chromosome distributions of each repeat sequence (Table 2) showed that signals of ten probes had similar chromosomal locations between diploid ssp. *falcata* (PI 631808) and tetraploid ssp. *falcata* (XiaNH-072X-824). However, signal distributions of E180, clone 68, and clone 87 probes were more abundant in tetraploid ssp. *falcata* than in diploid ssp. *falcata*. The signal numbers of *Ms*CR-1, *Ms*CR-2, *Ms*CR-4, *Ms*CR-5, E180, clone 68, and clone 87 probes were nearly twice between diploid ssp. *falcata* (PI 631808) and tetraploid ssp. *falcata* (XiaNH-072X-824). However, the signals of *Ms*CR-3, *Ms*TR-1, clone 65, and clone 74 probes was only located on one chromosome in diploid ssp. *falcata* rather than on many chromosomes in tetraploid ssp. *falcata*. Moreover, the cocktail

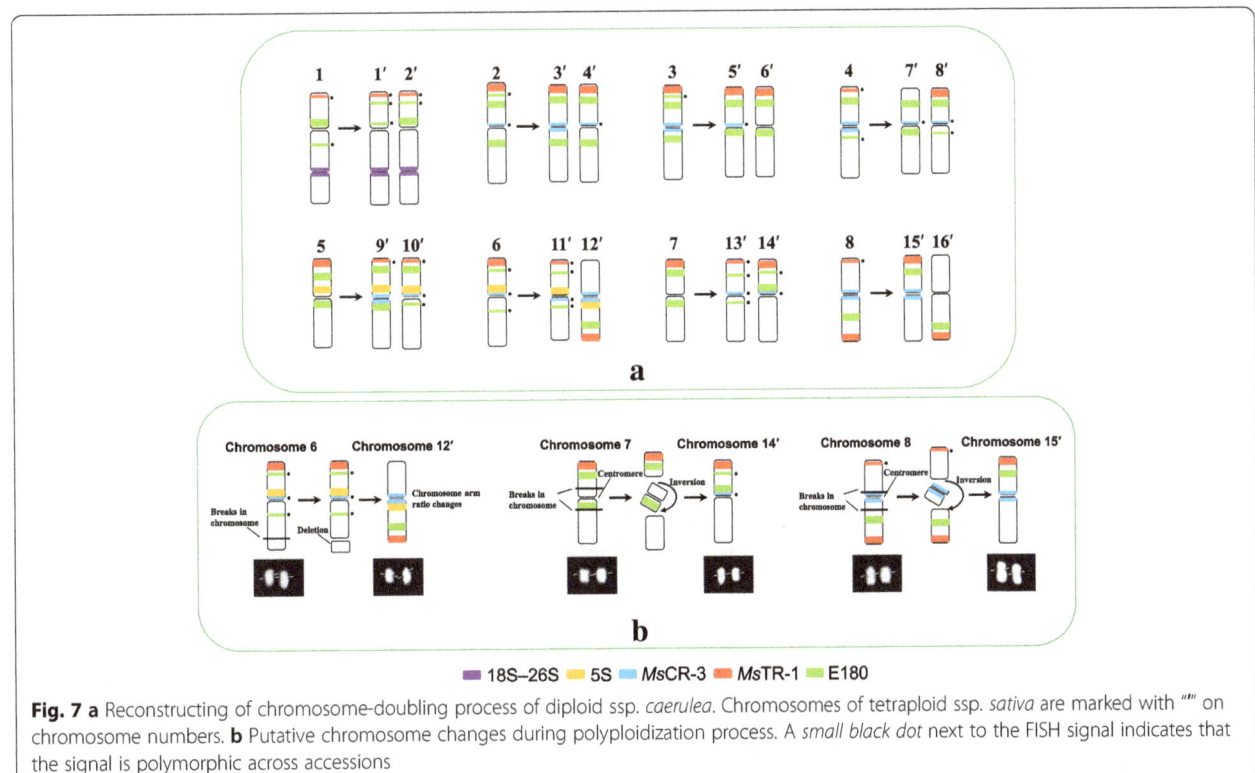

Fig. 7 a Reconstructing of chromosome-doubling process of diploid ssp. *caerulea*. Chromosomes of tetraploid ssp. *sativa* are marked with "'" on chromosome numbers. **b** Putative chromosome changes during polyploidization process. A *small black dot* next to the FISH signal indicates that the signal is polymorphic across accessions

FISH results revealed highly variable karyotypes across different tetraploid ssp. *falcata* accessions. Thus, chromosome collinearity analysis between diploid and tetraploid ssp. *falcata* could not be conducted as did between ssp. *caerulea* and ssp. *sativa*.

Discussion

Genomic differentiation of *M. sativa* ssp. *caerulea* and diploid ssp. *falcata*

Medicago sativa ssp. *caerulea* and diploid ssp. *falcata* are sympatrically distributed, with naturally occurring hybrids recorded between them [13, 14, 36]. The genetic affinity of the two species has been demonstrated by cytological research [15]. In addition, chromosomal differentiation between *M. sativa* ssp. *caerulea* and diploid ssp. *falcata* has been well described by analyses of both C- and N-banded chromosomes [20, 21]. C- and N-banding has revealed that chromosomes of diploid ssp. *falcata* possess only centromeric bands. In contrast, all chromosomes of ssp. *caerulea* have a centromeric band and a telomeric band in the short arm; in addition, most of the chromosomes of this subspecies have interstitial bands in the short arm, with a few chromosomes featuring prominent interstitial bands in the long arm.

C- and N-bands reflect constitutive heterochromatic DNA in chromosomes [23]. Our molecular cytogenetic analysis revealed the heterogeneous nature of the constitutive heterochromatin among centromeric, interstitial, and subtelomeric regions. In both ssp. *caerulea* and diploid ssp. *falcata*, centromeric bands were revealed to be a heterogeneous mix of *Ms*CR-1, *Ms*CR-2, *Ms*CR-3, *Ms*CR-4, *Ms*CR-5, clone 68, clone 87, and E180 sequences, along with a few 5S rDNA sites. The interstitial bands comprised E180 sequences along with 18S–26S rDNA and 5S rDNA sites, and the subtelomeric bands were represented by *Ms*TR-1, clone 68, clone 87, and E180.

Chromosomal differences between ssp. *caerulea* and diploid ssp. *falcata* as revealed by FISH were similar to those uncovered by C- or N-banding. The repetitive sequences were physically mapped onto centromeric, subtelomeric, or interstitial regions in ssp. *caerulea*, whereas the mapped sequences were mainly on centromeric regions in diploid ssp. *falcata*. Furthermore, conspicuous differences in distribution patterns were observed between ssp. *caerulea* and diploid ssp. *falcata*, even though the repetitive sequences detected in centromeric regions of both species displayed similar levels of heterogeneity. Unlike ssp. *caerulea*, more than half of the chromosomes of ssp. *falcata* contained E180 sequences in centromeric regions. In addition, centromeric sequences of *Ms*CR-3 were detected on one or no chromosomes of diploid ssp. *falcata*, whereas they were found on 3–5 pairs of chromosomes in ssp. *caerulea*.

Genetic differentiation between ssp. *caerulea* and diploid ssp. *falcata* has been previously revealed by nuclear markers [37–39]. Moreover, relationships uncovered among diploid members of the *M. sativa* species complex based on chloroplast DNA sequence analysis supports the recognition of ssp. *caerulea* and diploid ssp. *falcata* as distinct taxa [40]. Our study has revealed distinct genomic differentiation between ssp. *caerulea* and diploid ssp. *falcata* and supports their taxonomic differentiation at the chromosome level.

Chromosome evolution after polyploidization of diploid ssp. *caerulea*

Violet flowered diploid ssp. *caerulea* is postulated to have given rise to tetraploid ssp. *sativa* (alfalfa) [13, 16, 17]. The identical C-banding patterns of tetraploid alfalfa and ssp. *caerulea* support tetraploid alfalfa as an autotetraploid derived from diploid ssp. *caerulea* [21, 23]. Sequencing of chloroplast DNA has demonstrated that the two taxa have very closely related chloroplast haplotypes, with most individuals sharing the same haplotype, and are thus undifferentiated genetically for this characteristic. Similar to the C-banding analysis, chloroplast data supports a simple autopolyploid origin for ssp. *sativa* from diploid ssp. *caerulea* [19]. In our study, a putative chromosome doubling process from diploid ssp. *caerulea* to tetraploid alfalfa was reconstructed according to similar FISH patterns. The results strongly supported the simple autotetraploid origin of ssp. *sativa* from diploid ssp. *caerulea*.

It is generally believed that polyploid plants may have unstable genomes in a long term due to a genome-wide gene redundancy [41]. Ma and Gustafson [42] summarized the evolution of an allopolyploid species is a process of both cytological and genetic diploidization. Rapid genomic rearrangement such as chromosome insertion, chromosome deletion, and chromosome rearrangement, which would lead to diploidization of genome structure, has been investigated in some allopolyploid plant species [41, 43]. However, the occurrence of similar changes remains to be studied in detail during the generation of autopolyploids [44]. The limited data available so far imply that autopolyploids experience less genome restructuring than allopolyploids [44]. In our study, putative genome changes were discovered after polyploidization of diploid ssp. *caerulea*. Elimination of repetitive DNA was detected in pericentromeric regions of chromosome 6′ and chromosome 16′ of tetraploid ssp. *sativa*. Increase of repetitive DNA was detected in pericentromeric regions of chromosome 9′ of tetraploid ssp. *sativa*. Chromosome deletion was postulated to occur in the long arm of chromosome 12′ of tetraploid ssp. *sativa*. Pericentric inversions were postulated to occur in chromosome 14′ and chromosome 15′ of tetraploid ssp. *sativa*. Furthermore, significant diversification was recognized in four groups of homologous chromo-some of tetraploid ssp. *sativa* including chromosome 7′ and chromosome 8′, chromosome 11′ and chromosome 12′, chromosome 13′ and chromosome 14′ and chromosome 15′ and chromosome 16′. Thus, we con-cluded that autotetraploid alfalfa had undergone a partial diploidization by the progressive accumulation of chromosome structural rearrangements during evolu-tion. A previous study of pachytene karyotype reported that at least four groups of the tetraploid chromosomes appear sufficiently alike to be able to form quadrivalents in ssp. *sativa*, and three of these were seen to form quadrivalents at pachytene [45]. Subsequently, Armstrong summarized the quadrivalent frequency at pachytene ranged from 0.89 to 2.93 in tetraploid ssp. *sativa* [46]. It was considerably below theoretical expectations "5.34 quadrivalents per cell" for an autotetraploid [46]. This chromosome behavior in meiotic has confused the origin of tetraploid ssp. *sativa*. Partial diploidization of homolo-gous chromosome groups in tetraploid *M. sativa* found in our study should be an explanation for the low quadriva-lent frequencies at meiotic.

Phylogenetic relationships between diploid and tetraploid forms of ssp. *falcata*

Diploid and tetraploid forms of ssp. *falcata* have been traditionally treated as a single species, *M. sativa* ssp. *falcata*. Diploid ssp. *falcata* and tetraploid ssp. *falcata* have been recognized as diploid and tetraploid cytotypes on the basis of chromosome counting and morphology. Diploid ssp. *falcata* is hypothesized to be the ancestor of autoploid tetraploid ssp. *falcata* [13, 14]. As revealed by C- and N-banding, centromeric bands are a distinct fea-ture of the chromosomes of diploid ssp. *falcata* [20, 21]. Because of this assumption of autopolyploid origin, the C-banding pattern of tetraploid ssp. *falcata* was ex-pected to be similar to that of diploid ssp. *falcata*. Thus, the results of a preliminary study of six accessions of tetraploid ssp. *falcata* were surprising. Most of the plants possessed chromosomes that had C-bands in addition to normal centromeric bands [12]. Highly vari-able C-banding patterns were detected in these acces-sions. The accession containing the fewest number of additional bands had four pairs of chromosomes with an extra telomeric band on their short arms, whereas the remaining chromosomes had only centromeric bands. At the other extreme, two accessions had multiple bands on each chromosome, similar to doubled-diploid ssp. *caerulea*. Even though the studied accessions had yellow flowers with sickle-shaped pods, the accessions were speculated to be the product of hybridization with ssp. *sativa* [12]. Similarly, the six tetraploid ssp. *falcata* accessions used in our study—three acquired from the NPGS USDA germplasm bank and three collected in situ in China on the basis of morphological identificatio-n—also showed highly variable molecular karyotypes. We found that hybridization sites of *Ms*TR-1 and E180, which frequently produce subtelomeric and interstitial bands, respectively, in C-banding analyses, were highly variable across different individuals of tetraploid ssp. *falcata*. Along with the results of C-banding analysis [12], our results suggested the actual genomic character-istics of tetraploid ssp. *falcata*.

In an earlier analysis of chloroplast DNA, morpho-logically identical diploid and tetraploid cytotypes of ssp. *falcata* were found to possess very different chloroplast haplotypes. The most common haplotype of tetraploid ssp. *falcata* was shared with *M. prostrata* rather than diploid ssp. *falcata*, suggesting past introgression from *M. prostrata* into the polyploid. The evolutionary trajec-tory of ssp. *falcata* does not appear to have involved a simple autopolyploid origin as seen in ssp. *sativa* [19]. A high variability in the number of chromosomes with multiple E180 sites, which are frequently lacking in dip-loid ssp. *falcata*, was uncovered in our study. Although information on the chromosomal distribution of E180 in *M. prostrata* is not available, multiple E180 hybridization signals have been detected in species of section *Medicago*, such as *M. glutinosa*, *M. hemicycla*, and *M. polychroa* [47]. This finding suggests that the variable chromosomes in tetraploid ssp. *falcata* could have been introduced from *M. prostrata*. Our data indicate that the origin of tetraploid ssp. *falcata* from diploid ssp. *falcata* is far from simple. Elucidation of the evolutionary history of ssp. *falcata* will require a large amount of additional data.

Conclusions

The comparative cytogenetic results provide reliable evi-dence that diploid subspecies *caerulea* is the direct progenitor of tetraploid subspecies *sativa*. And autotet-raploid ssp. *sativa* has been suggested to undergo a par-tial diploidization by the progressive accumulation of chromosome structural rearrangements during evolu-tion. However, the tetraploid subspecies *falcata* is far from a simple autopolyploid from diploid subspecies *falcata* although no obvious morphological change was observed between these two subspecies.

Abbreviations
FISH: Fluorescence in situ hybridization; NPGS: National Plant Germplasm System; USDA: United States Department of Agriculture

Acknowledgements
We thank Professor Fujiang Hou (School of Pastoral Agriculture Science and Technology, Lanzhou University, China) for providing a few *M*. ssp. *caerulea* materials.

Funding

This work was supported by the Joint Scholars project of The Dawn of West China Talent Training Program of the Chinese Academy of Sciences and Natural Science Foundation of Qinghai Province (Nos. 2015-ZJ-903).

Authors' contributions

QD conceived of the study, and participated in its design and coordination and helped to draft the manuscript. FY carried out the molecular cytogenetic studies, performed the data analysis and drafted the manuscript. YZ and RL participated in the plant materials preparation, and helped in experiment. HW participated in the design of study and the language correction. JD and TW participated in data analysis, and helped the language correction. All authors read and approved the final manuscript.

Competing interests

The authors declare that they have no competing interests.

Author details

[1]Key Laboratory of Adaptation and Evolution of Plateau Biota, Northwest Institute of Plateau Biology, Chinese Academy of Sciences, Xining 810008, China. [2]University of Chinese Academy of Sciences, Beijing 100049, China. [3]State Key Laboratory of Agrobiotechnology, College of Biological Sciences, China Agricultural University, Beijing 100193, China.

References

1. Soltis PS, Soltis DE. The role of hybridization in plant speciation. Annu Rev Plant Biol. 2009;60:561–88.
2. Srisuwan S, Sihachakr D, Siljak-Yakovlev S. The origin and evolution of sweet potato (*Ipomoea batatas* Lam.) and its wild relatives through the cytogenetic approaches. Plant Sci. 2006;171(3):424–33.
3. Lim KY, Matyášek R, Kovarik A, Leitch AR. Genome evolution in allotetraploid *Nicotiana*. Biol J Linn Soc. 2004;82(4):599–606.
4. Lim KY, Matyášek R, Lichtenstein CP, Leitch AR. Molecular cytogenetic analyses and phylogenetic studies in the *Nicotiana* section Tomentosae. Chromosoma. 2000;109(4):245–58.
5. Chester M, Gallagher JP, Symonds VV, da Silva AVC, Mavrodiev EV, Leitch AR, Soltis PS, Soltis DE. Extensive chromosomal variation in a recently formed natural allopolyploid species, *Tragopogon miscellus* (Asteraceae). Proc Natl Acad Sci U S A. 2012;109(4):1176–81.
6. Wang K, Guo W, Zhang T. Detection and mapping of homologous and homoeologous segments in homoeologous groups of allotetraploid cotton by BAC-FISH. BMC Genomics. 2007;8(1):1.
7. Hanson RE, Islam-Faridi MN, Percival EA, Crane CF, Ji Y, McKnight TD, Stelly DM, Price HJ. Distribution of 5S and 18S–28S rDNA loci in a tetraploid cotton (*Gossypium hirsutum* L.) and its putative diploid ancestors. Chromosoma. 1996;105(1):55–61.
8. Maluszynska J, Hasterok R. Identification of individual chromosomes and parental genomes in *Brassica juncea* using GISH and FISH. Cytogenet Genome Res. 2005;109(1–3):310–4.
9. Xiong Z, Gaeta RT, Pires JC. Homoeologous shuffling and chromosome compensation maintain genome balance in resynthesized allopolyploid *Brassica napus*. Proc Natl Acad Sci U S A. 2011;108(19):7908–13.
10. Talukdar D. Meiotic consequences of selfing in grass pea (*Lathyrus sativus* L.) autotetraploids in the advanced generations: Cytogenetics of chromosomal rearrangement and detection of aneuploids. Nucleus. 2012;55(2):73–82.
11. Weiss H, Maluszynska J. Chromosomal rearrangement in autotetraploid plants of *Arabidopsis thaliana*. Hereditas. 2001;133(3):255–61.
12. Bauchan GR, Hossain MA. Advances in alfalfa cytogenetics. In: Bingham ET, editor. The Alfalfa Genome: 100 year of Alfalfa Genetics; 1999. http://www.naaic.org/TAG/TAGpapers/Bauchan/advcytog.html. Accessed 14 Apr 2017.
13. Small E, Jomphe M. A synopsis of the genus *Medicago* (Leguminosae). Can J Bot. 1989;67(11):3260–94.
14. Quiros CF, Bauchan GR. The genus *Medicago* and the origin of the *Medicago sativa* Complex. In: Hanson AA, Barnes DK, Hill RR, editors. Alfalfa and Alfalfa improvement. Madison: American Society of Agronomy, Crop Science Society of America, Soil Science Society of America; 1988. p. 93–124.
15. Gillies CB. Pachytene chromosomes of perennial *Medicago* species I. Species closely related to *M. sativa*. Hereditas. 1972;72(2):277–88.
16. Stanford EH. Tetrasomic inheritance in alfalfa. Agron J. 1951;43(5):222–5.
17. Quiros CF. Tetrasomic segregation for multiple alleles in alfalfa. Genetics. 1982;101(1):117–27.
18. Gillies CB, Bingham ET. Pachytene karyotypes of 2X haploids derived from tetraploid alfalfa (*Medicago sativa*)-evidence for autotetraploidy. Can J Genet Cytol. 1971;13(3):397–403.
19. Havananda T, Brummer EC, Doyle JJ. Complex patterns of autopolyploid evolution in alfalfa and allies (*Medicago sativa*; Leguminosae). Am J Bot. 2011;98(10):1633–46.
20. Bauchan GR, Hossain MA. Karyotypic analysis of N-banded chromosomes of diploid alfalfa: *Medicago sativa ssp. caerulea* and ssp. *falcata* and their hybrid. J Hered. 1998;89(2):533–7.
21. Bauchan GR, Hossain MA. Karyotypic analysis of C-banded chromosomes of diploid alfalfa: *Medicago sativa ssp. caerulea* and ssp. *falcata* and their hybrid. J Hered. 1997;88(6):533–7.
22. Bauchan GR, Hossain AM. Cytogenetic studies of the nine germplasm sources of alfalfa. In: Bouton J, Bauchan GR, editors. Symposium proceedings of the North American Alfalfa Improvement Conference, 36th. Bozeman; 1998.
23. Bauchan GR, Hossain MA. Distribution and characterization of heterochromatic DNA in the tetraploid African population alfalfa genome. Crop Sci. 2001;41(6): 1921–6.
24. Falistocco E, Falcinelli M, Veronesi F. Karyotype and C-banding pattern of mitotic chromosomes in alfalfa, *Medicago sativa* L. Plant Breed. 1995;114(5):451–3.
25. Bauchan GR, Hossain AM. Constitutive heterochromatin DNA polymorphisms in diploid *Medicago sativa ssp. falcata*. Genome. 1999;42(5):930–5.
26. De Jong JH, Fransz P, Zabel P. High resolution FISH in plants-techniques and applications. Trends Plant Sci. 1999;4(7):258–63.
27. Jiang J, Gill BS. Current status and the future of fluorescence in situ hybridization (FISH) in plant genome research. Genome. 2006;49(9):1057–68.
28. Kato A, Lamb JC, Birchler JA. Chromosome painting using repetitive DNA sequences as probes for somatic chromosome identification in maize. Proc Natl Acad Sci U S A. 2004;101(37):13554–9.
29. She CW, Jiang XH, Ou LJ, Liu J, Long KL, Zhang LH, Duan WT, Zhao W, Hu JC. Molecular cytogenetic characterisation and phylogenetic analysis of the seven cultivated *Vigna* species (Fabaceae). Plant Biol. 2015;17(1):268–80.
30. Paesold S, Borchardt D, Schmidt T, Dechyeva D. A sugar beet (*Beta vulgaris* L.) reference FISH karyotype for chromosome and chromosome-arm identification, integration of genetic linkage groups and analysis of major repeat family distribution. Plant J. 2012;72(4):600–11.
31. Leitch I, Hanson L, Lim K, Kovarik A, Chase M, Clarkson J, Leitch A. The ups and downs of genome size evolution in polyploid species of *Nicotiana* (Solanaceae). Ann Bot. 2008;101(6):805–14.
32. Dou Q, Wang RR-C, Lei Y, Yu F, Li Y, Wang H, Chen Z. Genome analysis of seven species of *Kengyilia* (Triticeae: Poaceae) with FISH and GISH. Genome. 2013;56(11):641–9.
33. Yu F, Lei Y, Li Y, Dou Q, Wang H, Chen Z. Cloning and characterization of chromosomal markers in alfalfa (*Medicago sativa* L.). Theor Appl Genet. 2013;126(7):1885–96.
34. Fukui K, Kamisugi Y, Sakai F. Physical mapping of 5s rDNA Loci by direct-cloned biotinylated probes in barley chromosomes. Genome. 1994;37(1): 105-11.
35. Fukui K. In situ hybridization. In: Fukui K, Nakayama S, editors. Plant chromosomes: laboratory method. Boca Raton: CRC press; 1996. p. 155–70.
36. Small E, Bauchan GR. Chromosome numbers of the *Medicago sativa* complex in Turkey. Can J Bot. 1984;62(4):749–52.
37. İlhan D, Li X, Brummer EC, Şakiroğlu M. Genetic diversity and population structure of tetraploid accessions of the *Medicago sativa–falcata* Complex. Crop Sci. 2016;56(3):1146–56.
38. Brummer E, Kochert G, Bouton J. RFLP variation in diploid and tetraploid alfalfa. Theor Appl Genet. 1991;83(1):89–96.
39. Brummer EC. Genomics research in alfalfa, *Medicago sativa* L. In: Wilson RF, Stalker HT, Brummer EC, editors. Legume crop genomics. Champaign: AOCS Press; 2004. p. 110–42.
40. Havananda T, Brummer EC, Maureira-Butler IJ, Doyle JJ. Relationships among diploid members of the *Medicago sativa* (Fabaceae) species complex based on chloroplast and mitochondrial DNA sequences. Syst Bot. 2010;35(1):140–50.

41. Hufton AL, Panopoulou G. Polyploidy and genome restructuring: a variety of outcomes. Curr Opin Genet Dev. 2009;19(6):600–6.

42. Ma XF, Gustafson J. Genome evolution of allopolyploids: a process of cytological and genetic diploidization. Cytogenet Genome Res. 2005;109(1–3):236–49.

43. Lim KY, Kovarik A, Matyasek R, Chase MW, Clarkson JJ, Grandbastien M, Leitch AR. Sequence of events leading to near-complete genome turnover in allopolyploid Nicotiana within five million years. New Phytol. 2007;175(4): 756–63.

44. Parisod C, Holderegger R, Brochmann C. Evolutionary consequences of autopolyploidy. New Phytol. 2010;186(1):5–17.

45. Gillies C. Alfalfa chromosomes. II. Pachytene karyotype of a tetraploid *Medicago sativa* L. Crop Sci. 1970;10:172–5.

46. Armstrong K. Chromosome associations at pachytene and metaphase in *Medicago sativa*. Can J Genet Cytol. 1971;13(4):697–702.

47. Rosato M, Galián JA, Rosselló JA. Amplification, contraction and genomic spread of a satellite DNA family (E180) in *Medicago* (Fabaceae) and allied genera. Ann Bot. 2011;109(4):773–82.

A passive mutualistic interaction promotes the evolution of spatial structure within microbial populations

Marie Marchal[1], Felix Goldschmidt[1,2], Selina N. Derksen-Müller[1,2], Sven Panke[3], Martin Ackermann[1,2*†] and David R. Johnson[1*†]

Abstract

Background: While mutualistic interactions between different genotypes are pervasive in nature, their evolutionary origin is not clear. The dilemma is that, for mutualistic interactions to emerge and persist, an investment into the partner genotype must pay off: individuals of a first genotype that invest resources to promote the growth of a second genotype must receive a benefit that is not equally accessible to individuals that do not invest. One way for exclusive benefits to emerge is through spatial structure (i.e., physical barriers to the movement of individuals and resources).

Results: Here we propose that organisms can evolve their own spatial structure based on physical attachment between individuals, and we hypothesize that attachment evolves when spatial proximity to members of another species is advantageous. We tested this hypothesis using experimental evolution with combinations of *E. coli* strains that depend on each other to grow. We found that attachment between cells repeatedly evolved within 8 weeks of evolution and observed that many different types of mutations potentially contributed to increased attachment.

Conclusions: We postulate a general principle by which passive beneficial interactions between organisms select for attachment, and attachment then provides spatial structure that could be conducive for the evolution of active mutualistic interactions.

Keywords: Mutualism, Cross-feeding, Cell aggregation, Experimental evolution, Microbial populations, Spatial structure

Background

Mutualistic interactions are pervasive in the natural environment and shape the assembly and functioning of nearly every ecological community [1, 2]. A mutualistic interaction occurs when two (or more) different organisms – referred to here as mutualistic partners - have reciprocal positive effects on each other's growth. Examples of mutualistic interactions include the associations between legumes and nitrogen-fixing bacteria, be-tween plants and pollinating insects, and between humans and members of their gut microbiota [1, 2]. Mutualistic interactions are also common between members of microbial communities and are important determinants of their ecological dynamics and processes [3]. A typical feature of mutualistic interactions within microbial communities is that they often require one or both mutualistic partners to excrete metabolites that positively affect the growth of others [4–8] (Fig. 1a). While we mainly focus on this type of mutualistic interaction in this manuscript, the main idea that we develop is potentially relevant for other types of mutualistic interactions and organisms.

How mutualistic interactions emerge and persist is not clear [9–11]. A first question pertains to the origin of

* Correspondence:

†Equal contributors martin.ackermann@env.ethz.ch; david.johnson@eawag.ch
[1]Department of Environmental Microbiology, Eawag, Überlandstrasse 133, 8600 Dübendorf, Switzerland
Full list of author information is available at the end of the article

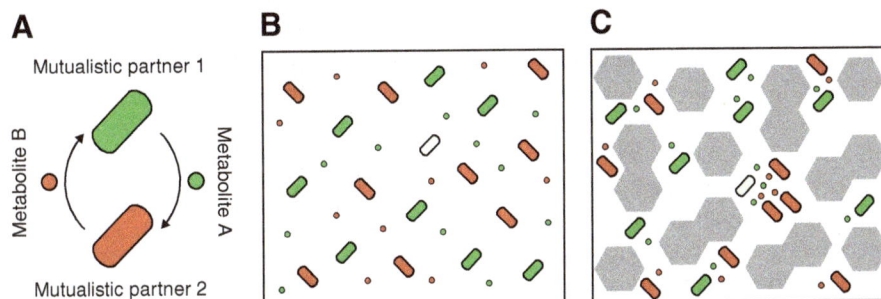

Fig. 1 Spatial structure promotes the evolution of active metabolite excretion. **a** A mutualistic interaction between two mutualistic partners. Partner 1 (*green cell*) excretes metabolite A (*green circle*) that promotes the growth of partner 2. Partner 2 (*red cell*) excretes metabolite B (*red circle*) that promotes growth of partner 1. **b** In a well-mixed environment, a mutant green cell that actively excretes more metabolites (*light green cell*) should decrease in frequency. This is because the mutant cell would pay all of the cost for increased metabolite excretion but would receive the same benefit as all the other green cells. **c** In a spatially structured environment, a mutant green cell that actively excretes more metabolites (*light green cell*) could potentially increase in frequency. This is because spatial structure might cause the mutant cell to receive a disproportionate amount of the benefits from increased metabolite excretion. These benefits originate from the positive effects of increased excretion of metabolite A on partner 2, thus resulting in increased excretion of metabolite B

mutualistic interactions. Starting from a situation where two microbial genotypes exist in the same environment but do not affect each other's growth, how could a mutant of one genotype emerge that excretes a metabolite that positively affects the growth of another genotype and increases in frequency, thus laying the foundation for a mutualistic interaction to evolve? In this scenario, the mutant might not receive an immediate benefit but would potentially carry a metabolic cost associated with excreting the metabolite. The mutant might therefore not be able to increase in frequency relative to its ancestor and a mutualistic interaction could not establish.

One possible solution to this problem is that mutually beneficial interactions could originally be *passive* in nature [12, 13]. We use the term 'passive' to refer to a behavior or property of an organism that, while potentially being beneficial to another organism, did not evolve because of its positive effects on that other organism. This is particularly evident for metabolic interactions between microorganisms, which often passively excrete metabolites that positively affect the growth of other microorganisms [12, 14–16]. For example, microorganisms might excrete metabolites via cell leakage or as side-products or end-products of their own metabolism [16–19], which could then be taken up by other microorganisms. If such metabolic interactions are reciprocal between two partners (Fig. 1a), they constitute as a mutually beneficial but passive mutualistic interaction.

While the passive scenario described above could be important for the origin of a mutualistic interaction, many mutualistic interactions in nature are not passive but are rather based on the *active* excretion of metabolites that positively affect the growth of a mutualistic partner (e.g. [7]). The active excretion of metabolites must, at least to some extent, divert cellular resources

away from the growth and reproduction of the excreting microorganism [20–22]. These metabolic costs then lead to a second fundamental question: starting from a situation where two microorganisms are coupled by an initially passive mutualistic interaction, how could investment of metabolic resources into each mutualistic partner evolve? Such an evolutionary transition requires a mutant that actively excretes metabolites to increase in frequency. Again, this is not trivial to explain. A mutant that actively excretes metabolites would increase the growth of its mutualistic partner, which in turn would lead to a benefit that is accessible to both the mutant and its ancestor (Fig. 1b). While the benefits of an increased investment are thus distributed uniformly, the costs are borne alone by the actively excreting mutant. The mutant might therefore have a growth disadvantage relative to its ancestor and decrease in frequency.

Ultimately, the evolutionary transition of a passive into an active mutualistic interaction requires that the mutant that actively invests metabolic resources into the growth of its mutualistic partner receive an exclusive benefit [9, 23]. Such coupling between investment and return arises if individuals of different mutualistic partners are spatially associated with each other for long periods of time [24, 25]. In this case, a mutant that invests metabolic resources into the growth of its mutualistic partner would receive an increased return if the partner's growth leads to an increased production of the benefit. This situation has been referred to as the "partner-fidelity model" [24]. A general scenario under which such a long-term association between individuals could arise is in the presence of spatial structure [11, 20, 21, 23, 25–29] (Fig. 1c); that is, in a situation where physical barriers constrain the movement of individuals and metabolites. In spatially structured environments, the benefits arising from an investment into another organism are not distributed evenly but are instead disproportionally directed back towards the investor, thus allowing those individuals to potentially increase in frequency and spread [20, 21, 23, 26–29] (Fig. 1c).

While there is accumulating theoretical and experimental evidence to support the importance of spatial structure on the evolution of mutualistic interactions, these studies have largely focused on imposing abiotic spatial structure on a mutualistic consortium and analyzing the evolutionary outcomes (e.g. [20, 21, 26, 30]). Mutualistic interactions, however, are also observed in habitats with relatively little abiotic spatial structure, for example between microorganisms that reside within the water columns of open oceans and lakes (e.g. [7, 31]). This then underscores an important gap in our knowledge: How can we explain the evolution of an active mutualistic interaction in the absence of extensive abiotic spatial structure?

The central idea that we are addressing in this manuscript is that microorganisms can readily evolve to create their own spatial structure based on physical attachment between individuals (i.e., cell aggregation) [23]. More specifically, we test the hypothesis that an initially passive mutualistic interaction selects for mutants that aggregate together with other individuals and thereby benefit from increased local concentrations of excreted metabolites. This scenario could have important consequences because cell aggregation leads to "partner fidelity" and could set the stage for the evolution of an active mutualistic interaction. Indeed, cell aggregates are prevalent within the mixed layers of oceans and non-stratified lakes and can harbor mutualistic interactions (e.g., [7, 31]). Whether a passive mutualistic interaction itself could promote the evolution of cell aggregation, however, is not clear.

To test this hypothesis, we experimentally created a passive mutualistic interaction between two auxotrophic strains of the bacterium *Escherichia coli* and tested for the evolution of cell aggregation. Each strain is defective in the biosynthesis of a different amino acid; they can only grow if the required amino acid is exogenously supplied from an abiotic source or if the strains are grown together in co-culture and passively excrete or release small amounts of the amino acid required by the other. This passive mutualistic interaction was based on a single genetic mutation in each mutualistic partner; it could therefore originate spontaneously via random mutation within large populations. We then propagated the mutualistic co-cultures in the absence of extensive abiotic spatial structure (i.e., in continuously-mixed batch reactors) and tested whether the passive mutualistic interaction itself promotes the evolution of cell aggregation. We are not investigating how spatial structure

influences the evolutionary transition to an active mutualistic interaction in this manuscript. Instead we ask how spatial structure itself evolves, and thereby focus on a process that has potentially profound implications for interactions both within and between populations of organisms.

Methods
Bacterial strains
We obtained amino acid-auxotrophic strains of *E. coli* BW25113 from the KEIO single-gene knockout library [32] (Table 1). Strain BW25113 ($\Delta proC$) contains a complete deletion of the open-reading frame of the *proC* gene and is predicted to be defective in proline biosynthesis. Strain BW25113 ($\Delta trpC$) contains a complete deletion of the open-reading frame of the *trpC* gene and is predicted to be defective in tryptophan biosynthesis. We refer to these strains as 'mutualistic partners' because they can grow together in co-culture but cannot grow alone; the ability to grow together is based on the passive excretion – presumably via cell leakage - of the amino acid that the other strain cannot biosynthesize.

We verified that the mutualistic partners are auxotrophic for the predicted amino acid by growing them in isolation with liquid minimal medium that was or was not supplemented with the required amino acid. The liquid minimal medium consisted of 6.8 g L^{-1} $Na_2HPO_4 \times 7H_2O$, 3 g L^{-1} KH_2PO_4, 0.5 g L^{-1} NaCl, 1 g L^{-1} NH_4Cl, 3.6 g L^{-1} glucose, 0.24 g L^{-1} $MgSO_4$, and 10 mg L^{-1} gentamycin (referred to as MM hereafter). We streaked each mutualistic partner onto a different lysogeny broth (LB) agar plate, picked three colonies from each LB agar plate, inoculated each colony into a different test tube containing 3 ml of MM, and incubated the test tubes for 24 h at 37 °C with continuous shaking (220 r. p. m.). As expected, neither mutualistic partner could grow in isolation with MM. However, strain BW25113 ($\Delta proC$) could grow in isolation when we supplemented MM with 50 mg L^{-1} L-proline while strain BW25113 ($\Delta trpC$) could grow in isolation when we supplemented MM with 20 mg L^{-1} L-tryptophan, thus verifying that each mutualistic partner is indeed auxotrophic for the predicted amino acid.

Genetic manipulations
We introduced a different plasmid into each mutualistic partner that carries a gene encoding for a different florescent protein, thus allowing us to distinguish and individually quantify each mutualistic partner when they are grown together in co-culture. To accomplish this, we constructed two derivatives of the pUC18T-mini-Tn7T-LAC-Gm conditionally replicative plasmid (Table 1) [33]. This plasmid contains an isopropyl-β-D-thiogalactopyranosid (IPTG)-inducible P_{lac} promoter located

Table 1 Bacterial strains and plasmids used in this study

Strain or plasmid	Relevant characteristics	Reference or source
E. coli strain		
BW25113 ($\Delta proC$)	BW25113 with $\Delta proC:Km^R$; Km^R	[32]
BW25113 ($\Delta trpC$)	BW25113 with $\Delta trpC:Km^R$; Km^R	[32]
DH5α/λpir	Used for replication of pUC18T derivatives; λpir80dlacZ ΔM15 Δ(lacZYA-argG)U169 recA1 hsdr17 deoR thi-1 supE44 gyrA96 relA	[34]
Plasmid		
pUC18T-mini-Tn7T-LAC-Gm	pUC18-based conditionally replicative delivery plasmid for mini-Tn7-LAC-Gm; Ap^R, Gm^R, mob^+	[33]
pUC18T-mini-Tn7T-LAC-Gm-*egfp*	pUC18T-mini-Tn7T-LAC-Gm containing *egfp* immediately downstream of P$_{lac}$; Ap^R, Gm^R, mob^+, $egfp^+$	this study
pUC18T-mini-Tn7T-LAC-Gm-*echerry*	pUC18T-mini-Tn7T-LAC-Gm containing *echerry* immediately downstream of P$_{lac}$; Ap^R, Gm^R, mob^+, $echerry^+$	this study

immediately upstream of a multiple cloning site (MCS). We first purified the pUC18T-mini-Tn7T-LAC-Gm plasmid from an overnight culture of *E. coli* DH5α/λpir (Table 1) [34]. We next used GoTaq DNA polymerase (Promega, Madison, WI, USA) to PCR amplify the *egfp* or *echerry* gene [35], which encode for green or red fluorescent protein respectively. The PCR amplification primers contain the *Bam*HI and *Kpn*I restriction sites that we used to clone the PCR products into the MCS of the pUC18T-mini-Tn7T-LAC-Gm plasmid (See Additional file 1). We then digested the PCR products and the pUC18T-mini-Tn7T-LAC-Gm plasmid with *Bam*HI and *Kpn*I (Thermo Fisher Scientific, Waltham, MA, USA) and ligated the PCR products into the pUC18T-mini-Tn7T-LAC-Gm plasmid. We designated the assembled derivative plasmids as pUC18T-mini-Tn7T-LAC-Gm-*egfp* and pUC18T-mini-Tn7T-LAC-Gm-*echerry* (Table 1). We finally replicated the assembled derivative plasmids in *E. coli* DH5α/λpir (Table 1) [34], introduced the derivative plasmids into the mutualistic partners via electroporation, and selected for transformants carrying the derivative plasmids by plating on LB agar plates supplemented with 10 μg ml^{-1} gentamycin and 1 mM IPTG. We introduced each derivative plasmid into each mutualistic partner, resulting in both *egfp*- and *echerry*-expressing variants of strains BW25113 ($\Delta proC$) and BW25113 ($\Delta trpC$).

Evolution experiment

We performed an evolution experiment with replicated co-cultures of the two mutualistic partners. We streaked the *egfp*-expressing BW25113 ($\Delta proC$), *echerry*-expressing BW25113 ($\Delta trpC$), *echerry*-expressing BW25113 ($\Delta proC$), and *egfp*-expressing BW25113 ($\Delta trpC$) strains onto different LB agar plates that were supplemented with 10 μg ml^{-1} gentamycin and 1 mM IPTG. We then picked one colony from each LB agar plate, inoculated

each colony into a different test tube containing 2.7 ml of MM supplemented with 300 μl of liquid LB medium, and incubated the test tubes for 24 h at 37 °C with continuous shaking (220 r. p. m.). After reaching stationary phase, we centrifuged the cultures, discarded the spent medium, washed the cells twice with H$_2$O containing 9 g L^{-1} NaCl, and suspended the washed cells in H$_2$O containing 9 g L^{-1} NaCl. We next prepared two binary mixes (i.e. co-cultures) of the mutualistic partners (50:50 ratio based on optical density measurements at 600 nm [OD$_{600}$]); one set of co-cultures consisted of the *egfp*-expressing BW25113 ($\Delta proC$) and *echerry*-expressing BW25113 ($\Delta trpC$) mutualistic partners while the other set of co-cultures consisted of the *echerry*-expressing BW25113 ($\Delta proC$) and *egfp*-expressing BW25113 ($\Delta trpC$) mutualistic partners. We finally inoculated 300 μl of each co-culture into eight replicated test tubes containing 2.7 ml of MM that was supplemented with 1 mM IPTG but not with amino acids, resulting in a total of 16 replicated mutualistic co-cultures. We designated the co-cultures consisting of the *egfp*-expressing BW25113 ($\Delta proC$) and *echerry*-expressing BW25113 ($\Delta trpC$) mutualistic partners as mutualistic co-cultures A1 to A8 and the co-cultures consisting of the *echerry*-expressing BW25113 ($\Delta proC$) and *egfp*-expressing BW25113 ($\Delta trpC$) mutualistic partners as mutualistic co-cultures B1 to B8. The initial OD$_{600}$ of each mutualistic co-culture was approximately 0.12. After incubating the mutualistic co-cultures for 3.5 days at 37 °C with continuous shaking (220 r. p. m.), we transferred 300 μl of each co-culture to a new test tube containing 2.7 ml of fresh MM that was supplemented with 1 mM IPTG but not with amino acids to achieve a 1:10 (volume: volume) dilution. We then repeated the incubation and transfer steps in the same MM supplemented with 1 mM IPTG for a total of 16 serial transfers. Immediately before each transfer, we measured the OD$_{600}$ of each mutualistic co-culture and archived a portion of each mutualistic co-culture in 20% glycerol at −80 °C for further analyses. We quantified the magnitude of cell aggregation within each mutualistic co-culture immediately before the fourteenth transfer as described below. If all cells in these co-cultures would have grown at the same rate, then the fourteen transfers with ten-fold dilution would correspond to approximately 46 generations of growth ($14 \times \log_2 10$); however, if only a portion of the cells would have grown efficiently under our experimental conditions (for example mutants that attach to other cells), then the number of cell generations during the evolution experiment could have been potentially much larger. We performed control experiments with the ancestral mutualistic partners using growth conditions identical to those described above, except that the MM was supplemented with 50 mg L^{-1} L-proline and 20 mg L^{-1} L-tryptophan.

Quantification of cell aggregation

We imaged each mutualistic co-culture using a Leica TCS SP5 confocal laser-scanning microscope (CLSM) with a 63 × (1.4 NA) oil-immersion lens (Leica Microsystems, Wetzlar, Germany). We removed 5-µl liquid aliquots from each mutualistic co-culture immediately after removal from the shaking incubator and deposited the liquid aliquots onto the surface of a glass slide. We imaged *egfp*-expressing cells using 488 nm excitation wavelength and 500–530 nm emission wavelengths. We imaged *echerry*-expressing cells using 633 nm excitation wavelength and 657–757 nm emission wavelengths. We collected images at a resolution of 1024 × 1024 using LAS AF v2.7 software (Leica Microsystems).

We quantified cell aggregation using the StatColoc plugin of the Icy software [36]. This algorithm computes the two-dimensional co-localization of different objects (in our case cells) using the Ripley's K function [37]. The resulting K-value measures the degree to which a set of objects deviates from spatial homogeneity. In our experiments with completely mixed batch reactors, a deviation from spatial homogeneity is most likely caused by cell aggregation, which we indeed confirmed by microscopy. We first detected *egfp*- and *echerry*-expressing cells using the Spot Detector plugin of Icy and translated the images into binary data using the NIH ImageJ analysis software (http://rsbweb.nih.gov/ij/). We then applied the StatColoc plugin using a radius of 0.48 µm to 4.8 µm. We analyzed between five and nine randomly selected microscope fields for each mutualistic co-culture and obtained ten K-values as a function of distance for each microscope field. We finally identified the maximum observed K-value for each microscope field and tested whether the maximum observed K-values are significantly different between test and reference mutualistic

co-cultures using the non-parametric Mann-Whitney U test. We performed the same statistical tests with the sum-of-the-ten K-values rather than the maximum observed K-value and obtained qualitatively identical results. We chose here to report the maximum observed K-values because they had better statistical properties. Namely, many of the microscopy fields did not contain any cell aggregates, which would be expected when aggregation is extensive and consists of a few sparsely distributed objects. This increases the variance when using the sum-of-the-ten K-values. We implemented the Mann-Whitney U test with the StatColoc plugin and considered a two-sided $P < 0.05$ to be statistically significant.

Isolation of evolved mutualistic partners

We isolated mutants emerging within each archived mutualistic co-culture that remained viable immediately before the fourteenth transfer of experimental evolution. We first streaked each archived mutualistic co-culture onto different LB agar plates that were supplemented with 10 µg ml^{-1} gentamycin and 1 mM IPTG. In many cases, the colonies expressed multiple fluorescent proteins, and we therefore had to perform a second streaking. After obtaining single colonies that only express one fluorescent protein, we qualitatively distinguished different evolved mutualistic partners within each co-culture based on colony morphology and the fluorescent gene that they expressed (*egfp* or *echerry*). For some co-cultures, we identified more than one colony morphology for each fluorescent gene. We finally picked one colony for each morphology, inoculated each colony into a different test tube containing 3 ml of liquid LB medium supplemented with 10 µg ml^{-1} gentamycin and 1 mM IPTG, incubated the test tubes for 24 h at 37 °C with continuous shaking (220 r. p. m.), and archived a portion of each culture in 20% glycerol at −80 °C for further analyses.

Genome sequencing of evolved mutualistic partners

We sequenced the genomes of all the isolated evolved mutualistic partners (see Additional file 2). In parallel, we sequenced the genomes of all the ancestral mutualistic partners (i.e., the *egfp*-expressing BW25113 (Δ*proC*), *echerry*-expressing BW25113 (Δ*trpC*), *echerry*-expressing BW25113 (Δ*proC*), and *egfp*-expressing BW25113 (Δ*trpC*) strains). We first streaked each mutualistic partner from the glycerol-archived samples onto different LB agar plates supplemented with 10 µg ml^{-1} gentamycin and 1 mM IPTG. We then picked one colony from each LB agar plate, inoculated each colony into a test tube containing 3 ml of liquid LB medium, incubated the test tubes for 24 h at 37 °C with continuous shaking (220 r. p. m.), and extracted genomic DNA from each culture using the ArchivePure DNA Purification kit (5prime, Hilden, Germany). We then prepared one sequence library for each mutualistic partner using 1 ng of genomic DNA, the Nextera XT DNA Sample Preparation kit (Illumina Inc., San Diego, CA, USA), and a different sample-specific multiplex adapter. We next pooled the libraries together, loaded the pool onto a single MiSeq flow cell (Ilumina Inc.), and sequenced the libraries using a MiSeq sequencer (Illumina Inc.) operated by the Genomic Diversity Center at ETH Zürich (Zürich, Switzerland.) We performed paired-end 150-cycle sequencing with the MiSeq Reagent Kit (version 2) (Illumina Inc.). All of the sequence reads are publically available in the NCBI Sequence Read Archive (http://www.ncbi.nlm.nih.gov/sra) under Bioproject ID number SUB2512304.

We analyzed the resulting sequence reads by first binning the raw sequence reads into libraries using the automated run protocol on the MiSeq sequencer (Illumina Inc.). We then performed quality control with FastQC version 0.10.1 and quality filtering using PrinSeq Lite version 0.20.4 software [38]. The parameters used for quality filtering are as follows: out_format, 3; min_qual_mean, 28; min_len, 50; range_gc 15–85; ns_max_n, 1; derep, 14; derep_min, 2; trim_ns_left, 1; trim_ns_right, 1; trim_qual_left, 28; trim_qual_right, 28; trim_left, 1. In summary, we trimmed or discarded all sequence reads with mean quality scores below 28 or had ambiguous nucleotides. We further discarded all sequence reads that were shorter than 50 bases. We finally applied the breseq pipeline (version 0.24rc1) and the utility program gdtools [39, 40] to predict genetic changes between each evolved mutualistic partner and its corresponding ancestral mutualistic partner. This included nucleotide polymorphisms, deletions, insertions, and multiplications. We used the genome sequence of E. coli K-12 MG1655 [41] as a reference for the mapping and assembly of the sequence reads. The parameters used for calling genetic changes are as follows: reference, NC_000913; base-quality-cutoff, 15; require-match-length, 30.

Crossing experiment

We randomly selected the B2 mutualistic co-culture to test whether one or more of the evolved mutualistic partners were responsible for the evolution of cell aggregation. Based on colony morphology, fluorescent protein production, and genome analyses, we identified one evolved mutualistic partner of strain BW25113 ($\Delta proC$) and two different evolved mutualistic partners of strain BW25113 ($\Delta trpC$) within the B2 mutualistic co-culture immediately before the fourteenth transfer of experimental evolution (see Additional file 2). We first streaked each of the evolved mutualistic partners and their corresponding ancestral mutualistic partners from

the archived isolate samples onto different LB agar plates that were supplemented with 10 μg ml^{-1} gentamycin and 1 mM IPTG. We then picked one colony from each LB agar plate, inoculated each colony into a different test tube containing 2.7 ml of MM supplemented with 300 μl of liquid LB medium, and incubated the test tubes for 24 h at 37 °C with continuous shaking (220 r. p. m.). After reaching stationary phase, we washed and suspended the cells in water containing 9 g L^{-1} NaCl as described above for the evolution experiment. We next prepared different binary (50:50 ratio based on OD$_{600}$ measurements) or ternary (33:33:33 ratio based on OD$_{600}$ measurements) mixes of different evolved and ancestral mutualistic partners as described in the results section. We finally inoculated 300 μl of each mix into replicated test tubes containing 2.7 ml of MM that was supplemented with 1 mM IPTG but not with amino acids. The initial OD$_{600}$ of each mutualistic co- or tri-culture was approximately 0.12. After incubating the mutualistic co- or tri-cultures for 7 days at 37 °C with continuous shaking (200 r. p. m.), we measured the OD$_{600}$ of each co-or tri-culture and quantified the magnitude of cell aggregation as described above.

Results and discussion
Experimental creation of a passive mutualistic interaction
We created a passive and obligate mutualistic interaction by inoculating pairs of the BW25113 ($\Delta proC$) and BW25113 ($\Delta trpC$) mutualistic partners (Table 1) together into MM that was not supplemented with amino acids. We found that the mutualistic partners could grow when they were inoculated together but could not grow when they were inoculated in isolation, thus demonstrating that we could indeed create the expected passive mutualistic interaction. We use the term 'passive' because the interaction emerges spontaneously when inoculating the two auxotrophic strains together that were neither engineered (e.g., as in [42]) nor evolved to actively excrete the amino acid that the other strain requires. We further tested whether access to these two amino acids limits the growth of the mutualistic co-cultures. We found that the co-cultures reached stationary phase approximately four-times more rapidly when we provided exogenous supplements of the required amino acids (50 mg L^{-1} L-proline and 20 mg L^{-1} L-tryptophan) (< 18 h) than when we did not (> 3.5 days), thus verifying that the supply of the required amino acids did indeed limit the growth of the mutualistic co-cultures.

Evolutionary dynamics
We next investigated the evolutionary dynamics of the mutualistic co-cultures. We performed an evolution experiment by serially transferring the 16 replicated

mutualistic co-cultures every 3.5 days into fresh MM that was not supplemented with amino acids, corresponding to 8 weeks of experimental evolution. Mutualistic co-cultures A1-A8 consisted of the *egfp*-expressing BW25113 (Δ*proC*) and *echerry*-expressing BW25113 (Δ*trpC*) mutualistic partners while mutualistic co-cultures B1 to B8 consisted of the *echerry*-expressing BW25113 (Δ*proC*) and *egfp*-expressing BW25113 (Δ*trpC*) mutualistic partners. We used two different combinations of fluorescent protein-encoding genes for two reasons. First, it allowed us to assess whether differences in the fluorescent protein-encoding genes themselves affect the outcome of the evolution experiment. Second, it allowed us to monitor for cross-contamination among the mutualistic co-cultures during the evolution experiment, which we never detected. We measured the OD_{600} of each mutualistic co-culture immediately before each subsequent transfer as a proxy of total cell numbers.

We observed three qualitatively different evolutionary dynamics. For all 16 mutualistic co-cultures, the OD_{600} decreased by about 2.3-fold between the first and second transfers (Mann-Whitney U test, two-sided $P = 1.5 \times 10^{-6}$), indicating a substantial reduction in total cell numbers. Thereafter, however, the mutualistic co-cultures exhibited different dynamics. For three of the mutualistic co-cultures (co-cultures B4, B5, and B6), the OD_{600} continued to decrease and fell below detection levels at the fourth transfer (Fig. 2), indicating a persistent reduction in total cell numbers and eventual extinction. For another three of the mutualistic co-cultures (co-cultures A5, B3, and B8), the OD_{600} increased after the second transfer but then decreased again and fell below detection levels at the sixth, fourteenth, or fifteenth transfer (Fig. 2), indicating an initial increase in total cell numbers followed by a rapid shift in growth dynamics and eventual extinction. Finally, for the remaining ten mutualistic co-cultures, the OD_{600} increased nearly continuously and significantly by about 2.2-fold between the second and final transfers (Mann-Whitney U test, two-sided $P = 1.8 \times 10^{-4}$), demonstrating a progressive and substantial increase in total cell numbers and avoidance of extinction. These results are qualitatively consistent with a previous evolution experiment that imposed a mutualistic interaction between a methanogenic and a sulfate-reducing microorganism [43]. In that experiment, the authors similarly found that some mutualistic co-cultures went extinct while others progressively increased in total cell numbers and avoided extinction.

One limitation of our analysis above is our use of OD_{600} measures as proxy measures of total cell numbers, as cell aggregation can prevent a linear correspondence between OD_{600} and total cell numbers. In general, however, cell aggregation tends to decrease OD_{600}. Thus, the fact that we observed a general increase in OD_{600}

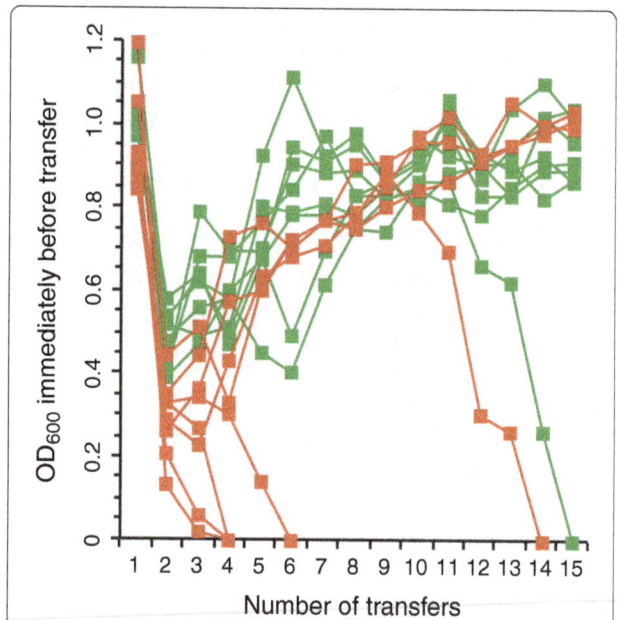

Fig. 2 Experimental evolution of the mutualistic consortia. Sixteen replicate mutualistic consortia were assembled from the *egfp*-expressing BW25113 (Δ*proC*) and *echerry*-expressing BW25113 (Δ*trpC*) mutualistic partners (red symbols) or from the echerry-expressing BW25113 (Δ*proC*) and *egfp*-expressing BW25113 (Δ*trpC*) mutualistic partners (green symbols). Total cell numbers were then measured as OD_{600} and tracked over time. After the mutualistic interaction was imposed, ten consortia improved in growth performance over evolutionary time while six consortia went extinct, suggesting that different consortia undergo different evolutionary trajectories

over evolutionary time as cell aggregation emerged suggests that total cell numbers also increased over the same evolutionary time. The increase in total cell numbers, however, should be interpreted as a qualitative observation rather than a quantitative measure.

Evolution of cell aggregation

We next examined whether the created passive mutualistic interaction promotes the subsequent evolution of cell aggregation. To accomplish this, we used a confocal microscope to analyze the spatial structure of the mutualistic co-cultures that avoided extinction immediately before the fourteenth transfer. We found that the cells within all of these mutualistic co-cultures were significantly more aggregated together than were the cells within their corresponding ancestral mutualistic co-culture (Fig. 3). All observed aggregates contained mixtures of both mutualistic partners. All of the evolved mutualistic co-cultures had significantly larger maximum observed K-values (which we used as a measure of cell aggregation as described in the Materials and Methods) than did their corresponding ancestral mutualistic co-cultures (Fig. 4 and Table 2; Mann-Whitney U test, two-sided $P \leq 0.0021$). This indicates that cells within the evolved mutualistic co-cultures were more

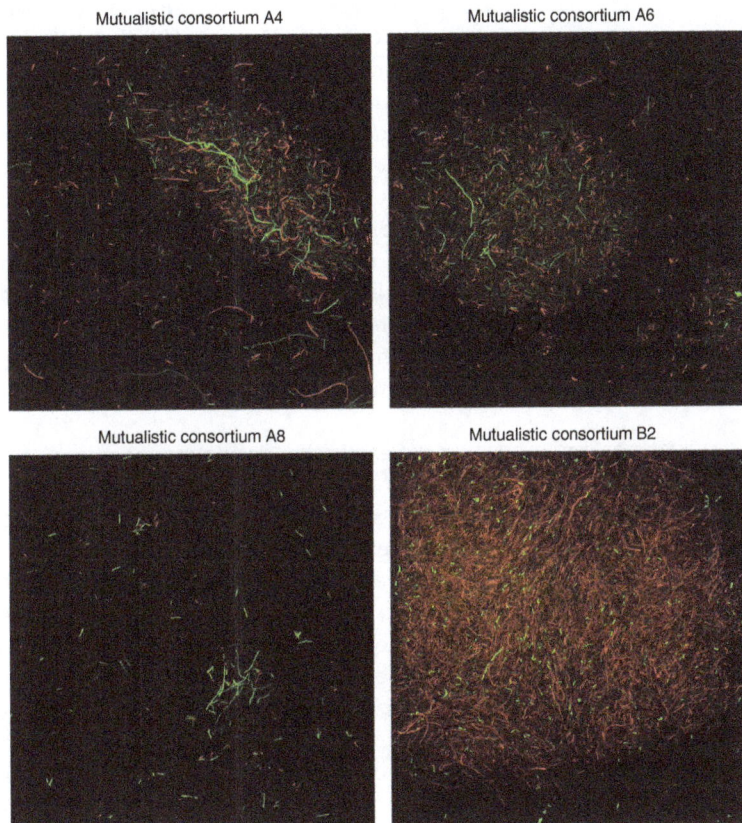

Fig. 3 Representative images of cell aggregation within the mutualistic consortia. Images were obtained immediately before the fourteenth transfer of experimental evolution. For the A4, A6, and A8 consortia, green cells are the *egfp*-expressing BW25113 (Δ*proC*) mutualistic partner and red cells are the *echerry*-expressing BW25113 (Δ*trpC*) mutualistic partner. For the B2 consortium, red cells are the *echerry*-expressing BW25113 (Δ*proC*) mutualistic partner and green cells are the *egfp*-expressing BW25113 (Δ*trpC*) mutualistic partner. Note that all cell aggregates contained both mutualistic partners, but the qualitative structure and organization of the cell aggregates varied across the different consortia

heterogeneously distributed in space via cell aggregation than were the cells within their corresponding ancestral mutualistic co-culture. In general, aggregation was qualitatively similar within lineages but varied across lineages (Fig. 3). We further found that cell aggregation only evolved in co-cultures consisting of strains coupled by the passive mutualistic interaction. Spatial structure never evolved when we prevented the mutualistic interaction from establishing by growing the ancestral mutualistic partners together in LB liquid medium or in MM supplemented with the required amino acids (note that the MM contained gentamycin and IPTG and the strains contained their respective plasmids in these controls). Finally, there was no statistically detectable effect of the combination of fluorescent proteins used to mark each strain on the maximum observed K-values reported in Fig. 4 (i.e. there was no difference whether the Δ*trpC* or Δ*proC* mutant expressed *egfp* or *echerry*) (Mann-Whitney U-test, two-sided $P > 0.05$). Taken together, our results demonstrate that the creation of the passive mutualistic

interaction was necessary for and promoted the subsequent evolution of cell aggregation.

One alternative explanation is that the cell aggregates consisted of dead cells rather than viable cells. To test this, we routinely plated the cultures onto agar plates. After incubation, we always observed individual colonies that expressed both fluorescent proteins (i.e., they contained mixtures of cells that express *gfp* or *echerry*). This indicates that the mutualistic partners were indeed physically attached to each other and that cells within the aggregates were viable. The cell aggregation phenotype therefore clearly contributed towards population-level phenotype. However, as shown in Fig. 3, individual planktonic cells remained. Thus, while aggregation significantly contributed to population-level phenotype, it remained incomplete.

While we observed the emergence of cell aggregation, we note that our results do not suggest that the emergence of cell aggregation was directly caused by the mutualistic interaction itself. This could occur via mechanisms such as specific partner recognition, where each

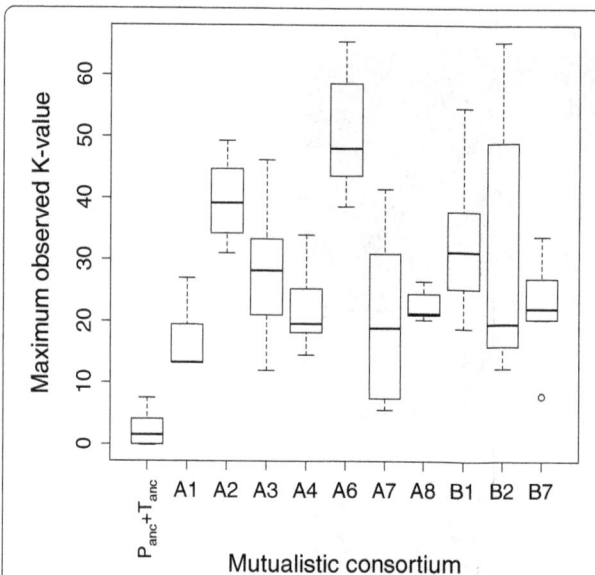

Fig. 4 Magnitude of cell aggregation within the ancestral and evolved mutualistic consortia. Cell aggregation was measured as the maximum observed K-value. Definitions: P_{anc}, ancestral strain of BW25113 ($\Delta proC$); T_{anc}, ancestral strain of BW25113 ($\Delta trpC$). The data are presented as Tukey box plots. All the mutualistic consortia had larger K-values than the ancestral consortia, indicating an increase in cell aggregation over the course of the evolution experiment

Table 2 Comparison of the K-values for the test and ancestral mutualistic consortia

Experiment	[a]Mutualistic consortium	[b]p
Evolution experiment	A1	**0.0013**
	A2	**0.0013**
	A3	**0.0013**
	A4	**0.0013**
	A6	**0.0013**
	A7	**0.0021**
	A8	**0.0013**
	B1	**0.0003**
	B2	**0.0003**
	B7	**0.0004**
[c,d]Crossing experiment	$P_{anc} + T_{evol,a}$	0.015
(mutualistic consortium B2)	$P_{anc} + T_{evol,b}$	0.028
	$P_{evol} + T_{anc}$	0.17
	$P_{evol} + T_{evol,a}$	**0.0002**
	$P_{evol} + T_{evol,b}$	**0.0009**
	$P_{evol} + T_{evol,a} + T_{evol,b}$	**0.0002**

[a]Only those consortia that avoided extinction were analyzed. [b]Bold font: statistically significant differences between the test and ancestral mutualistic consortia ($P < 0.01$). [c]Definitions: P_{anc}, ancestral strain of BW25113 ($\Delta proC$); T_{anc}, ancestral strain of BW25113 ($\Delta trpC$); P_{evol}, evolved strain of BW25113 ($\Delta proC$); $T_{evol,a}$, evolved strain of BW25113 ($\Delta trpC$); $T_{evol,b}$, evolved strain of BW25113 ($\Delta trpC$). [d]$T_{evol,a}$ and $T_{evol,b}$ are two genetically different strains of BW25113 ($\Delta trpC$) that emerged within the B2 mutualistic consortium

mutualistic type aggregates with its partner. However, such adaptations would likely require long evolutionary times. Instead, the emergence of cell aggregation in our study is likely an indirect consequence of the small amounts of amino acids that leak or are released from the mutualistic partners. In other words, the mutualistic interaction creates an environment with low concentrations of amino acids, which then promotes the evolution of non-specific cell aggregation. Regardless, non-specific cell aggregation might then set the stage for further evolutionary changes, such as the emergence of partner recognition and specific attachment between the different mutualistic partners.

Genetic changes during emergence of cell aggregation

We investigated the genetic changes that occurred during the evolution experiment. The goal here was not to identify the specific genetic changes that cause cell aggregation, but rather to investigate whether each lineage accumulated similar or different genetic changes during the acquisition of the cell aggregation phenotype. This then allows us to hypothesize whether there is a single or multiple evolutionary pathways to cell aggregation. To accomplish this, we reconstructed the ancestral and evolved mutualistic co-cultures from isolates. We first grew each ancestral or evolved mutualistic partner on amino acid-rich LB agar plates, assembled the mutualistic partners together into mutualistic co-cultures, and inoculated the co-cultures into MM that was not supplemented with amino acids. We found that the reconstructed evolved mutualistic co-cultures formed cell aggregates after 7 days of incubation while the ancestral mutualistic co-cultures did not (i.e., the ancestral cells were completely planktonic in microscopy images). Thus, cell aggregation was heritable and therefore likely had a genetic basis.

We next sequenced the genomes of the ancestral and evolved isolates and identified genetic changes in the evolved mutualistic partners that might have caused the observed emergence of cell aggregation. We provide a complete list of all the observed genetic changes in Additional file 2. We excluded mutualistic co-culture A2 from the analysis because we could not separate the mutualistic partners into individual cells (i.e., they continued to form mixed-strain colonies after repeated streaking on LB agar plates, indicating very strong aggregation). Each evolved mutualistic co-culture contained a mutualistic partner that fixed at least one genetic change located within or immediately upstream of a gene known to be involved with biofilm formation (Table 3). However, as opposed to comparable experimental evolution studies and analytical methods [40, 44–47], we observed relatively limited evolutionary parallelism of the genetic changes. More than 75% (13/17) of the genes or

Table 3 Genetic changes within or upstream of genes that have experimentally verified roles in *E. coli* biofilm formation in other studies

[a]Mutualistic consortium	[b]Evolved mutualistic partner	Mutation type	Gene(s) or intergenic region	Role of gene(s) in *E. coli* biofilm formation	Reference
A1	BW25113 (Δ*proC*)	non-synonymous point mutation, A- > G	*spoT*[c]	c-di-GMP regulation	[51, 52]
A3	BW25113 (Δ*proC*)	Δ16 bp, intergenic region	upstream of *flhD*	motility regulation	[57, 58]
	BW25113 (Δ*trpC*)a	Δ1 bp, coding region	*hdfR*	motility regulation	[58, 59]
	BW25113 (Δ*trpC*)b	Δ1 bp, coding region	*hdfR*	motility regulation	[58, 59]
A4	BW25113 (Δ*proC*)	non-synonymous point mutation, G- > A	*flhC*	motility regulation	[60]
	BW25113 (Δ*trpC*)a	Δ627 bp, coding and intergenic regions	*spoT*[c]-*trmH*	c-di-GMP regulation	[51, 52]
	BW25113 (Δ*trpC*)b	Δ627 bp, coding and intergenic regions	*spoT*[c]-*trmH*	c-di-GMP regulation	[51, 52]
A6	BW25113 (Δ*trpC*)	non-synonymous point mutation, C- > T	*glmU*[c]	extracellular matrix	[53, 54]
	BW25113 (Δ*trpC*)	point mutation, intergenic region	upstream of *yqcC*	biofilm maturation	[61]
A7	BW25113 (Δ*trpC*)a	Δ15 bp, coding region	*spoT*[c]	c-di-GMP regulation	[51, 52]
	BW25113 (Δ*trpC*)b	Δ15 bp, coding region	*spoT*[c]	c-di-GMP regulation	[51, 52]
	BW25113 (Δ*trpC*)c	Δ15 bp, coding region	*spoT*[c]	c-di-GMP regulation	[51, 52]
A8	BW25113 (Δ*proC*)	Δ3 bp, coding region	*glmU*[c]	extracellular matrix	[53, 54]
	BW25113 (Δ*proC*)	Δ3 bp, coding region	*gspA*	biofilm maturation	[62]
	BW25113 (Δ*proC*)	Δ3 bp, coding region	*rcsF*	EPS regulation	[63, 64]
	BW25113 (Δ*trpC*)c	Δ6 bp, coding region	*bamA*	extracellular matrix	[65, 66]
B1	BW25113 (Δ*trpC*)	Δ11 bp, coding region	*nlpI*	extracellular matrix	[67]
B2	BW25113 (Δ*proC*)	Δ10 bp, intergenic region	upstream of *bluF* and *ycg*	EPS regulation	[61, 62]
	BW25113 (Δ*proC*)	Δ5 bp, coding region	*dksA*	c-di-GMP regulation	[61, 63, 68]
	BW25113 (Δ*proC*)	Δ3 bp, intergenic region	upstream of *yliE*	c-di-GMP regulation	[52]
	BW25113 (Δ*proC*)	113 bp duplication of coding region	*yeaP*	fimbriae regulation	[69]
	BW25113 (Δ*trpC*)a	non-synonymous point mutation, G- > T	*gpp*	c-di-GMP regulation	[52, 70]
B7	BW25113 (Δ*proC*)	Δ1 bp, coding region	*bluR*	EPS regulation	[56]

[a]Only those consortia that avoided extinction were analyzed. [b]Subscripts indicate that more than one clone of that mutualistic partner was sequenced from the corresponding mutualistic consortium. [c]Genetic changes in *spoT* and *glmU* have been observed in other studies where cell aggregation did not emerge

upstream regions that contained genetic changes were changed in only one co-culture (Table 3). Thus, there appears to be a large number of possible genetic targets that could result in the evolution of cell aggregation. This is not unexpected given the large number of gene products required for the regulation, initiation, development and maturation of *E. coli* biofilms [48–50].

While we observed limited evolutionary parallelism in the genetic changes, a few genes or upstream regions were changed in mutualistic partners from more than one co-culture. Genetic changes in *spoT* or its upstream region occurred in mutualistic partners from three co-cultures (Table 3). *spoT* affects biofilm formation by modifying levels of (p) ppGpp [51, 52] and, under certain conditions, the inactivation of *spoT* can enhance biofilm formation [51]. Genetic changes in *glmU*

occurred in mutualistic partners from two co-cultures (Table 3). *glmU* affects biofilm formation by controlling the biosynthesis of surface adhesion molecules [53, 54]. We note here, however, that genetic changes in *spoT* and *glmU* have been reported in other evolution experiments where cell aggregation did not emerge [40], and that further molecular work would therefore be required to test their role here. Genetic changes in three flagellar genes (*flhC*, *flhD*, and *hdfR*) occurred in mutualistic partners from two co-cultures (Table 3). These genes affect biofilm formation and surface attachment by regulating the biosynthesis of flagella [55]. Finally, genetic changes in three other genes or upstream regions (*bluR*, *bluF*, *ycgG*) occurred in mutualistic partners from two co-cultures (Table 3). These genes affect biofilm formation by activating the Rcs system, which regulates the biosynthesis of surface adhesion molecules and curli fimbre [56]. We did observe some mutations in *lacI,* which is involved with the transcriptional regulation of the fluorescent proteins. These mutations were not unique to one mutualistic partner or another, and they are therefore unlikely to have created confounding factors that compromise our main conclusions.

Both mutualistic partners are required for cell aggregation

The genome sequences demonstrate that both mutualistic partners could acquire mutations in biofilm-related genes or in regions immediately upstream of those genes. This observation then leads to the following hypothesis: both mutualistic partners could contribute towards the observed evolution of cell aggregation. To test this hypothesis, we performed a crossing experiment with the B2 mutualistic co-culture. Based on colony morphology and genome sequences, we determined that this mutualistic co-culture had one evolved mutualistic partner of BW25113 ($\Delta proC$) (designated as P_{evol}) and two different evolved mutualistic partners of BW25113 ($\Delta trpC$) (designated as $T_{evol,a}$ and $T_{evol,b}$) (see Additional file 2). $T_{evol,a}$ contains one genetic change not present in $T_{evol,b}$ while $T_{evol,b}$ contains seven genetic changes not present in $T_{evol,a}$ (see Additional file 2). We first grew each mutualistic partner in isolation and then mixed the evolved mutualistic partners together or mixed each evolved mutualistic partner with its corresponding ancestral mutualistic partner of BW25113 ($\Delta proC$) (designated as P_{anc}) or BW25113 ($\Delta trpC$) (designated as T_{anc}). We finally quantified cell aggregation of the resulting mutualistic consortia using a confocal microscope (Fig. 5). We acknowledge here that we only performed these experiments for one lineage; the results may therefore be lineage-specific and we cannot make general statements. Instead, our objective here was to investigate a single lineage in order to provide initial insight into the observed phenomena and set the stage for future investigations (e.g., to investigate evolutionary parallelism, etc.).

We found that heritable changes in both evolved mutualistic partners are required to maximize cell aggregation. Mutualistic consortia of P_{anc} and $T_{evol,a}$ or $T_{evol,b}$ produced significantly more cell aggregation than mutualistic consortia of P_{anc} and T_{anc} (Fig. 5 and Table 2; Mann-Whitney U test, two-sided $P < 0.05$). In contrast, mutualistic consortia of P_{evol} and T_{anc} did not form significantly more cell aggregation than mutualistic consortia of P_{anc} and T_{anc} (Fig. 5 and Table 2; Mann-Whitney U test, two-sided $P > 0.05$). However, mutualistic consortia of P_{evol} and $T_{evol,a}$ or $T_{evol,b}$ produced the most significant increase in cell aggregation (Fig. 5 and Table 2; Mann-Whitney U test, two-sided $P < 0.05$). Thus, within the B2 mutualistic consortium, the most

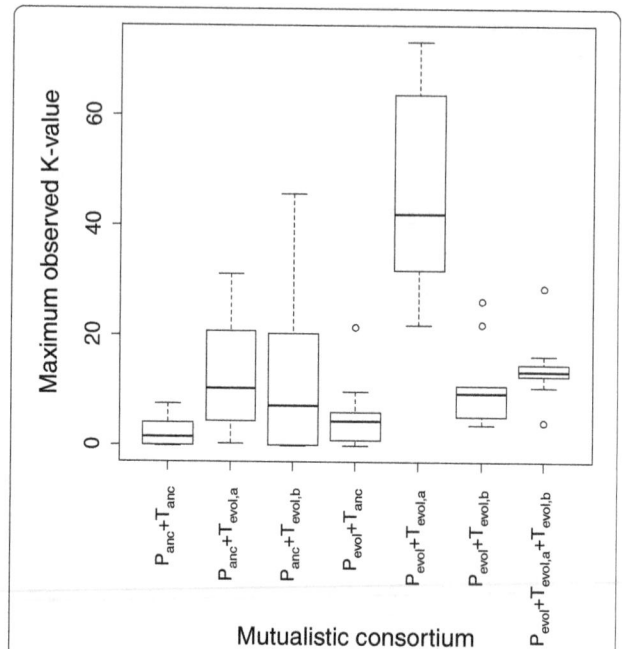

Fig. 5 Magnitude of cell aggregation for different combinations of ancestral and evolved mutualistic partners from the B2 mutualistic consortium. Cell aggregation was measured as the maximum observed K-value. Definitions: P_{anc}, ancestral strain of BW25113 ($\Delta proC$); T_{anc}, ancestral strain of BW25113 ($\Delta trpC$); P_{evol}, evolved strain of BW25113 ($\Delta proC$); $T_{evol,a}$, evolved strain of BW25113 ($\Delta trpC$); $T_{evol,b}$, evolved strain of BW25113 ($\Delta trpC$). $T_{evol,a}$ and $T_{evol,b}$ are two genetically different strains of BW25113 ($\Delta trpC$) that emerged within the B2 mutualistic consortium. The data are presented as Tukey box plots. Note that the K-values (and thus cell aggregation) were the largest for pairs of evolved mutualistic partners, suggesting that genetic changes in both partners contributed to the observed cell aggregation

substantial increase in cell aggregation occurred when mutations in both mutualistic partners were present together within the consortium.

Only one mutualistic partner contributes towards increased cell numbers

We next asked whether both evolved mutualistic partners contribute towards increased cell numbers of the mutualistic co-cultures (measured as the OD_{600} after 7 days). To test this, we performed the same crossing experiment with the B2 mutualistic consortium as described above. We found that mutations in only one evolved mutualistic partner contributed towards increased cell numbers. Mutualistic co-cultures of P_{anc} and $T_{evol,a}$ or $T_{evol,b}$ had significantly higher OD_{600} than mutualistic co-cultures of P_{anc} and T_{anc} and could account for all or a substantial portion of the observed increase in growth of the evolved mutualistic co-culture (Fig. 6; Welch test, two-sided $P < 0.02$). In contrast, mutualistic co-cultures of P_{evol} and T_{anc} had significantly lower OD_{600} than mutualistic co-cultures of P_{anc} and T_{anc} (Fig. 6; Welch test, two-sided $P = 0.0382$). Thus,

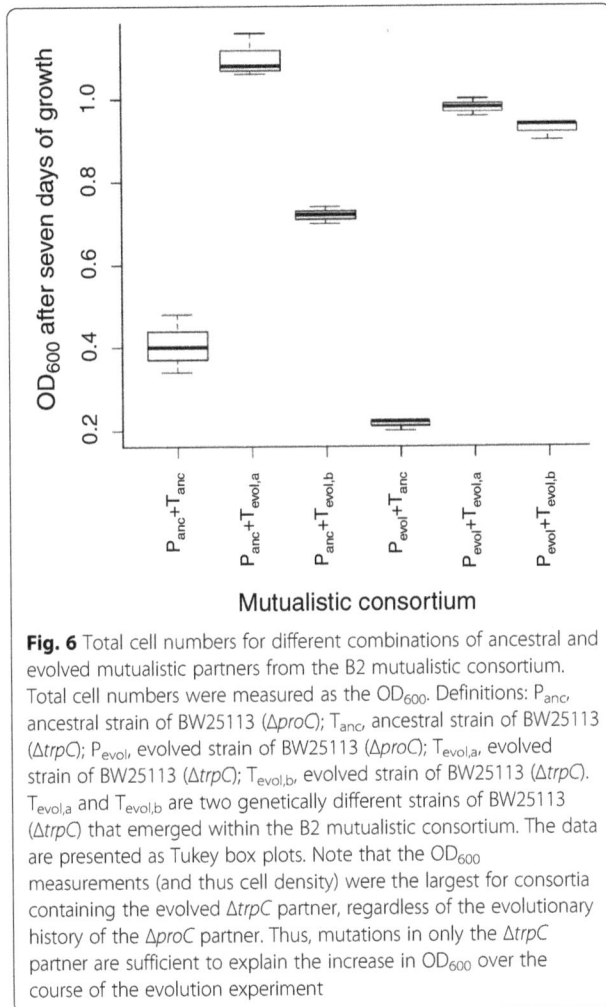

Fig. 6 Total cell numbers for different combinations of ancestral and evolved mutualistic partners from the B2 mutualistic consortium. Total cell numbers were measured as the OD_{600}. Definitions: P_{anc}, ancestral strain of BW25113 ($\Delta proC$); T_{anc}, ancestral strain of BW25113 ($\Delta trpC$); P_{evol}, evolved strain of BW25113 ($\Delta proC$); $T_{evol,a}$, evolved strain of BW25113 ($\Delta trpC$); $T_{evol,b}$, evolved strain of BW25113 ($\Delta trpC$). $T_{evol,a}$ and $T_{evol,b}$ are two genetically different strains of BW25113 ($\Delta trpC$) that emerged within the B2 mutualistic consortium. The data are presented as Tukey box plots. Note that the OD_{600} measurements (and thus cell density) were the largest for consortia containing the evolved $\Delta trpC$ partner, regardless of the evolutionary history of the $\Delta proC$ partner. Thus, mutations in only the $\Delta trpC$ partner are sufficient to explain the increase in OD_{600} over the course of the evolution experiment

genetic changes in the P_{evol} mutualistic partner resulted in a growth defect when combined with its ancestral mutualistic partner. This suggests that the evolutionary changes in P_{evol} were acquired in direct response to evolutionary changes in $T_{evol,a}$ or $T_{evol,b}$, and are therefore a result of co-evolutionary dynamics between the mutualistic partners. Thus, within the B2 mutualistic co-culture, mutations in only one mutualistic partner was sufficient to explain the observed increase in cell numbers – but not necessarily the growth rate – of the B2 mutualistic co-culture over evolutionary time. After these initial mutants emerge, this may then set the stage for future mutants to emerge that reinforce the mutualistic interaction. We believe these dynamics are generalizable, in that the origin of any mutualistic interaction must necessarily emerge via single genetic changes. The dynamics with respect to the further strengthening of the mutualistic interaction, where mutants of both cell-types increase in frequency, would require longer evolutionary times than investigated in this study. We acknowledge here that we are not investigating the effects of individual genetic changes, which would require introducing those genetic changes into the ancestral background. It is possible that the presence of multiple genetic changes may be critically important for our observations, such as reciprocal adaptations between the mutualistic partners.

Conclusions

We propose that the scenario we investigated here might be applicable to different types of organisms and interactions: for many organisms, proximity to members of the same or another genotype might be advantageous, for example because these other individuals provide resources or protection. This is expected to impose selection for biological traits that ensure proximity, through physical attachment, behavior, or by other means. This would lead to a situation that is equivalent to spatial structure; that is, a situation where individuals within or between genotypes are associated with each other for extended periods of time, as assumed in the "partner fidelity" model [24]. According to this scenario, mutually beneficial interactions within and between organisms are expected to evolve more readily than often assumed, because the organisms themselves can readily generate the spatial structure that is necessary for these beneficial interactions to emerge and be stable. While this scenario is consistent with our results, a conclusive test would require tracking the spatial positioning of individual cells over evolutionary time, which is not possible with our current data set. However, recent developments in analytical microbiology should now allow for such tracking and for explicit tests of this hypothesis.

Acknowledgements
We thank Jan Roelof van der Meer and Victor de Lorenzo for generously providing the plasmids used in this study. We thank the Genetic Diversity Center at ETH Zurich for technical support with the genomic analyses.

Funding
This work was supported by grants from Eawag (Category: Seed), the Swiss National Science Foundation (31003A_149304), and SystemsX.ch (Category: IPP). The authors declare no conflict of interest related to this work.

Authors' contributions
All authors conceived and designed the research project and wrote the manuscript. MM performed all the experiments. All authors read and approved the final manuscript.

Competing interests
The authors declare that they have no competing interests.

Author details
[1]Department of Environmental Microbiology, Eawag, Überlandstrasse 133, 8600 Dübendorf, Switzerland. [2]Department of Environmental Systems Science, ETH Zürich, 8092 Zürich, Switzerland. [3]Department of Biosystems Science and Engineering, ETH Zürich, 4058 Basel, Switzerland.

References

1.   Boucher DH, James S, Keeler KH. The ecology of mutualism. Annu Rev Ecol Syst. 1982;13:315–47.

2.   Mougi A, Kondoh M. Diversity of interaction types and ecological community stability. Science. 2012;337:349–51.

3.   Little AE, Robinson CJ, Peterson SB, Raffa KF, Handelsman J. Rules of engagement: interspecies interactions that regulate microbial communities. Annu Rev Microbiol. 2008;62:375–401.

4.   Schink B. Energetics of syntrophic cooperation in methanogenic degradation. Microbiol Mol Biol Rev. 1997;61:262–80.

5.   Pernthaler A, Dekas AE, Brown CT, Goffredi SK, Embaye T, Orphan VJ. Diverse syntrophic partnerships from deep-sea methane vents revealed by direct cell capture and metagenomics. Proc Natl Acad Sci U S A. 2008;105:7052–7.

6.   Men Y, Feil H, Verberkmoes NC, Shah MB, Johnson DR, Lee PK, et al. Sustainable syntrophic growth of *Dehalococcoides ethenogenes* strain 195 with *Desulfovibrio vulgaris* Hildenborough and *Methanobacterium congolense*: global transcriptomic and proteomic analyses. ISME J. 2012;6:410–21.

7.   Thompson AW, Foster RA, Krupke A, Carter BJ, Musat N, Vaulot D, et al. Unicellular cyanobacterium symbiotic with a single-celled eukaryotic alga. Science. 2012;337:1546–50.

8.   Zelezniak A, Andrejev S, Ponomarova O, Mende DR, Bork P, Patil KR. Metabolic dependencies drive species co-occurrence in diverse microbial communities. Proc Natl Acad Sci U S A. 2015;112:6449–54.

9.   Doebeli M, Knowlton N. The evolution of interspecific mutualisms. Proc Natl Acad Sci U S A. 1998;95:8676–80.

10. Ferriere R, Bronstein JL, Rinaldi S, Law R, Gauduchon M. Cheating and the evolutionary stability of mutualisms. Proc Biol Sci. 2002;269:773–80.

11. Nahum JR, Harding BN, Kerr B. Evolution of restraint in a structured rock-paper-scissors community. Proc Natl Acad Sci U S A. 2011;108:10831–8.

12. Sachs JL, Hollowell AC. The origins of cooperative bacterial communities. MBio. 2012;3:e00099–12.

13. Estrela S, Morris JJ, Kerr B. Private benefits and metabolic conflicts shape the emergence of microbial interdependencies. Environ Microbiol. 2015; doi:10.1111/1462-2920.13028.

14. Duan K, Sibley CD, Davidson CJ, Surette MG. Chemical interactions between organisms in microbial communities. Contrib Microbiol. 2009;16:1–17.

15. Zengler K, Palsson BO. 2012. A road map for the development of community systems (CoSy) biology. Nat Rev Microbiol. 2012;10:366–72.

16. Morris JJ. Black queen evolution: the role of leakiness in structuring microbial communities. Trends Genet. 2015;31:475–82.

17. Ovádi J. Physiological significance of metabolic channelling. J Theor Biol. 1991;152:1–22.

18. Wolfe AJ. The acetate switch. Microbiol Mol Biol Rev. 2005;69:12–50.

19. Lilja EE, Johnson DR. Segregating metabolic processes into different microbial cells accelerates the consumption of inhibitory substrates. ISME J. 2015; doi:10.1038/ismej.2015.243.

20. Bever JD, Richardson SC, Lawrence BM, Holmes J, Watson M. Preferential allocation to beneficial symbiont with spatial structure maintains mycorrhizal mutualism. Ecol Lett. 2009;12:13–21.

21. Harcombe W. Novel cooperation experimentally evolved between species. Evolution. 2010;64:2166–72.

22. Johnson DR, Goldschmidt F, Lilja EE, Ackermann M. Metabolic specialization and the assembly of microbial communities. ISME J. 2012;6:1985–91.

23. Werner GD, Strassmann JE, Ivens AB, Engelmoer DJ, Verbruggen E, Queller DC, et al. Evolution of microbial markets. Proc Natl Acad Sci U S A. 2014;111:1237–44.

24. Bull JJ, Rice WR. Distinguishing mechanisms for the evolution of co-operation. J Theor Biol. 1991;149:63–74.

25. Foster KR, Wenseleers T. A general model for the evolution of mutualisms. J Evol Biol. 2006;19:1283–93.

26. Kim HJ, Boedicker JQ, Choi JW, Ismagilov RF. Defined spatial structure stabilizes a synthetic multispecies bacterial community. Proc Natl Acad Sci U S A. 2008;105:18188–93.

27. Wakano JY, Nowak MA, Hauert C. Spatial dynamics of ecological public goods. Proc Natl Acad Sci U S A. 2009;106:7910–4.

28. Allen B, Nowak MA. Cooperation and the fate of microbial societies. PLoS Biol. 2013;11:e1001549.

29. Pande S, Kaftan F, Lang S, Svatoš A, Germerodt S, Cost C. Privatization of cooperative benefits stabilizes mutualistic cross-feeding interactions in spatially structured environments. 2015;ISME J. doi:10.1038/ismej.2015.212.

30. Hol FJ, Galajda P, Nagy K, Woolthuis RG, Dekker C, Keymer JE. Spatial structure facilitates cooperation in a social dilemma: empirical evidence from a bacterial community. PLoS One. 2013;8:e77042.

31. Liu Z, Müller J, Li T, Alvey RM, Vogl K, Frigaard NU, et al. Genomic analysis reveals key aspects of prokaryotic symbiosis in the phototrophic consortium "*Chlorochromatium aggregatum*". Genome Biol. 2013;14:R127.

32. Baba T, Ara T, Hasegawa M, Takai Y, Okumura Y, Baba M, et al. Construction of *Escherichia coli* K-12 in-frame, single-gene knockout mutants: the Keio collection. Mol Syst Biol. 2006;2:2006.0008.

33. Choi KH, Gaynor JB, White KG, Lopez C, Bosio CM, Karkhoff-Schweizer RR, et al. A Tn7-based broad-range bacterial cloning and expression system. Nat Methods. 2005;2:443–8.

34. Miller VL, Mekalanos JJ. A novel suicide vector and its use in construction of insertion mutations: osmoregulation of outer membrane proteins and virulence determinants in *Vibrio cholerae* requires toxR. J Bacteriol. 1998;170:2575–83.

35. Minoia M, Gaillard M, Reinhard F, Stojanov M, Sentchilo V, van der Meer JR. Stochasticity and bistability in horizontal transfer control of a genomic island in *Pseudomonas*. Proc Natl Acad Sci U S A. 2008;105:20792–7.

36. Lagache T, Lang G, Sauvonnet N, Olivo-Marin JC. Analysis of the spatial organization of molecules with robust statistics. PLoS One. 2013;8:e80914.

37. Ripley BD. The second-order analysis of stationary point processes. J Appl Prob. 1976;13:255–66.

38. Schmieder R, Edwards R. Quality control and preprocessing of metagenomic datasets. Bioinformatics. 2011;27:863–4.

39. Barrick JE, Lenski RE. Genome-wide mutational diversity in an evolving population of *Escherichia coli*. Cold Spring Harb Symp Quant Biol. 2009;74:119–29.

40. Barrick JE, Yu DS, Yoon SH, Jeong H, Oh TK, Schneider D, et al. Genome evolution and adaptation in a long-term experiment with *Escherichia coli*. Nature. 2009;461:1243–7.

41. Hayashi K, Morooka N, Yamamoto Y, Fujita K, Isono K, Choi S, et al. Highly accurate genome sequences of *Escherichia coli* K-12 strains MG1655 and W3110. Mol Syst Biol. 2006;2:2006.0007.

42. Pande S, Merker H, Bohl K, Reichelt M, Schuster S, de Figueiredo LF, et al. Fitness and stability of obligate cross-feeding interactions that emerge upon gene loss in bacteria. ISME J. 2014;8:953–62.

43. Hillesland KL, Stahl DA. Rapid evolution of stability and productivity at the origin of a microbial mutualism. Proc Natl Acad Sci U S A. 2010;107:2124–9.

44. Gerstein AC, Lo DS, Otto SP. Parallel genetic changes and nonparallel gene-environment interactions characterize the evolution of drug resistance in yeast. Genetics. 2012;192:241–52.

45. Tenaillon O, Rodríguez-Verdugo A, Gaut RL, McDonald P, Bennett AF, Long AD, et al. The molecular diversity of adaptive convergence. Science. 2012;335:457–61.

46. Le Gac M, Cooper TF, Cruveiller S, Edigue CM, Schneider D. Evolutionary history and genetic parallelism affect correlated responses to evolution. Mol Ecol. 2013;22:3292–303.

47. Blank D, Wolf L, Ackermann M, Silander OK. The predictability of molecular evolution during functional innovation. Proc Natl Acad Sci U S A. 2014;111:3044–9.

48. Pratt LA, Kolter R. Genetic analyses of bacterial biofilm formation. Curr Opin Microbiol. 1999;2:598–603.

49. Wood TK. Insights on *Escherichia coli* biofilm formation and inhibition from whole-transcriptome profiling. Environ Microbiol. 2009;11:1–15.

50. Hung C, Zhou Y, Pinkner JS, Dodson KW, Crowley JR, Heuser J, et al. *Escherichia coli* Biofilms have an organized and complex extracellular matrix structure. MBio. 2013;4:e00645–13.

51. Balzer GJ, McLean RJ. The stringent response genes *relA* and *spoT* are important for *Escherichia coil* biofilms under slow-growth conditions. Can J Microbiol. 2002;48:675–80.
52. Boehm A, Steiner S, Zaehringer F, Casanova A, Hamburger F, Ritz D, et al. Second messenger signalling governs *Escherichia coli* biofilm induction upon ribosomal stress. Mol Microbiol. 2009;72:1500–16.
53. Itoh Y, Wang X, Hinnebusch BJ, Preston JF, Romeo T. Depolymerization of beta-1, 6-N-acetyl-D-glucosamine disrupts the integrity of diverse bacterial biofilms. J Bacteriol. 2005;187:382–7.
54. Burton E, Gawande PV, Yakandawala N, LoVetri K, Zhanel GG, Romeo T, et al. Antibiofilm activity of GlmU enzyme inhibitors against catheter-associated uropathogens. Antimicrob Agents Chemother. 2006;50:1835–40.
55. Pratt LA, Kolter R. Genetic analysis of *Escherichia coli* biofilm formation: roles of flagella, motility, chemotaxis and type I pili. Mol Microbiol. 1998;30:285–93.
56. Tschowri N, Lindenberg S, Hengge R. Molecular function and potential evolution of the biofilm-modulating blue light-signalling pathway of *Escherichia coli*. Mol Microbiol. 2012;85:893–906.
57. Claret L, Hughes C. Interaction of the atypical prokaryotic transcription activator FlhD2C2 with early promoters of the flagellar gene hierarchy. J Mol Biol. 2002;321:185–99.
58. Prüss BM, Besemann C, Denton A, Wolfe AJ. A complex transcription network controls the early stages of biofilm development by *Escherichia coli*. J Bacteriol. 2006;188:3731–9.
59. Ko M, Park C. H-NS-dependent regulation of flagellar synthesis is mediated by a LysR family protein. J Bacteriol. 2000;182:4670–2.
60. Sule P, Horne SM, Logue CM, Prüss BM. Regulation of cell division, biofilm formation, and virulence by FlhC in *Escherichia coli* O157:H7 grown on meat. Appl Environ Microbiol. 2011;77:3653–62.
61. Beloin C, Valle J, Latour-Lambert P, Faure P, Kzreminski M, Balestrino D, et al. Global impact of mature biofilm lifestyle on *Escherichia coli* K-12 gene expression. Mol Microbiol. 2004;51:659–74.
62. Tenorio E, Saeki T, Fujita K, Kitakawa M, Baba T, Mori H, et al. Systematic characterization of *Escherichia coli* genes/ORFs affecting biofilm formation. FEMS Microbiol Lett. 2003;225:107–14.
63. Majdalani N, Heck M, Stout V, Gottesman S. Role of RcsF in signaling to the Rcs phosphorelay pathway in *Escherichia coli*. J Bacteriol. 2005;187:6770–8.
64. Ferrières L, Clarke DJ. The RcsC sensor kinase is required for normal biofilm formation in *Escherichia coli* K-12 and controls the expression of a regulon in response to growth on a solid surface. Mol Microbiol. 2003;50:1665–82.
65. Wu T, Malinverni J, Ruiz N, Kim S, Silhavy TJ, Kahne D. Identification of a multicomponent complex required for outer membrane biogenesis in *Escherichia coli*. Cell. 2005;121:235–45.
66. Ma Q, Wood TK. OmpA influences *Escherichia coli* biofilm formation by repressing cellulose production through the CpxRA two-component system. Environ Microbiol. 2009;11:2735–46.
67. Sanchez-Torres V, Maeda T, Wood TK. Global regulator H-NS and lipoprotein NlpI influence production of extracellular DNA in *Escherichia coli*. Biochem Biophys Res Commun. 2010;401:197–202.
68. Magnusson LU, Gummesson B, Joksimović P, Farewell A, Nyström T. Identical, independent, and opposing roles of ppGpp and DksA in *Escherichia coli*. J Bacteriol. 2007;189:5193–202.
69. Sommerfeldt N, Possling A, Becker G, Pesavento C, Tschowri N, Hengge R. Gene expression patterns and differential input into curli fimbriae regulation of all GGDEF/EAL domain proteins in *Escherichia coli*. Microbiology. 2009;155:1318–31.
70. Hara A, Sy J. Guanosine 5'-triphosphate, 3'-diphosphate 5'-phosphohydrolase. Purification and substrate specificity. J Biol Chem. 1983;258:1678–83.

Evolution of group I introns in Porifera: new evidence for intron mobility and implications for DNA barcoding

Astrid Schuster[1], Jose V. Lopez[2], Leontine E. Becking[3,4], Michelle Kelly[5], Shirley A. Pomponi[6], Gert Wörheide[1,7,8], Dirk Erpenbeck[1,8*] (iD) and Paco Cárdenas[9*]

Abstract

Background: Mitochondrial introns intermit coding regions of genes and feature characteristic secondary structures and splicing mechanisms. In metazoans, mitochondrial introns have only been detected in sponges, cnidarians, placozoans and one annelid species. Within demosponges, group I and group II introns are present in six families. Based on different insertion sites within the *cox1* gene and secondary structures, four types of group I and two types of group II introns are known, which can harbor up to three encoding homing endonuclease genes (HEG) of the LAGLIDADG family (group I) and/or reverse transcriptase (group II). However, only little is known about sponge intron mobility, transmission, and origin due to the lack of a comprehensive dataset. We analyzed the largest dataset on sponge mitochondrial group I introns to date: 95 specimens, from 11 different sponge genera which provided novel insights into the evolution of group I introns.

Results: For the first time group I introns were detected in four genera of the sponge family Scleritodermidae (*Scleritoderma, Microscleroderma, Aciculites, Setidium*). We demonstrated that group I introns in sponges aggregate in the most conserved regions of *cox1*. We showed that co-occurrence of two introns in *cox1* is unique among metazoans, but not uncommon in sponges. However, this combination always associates an active intron with a degenerating one. Earlier hypotheses of HGT were confirmed and for the first time VGT and secondary losses of introns conclusively demonstrated.

Conclusion: This study validates the subclass Spirophorina (Tetractinellida) as an *intron hotspot* in sponges. Our analyses confirm that most sponge group I introns probably originated from fungi. DNA barcoding is discussed and the application of alternative primers suggested.

Keywords: Porifera, Tetractinellida, *cox1*, HGT, VGT, homing endonuclease gene (HEG), LAGLIDADG, group I intron, DNA barcoding

Background

Mobile introns are self-splicing DNA sequences that play a major role in genome evolution. Group I and group II introns are distinguished based on their splicing mechanisms and secondary structures. Apart from unique splicing mechanisms, differences between group I and group II introns were observed within the core regions of their secondary structures. Depending on these structural characteristics, group I introns have been further categorized into IA-IE classes. Group II introns constitute up to six stem-loop domains and are classified I-VI respectively (e.g., [35]). Group I and group II introns often contain open reading frames (ORFs) in their loop regions [70], which can encode for different site-specific homing endonuclease genes (HEGs). The majority of group I introns include HEGs, which have a conservative single or a double motif of the amino-acid sequence LAGLIDADG. In contrast group II introns encode in most cases a

* Correspondence: erpenbeck@lmu.de; paco.cardenas@fkog.uu.se
[1]Department of Earth- & Environmental Sciences, Palaeontology and Geobiology, Ludwig-Maximilians-Universität München, Richard-Wagner-Str. 10, 80333 Munich, Germany
[9]Department of Medicinal Chemistry, Division of Pharmacognosy, BioMedical Center, Uppsala University, Husargatan 3, 75123 Uppsala, Sweden
Full list of author information is available at the end of the article

reverse transcriptase-like (RT) ORF (e.g., [36]). Group I and group II introns are found in all domains of life: group I introns are present in bacterial, organellar, bacteriophage and viral genomes as well as in the nuclear rDNA of eukaryotes. Group II introns have a similar distribution, but are not known from the nuclear rDNA (e.g., [33]). More specifically, group I and/or group II introns are found, e.g., in eukaryotic viruses [92], slime molds [45], choanoflagellates [7], the annelid *Nephtys* sp. [84], red algae [8], brown algae [25] and plants: green algae [85], liverworts [58, 66, 67] and different angiosperms [58, 66, 67]. Group II introns seem to thrive especially in plants [60], whereas the largest abundance of group I introns currently occurs within fungi [23, 54, 68]. As an example, the mitochondrial (mt) genome of the fungus *Ophioscordyceps sinensis* harbors 44 group I introns and six group II introns, accounting for 68.5% of its mt genome nucleotides. Here, 12 out of 44 group I introns and only one out of six group II introns are located in the cytochrome c oxidase subunit 1 (*cox1*) gene [54], an acknowledged insertion hotspot for mt group I introns [23].

More recently, group I introns have been discovered in the *cox1* of early branching metazoan phyla: Placozoa [9, 17, 76], Cnidaria [27, 31] as well as Porifera [21, 29, 65, 82, 90]. Group II introns are rarer, and found in the *cox1* of Placozoa [17, 76], and in one demosponge species of the order Axinellida (referred to as *Cymbaxinella verrucosa*) [43]. In Porifera, group I introns have only been recorded from Demospongiae and Homoscleromorpha and, like in group II introns, always in the *cox1*

gene, with only occasional double insertions (Fig. 1). The current nomenclature of sponge group I and group II introns is based on the intron insertion site positions in reference to the *Amphimedon queenslandica cox1* gene (DQ915601) [82]. In Homoscleromorpha, three different intron positions (714, 723 and 870; Fig. 1) are known for three species of the family Plakinidae [29]. Within the demosponge subclass Verongimorpha intron 723 is detected in one species (*Aplysinella rhax*) of the order Verongiida [21]. Most intron insertions have been found within the demosponge subclass Heteroscleromorpha, in the orders Agelasida, Axinellida [43] and especially Tetractinellida [82]. In the Tetractinellida, group I introns are currently known in five sponge species belonging to three genera (*Cinachyrella*, *Tetilla* and *Stupenda*) and inserted at four mtDNA intron positions: 387, 714, 723 and 870 [47, 65, 82] (Fig. 1). To this date, all sponge group I introns encode a HEG with two LAGLIDADG motifs [21, 29, 47, 82] with the exception of intron 714 in *Plakinastrella* sp. and intron 870 in *Agelas oroides* and *Axinella polypoides,* in which no ORF was detected [29, 43, 52, 90]. Intriguingly, Tetractinellida introns are currently only detected in the families Tetillidae [82] and Stupendidae [47].

Fungi and Placozoa have been proposed as possible donors for group I introns among sponges [43, 47, 65]. However, these findings await corroboration with a broader and more comprehensive taxon set. Intron/HEG phylogenetic analyses and group I/II intron secondary structures are the basis for different scenarios on the

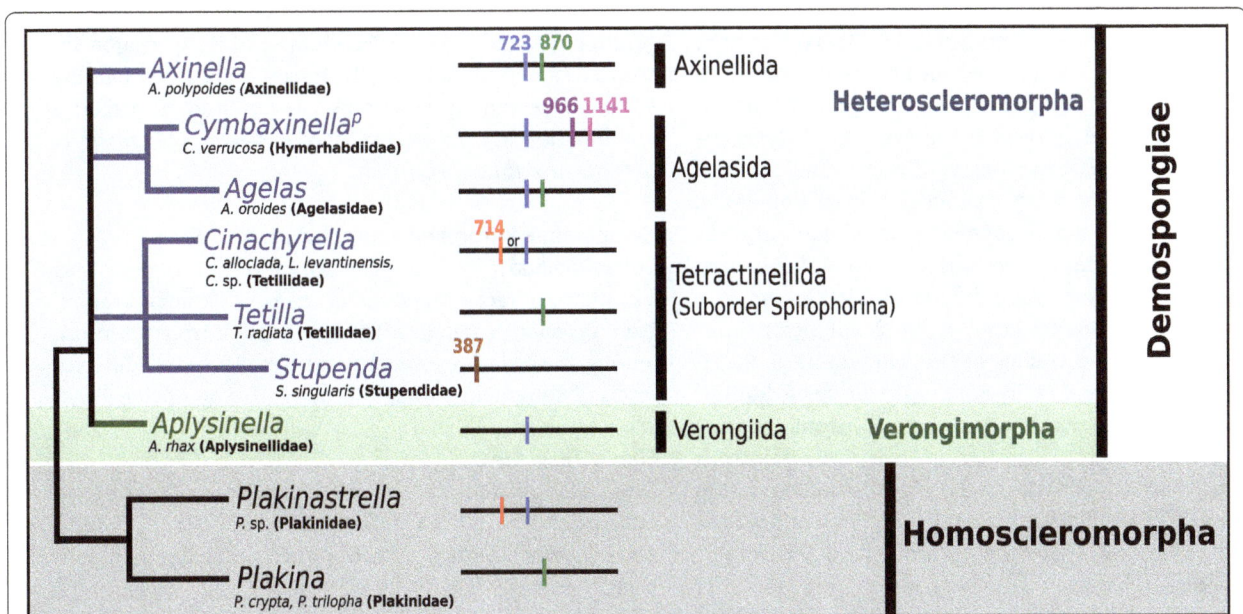

Fig. 1 Simplified sponge phylogeny highlighting currently known group I and II intron insertion sites. Horizontal *black* lines with colored vertical bars and numbers (*blue*: intron 723, *green*: intron 870, *red*: intron 714, *brown*: intron 387, *purple*: intron 966, *pink*: intron 1141) behind the taxa names represent different intron locations within *cox1*

origin of introns within sponges [43, 47, 82]. The presence of independent horizontal gene transfers (HGT) for introns is supported by their haphazard distribution over phylogenetically distant sponge groups [21, 43, 82]. Vertical gene transfer (VGT) of introns is assumed among closely related taxa, but never confirmed due to the lack of comprehensive taxon sampling [82].

To gain new insights into the evolution of sponge introns we required an intron-rich taxonomic group. Based on earlier studies on sponge introns, the order Tetractinellida represents an obvious target. Other lines of evidence support this choice, such as unsuccessful attempts to amplify *cox1* in this group with standard protocols [10, 72, 81], potentially due to introns in the relevant primer regions [82]. Consequently, this study focuses on tetractinellid *cox1* mitochondrial data to broaden our knowledge on mt intron evolution in this early-branching metazoan phylum.

The data from this "intron-hotspot taxon" presented here constitutes the most representative dataset to target specific questions pivotal to understand intron structure and distribution including activity and mobility. Importance of HGT or VGT or a combination of both will be addressed. Additionally, current hypotheses on the origin of sponge mitochondrial introns will be discussed by comparing intron data across other phyla.

Results
Mitochondrial intron diversity and characteristics in tetractinellid sponges

The current study comprises the largest dataset of sponge mitochondrial introns to date (95 sequences of which 72 are new), encompassing 13 different sponge genera. All 72 newly sequenced introns were group I introns of the class IB, and all encoded a HEG of the LAGLIDADG family, except for intron 723 of *Aciculites* sp.1, where no HEG was observed. A double motif of the LAGLIDADG domain was located in all introns, if the sequence was not degenerated or without a HEG. Different intron lengths were observed for different species, and an overview of the different initiation and stop codons of all HE ORFs is given in Fig. 2. All introns possessed start and stop codons in the same frame as the 5′ exon, except intron 714 in *Plakinastrella* sp. Additionally, uninterrupted ORFs in the same 5′ exonic reading frame were observed for all sequences unless introns were degenerated or without HEG. Initiation and stop codons varied among the intron HEGs. For example the HEG of intron 387 potentially starts with a GTG initiation codon at position 19 of the intron, and not TTA (position 16) as suggested previously [47]. The HEG of intron 714 potentially starts with TTG as its initial codon (position 27). Instead of TTA (position 1) as suggested by Rot et al. [65], all intron 723 HEGs potentially

start with ATT (position 10). In the intron HEG 870 we have ATT (position 24 for all) and GTG (position 9 for *Plakina* and 21 for *Tetilla*) as initiation codons. The stop codon for most HE ORFs was TAG or TAA except for *Cinachyrella* sp. 2 and *Setidium* sp.1 (intron 723), where truncated HEGs were found.

We discovered more introns in the Spirophorina at positions 714, 723 and 870 (Fig. 2); no intron at position 387 was found. As an example, intron 714 sequences were generated for five more *Cinachyrella* sp. 2 taxa; four from the Indian Ocean (Kenya, Myanmar) and one from the southwest-Pacific (Indonesia). *Cinachyrella* species are previously known to have only one intron insertion at a time (either 714 or 723). However, our study reveals that both introns 714 and 723 can occur together in *cox1*, e.g., in *Cinachyrella* sp. 2 from marine lakes (RMNH POR11161) and mangroves (RMNH POR11187). Intron 723 was sequenced from 11 different *Cinachyrella* species, and it is particularly present in the *Cinachyrella alloclada* complex. In total this study contains 42 sequences of *Cinachyrella alloclada* (intron 723) from the western Atlantic, the Caribbean Sea and the Gulf of Mexico. We added six additional intron 723 sequences including *C.* cf. *anomala*, *C.* cf. *providentiae*, *C. porosa* (all Indonesia), *C.* sp. 3 (Red Sea), *C.* sp. 4 (Morocco) and *C.* sp. 5 (Taiwan). The resulting *Cinachyrella* dataset covers subtropical-tropical areas from 1 to 90 m depth.

For the first time we discovered intron 723 in the Scleritodermidae (*Microscleroderma, Aciculites, Setidium* and *Scleritoderma*). Intron 870 was found in *Tetilla quirimure* from Brazil and *Microscleroderma herdmani* from the Indian Ocean (Mauritius), and the Pacific (Philippines and Hawaii). Huchon et al. [43] located intron 723 in combination with intron 870 in two families (Axinellidae and Agelasidae), while our study reveals this combination in three scleritodermid genera (*Microscleroderma, Aciculites, Setidium*).

Comparative intron and exon phylogenies of Tetractinellida

Phylogenetic reconstructions of the *cox1* exon and the intron revealed a patchy distribution of intron insertions among the Scleritodermidae and Tetillidae and different levels of congruence among intron and exon phylogenies.

Family Tetillidae

The relationships of major clades (Fig. 3) were in concordance with a previous study [81]. Unlike *Cinachyrella*, *Tetilla* appeared monophyletic in reconstructions with five species (*T. radiata, T. japonica, T. quirimure, T. dactyloidea* and *T. muricyi*). *Tetilla radiata* (intron 870) is sister to the intron-lacking *Tetilla japonica* (posterior probability [PP] = 1.00 / bootstrap support [BS] = 98). As shown by Szitenberg et al. [81],

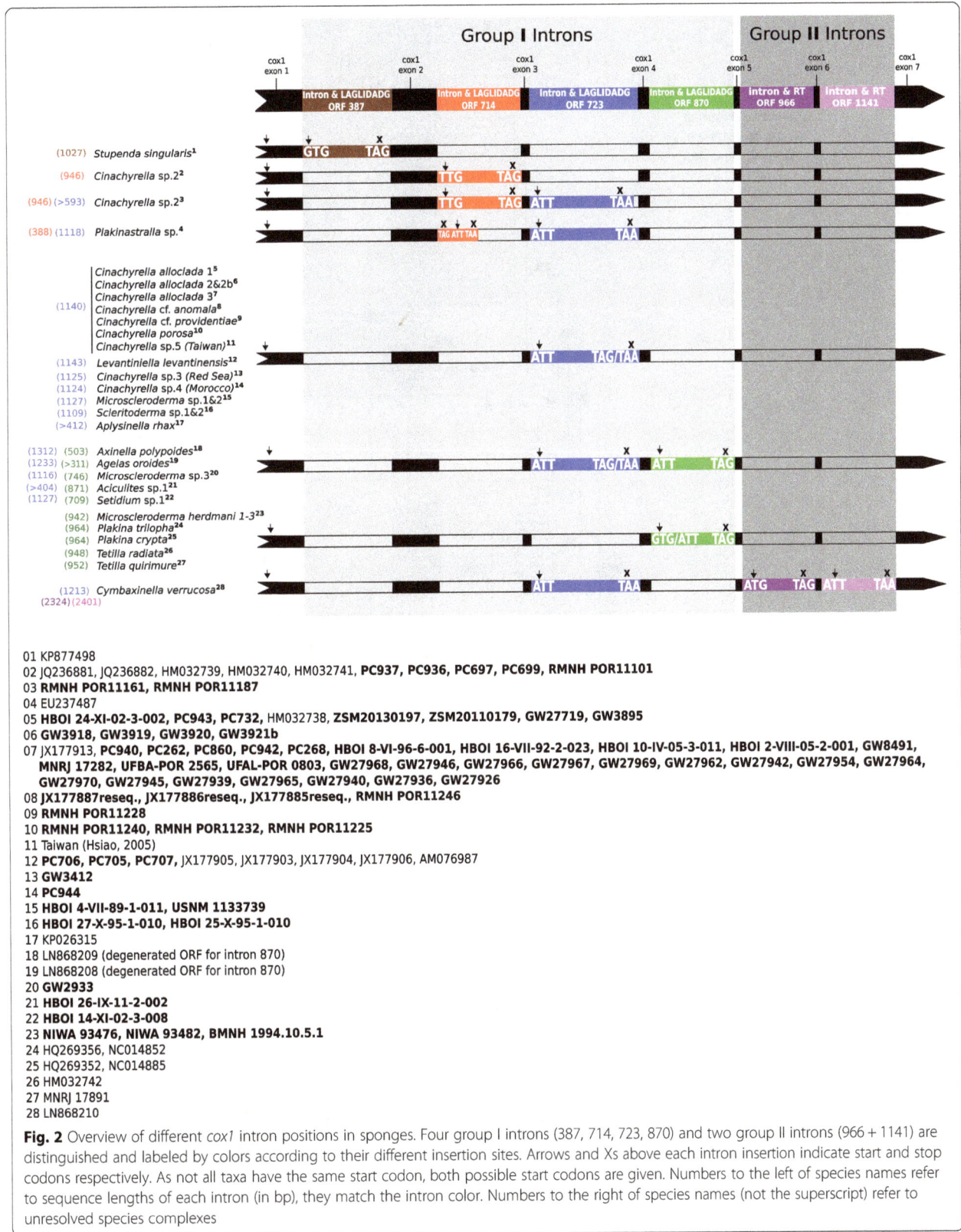

01 KP877498
02 JQ236881, JQ236882, HM032739, HM032740, HM032741, **PC937, PC936, PC697, PC699, RMNH POR11101**
03 **RMNH POR11161, RMNH POR11187**
04 EU237487
05 **HBOI 24-XI-02-3-002, PC943, PC732,** HM032738, **ZSM20130197, ZSM20110179, GW27719, GW3895**
06 **GW3918, GW3919, GW3920, GW3921b**
07 JX177913, **PC940, PC262, PC860, PC942, PC268, HBOI 8-VI-96-6-001, HBOI 16-VII-92-2-023, HBOI 10-IV-05-3-011, HBOI 2-VIII-05-2-001, GW8491, MNRJ 17282, UFBA-POR 2565, UFAL-POR 0803, GW27968, GW27946, GW27966, GW27967, GW27969, GW27962, GW27942, GW27954, GW27964, GW27970, GW27945, GW27939, GW27965, GW27940, GW27936, GW27926**
08 **JX177887reseq., JX177886reseq., JX177885reseq., RMNH POR11246**
09 **RMNH POR11228**
10 **RMNH POR11240, RMNH POR11232, RMNH POR11225**
11 Taiwan (Hsiao, 2005)
12 **PC706, PC705, PC707,** JX177905, JX177903, JX177904, JX177906, AM076987
13 **GW3412**
14 **PC944**
15 **HBOI 4-VII-89-1-011, USNM 1133739**
16 **HBOI 27-X-95-1-010, HBOI 25-X-95-1-010**
17 KP026315
18 LN868209 (degenerated ORF for intron 870)
19 LN868208 (degenerated ORF for intron 870)
20 **GW2933**
21 **HBOI 26-IX-11-2-002**
22 **HBOI 14-XI-02-3-008**
23 **NIWA 93476, NIWA 93482, BMNH 1994.10.5.1**
24 HQ269356, NC014852
25 HQ269352, NC014885
26 HM032742
27 MNRJ 17891
28 LN868210

Fig. 2 Overview of different *cox1* intron positions in sponges. Four group I introns (387, 714, 723, 870) and two group II introns (966 + 1141) are distinguished and labeled by colors according to their different insertion sites. Arrows and Xs above each intron insertion indicate start and stop codons respectively. As not all taxa have the same start codon, both possible start codons are given. Numbers to the left of species names refer to sequence lengths of each intron (in bp), they match the intron color. Numbers to the right of species names (not the superscript) refer to unresolved species complexes

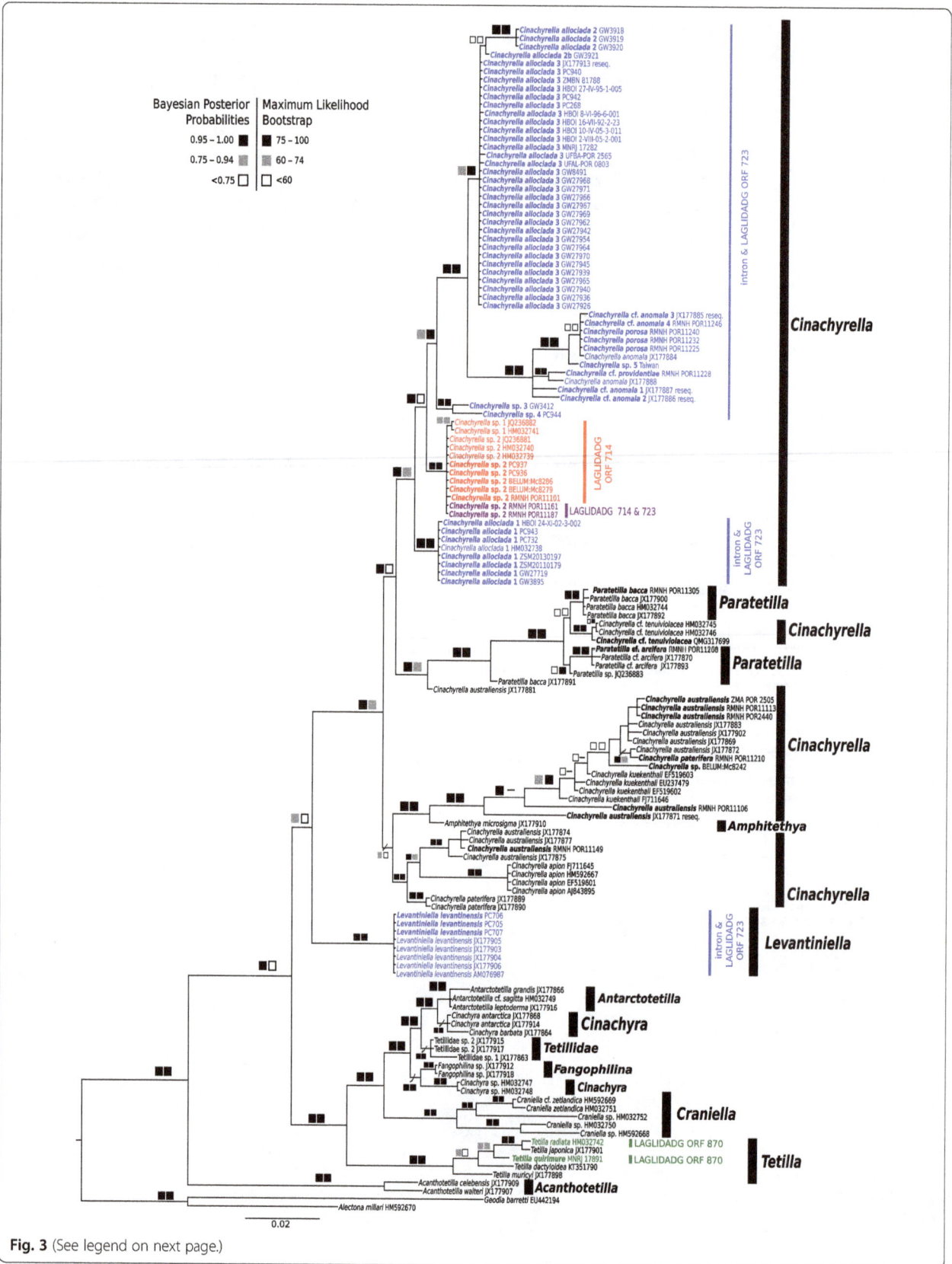

Fig. 3 (See legend on next page.)

our analysis supported the early-branching position of the intron 723-bearing *Levantiniella levantinensis* with respect to the *Cinachyrella/Paratetilla/Amphitethya* clade. Intron 723-bearing Tetillidae did not form a monophyletic group due to the position of *L. levantinensis* and the presence of a highly supported (PP 1.00/BS 100) clade of intron 714 bearing species consisting of *C.* sp. 1 (intron 714 only) and *C.* sp. 2 (with intron 714 only or in combination with intron 723). There were no genetic differences in *cox1* between *Cinachyrella* sp. 2 bearing one (714) or two (714 *and* 723) introns.

Family Scleritodermidae
The *cox1* exon phylogeny (Fig. 4) corroborated the sister group relationship Scleritodermidae/Stupdendidae, previously suggested with 18S rDNA data [47]. Results also supported the monophyly of Scleritodermidae and its genera *Microscleroderma*, *Aciculites* and *Scleritoderma* as previously suggested by 18S and 28S rDNA phylogenies [61, 72]. Species in these genera displayed different intron distributions (Fig. 4).

Intron + LAGLIDADG Phylogeny
The intron 723 phylogeny (intron + LAGLIDADG) (Fig. 5) broadly agreed with the corresponding exon phylogeny (Fig. 3), but also displayed several differences crucial for the understanding of sponge intron evolution. Notably, we recovered intron 723 of *Cinachyrella* sp. 2 in a different clade than for the *cox1* of all *Cinachyrella* sp. 2, whether they have intron 714 or both 714 + 723 (Fig. 3). The clade of introns 723 of *Cinachyrella* sp. 3 and sp. 4, from the Red Sea and Morocco respectively, were in a sister-group relationship with *L. levantinensis* whereas in the exon phylogeny these two species branched within the *Cinachyrella* clade (Fig. 3). These incongruences between the exon and the intron phylogenies were shown to be significant ($p < 0.01$, Shimodaira-Hasegawa (SH)-test). One single *C. alloclada* 1 sequence falls within the *C. alloclada* 3 clade. This position is regarded as artifactual due to an incomplete intron sequence as retrieved from degraded DNA.

Interestingly, *Agelas oroides* and *Cymbaxinella verrucosa* intron sequences grouped within Spirophorina in a highly supported sister group to *Cinachyrella* introns. The intron + LAGLIDADG phylogenies for 870 and 714 were congruent with the exon phylogeny for all supported clades (Additional file 1).

Secondary structure analyses of introns 723 and 870
The secondary structures of intron 723 and intron 870 presented the typical RNA fold of a group I intron structure [88], consisting of a P1-P2-P10 substrate domain, a P4-P5-P6 scaffold domain and a P3-P7-P8 catalytic domain (Fig. 6). The conserved regions Q, P, S, and R, building the core, were found in all of the structures.

Our current study expands our knowledge on *Cinachyrella* intron 723 [82] thanks to five additional structures predicted for *Cinachyrella porosa* (RMNH POR11225), *Cinachyrella* sp. 4 (PC944), *Cinachyrella alloclada* 2 (GW3920), *Cinachyrella* cf. *providentiae* (RMNH POR11228), and *Cinachyrella* cf. *anomala* (JX177887) (Additional file 2). The structures of introns 714 and 723 only have a single-stranded P2 region, whereas intron 870 has a double stranded P2 region (Fig. 6, [82]). The LAGLIDADG ORF is always located in the loop of the P8 helix (Fig. 6, Additional file 2).

Intron 723 structural differences between the species are in the P6 and P9 regions. In particular, *Cinachyrella* sp. 4 from Morocco has reduced helices P9.1c and P9.1d compared to all others. *Cinachyrella alloclada* 2 differs slightly in the P6, P6a, P6b and P6d regions to other *Cinachyrella* species. We generated for the first time secondary structures of scleritodermid intron 723 in *Microscleroderma* sp. 2 (USNM 1133739), *Setidium* sp. 1 (HBOI 14-XI-02-3-008) and *Scleritoderma* sp. 2 (HBOI 25-X-95-1-010) (Fig. 6). All three species show a high variability in loops and helices within the P9 region. Only a few differences were observed in the P6 region between the species. The main difference between *Cinachyrella* and Scleritodermidae intron 723 is the absence of the P6d region in the latter (Fig. 6a, b and c).

The secondary structure of intron 870 was reconstructed for *Tetilla quirimure* (MNRJ 17891) and *Microscleroderma herdmani* 3 (BMNH 1994.10.5.1) (Fig. 6d and e). Both taxa contain the known core helices and conserved structures of Q, P, S, and R. *Tetilla quirimure* intron structure is very similar regarding P6 and P9 regions to the one from *Tetilla radiata* [82]. The intron of *Microscleroderma herdmani*, in turn, has a reduced P6a helix and a different P5a region compared to that found in *Tetilla*.

The LAGLIDADG protein phylogeny
The sponge LAGLIDADG sequences displayed phylogenetic affinities to four different clades (Fig. 7). LAGLIDADG

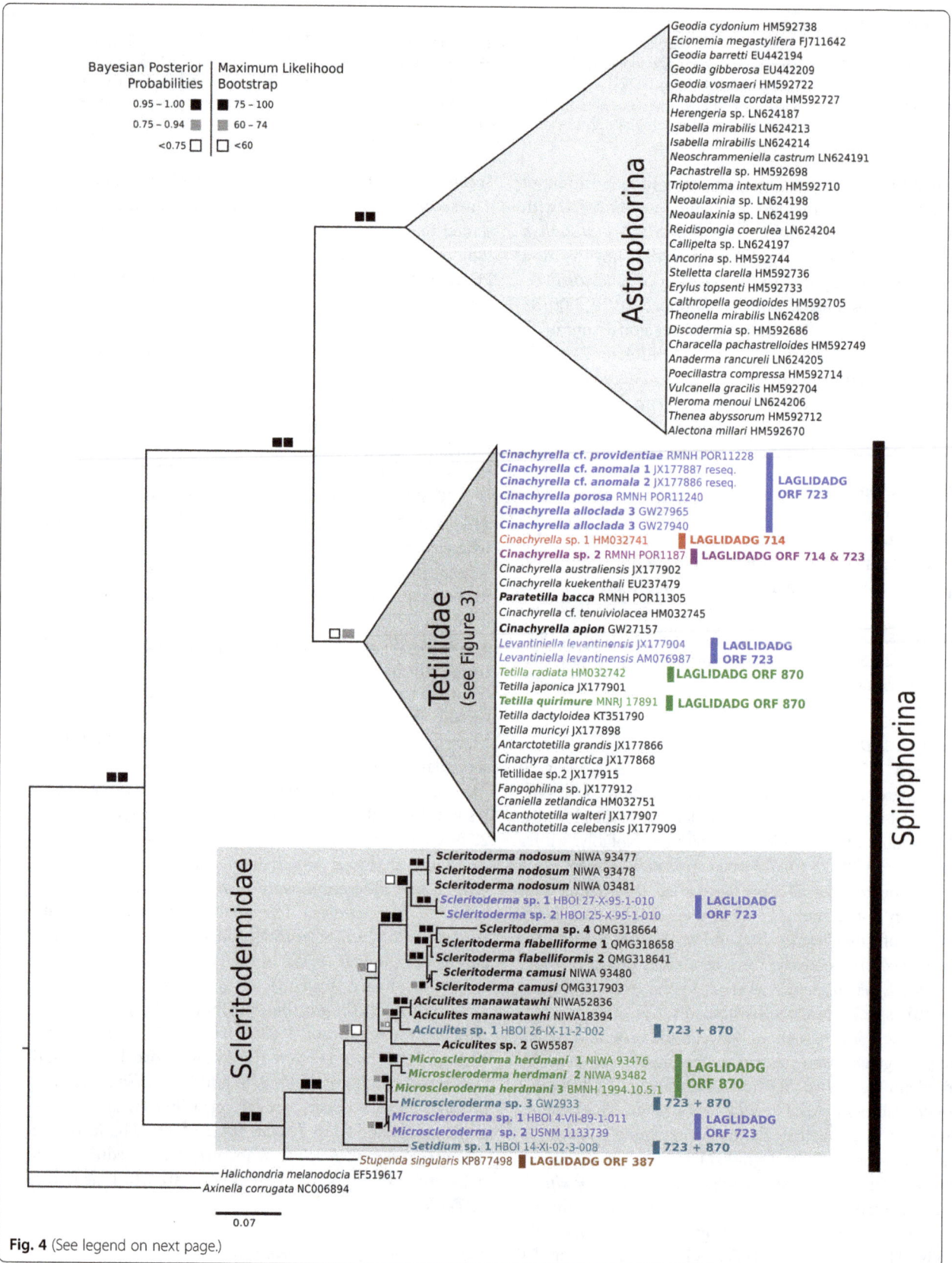

Fig. 4 (See legend on next page.)

(See figure on previous page.)
Fig. 4 *Cox1* (introns excluded) bayesian inference (MrBayes, GTR + G + I model) phylogeny of the Tetractinellida with focus on the family Scleritodermidae. The maximum likelihood (RAxML) tree is congruent. Squares represent node supports. *Black* squares: PP = 0.95–1.00, BP = 75–100. *Dark gray* squares: PP = 0.75–0.94, BP = 60–74. *White* squares: PP < 0.75, BP < 60. Numbers after each taxa are GenBank accession numbers or voucher numbers (for new sequences). Sequences generated in this study are in bold (32 sequences). Color code follows Fig. 2, except for three Scleritodermidae taxa, which possess two intron insertions (723 + 870) and marked in *dark-cyan* respectively

(intron 387) of *Stupenda singularis* forms a highly supported sister group relationship with a Fungi/Marchantiophyta clade. Cnidarian LAGLIDADG encoding sequences are only present in Hexacorallia and never revealed in the other subclasses Octocorallia and Ceriantharia. Within Hexacorallia, the sequences of intron 888 are monophyletic and include the orders Actiniaria (sea anemones), Scleractinia (stony corals), Corallimorpharia (corallimorphs) and Antipatharia (black corals) [31]. Notably, there are no sponge LAGLIDADG (intron 888) sequences. In comparison, several scleractinian sequences form a clade with the sponge LAGLIDADG intron 723 sequences.

LAGLIDADG (intron 870) of *Plakina/Tetilla* and the Scleritodermidae are closely related to Hexacorallia (Zoantharia) LAGLIDADG sequences, but with considerable genetic distance. The genealogical affinities of intron 714 appear unresolved.

Discussion

Characterisation and mobility mechanisms of group I intron

In sponges, group I and II introns occur coincidentally and exclusively in the *cox1* gene in a few sponge groups (Fig. 1). The current study specifically targeting sponge mitochondrial *cox1* introns represents the largest sponge mt intron dataset (95 specimens), encompassing 11 different sponge genera. In accordance with previous studies [21, 43, 47, 65, 82], double or single intron insertion sites within the *cox1* gene were observed among different taxa (Fig. 2). Sponge group I intron sequences in most cases contain a putative homing endonuclease (HE) ORF, which encode for a LAGLIDADG-type protein, unless they are degenerated (Fig. 2). Intron mobility is facilitated by those site specific HEs, which conduct double-strand breaks (DSBs) in alleles that lack introns, hence activating intron mobility via a DSB-repair process [2]. Those HEs are known to promote their mobility towards conserved regions [35]. However, only a few studies have investigated the conservation of those introns in their host genes. Swithers et al. [80] analysed the conservation of group I and group II introns in the host genes of vascular plants, protists, fungi, green algae, liverworts and amoeba, but not in animals. In the former, group I introns are preferentially located within conserved regions, whereby group II introns were not shown to remain particularly in conserved sites [80]. In comparison, our conservation analysis (Fig. 8) corroborated these

findings by demonstrating that even in early branching metazoans like sponges, group I introns are located in the most conserved regions of their host proteins. At present, sponge group II intron insertions are only known from a single demosponge (*Cymbaxinella verrucosa*) [43], with two group II introns in the 3′ region of *cox1*. Our conservation analysis showed that these two group II introns were located in conserved regions (Fig. 8), however, additional data are needed for a better understanding of group II intron conservation and mobility in sponges.

The *cox1* gene in demosponges has the lowest substitution rate of all mt protein coding genes [90]. In fact, the mitochondrial genomes of the sponge classes Homoscleromorpha and Demospongiae possess features shared with non-metazoan opisthokonts rather than Bilateria such as the presence of intergenic regions, genes of foreign origin, a low substitution rate, selfish elements and introns [50]. Mt introns are also found in plants [28], fungi [44], Placozoa [76], and Hexacorallia [13, 37, 74] all known to have slow rates of evolution. It is assumed that this lower substitution rate slows down the elimination of ribozyme activities within group I introns, therefore HEGs would degenerate slower in most fungi [23], anthozoans [27, 31] and placozoans [9]. Intron mobility is particularly dependant on secondary structure and therefore mutation pressure, so sponge introns survive in the most conserved mt gene (*cox1*), and the most conserved regions of this gene (Fig. 8), where their HEG is most likely to degenerate slowly. On the other hand, hexactinellid and particularly calcareous sponges possess an accelerated substitution rate [51, 52] and no known mt introns to this date. This correlation between higher mutation rate and absence of mt intron is shared by Cubozoa [77], Ceriantharia [79]; Ctenophora [48, 59], Hydrozoa and Scyphozoa [42]. One exception to this pattern is the apparent lack of introns in Octocorallia, despite their lower substitution rates compared to the intron-bearing Hexacorallia [37, 74]. This might be due to the presence of a unique *MutS* gene, which encodes a DNA mismatch repair machinery [4], which prevents intron insertions. The mt DNA mismatch repair machinery in sponges remains unknown.

Sponge group I introns consist of complex catalytic ribozymes (RNAs) that fold into a conserved three-dimensional core structure of ten helices. Within this structure, the sponge HEGs are found to be always

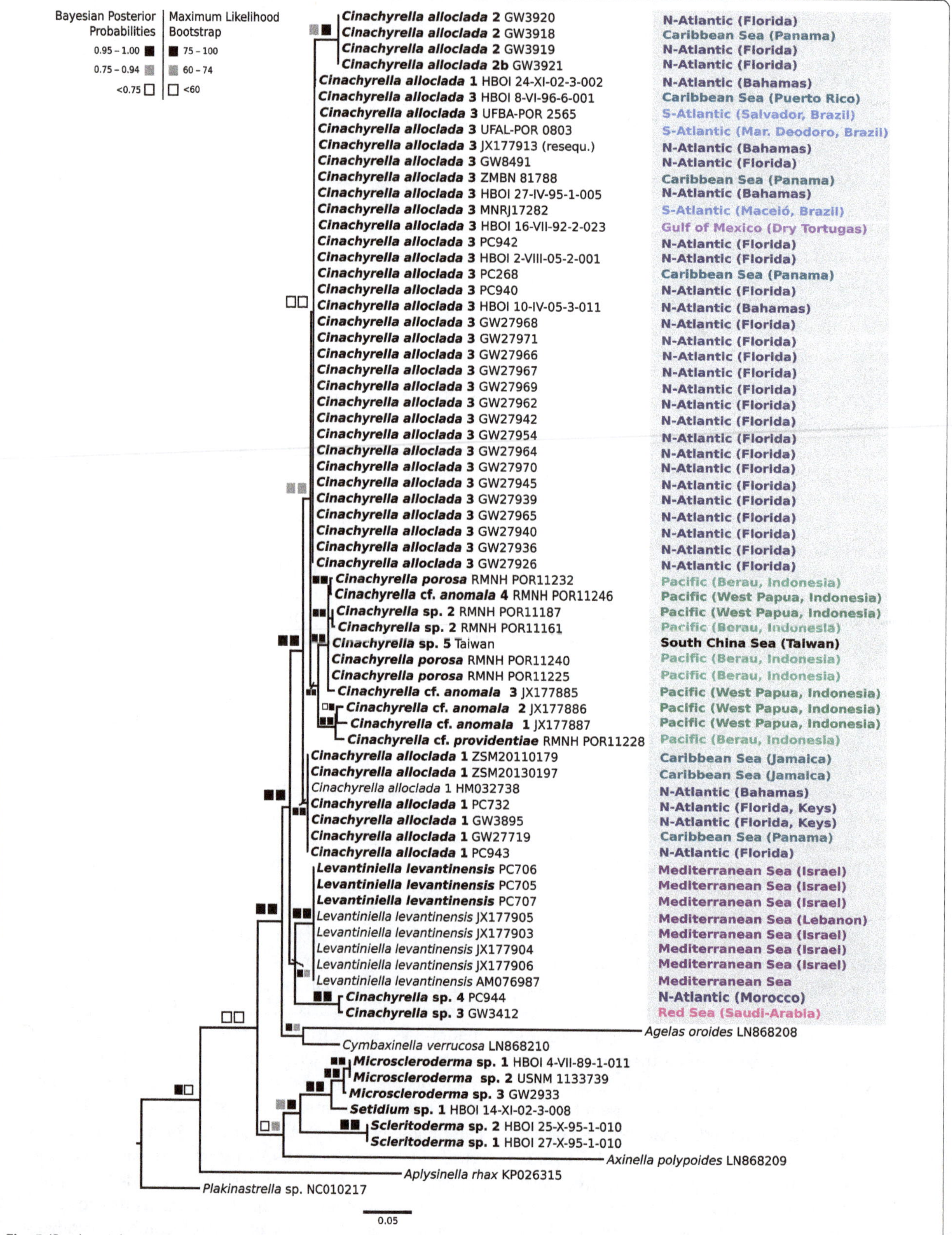

Fig. 5 (See legend on next page.)

(See figure on previous page.)
Fig. 5 Intron 723 bayesian inference (MrBayes, GTR + G + I model) phylogeny in sponges. The maximum likelihood (RAxML) tree is congruent. Squares represent node supports. *Black squares:* PP = 0.95–1.00, BP = 75–100. *Dark gray* squares: PP = 0.75–0.94, BP = 60–74. *White squares:* PP < 0.75, BP < 60. Numbers after each taxa are GenBank accession numbers or voucher numbers (cf. Additional file 1). Sequences generated in this study are in bold (63 sequences). Sampling localities for the subtropical-tropical *Cinachyrella* taxa are given

located within the loop region of the catalytic domain (helix P8, Fig. 6) as in Hexacorallia [31]. HEGs and their intron partners are thought to move either independently from each other [73] or as a single unit [34]. Whether those HEGs are actively expressed or not often depends on their functionality. The functional expression of HEG group I introns and the resulting gains and/or losses are considered as a cyclical process of different stages [30]. Emblem et al. [20] applied this into an evolutionary model for a group I intron in sea anemones and reported five stages: 1) Intron with HEG expressed and fused in frame with the upstream host gene exon; 2) Intron with expressed free-standing HEG; 3) Intron with shortened/degenerated HEG; 4) Intron without a conserved HEG and 5) Exon *cox1* without intron. Until now, only a few insights into this evolutionary model were given for sponges. Different stages are observed for intron 723 and intron 870 in different sponge species [43]. However, no detailed information has been provided yet on the potential start and stop codons, which are crucial diagnostic features for their categorisation. The potential start and stop codons, observed in all group I introns (Fig. 2), in addition to the predicted secondary structures (Fig. 6, Additional file 2), provide insights into the respective evolutionary stages of all sponge group I introns. In detail, we classified intron 387 of *Stupenda singularis* in stage 1. Intron 714 of all *Cinachyrella* sp. 2 appear in stage 1, and in stage 4 for *Plakinastrella* sp. due to several start and stop codons and no HEG. Intron 723 is found to be in stage 1 among all *Cinachyrella* species except *C.* sp. 2 (see below), which is in concordance with the already published data [43]. Intron 723 in *Microscleroderma* sp. 1 & 2 and *Scleritoderma* sp. 1 & 2 are also found to be in stage 1. A study comparing the length of DNA and RNA in combination with RT-PCR on a *Cinachyrella* intron 723 from Taiwan (probably *Cinachyrella* sp. 5) suggests that it can self-splice in vivo or in vitro [12]. It confirms that this particular stage 1 intron 723 is active. We observe intron 723 also in stages 3 and 4 in *Aplysinella rhax*, *Microscleroderma* sp. 3, *Cinachyrella* sp. 2, *Setidium* sp.1 (degenerated HEG) and *Aciculites* sp. 1 (short sequence and no HEG) respectively, which rebuts the suggested recent infection of intron 723 in sponges [43]. Intron 870 was at stage 1 for *Tetilla radiata*, *Plakina crypta* and *Plakina trilopha* [43] and now shown for *Microscleroderma herdmani* 1–3, *Tetilla quirimure* and *Setidium* sp. 1. Interestingly, both stage 3–4 intron 870 (*A.*

polypoides, *A. oroides*) previously described [43] co-occur with stage 1 intron 723. Also, the only stage 4 intron 714 (*Plakinastrella* sp.) co-occurs with a stage 1 intron 723. Similarly, all of the stage 3–4 intron 723 (*Microscleroderma* sp. 3, *Cinachyrella* sp. 2, *Aciculites* sp. 1, *Setidium* sp. 1) co-occur with stage 1 introns (either 714 or 870). Overall, two stage 1 introns never co-occur, one of the two is always degenerating. We can therefore hypothesize that the presence of two group I introns is unstable or that maybe the degeneration of one somehow enables the insertion of a different intron. More double-intron-bearing *cox1* sequences are needed to study this further. Moreover, we noted that although Scleractinia (Hexacorallia) possesses intron 723 or 888, no evidence of double-intron *cox1* sequences in this group is given, which applies for Cnidaria in general. Since double-intron *cox1* sequences are also absent in Placozoa, sponges (e.g., demosponges and homoscleromorphs) are to date the only metazoans with double-intron *cox1* sequences.

HGT versus VGT of group I introns

The sporadic detections and patchy distributions of group I introns not only among sponges, but also among other Metazoa in e.g., scleractinian corals [27, 31], plants [67] and fungi [45] are the main arguments for HGT.

HGT events for group I introns in sponges were first hypothesized by Rot et al. [65] and later corroborated by other studies, based on major differences between *cox1* and intron phylogeny topologies [82] as well as the occurrence of homologous introns in phylogenetically distantly related sponge groups e.g., homoscleromorphs [90] and demosponges (Verongimorpha and Heteroscleromorpha) [21, 43, 47] (Fig. 1). Our results reveal new cases of intron HGT, this time within the *Cinachyrella* species (Fig. 9). As an example *Cinachyrella* sp. 3 (GW3412, Red Sea) and *Cinachyrella* sp. 4 (PC944, Morocco) are sister to *Cinachyrella alloclada* 2–3/ *Cinachyrella* spp. from the Pacific whereas their intron 723 are sister to the intron of *Levantiniella levantinensis* (Fig. 9). One can therefore hypothesize that an intron 723 of *L. levantinensis* (or one of its ancestors) invaded the ancestor of *Cinachyrella* sp. 3 and *Cinachyrella* sp. 4. In addition, two *Cinachyrella* sp. 2 specimens from the Pacific (RMNH POR11161 and RMNH POR11187) group together with other conspecifics in the *cox1* phylogeny; however, their introns are closely related to *Cinachyrella* species from the Pacific. Therefore, we

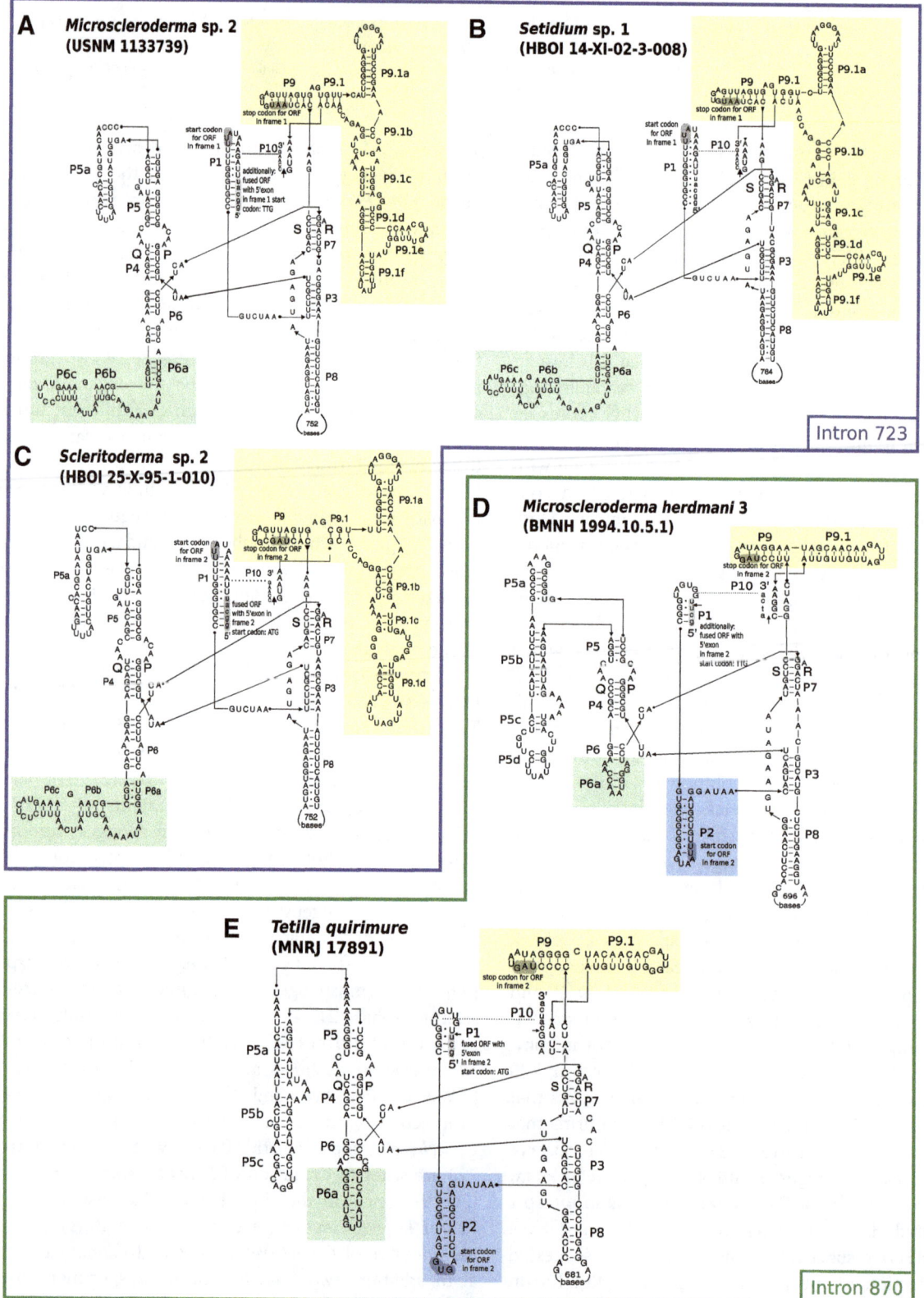

Fig. 6 (See legend on next page.)

Fig. 6 Predicted secondary structures of *Microscleroderma* sp. 2 (**a**), *Setidium* sp. 1 (**b**), *Scleritoderma* sp. 2 (**c**) (group I, IB, intron 723), *Microscleroderma herdmani* 3 (**d**) and *Tetilla quirimure* (**e**) (group I, IB, intron 870). Exon bases are in lower-case letters and intron bases in upper-case letters. Paired P1-P10 helices and their conserved sequences (P, Q, R, S) are labeled according to the standard group I intron scheme [88]. The HEGs are present in the loops of their respective P8 helix. For a better comparison the same color scheme as in Szitenberg et al. [82] was used to highlight differences in the P2, P6 and P9 regions. Potential start and stop codons are highlighted in *light gray*

hypothesize that those two *Cinachyrella* sp. 2 specimens from marine lakes and mangroves were reinfested by intron 723 after it was primarily lost, which would indicate a "secondary" HGT between *C.* sp. 2 and the Pacific *Cinachyrella* spp. (Fig. 9). It also seems that these two specimens are losing their intron 723 again, which is at stage 3. These hypotheses are significantly corroborated by the SH-test ($p < 0.01$). Another HGT event was found within the family Sceritodermidae: while *Setidium* sp. 1 is sister to *Microscleroderma* taxa in the intron phylogeny (Fig. 5), it branches off first of all other Scleritodermidae taxa in the *cox1* phylogeny (Fig. 4). However, the *cox1* topology is poorly supported around *Setidium* sp., therefore this HGT event is less obvious than in the two previous cases. Although the introns 723 found in the Agelasida (*A. oroides* and *C. verrucosa*) are fairly divergent, the intron and the LAGLIDADG phylogenies both suggest that they are phylogenetically related to the *Cinachyrella/Levantiniella* introns. Indeed, the Agelasida introns share a more recent common ancestor with the *Cinachyrella/Levantiniella* introns, than with the Scleritodermidae (Fig. 5). A HGT from an ancestor of *Cinachyrella/Levantiniella* species to some Agelasida could account for this result. In all these examples it can therefore be hypothesised that a HGT occurred between distantly related sponge groups. Although the mechanism of intron HGT is unknown at this point, we noted that these donor/receiver species originate from the same regions and share the same habitats (reef, lake or mangrove, see Additional file 3), which is expected to make HGT possible.

Similarities in intron secondary structures of distantly related sponges are further evidence for HGT [43]. Hence, independent insertion events in Tetillidae, Axinellidae and Agelasida were proposed for intron 723 [43]. This is confirmed by secondary structure differences we observed in closely related families (Tetillidae and Scleritodermidae) (Fig. 6, Additional file 2). Additional loops (P9.1e,f), reduced stems (e.g., P9.1d) and the absence of the P6d region in Sceritodermidae (Fig. 6) result in a higher structure similarity to e.g., *Axinella polypoides* [43], rather than to other Tetillidae structures ([65]; Additional file 2, [82]), which confirms independent insertions of intron 723 in Scleritodermidae and Tetillidae. No major structural differences of intron 723 in the P9 and P6 regions were observed within different species of *Cinachyrella* (Additional file 2) except for *Cinachyrella* sp. 4 from

Morocco, which showed reduced P9.1c and P9.1d helices. A few minor differences were also noted between *Cinachyrella* and *L. levantinensis* structures (Fig. 2 in [65]): the latter had an additional loop in P5a, a reduced P9.1d and a loop at the end of P6d. Interestingly, the latter two features are also observed in *C.* sp. 4 (Additional file 2), which could be explained by their common origin resulting from a HGT (Fig. 9). The relative similarity of intron 723 between *Cinachyrella* and *Levantiniella* is a strong argument in favor of a single insertion event in this clade, which therefore implies at least two losses of intron 723 to account for the two major *Cinachyrella/Paratetilla/Amphitethya* clades without any intron (Fig. 9). These would be the first reported cases of mt intron secondary loss in sponges.

For intron 870 no structure differences were observed between *Tetilla quirimure* (MNRJ 17891) (Fig. 6) and *Tetilla radiata* (HM032742) [82]. Remarkably, *Tetilla japonica* (JX177901), which is sister to *Tetilla radiata* (with a strongly supported node) does not posses intron 870. We therefore assume that *Tetilla japonica* secondarily lost intron 870, which would represent another case of mt intron loss in sponges. The structure of intron 870 in *Microscleroderma herdmani* 3 displays a reduced P6a and an additional P5d region (Fig. 6d) compared to *Tetilla radiata / quirimure*, which suggests an independent insertion of intron 870 as the most plausible explanation. This is further corroborated by the distant phylogenetic relationship between *Tetilla* and Scleritodermidae (Fig. 4) and the LAGLIDADG phylogeny (Fig. 7).

Until now VGT was only assumed within sponges [82], but awaited proof with a wider sampling. For the first time our study on 63 *Cinachyrella* sequences provides conclusive evidence that introns were vertically transmitted due to 1) mostly congruent *cox1* versus intron phylogenies and 2) similarity of secondary structures among closely related species. Introns 714, 723 and 870 have all undergone VGT, but this is especially apparent for intron 723 for which we have the largest sampling (Figs. 5 and 9). VGT for group I introns are also known e.g., from hexacorals (nad5-717 intron, [19]), but is often difficult to ascertain due to the patchy distribution of introns. To conclude, our results demonstrate that introns 714, 723 and 870 undergo VGT, HGT and secondary loss events, and that both VGT and HGT can occur within one genus (e.g., *Cinachyrella*) (Fig. 9).

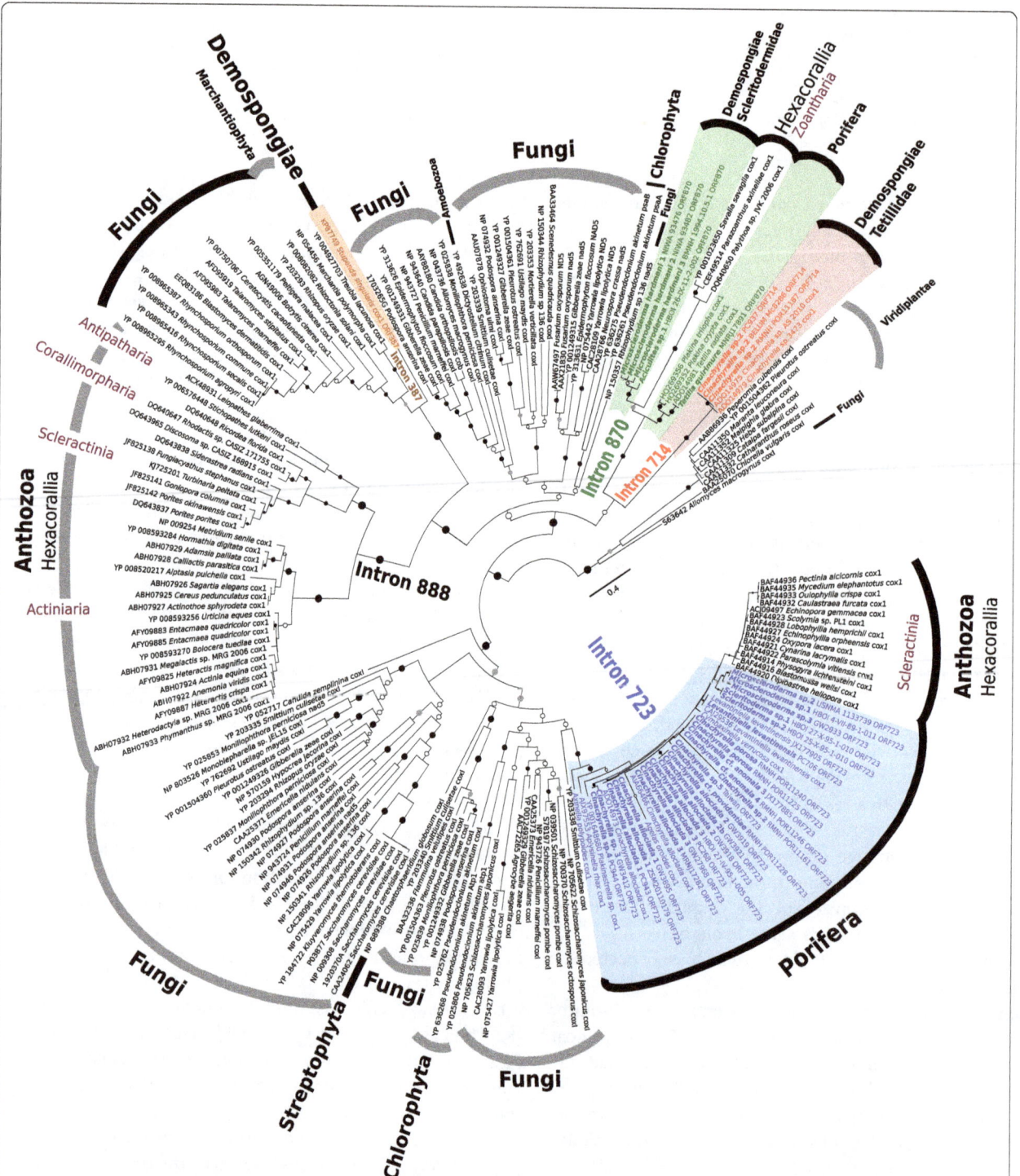

Fig. 7 Maximum Likelihood phylogeny based on LAGLIDADG protein sequences of representative taxa from diverse groups. Circles on the branches indicate support values. *Black* circles: BP = 75–100. *Dark gray* circles: BP = 60–74. *White* circles: BP < 60. Numbers behind each taxa are GenBank accession numbers or voucher numbers. Sequences generated in this study are in bold. Sponge LAGLIDADG clades for introns 387, 714, 723 and 870 are indicated with *brown*, *red*, *blue* and *green* background respectively

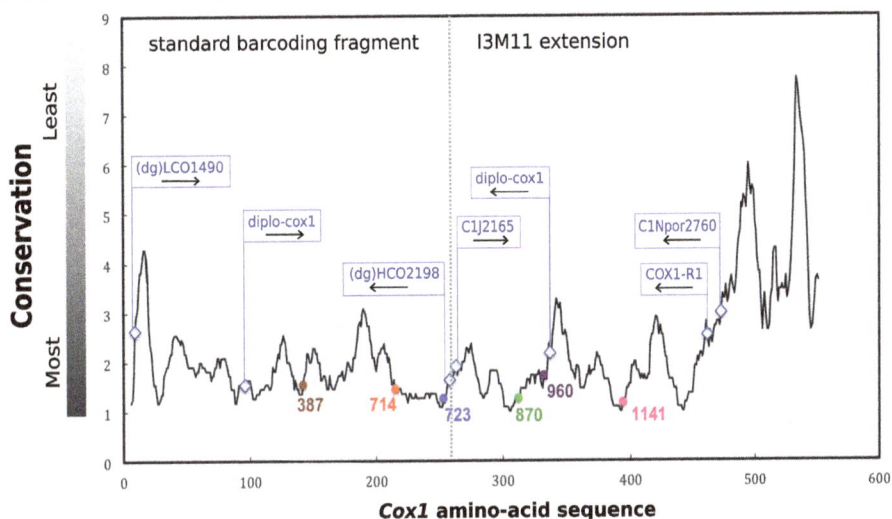

Fig. 8 Conservation profile of the complete translated *cox1* gene for sponges. X-axis indicates the amino acid position along the alignment. Y-axis assigns the number of amino-acid substitutions over a range of 11 aligned positions throughout the alignment. Different intron insertion positions of group I introns (387, 714, 723 and 870) and group II introns (966 and 1141) are plotted on the profile line and color-coded with respect to Fig. 2. Additionally, the 5′ position of commonly used sponge barcoding markers is indicated as *light blue* rhombuses. A detailed list of primers is provided in Additional file 4

Origin of group I introns

The origins of group I introns has been debated for many eukaryotic organisms (e.g., [63]) including sponges [43]. Fungi are proposed as the primary donor of mt group I introns not only in plants (e.g., [14]), but also in cnidarians [31] and sponges [47, 65, 82]. Placozoa have also been suggested as possible donors in sponges, but only for intron 387 in one species [48]. Sponge-fungal associations, pivotal for such HGT, are well-known for sponges (e.g., [40, 71]). However, only little is known about the specific fungal lineages associated with intron-bearing sponge taxa. For *Cinachyrella*, however, deep sequencing analysis recently identified Ascomycota as a dominant fungal phylum (*Cinachyrella* cf. *australiensis* and *Cinachyrella* sp. from the China Sea [38]). This is corroborated by data for *Cinachyrella alloclada* from the Caribbean that showed that the cosmopolitan *Phoma* sp. (Ascomycota) is the dominant sponge-associated fungus, while nine more ascomycete species were found [5]. Indeed, in the reconstructed LAGLIDADG protein phylogeny (Fig. 7) all intron 723 sequences of *Cinachyrella* and other sponge taxa display a close relationship to ascomycete intron sequences, but also Viridiplantae (Chlorophyta and Streptophyta). The latter is not surprising, as intron of chlorophytes and Viridiplantae similarly have their origin from Ascomycota [53, 87]. The huge assemblage of group I introns described in fungi increases the chance of a HGT event. When metazoans and plants host one or two group I introns in their *cox1*, fungus like *Ophiocordyceps sinensis*

contains 21 group I introns within its *cox1* region alone [54].

The non-bilaterian LAGLIDADG-protein sequence dataset (Fig. 7) identifies Scleractinia (stony corals) and Zoantharia (zoanthids) sequences as the only sequences respectively homologous to the sponge introns 723 and 870. Although this is not very well supported, our results show a sister relationship between LAGLIDADG in Scleritodermidae and Zoantharia, which suggests they may have contaminated each other (the direction of the HGT is unclear at this point); alternatively they were contaminated by the same donor, un-sequenced as of today. Because the Scleractinia intron 723 LAGLIDADG sequences are nested within the sponge sequences (Fig. 7), Fukami et al. [27] suggested two alternative scenarios: 1) Scleractinia and sponges have a similar fungi donor which independently transferred intron 723 in each group or 2) HGT events from each other (sponges to coral, or vice versa). However, although intron 723 has a patchy distribution in the Scleractinia [27], the Scleractinia LAGLIDADG sequences form a well supported clade (Fig. 7), which suggests that the origin and therefore the donor must have been the same. Different sponges can be excluded as donors, because otherwise the Scleractinia sequences would be partly mixed with the sponge sequences. Also, the possibility of one single sponge donor is unlikely, because there is no sponge species living in close contact with all these Scleractinia species from the Indo-Pacific and the Atlantic. Thus, we are in favor of a donor, most probably a fungus, which transferred intron 723 to different Scleractinia, while

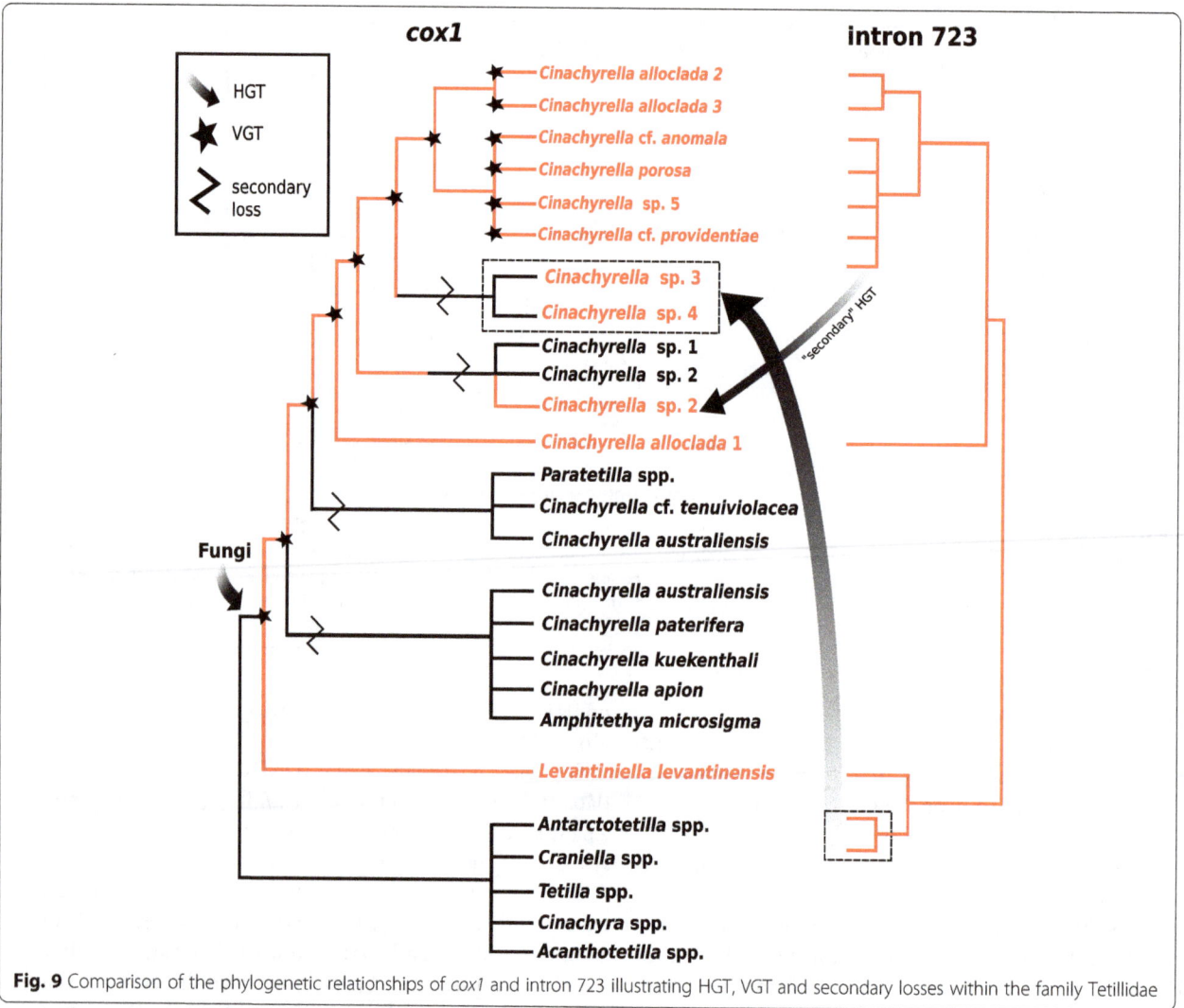

Fig. 9 Comparison of the phylogenetic relationships of *cox1* and intron 723 illustrating HGT, VGT and secondary losses within the family Tetillidae

similar donors transferred intron 723 to different sponges. These donors can probably also act as vectors, thereby enabling HGT between *Cinachyrella* species, as shown above. Intron 387 of *Stupenda singularis* is the only sponge LAGLIDADG sequence apparently unrelated to any coral LAGLIDADG in the data set, but is closely related to LAGLIDADG found in fungi and Marchantiophyta (liverworts). Liverworts are thought to have received their introns from fungi [58], and the close relationship of the *Stupenda singularis* LAGLIDADG 387 likewise suggests a fungal origin (see also [47]).

Unfortunately, the origin of sponge intron 714 remains unresolved, since no supported relationship to any of the included taxonomic groups is given. Therefore, we cannot exclude the possibility that group I introns in sponges may originate from an undiscovered and/or un-sequenced sponge-associated symbionts, e.g., fungi, Archaea, Bacteria or dinoflagellates, since all are known to posses group I introns. In particular sponge bacterial symbionts, which can

contribute to over 50% of the sponge biomass [39, 69], may play an essential role as potential intron donors. However, according to our results (Fig. 7) it is unlikely that bacteria, archaea or dinoflagellates are donors, because of the absence of a homologous LAGLIDADG motif. Blastx of intron 714 specified fungi as best hits, therefore fungi remains a good donor candidate for this intron. According to the diversity of habitats of intron-bearing sponges, from shallow water (in reefs, marine lakes and mangrove) to the deep sea (Additional file 3), we can hypothesize that the putative intron donor may also be ubiquitous, and present in all these different environments.

Implications for DNA barcoding

Intron insertion in highly conserved *cox1* regions decreases the possibility of intron elimination, because the removal must be specific in order to avoid any disruption of the protein function. The most widespread intron

723 is located at the most conserved site in the *cox1* gene (Fig. 8), suggesting this position as an "intron hotspot". In addition to the high number of intron 723 in Tetillidae, we discovered several more intron 714 (*Cinachyrella*) and intron 870 (Scleritodermidae, *Tetilla*), both in conserved *cox1* regions. Intron presence in conserved *cox1* gene regions has major consequences for other fields of science such as molecular taxonomy. In order to gain a better understanding of locations and conservation of the currently recommended barcoding primers [22, 24, 91], we plotted the 5′ site of each primer on the *cox1* conservation profile line (Fig. 8). Interestingly, for the standard barcoding fragment and the I3M11 extension our analysis shows that all previously applied sponge barcoding primers are located in comparatively less conserved regions (Fig. 8). Moreover, our results indicate that group I intron 723 and group II intron 960 are in close proximity to barcoding reverse primer sites (HCO2198 and diplo-cox1) or interrupt the priming regions (intron 723), which corroborates earlier findings from Szitenberg (2010). These findings may partly explain the low (~25% mean) amplification success reported for barcoding museum samples using standard barcoding primers [86]. We therefore recommend for future sponge barcoding studies to test the reverse COX1-R1 primer, which is more distant to the intron insertions, instead of the HCO2198 primer (Fig. 8). The COX1-R1 primer, originally designed to amplify the *cox1* of Tetillidae [65], has been shown to successfully amplify *cox1* in Poecilosclerida [62], Agelasida and Axinellida [43], Chondrosiida and Dictyoceratida [3], some Astrophorina (P. Cárdenas, unpublished results) and Spirophorina (this study).

Conclusion

This study provides novel insights into the taxonomic distribution, diversity and mobility of mitochondrial group I introns in sponges, and validates the subclass Spirophorina (Tetractinellida), as an *intron hotspot* in sponges, notably by increasing the number of Tetillidae introns known by a factor of 5. We wonder whether this could be linked to a lower mt mutation rate in the Spirophorina with respect to other sponges, as suggested for some intron hotspot fungi groups [44]. We show that co-occurrence of two introns in *cox1* is unique among metazoans, but not uncommon in sponges. However, this combination always associates a potentially active intron with a degenerating one. Earlier hypotheses of HGT were confirmed and for the first time VGT and secondary losses of introns conclusively demonstrated. Consequently, such a high level of HGT in combination with the relative low variation in case of VGT (e.g., intron 723, Fig. 9), rejects any alternative use of mt introns as phylogeographic markers. Since the majority of

sponge introns encode a HEG in frame with the 5′ exon, activity of those introns is assumed. We further demonstrate that introns are not restricted to shallow water sponge species, but also occur in species from deeper (~500 m) habitats and extreme environments (mangroves and marine lakes). Conservation profile analysis reveals that all group I and possibly also group II intron insertions in sponges are located within the most conserved regions of their host protein, which may partly explain why they persist in their host genes. At the same time, we show that the currently used sponge barcoding primers are usually located in less conserved regions compared to the introns, but can also overlay intron insertion sites. Therefore, we recommend applying different primers (in particular reverse primers) when standard barcoding primers fail to amplify the *cox1* gene. Finally, our study enhances the support for a fungal origin for the majority of introns in sponges.

Methods
Sampling and identification of specimens

Cinachyrella samples were collected in Florida (U.S.A.) by snorkeling in the seagrass meadows adjacent to the Mote Marine Laboratory/Tropical Research Laboratory (Summerland Key, Florida U.S.A.) and by scuba-diving on the Broward County reef located off Fort Lauderdale (26° 10.498, −80° 05.632). More *Cinachyrella* spp. were collected in Indonesia by diving on reefs and snorkeling in mangroves and marine lakes in West Papua and East Kalimantan, Indonesia. The remaining material was obtained through collaborators or sampled in several museum collections (Additional file 3). Because of ambiguous sequences or missing data, some *Cinachyrella* specimens from Szitenberg et al. [81] were successfully re-sequenced (JX177885, JX177886, JX177887 and JX177913). Taxonomic identification to genus and species level was performed by the authors and follows the findings of Carella et al. [11] on Tetillidae. The species *Craniella quirimure* from Brazil was re-assigned to the genus *Tetilla* based on the absence of a clear double-layered cortex. In some cases identification of species was adopted from collections and earlier publications. Numbers were added for lineages of species that could not be recovered as monophyletic and await revision (e.g., *C. alloclada* 1–3). A detailed list of species origin including collector, voucher numbers and accession numbers, location and depth are provided in Additional file 3. A *Cinachyrella cox1* sequence (including a group I intron) from Taiwan was manually copied from Hsiao [41]. This species was first identified as *Cinachyrella australiensis* and is identified as *Cinachyrella* sp. 5 based on our *cox1* CDS phylogeny. One complete *cox1* sequence of *Microscleroderma* sp. (USNM 1133739) with an intron was kindly provided by D. V. Lavrov

(Department of Ecology, Evolution, and Organismal Biology, Iowa State University, USA). The higher level demosponge classification follows Morrow & Cárdenas [56].

Molecular approach

Genomic DNA was isolated from the choanosome of the sponge tissue by using the NucleoSpin (Machery-Nagel) or the DNeasy (Qiagen) Blood and Tissue Kit according to the manufacturer's protocol. An additional centrifugation step was added before transferring the lysate to the Spin Column in order to avoid any clogging of the membrane, caused by sponge spicules. Quantification of the isolated genomic DNA was performed using a Nano-Drop 1000 Spectrophotometer (Thermo Scientific).

Amplification of the partial *cox1* was performed by using different primers and PCR conditions. Detailed information of primers used for each sample is provided in Additional file 4. For most Tetillidae the *cox1* fragment was amplified using the primers LCO1490 [24] and COX1-R1 [65] and for most Scleritodermidae we used the primers diplo-cox1-f1 and diplo-cox1-r1 [52]. For both primer pairs the PCR settings were: 94 °C, 5 min; (94 °C, 1 min; 50–52 °C, 1:30 min; 72 °C, 1:30 min) × 40 cycles; 72 °C, 10 min. Amplified fragments were visually checked for introns by length on a 1.5% agarose gel. For the majority of the *Cinachyrella* samples with introns, we observed an additional non-specific band at position ~600 bp of bacteria and fungi *cox1* fragments. Separation of double bands and PCR clean-up was performed using a modified freeze-squeeze method [83] in which 20 μl of the PCR product were cut from the gel and stored at −80 °C for one hour, followed by a 40 min centrifugation step at 14,000 rpm. The supernatant (6 μl) was used for cycle sequencing with different and multiple sequencing primers (Additional file 4) together with BigDye Terminator v3.1 (Applied Biosystems, Forster City, CA, USA) chemicals and sequenced by an ABI 3730 Genetic Analyzer at the Sequencing Service of the Department of Biology (LMU München), or by Macrogen (South Korea).

Positions and secondary structures of group I introns within Tetractinellida

Insertion sites for each intron were ascertained in an alignment including other intron-bearing sequences [21, 43, 82]. Intron specific positions were defined according to the *cox1* sequence of the sponge *Amphimedon queenslandica* following Szitenberg et al. [82]. Blast hits and sequence similarity to already published group I intron insertions were used to distinguish between different insertion sites and group I and group II introns. An overview of the different group I (intron 387, 714, 723, 870) insertion sites as well as group II (intron 966 & 1141) insertion sites is given in Fig. 2. Identification of the HEG

for each ORF was conducted by blastp against NCBI Genbank [1]. The class of group I introns (IA, IB, IC, ID or IE) was obtained using the RNAweasel Website http://megasun.bch.umontreal.ca/RNAweasel/ [49]. Initiation and stop codons of the HEG ORFs were located using the ORF finder as implemented in Geneious v.8.1.8 (www.geneious.com) with the following settings: translation Table 4 (Mold and Protozoan mitochondrial) with start codons ATG, GTG, TTG and ATT [90], minimum size 100 bp, including interior ORFs. Although considered as potential start codon in sponge group I introns [47, 65], there is actually no evidence that TTA is used in sponges as start codon; it has only been found so far in *Trypanosoma* [26]. We therefore excluded the TTA start codon in our searches and also revisited the ORFs of previously reported sponge group I introns.

In order to predict the secondary structures of group I introns, we manually converted the given secondary core structures into a dot-bracket notation including pseudoknot informations in square brackets. As secondary structure references, we used *Cinachyrella alloclada* (HM032738) for intron 723 and *Tetilla radiata* (HM032742) for intron 870 [82]. In order to ensure the right structure annotation for short variable (mainly P6 and P9) domains, we used Mfold http://unafold.rna.albany.edu/ [93] under the general settings, presupposing the exclusion of additional pseudoknots, which cannot be predicted by this program. Those Mfold structures were then manually converted into a dot-bracket notation and implemented to the already established core structure sequence. SeaView v4 [32] was used to align the sequences to their structure annotation. The LAGLIDADG regions were removed from the sequences for further analysis. The rest of the intron sequence together with its structure information, was converted to a ct-format using the Perl-script (2ct.zip) of Voigt et al. [89] (available at http://www.palaeontologie.geo.lmu.de/molpal/RRNA/index.htm). All secondary structures were visualized in RNAViz 2.0.3 http://rnaviz.sourceforge.net/ [18]. Helix names follow Szitenberg et al. [82].

Tetractinellida phylogenies predicted by *cox1* CDS
Sequence alignments and outgroup choice

Newly generated sequences as well as additional GenBank sequences were manually aligned to the datasets from Szitenberg et al. [81, 82]. Aligned sequences were subsequently controlled for discrepancies and corrected by eye. Two Astrophorina species (*Geodia barretti* and *Alectona millari*) were used as outgroups in Tetillidae phylogenetic analyses. Astrophorina has been established as the sister clade of Spirophorina in previous studies [6, 57]. For the analysis of the tetractinellid phylogeny we chose *Halichondria melanodocia* and *Axinella corrugata*, which were already successfully used as outgroups in previous studies on the molecular phylogeny of the Tetractinellida (e.g., [47]).

The final *cox1* alignment (excluding intron(s)) of the Tetillidae phylogeny comprised 133 sequences (including the two outgroups), of which 76 were newly generated from this study. The alignment was 1177 bp long, of which 829 bp were constant, 62 bp were parsimony uninformative and 286 bp were parsimony informative. The final *cox1* alignment of the Tetractinellida phylogeny constituted 82 sequences (including the two outgroups) of which 33 were newly generated from this study. In total the alignment comprised 1118 bp, of which 642 bp were invariant, 77 bp parsimony uninformative and 399 bp were parsimony informative.

Phylogenetic reconstructions

Phylogenetic tree reconstructions for both analyses were performed on a parallel version of MrBayes v3.2.4 [64] and RAxML v8.0.26 [78] on a Linux cluster. Bayesian analyses were conducted under the most generalized GTR + G + I evolutionary model, as resulted from jModelTest v.2.1.7 [16]. Analyses were run in two concurrent runs of four Metropolis-coupled Markov-chains (MCMC) for 100,000,000 generations and stopped when the average standard deviation of split frequencies reached below 0.01. The first 25% (burn-in) of the sampled trees were removed for further analysis. For both datasets, Maximum Likelihood (ML) and bootstrap analyses (1,000 replicates) under the GTR + G model as resulted from jModelTest v.2.1.7 [16] were performed. Tree topologies from Bayesian and ML analyses were compared and visualized using Figtree v1.4.2 http://tree.bio.ed.ac.uk/software/figtree/.

Phylogenetic inference based on intron + LAGLIDADG sequences

In order to test for vertical transmission of group I introns (including both LAGLIDADG and the non-coding regions) in the genus *Cinachyrella*, we conducted phylogenetic analyses on separate datasets respectively including all sponge introns 723, 714 and 870. For the analysis of intron 723, we included 74 sequences of which 63 belong to the genus *Cinachyrella*. One taxon *Aciculites* sp. 1 (HBOI 26-IX-11-2-002), was excluded from this analysis, as no putative HEG were detected in the intron. The final intron 723 alignment was 1167 bp long, of which 488 bp were constant, 307 bp parsimony uninformative and 372 bp parsimony informative. The final intron 714 dataset included 13 sequences and was 946 bp long, of which 891 bp were constant, 49 bp were parsimony uninformative and 6 bp were parsimony informative. As an outgroup for both analysis we used the introns of *Plakinastrella* sp. (NC 010217), a species that belongs to a different sponge class (Fig. 1). The final

alignment of intron 870 contained 12 taxa and was 974 bp long, of which 615 pb were constant, 46 bp were parsimony uninformative and 313 bp parsimony informative. *Plakina trilopha* (HQ269356) and *P. crypta* (HQ269352) which belong to a different sponge class (Fig. 1) were used as outgroups. Phylogenetic tree reconstructions were performed as described above for the *cox1* exon phylogeny.

In order to test whether the incongruencies between the exon and the intron/HE phylogeny were significant, we performed a series of Shimodaira-Hasegawa (SH) tests [75] as implemented in RAxML [78] on the exon tree against ML topologies constrained towards the intron tree topology. Constraints were inferred with Mesquite v.3.10 [55].

Phylogenetic reconstructions based on LAGLIDADG protein sequences of group I introns

In order to investigate the evolutionary origins of the putative LAGLIDADG encoding introns (387, 714, 723 and 870) in sponges the newly generated sequences were added to the LAGLIDADG dataset by Huchon et al. [43]. Additionally, we included 12 fungal and two Marchantiophyta LAGLIDADG sequences resulting from Blastp hits of the sponge LAGLIDADG for intron 387 (Table 2, [47]). Subsequently, MAFFT v.7 [46] under the L-INS-I algorithm was used to generate the protein alignment. The resulting alignment contains sequences of fungi, plants, cnidarians and sponges. Here, 291 amino-acids (aa) out of 1278 aa were parsimony-uninformative variable characters, 729 aa were constant and 708 parsimony-informative. As a result, we manually corrected the LAGLIDADG alignment. Parts with more than approximately 50% of missing data were removed manually using the custom site set selection tool in SeaView. The final alignment was 317 amino-acids long, of which one character was constant and two variable characters were parsimony-uninformative. The rest of the 314 characters were phylogenetically informative. The maximum likelihood (ML) analysis was performed using RAxML v8.0.26 [78] on a Linux cluster with 1,000 bootstrap repeats. Using ProtTest 3.4 [15] the best evolutionary model was found to be VT + I + Gamma + F. However, for the RAxML analysis we excluded the invariant parameter (I) from the model, as it is not recommended to use both gamma (G) and invariant (I) parameters among site-rate variations according to the RAxML manual. No root for the tree was specified, as it was not needed for our purpose.

Compilation of the conservation profile

A conservation profile was calculated from a *cox1* protein alignment dataset compiled from the demosponge sequences from [90], complemented by Homoscleromorpha sequences from [29]. The final protein alignment consisted of 58 sequences and 556 characters. The conservation profile was made following Swithers et al. [80] using the same perl script (made available in the supplementary material of Swithers et al. [80]) but with a slightly modification to allow 'X' characters in the alignment and calculation. The 5′ position of the common barcoding markers as well as all sponge intron insertion positions were plotted on the profile line.

Additional files

Additional file 1: Intron 714 and 870 phylogenies. (PDF 38 kb)

Additional file 2: Predicted secondary structure of introns from different *Cinachyrella* species. (PDF 86 kb)

Additional file 3: Metadata of samples used in this study. (XLSX 85 kb)

Additional file 4: PCR and sequencing primers used in this study. (XLS 80 kb)

Acknowledgements

We greatly acknowledge Dennis V. Lavrov (Iowa State University, USA) for sharing the *Microscleroderma* sp. 2 *cox1* intron sequence and Oliver Voigt (Dept. of Earth- & Environmental Sciences, LMU Munich, Germany) for providing scripts and support for secondary structure analysis. We greatly thank Eduardo Hajdu and Cristina Castello-Branco (Universidade Federal Do Rio De Janeiro, Brasil), Carsten Lüter (Museum für Naturkunde Berlin, Germany), Helmut Lehnert (Zoologische Staatssammlung München, Germany), Sigal Shefer and Yaniv Aluma (Tel Aviv University), Sadie Mills and Kareen Schnabel (National Institute of Water and Atmospheric Research, Wellington, New Zealand), Sara Griffiths (University of Manchester, UK), Cécile Debitus (IRD, Institut de Recherche pour le Développement, Marseille, France)France), Christopher J. Freeman (Smithsonian Marine Station, Fort Pierce, USA), Belkassem El Amraoui (University Ibn Zohr, Taroudant, Morocco) , Nadia Santodomingo (Natural History Museum, London), Nicole J. De Voogd (Naturalis Biodiversity Center), Yosephine Tuti (Indonesian National Institute of Sciences) and the governments of Ecuador, Commonwealth of the Bahamas, Panama, and Martinique (France) for sampling, sharing material and help in the collections. We thank Kyle Roebuck, Nidhi Vijayan and Marissa Wickes with help in sample preparation and shipping from Florida. We thank the Systematic Biology lab (Dept. of Organismal Biology, Uppsala University, Sweden) and Gabrielle Büttner and Simone Schätzle (Dept. of Earth- & Environmental Sciences, LMU Munich, Germany) for sequencing assistance. Finally, we want to thank the three anonymous reviewers for their helpful comments and suggestions.

Funding

This work was funded by the German Science Foundation (DFG) (grant number DFG ER 611/3-1, DFG Wo896/15-1); LMUMentoring Program; HELGE AX:Son JOHNSON STIFTELSE (Sweden), Inez Johanssons HT2012 (Uppsala University, Sweden) and Netherlands Organisation for Scientific Research (NWO) Veni#863.14.020.

Authors' contributions

AS and PC conceived and designed the study. AS carried out PCR, sequencing, phylogenetic analyses and predicted the secondary structures. AS, LEB, MK, PC, and SAP identified the specimens. All authors contributed with samples and reagents. AS drafted the manuscript and figures. PC, JVL, SAP, GW and DE assisted in revising the MS. All authors approved the final version of the manuscript.

Competing interests

The authors declare that they have no competing interests.

Author details

[1]Department of Earth- & Environmental Sciences, Palaeontology and Geobiology, Ludwig-Maximilians-Universität München, Richard-Wagner-Str. 10, 80333 Munich, Germany. [2]Halmos College of Natural Sciences and Oceanography, Nova Southeastern University, Dania Beach, FL 33004, USA. [3]Marine Animal Ecology, Wageningen University & Research Centre, P.O. Box 3700AH, Wageningen, The Netherlands. [4]Naturalis Biodiversity Center, Marine Zoology Department, PO Box 95172300RA, Leiden, The Netherlands. [5]National Centre for Aquatic Biodiversity and Biosecurity, National Institute of Water and Atmospheric Research, P.O. Box 109–695Newmarket, Auckland, New Zealand. [6]Harbor Branch Oceanographic Institute-Florida Atlantic University, 5600 U.S. 1 North, Ft Pierce, FL 34946, USA. [7]SNSB - Bavarian State Collections of Palaeontology and Geology, Richard-Wagner Str. 10, 80333 Munich, Germany. [8]GeoBio-CenterLMU, Ludwig-Maximilians-Universität München, Richard-Wagner Str. 10, 80333 Munich, Germany. [9]Department of Medicinal Chemistry, Division of Pharmacognosy, BioMedical Center, Uppsala University, Husargatan 3, 75123 Uppsala, Sweden.

References

1. Altschul SF, Madden TL, Schaffer AA, Zhang JH, Zhang Z, Miller W, Lipman DJ. Gapped BLAST and PSI-BLAST: a new generation of protein database search programs. Nucleic Acids Res. 1997;25:3389–402.
2. Belfort M, Perlman PS. Mechanisms of intron mobility. J Biol Chem. 1995; 270:30237–40.
3. Belinky F, Szitenberg A, Goldfarb I, Feldstein T, Wörheide G, Ilan M, Huchon D. ALG11 - A new variable DNA marker for sponge phylogeny: Comparison of phylogenetic performances with the 18S rDNA and the COI gene. Mol Phylogenet Evol. 2012;63:702–13.
4. Bilewitch JP, Degnan SM. A unique horizontal gene transfer event has provided the octocoral mitochondrial genome with an active mismatch repair gene that has potential for an unusual self-contained function. BMC Evol Biol. 2011;11:228.
5. Bolaños J, Fernando De León L, Ochoa E, Darias J, Raja HA, Shearer CA, Miller AN, Vanderheyden P, Porras-Alfaro A, Caballero-George C. Phylogenetic diversity of sponge-associated fungi from the Caribbean and the Pacific of Panama and their in vitro effect on angiotensin and endothelin receptors. Mar Biotechnol. 2015;17:533–46.
6. Borchiellini C, Chombard C, Manuel M, Alivon E, Vacelet J, Boury-Esnault N. Molecular phylogeny of Demospongiae: implications for classification and scenarios of character evolution. Mol Phylogenet Evol. 2004;32:823–37.
7. Burger G, Forget L, Zhu Y, Gray MW, Lang BF. Unique mitochondrial genome architecture in unicellular relatives of animals. Proc Natl Acad Sci U S A. 2003;100:892–7.
8. Burger G, Saint-Louis D, Gray MW, Lang BF. Complete sequence of the mitochondrial DNA of the red alga *Porphyra purpurea*. Cyanobacterial introns and shared ancestry of red and green algae. Plant Cell. 1999;11: 1675–94.
9. Burger G, Yan Y, Javadi P, Lang BF. Group I-intron trans-splicing and mRNA editing in the mitochondria of placozoan animals. Trends Genet. 2009;25:381–6.
10. Cárdenas P, Menegola C, Rapp HT, Cristina DM. Morphological description and DNA barcodes of shallow-water *Tetractinellida* (Porifera: Demospongiae) from Bocas del Toro, Panama, with description of a new species. Zootaxa. 2009;2276:1–39.
11. Carella M, Agell G, Cárdenas P, Uriz MJ. Phylogenetic reassessment of Antarctic Tetillidae (Demospongiae, Tetractinellida) reveals new genera and genetic similarity among morphologically distinct species. PLoS One. 2016; 11(8):e0160718.
12. Chan H-M. Self-splicing of Group I Intron of the Mitochondrial Genome of the Sponge, *Cinachyrella australiensis*. Kahosiung: Master Thesis: National Sun Yat-sen University; 2009. http://etd.lib.nsysu.edu.tw/ETD-db/ETD-search/ view_etd?URN=etd-0819109-130742 Accessed 20 Mar 2016.
13. Chen I-P, Tang C-Y, Chiou C-Y, Hsu J-H, Wei NV, Wallace CC, Muir P, Wu H, Chen CA. Comparative analyses of coding and noncoding DNA regions indicate that *Acropora* (Anthozoa: Scleractina) possesses a similar evolutionary tempo of nuclear vs. mitochondrial genomes as in plants. Mar Biotechnol. 2009;11:141–52.
14. Cho Y, Qiu YL, Kuhlman P, Palmer JD. Explosive invasion of plant mitochondria by a group I intron. Proc Natl Acad Sci U S A. 1998;95:14244–9.

15. Darriba D, Taboada GL, Doallo R, Posada D. ProtTest 3: fast selection of best-fit models of protein evolution. Bioinformatics. 2011;27:1164–5.

16. Darriba D, Taboada GL, Doallo R, Posada D. jModelTest 2: more models, new heuristics and parallel computing. Nat Methods. 2012;9:772.

17. Dellaporta SL, Xu A, Sagasser S, Jakob W, Moreno MA, Buss LW, Schierwater B. Mitochondrial genome of Trichoplax adhaerens supports Placozoa as the basal lower metazoan phylum. Proc Natl Acad Sci U S A. 2006;103:8751–6.

18. De Rijk P, Wuyts J, De Wachter R. RnaViz 2: an improved representation of RNA secondary structure. Bioinformatics. 2003;19:299–300.

19. Emblem Å, Karlsen BO, Evertsen J, Johansen SD. Mitogenome rearrangement in the cold-water scleractinian coral Lophelia pertusa (Cnidaria, Anthozoa) involves a long-term evolving group I intron. Mol Phylogenet Evol. 2011;61:495–503.

20. Emblem Å, Okkenhaug S, Weiss ES, Denver DR, Karlsen BO, Moum T, Johansen SD. Sea anemones possess dynamic mitogenome structures. Mol Phylogenet Evol. 2014;75:184–93.

21. Erpenbeck D, Aryasari R, Hooper JNA, Wörheide G. A Mitochondrial Intron in a Verongid Sponge. J Mol Evol. 2014;80:13–7.

22. Erpenbeck D, Hooper JNA, Wörheide G. CO1 phylogenies in diploblasts and the "Barcoding of Life"- are we sequencing a suboptimal partition? Mol Ecol Notes. 2006;6:550–3.

23. Férandon C, Moukha S, Callac P, Benedetto J-P, Castroviejo M, Barroso G. The Agaricus bisporus cox1 Gene: The Longest Mitochondrial Gene and the Largest Reservoir of Mitochondrial Group I Introns. PLoS One. 2010;5:e14048.

24. Folmer O, Black M, Hoeh W, Lutz R, Vrijenhoek R. DNA primers for amplification of mitochondrial cytochrome c oxidase subunit I from diverse metazoan invertebrates. Mol Mar Biol Biotechnol. 1994;3:294–9.

25. Fontaine J-M, Rousvoal S, Leblanc C, Kloareg B, Loiseau-de GS. The Mitochondrial LSU rDNA of the Brown Alga Pylaiella littoralis Reveals α- Proteobacterial Features and is Split by Four Group IIB Introns with an Atypical Phylogeny. J Mol Biol. 1995;251:378–89.

26. Fox TD. Natural variation in the genetic code. Annu Rev Genet. 1987;21:67–91.

27. Fukami H, Chen CA, Chiou CY, Knowlton N. Novel group I introns encoding a putative homing endonuclease in the mitochondrial cox1 gene of Scleractinian corals. J Mol Evol. 2007;64:591–600.

28. Galtier N. The intriguing evolutionary dynamics of plant mitochondrial DNA. BMC Biol. 2011;9:1–3.

29. Gazave E, Lapébie P, Renard E, Vacelet J, Rocher C, Ereskovsky AV, Lavrov DV, Borchiellini C. Molecular Phylogeny Restores the Supra-Generic Subdivision of Homoscleromorph Sponges (Porifera, Homoscleromorpha). PLoS One. 2010;5:e14290.

30. Goddard MR, Burt A. Recurrent invasion and extinction of a selfish gene. Proc Natl Acad Sci U S A. 1999;96:13880–5.

31. Goddard MR, Leigh J, Roger AJ, Pemberton AJ. Invasion and persistence of a selfish gene in the Cnidaria. PLoS One. 2006;1:e3.

32. Gouy M, Guindon S, Gascuel O. SeaView version 4: A multiplatform graphical user interface for sequence alignment and phylogenetic tree building. Mol Biol Evol. 2010;27:221–4.

33. Hafez M, Hausner G, Bonen L. Homing endonucleases: DNA scissors on a mission. Genome. 2012;55:553–69.

34. Haugen P, Bhattacharya D. The spread of LAGLIDADG homing endonuclease genes in rDNA. Nucleic Acids Res. 2004;32:2049–57.

35. Haugen P, Simon DM, Bhattacharya D. The natural history of group I introns. Trends Genet. 2005;21:111–9.

36. Hausner G. Introns, Mobile Elements, and Plasmids. In: Organelle Genetics. Heidelberg: Springer Berlin Heidelberg; 2012. p. 329–57.

37. Hellberg ME. No variation and low synonymous substitution rates in coral mtDNA despite high nuclear variation. BMC Evol Biol. 2006;6:24.

38. He L, Liu F, Karuppiah V, Ren Y, Li Z. Comparisons of the fungal and protistan communities among different marine sponge holobionts by pyrosequencing. Microb Ecol. 2014;67:951–61.

39. Hentschel U, Piel J, Degnan SM, Taylor MW. Genomic insights into the marine sponge microbiome. Nat Rev Microbiol. 2012;10:641–54.

40. Höller U, Wright AD, Matthee GF, Konig GM, Draeger S, Aust H-J, Schulz B. Fungi from marine sponges: diversity, biological activity and secondary metabolites. Mycol Res. 2000;104:1354–65.

41. Hsiao S-T. Genomic Analyses of the Complete Mitochondrial DNA Sequence of the Sponge, Cinachyrella australiensis (Demospongiae: Spirophorida). Kahosiung: Master Thesis: National Sun Yat-sen University; 2005. http://etd.lib.nsysu.edu.tw/ETD-db/ETD-search/view_etd?URN=etd-0707105-161158. Accessed 20 Mar 2016.

42. Huang D, Meier R, Todd PA, Chou LM. Slow mitochondrial COI sequence evolution at the base of the metazoan tree and its implications for DNA barcoding. J Mol Evol. 2008;66:167–74.

43. Huchon D, Szitenberg A, Shefer S, Ilan M, Feldstein T. Mitochondrial group I and group II introns in the sponge orders Agelasida and Axinellida. BMC Evol Biol. 2015;15:278.

44. Jalalzadeh B, Saré IC, Férandon C, Callac P, Farsi M, Savoie J-M, Barroso G. The intraspecific variability of mitochondrial genes of Agaricus bisporus reveals an extensive group I intron mobility combined with low nucleotide substitution rates. Curr Genet. 2015;61:87–102.

45. Johansen S, Johansen T, Haugli F. Structure and evolution of myxomycete nuclear group I introns: a model for horizontal transfer by intron homing. Curr Genet. 1992;22:297–304.

46. Katoh K, Standley DM. MAFFT Multiple sequence alignment software version 7: Improvements in performance and usability. Mol Biol Evol. 2013;30:772–80.

47. Kelly M, Cárdenas P. An unprecedented new genus and family of Tetractinellida (Porifera, Demospongiae) from New Zealand's Colville Ridge, with a new type of mitochondrial group I intron. Zool J Linn Soc. 2016;177:335–52.

48. Kohn AB, Citarella MR, Kocot KM, Bobkova YV, Halanych KM, Moroz LL. Rapid evolution of the compact and unusual mitochondrial genome in the ctenophore, Pleurobrachia bachei. Mol Phylogenet Evol. 2012;63:203–7.

49. Lang BF, Laforest MJ, Burger G. Mitochondrial introns: a critical view. Trends Genet. 2007;23:119–25.

50. Lavrov DV. Mitochondrial Genomes of Two Demosponges Provide Insights into An Early Stage of Animal Evolution. Mol Biol Evol. 2005;22:1231–9.

51. Lavrov DV, Adamski M, Chevaldonné P, Adamska M. Extensive Mitochondrial mRNA Editing and Unusual Mitochondrial Genome Organization in Calcaronean Sponges. Curr Biol. 2016;26:86–92.

52. Lavrov D, Wang X, Kelly M. Reconstructing ordinal relationships in the Demospongiae using mitochondrial genomic data. Mol Phylogenet Evol. 2008;49:111–24.

53. Lindstrom SC, Pistolic J. Detection of a group I (IE) fungal intron in the green algal genus Urospora (Ulvophyceae). J Phycol. 2005;41:359–65.

54. Li Y, Hu XD, Yang R-H, Hsiang T, Wang K, Liang D-Q, Liang F, Cao D-M, Zhou F, Wen G, et al. Complete mitochondrial genome of the medicinal fungus Ophiocordyceps sinensis. Sci Rep. 2015;5:13892.

55. Maddison WP, Maddison DR. Mesquite: a modular system for evolutionary analyses. 2016. Version 3.10 http://mesquiteproject.org.

56. Morrow C, Cárdenas P. Proposal for a revised classification of the Demospongiae (Porifera). Front. Zool. 2015;12:1–27.

57. Nichols S. An evaluation of support for order-level monophyly and interrelationships within the class Demospongiae using partial data from the large subunit rDNA and cytochrome oxidase subunit I. Mol Phylogenet Evol. 2005;34:81–96.

58. Ohta E, Oda K, Yamato K, Nakamura Y, Takemura M, Nozato N, Akashi K, Ohyama K, Michel F. Group I introns in the liverwort mitochondrial genome: the gene coding for subunit 1 of cytochrome oxidase shares five intron positions with its fungal counterparts. Nucleic Acids Res. 1993;21:1297–305.

59. Osigus HJ, Eitel M, Bernt M, Donath A, Schierwater B. Mitogenomics at the base of Metazoa. Mol Phylogenet Evol. 2013;69:339–51.

60. Palmer JD, Adams KL, Cho Y, Parkinson CL, Qiu YL, Song K. Dynamic evolution of plant mitochondrial genomes: mobile genes and introns and highly variable mutation rates. Proc Natl Acad Sci U S A. 2000;97:6960–6.

61. Redmond NE, Morrow CC, Thacker RW, Diaz MC, Boury-Esnault N, Cárdenas P, Hajdu E, Lobo-Hajdu G, Picton BE, Pomponi SA, et al. Phylogeny and Systematics of Demospongiae in Light of New Small-Subunit Ribosomal DNA (18S) Sequences. Integr Comp Biol. 2013;53:388–415.

62. Reveillaud J, van Soest R, Derycke S, Picton B, Rigaux A, Vanreusel A. Phylogenetic relationships among NE Atlantic Plocamionida Topsent (1927) (Porifera, Poecilosclerida): Under-estimated diversity in reef ecosystems. PLoS One. 2011;6:e16533.

63. Rodríguez-Trelles F, Tarrío R, Ayala FJ. Origins and evolution of spliceosomal introns. Annu Rev Genet. 2006;40:47–76.

64. Ronquist F, Teslenko M, van der Mark P, Ayres DL, Darling A, Höhna S, Larget B, Liu L, Suchard MA, Huelsenbeck JP. MrBayes 3.2: Efficient Bayesian Phylogenetic Inference and Model Choice across a Large Model Space. Syst Biol. 2012;61:539–42.

65. Rot C, Goldfarb I, Ilan M, Huchon D. Putative cross-kingdom horizontal gene transfer in sponge (Porifera) mitochondria. BMC Evol Biol. 2006;6:71.

66. Sanchez-Puerta MV, Abbona CC, Zhuo S, Tepe EJ, Bohs L, Olmstead RG, Palmer JD. Multiple recent horizontal transfers of the cox1 intron in Solanaceae and extended co-conversion of flanking exons. BMC Evol Biol. 2011;11:277.

67. Sanchez-Puerta MV, Cho Y, Mower JP, Alverson AJ, Palmer JD. Frequent, phylogenetically local horizontal transfer of the cox1 group I Intron in flowering plant mitochondria. Mol Biol Evol. 2008;25:1762–77.

68. Santamaria M, Vicario S, Pappadà G, Scioscia G, Scazzocchio C, Saccone C. Towards barcode markers in Fungi: an intron map of Ascomycota mitochondria. BMC Bioinformatics. 2009;10:1–13.

69. Santavy DL, Willenz P, Colwell RR. Phenotypic study of bacteria associated with the caribbean sclerosponge, *Ceratoporella nicholsoni*. Appl Environ Microbiol. 1990;56:1750–62.

70. Schäfer B. Genetic conservation versus variability in mitochondria: the architecture of the mitochondrial genome in the petite-negative yeast *Schizosaccharomyces pombe*. Curr Genet. 2003;43:311–26.

71. Schmitt S, Tsai P, Bell J, Fromont J, Ilan M, Lindquist N, Perez T, Rodrigo A, Schupp PJ, Vacelet J, et al. Assessing the complex sponge microbiota: core, variable and species-specific bacterial communities in marine sponges. ISME J. 2012;6:564–76.

72. Schuster A, Erpenbeck D, Pisera A, Hooper J, Bryce M, Fromont J, Wörheide G. Deceptive Desmas: Molecular Phylogenetics Suggests a New Classification and Uncovers Convergent Evolution of Lithistid Demosponges. PLoS One. 2015;10:e116038.

73. Sellem CH, Belcour L. Intron open reading frames as mobile elements and evolution of a group I intron. Mol Biol Evol. 1997;14:518–26.

74. Shearer TL, Van Oppen MJH, Romano SL, Wörheide G. Slow mitochondrial DNA sequence evolution in the Anthozoa (Cnidaria). Mol Ecol. 2002;11:2475–87.

75. Shimodaira H, Hasegawa M. Multiple comparisons of log-likelihoods with applications to phylogenetic inference. Mol Biol Evol. 1999;16:1114–6.

76. Signorovitch AY, Buss LW, Dellaporta SL. Comparative genomics of large mitochondria in placozoans. PLoS Genet. 2007;3:e13.

77. Smith DR, Kayal E, Yanagihara AA, Collins AG, Pirro S, Keeling PJ. First complete mitochondrial genome sequence from a box jellyfish reveals a highly fragmented linear architecture and insights into telomere evolution. Genome Biol Evol. 2012;4:52–8.

78. Stamatakis A. RAxML version 8: a tool for phylogenetic analysis and post-analysis of large phylogenies. Bioinformatics. 2014;30:1312–3.

79. Stampar SN, Maronna MM, Kitahara MV, Reimer JD, Morandini AC. Fast-Evolving Mitochondrial DNA in Ceriantharia: A Reflection of Hexacorallia Paraphyly? PLoS One. 2014;9:e86612.

80. Swithers KS, Senejani AG, Fournier GP, Gogarten JP. Conservation of intron and intein insertion sites: implications for life histories of parasitic genetic elements. BMC Evol Biol. 2009;9:303.

81. Szitenberg A, Becking LE, Vargas S, Fernandez JCC, Santodomingo N, Wörheide G, Ilan M, Kelly M, Huchon D. Phylogeny of Tetillidae (Porifera, Demospongiae, Spirophorida) based on three molecular markers. Mol Phylogenet Evol. 2013;67:509–19.

82. Szitenberg A, Rot C, Ilan M, Huchon D. Diversity of sponge mitochondrial introns revealed by cox 1 sequences of Tetillidae. BMC Evol Biol. 2010;10:288.

83. Tautz D, Renz M. An optimized freeze-squeeze method for the recovery of DNA fragments from agarose gels. Anal Biochem. 1983;132:14–9.

84. Vallès Y, Halanych KM, Boore JL. Group II introns break new boundaries: presence in a bilaterian's genome. PLoS One. 2008;3:e1488.

85. Van Oppen MJ, Olsen JL, Stam WT. Evidence for independent acquisition of group I introns in green algae. Mol Biol Evol. 1993;10:1317–26.

86. Vargas S, Schuster A, Sacher K, Büttner G, Schätzle S, Läuchli B, Hall K, Hooper JNA, Erpenbeck D, Wörheide G. Barcoding Sponges: An Overview Based on Comprehensive Sampling. PLoS One. 2012;7:e39345.

87. Vaughn JC, Mason MT, Sper-Whitis GL, Kuhlman P, Palmer JD. Fungal origin by horizontal transfer of a plant mitochondrial group I intron in the chimeric *cox1* gene of *Peperomia*. J Mol Evol. 1995;41:563–72.

88. Vicens Q, Cech TR. Atomic level architecture of group I introns revealed. Trends Biochem Sci. 2006;31:41–51.

89. Voigt O, Erpenbeck D, Wörheide G. Molecular evolution of rDNA in early diverging Metazoa: First comparative analysis and phylogenetic application of complete SSU rRNA secondary structures in Porifera. BMC Evol Biol. 2008;8:69.

90. Wang X, Lavrov D. Seventeen new complete mtDNA sequences reveal extensive mitochondrial genome evolution within the Demospongiae. PLoS One. 2008;3:e2723.

91. Wörheide G, Erpenbeck D, Menke C. The Sponge Barcoding Project: aiding in the identification and description of poriferan taxa. In: Custódio MR, Lôbo-Hajdu G, Hajdu E, Muricy G, editors. Porifera research: biodiversity, innovation and sustainability, vol. 28. Série Livros: Rio de Janeiro; 2007. p. 123–8.

92. Yamada T, Tamura K, Aimi T, Songsri P. Self-splicing group I introns in eukaryotic viruses. Nucleic Acids Res. 1994;22:2532–7.

93. Zuker M. Mfold web server for nucleic acid folding and hybridization prediction. Nucleic Acids Res. 2003;31:3406–15.

The protein subunit of telomerase displays patterns of dynamic evolution and conservation across different metazoan taxa

Alvina G. Lai*, Natalia Pouchkina-Stantcheva, Alessia Di Donfrancesco, Gerda Kildisiute, Sounak Sahu and A. Aziz Aboobaker*

Abstract

Background: Most animals employ telomerase, which consists of a catalytic subunit known as the telomerase reverse transcriptase (TERT) and an RNA template, to maintain telomere ends. Given the importance of TERT and telomere biology in core metazoan life history traits, like ageing and the control of somatic cell proliferation, we hypothesised that TERT would have patterns of sequence and regulatory evolution reflecting the diverse life histories across the Animal Kingdom.

Results: We performed a complete investigation of the evolutionary history of TERT across animals. We show that although TERT is almost ubiquitous across Metazoa, it has undergone substantial sequence evolution within canonical motifs. Beyond the known canonical motifs, we also identify and compare regions that are highly variable between lineages, but show conservation within phyla. Recent data have highlighted the importance of alternative splice forms of *TERT* in non-canonical functions and although animals may share some conserved introns, we find that the selection of exons for alternative splicing appears to be highly variable, and regulation by alternative splicing appears to be a very dynamic feature of *TERT* evolution. We show that even within a closely related group of triclad flatworms, where alternative splicing of *TERT* was previously correlated with reproductive strategy, we observe highly diverse splicing patterns.

Conclusions: Our work establishes that the evolutionary history and structural evolution of TERT involves previously unappreciated levels of change and the emergence of lineage specific motifs. The sequence conservation we describe within phyla suggests that these new motifs likely serve essential biological functions of TERT, which along with changes in splicing, underpin diverse functions of TERT important for animal life histories.

Keywords: Telomerase, TERT, Evolution, Alternative splicing, Planarian, Metazoa

Background

In linear eukaryotic chromosomes, telomeres function as protective nucleoprotein structures that guard chromosome ends against the action of nucleases, DNA damage signalling and nucleolytic degradation [1–3]. Telomeric DNA, consisting entirely of guanine-rich tandem repeats, is maintained by the action of the enzyme

* Correspondence: Alvina.Lai@zoo.ox.ac.uk; alvinagracelai@gmail.com; Aziz.Aboobaker@zoo.ox.ac.uk
Department of Zoology, University of Oxford, South Parks Road, Oxford OX1 3PS, UK

telomerase. The activity of telomerase was first eluded with the framing of the "end replication problem" by Watson [4] and Olovnikov [5], which noted that the lagging strand nature of DNA replication would lead to progressive shortening of linear eukaryotic chromosomes. In humans, cells in culture have a finite replicative lifespan limiting the number of times a cell can divide, and eventually senesce [6]. This has since been shown to be a key mechanism in protecting against cancer and may also contribute to causing a decline in physiological functions and organismal ageing [7–12].

While together the shortening of chromosome ends and the observation of the "Hayflick limit" led to the notion that telomeres could act as timers of cellular replicative age, it also raised the question as to how chromosomes maintain their ends across generations, through reproduction, development and during the lifespan of long-lived individuals [2, 13–15]. This led to intensive research on both the structure of chromosome ends and a search for the cellular activities that could maintain these structures in light of the "end replication problem". Eventually, the ribonucleoprotein telomerase was discovered in the ciliated protozoa *Tetrahymena thermophila* [16–20]. Soon after, the catalytic subunit of this enzyme, telomerase reverse transcriptase (TERT) was also discovered and then studied in human cells [21–27].

However, despite telomeres and telomerase providing the solution for the existence of linear chromosomes, it is clear that the huge variation in life histories, particularly amongst metazoans, would have different demands on an end chromosome maintenance system. For example, short-lived and long-lived metazoans would have very different needs for telomerase and telomere maintenance. This is exemplified by comparing ourselves with single celled protists, where human life history in an evolutionary context means that signalling from chromosome ends must achieve a balance between allowing growth and preventing inevitable somatic mutations from causing cancer and death before reproductive age [28, 29]. Subsequent pleiotropic effects of signalling mechanisms associated with telomeres now appear to be central to the human ageing process and age associated pathologies, including processes that are unconnected to telomere length [30–33]. On this basis we would expect to observe telomere biology adaptations to reflect life history specific requirements and therefore variation between different groups. For instance, telomerase levels in the long-lived rodent, the naked-mole rat show stable expression across different life stages [34]. In long-lived birds, telomerase activity is correlated with proliferative potential and the demands of specific organ types, for example birds with long lifespans have high levels of telomerase in their bone marrow [35]. In other species telomerase does not appear to be related to homeostasis and longevity, for example studies in laboratory mice have demonstrated that ageing is unlikely to be caused by telomere shortening alone as aged mice have long telomeres and loss of telomerase has no immediate effect on ageing [11, 36–39]. Unlike mammals that stop growing upon reaching adulthood, certain animals with indeterminate growth such as some fish [40–42], bivalve molluscs [43], sea urchins [44], lobsters [45] and planarians [46] express telomerase through the life cycle.

In addition to its canonical role in telomere maintenance, TERT, the protein subunit of the telomerase enzyme has also been shown to have non-canonical functions [47–50]. Not much is known about whether these functions are conserved or lineage specific. Additionally, *TERT* transcripts in many animal species, including vertebrates, insects, nematodes and planarians are alternatively spliced [46, 51–55] and spliced variants have also been shown to have non-canonical roles, for example in cell proliferation [56, 57]. *TERT* splicing dysregulation is present in many cancers and it is thought that human *TERT (hTERT)* splicing could be manipulated for therapeutic purposes [58, 59]. While only a little is known about *TERT* alternatively spliced (AS) variants in animals, this represents a potential evolutionary mode for lineage specific life history adaptation, and thus broader assessment of *TERT* alternate splicing is required particularly in lineages spanning short evolutionary time frames.

Given the importance of telomerase function in lineage-specific life history traits, it is surprising that an in depth understanding of the evolutionary history of TERT in the Metazoa is currently lacking from the literature. Here, we make a detailed study of TERT across the Animal Kingdom using extant data to look for evidence (or lack) of patterns of TERT protein and alternative splicing evolution. We identify TERTs across animal phylogeny and perform within and between phyla comparison of the structure of this gene. We are able to confirm the domain content of the ancestral TERT at the base of the Animal Kingdom and then we are able to show dynamic evolution of the protein in different lineages. This includes both the loss of canonical motifs and the invention of lineage specific domains, for example the invention of novel C-terminal extension and N-terminal linker motifs specific to the vertebrate lineage. In some cases we hypothesized that novel motifs may simply compensate for lost canonical motifs, and in others they may relate to lineage specific activities and interactions that remain to be discovered. Next we use available data to assess the levels of alternative splicing, implicated in non-canonical TERT function in humans and described these in different taxa. We find evidence that alternative splicing is likely to be common feature of some lineages, but is not broadly conserved with respect to splicing patterns. Finally, we study the evolution of splicing in one particular related group of animals, the free-living Tricladida planarians [60–62]. We find that it evolves relatively rapidly in this group of highly regenerative animals, suggesting that study of alternative splicing over shorter evolutionary timescales may be required to understand adaptive non-canonical roles of TERT. Taken together our work highlights a previously unappreciated evolutionary and likely functional diversity in this core eukaryotic protein. We conclude that telomere biology, a core synapomorphy of eukaryotes, has undergone significant changes in different lineages and some of this is through the evolution of the TERT protein subunit of telomerase itself.

Results and discussion

TERT distribution across the animal kingdom and reconstruction of the ancestral TERT domain structure

While TERT proteins perform core functions across eukaryotes, not much is known about how conserved or not TERT protein structure is between different metazoan lineages. In order to address this, we set out to define the structure of TERT in the metazoan ancestor and use this as a basis for comparison amongst animals. We searched for TERT orthologs from publicly available genomic and transcriptomic datasets and retrieved 135 *TERT* sequences representative of different metazoans and unicellular eukaryotes. We constructed a maximum-likelihood tree with the TERT sequence from Euglenozoa as outgroup. In many cases but not all, we observed that tree topology reflected the appropriate phylogenetic relationships between phyla (Fig. 1).

TERT is found in unicellular relatives of metazoans and early-branching metazoans

Metazoans are related to unicellular lineages from the holozoan clade: filastereans, ichthyosporeans and choanoflagellates [63–67]. While we could identify *TERT* from *Capsaspora owczarzaki*, we could not find evidence for *TERT* orthologs in either the choanoflagellates *Monosiga brevicollis* or *Salpingoeca rosetta*. Despite this, both *M. brevicollis* [68] and *S. rosetta* [69] have the 5′-TTAGGG(n)-3′ telomeric repeat. Based on this finding, a canonical TERT is predicted to be present [68]. The observed absence could reflect incomplete genome sequences or, more likely given the absence from two different choanoflagellate genomes, that TERT primary sequence has evolved beyond detection or a related reverse transcriptase activity has instead taken up the role of TERT. From the genome sequences of other unicellular parasitic protozoans, putative *TERT* sequences have been predicted [70–77] and our phylogenetic analyses confirms that these predictions are all canonical *TERT* proteins, so it seems likely that loss of a recognisable TERT in choanoflagellates is a derived character rather than the alternative hypothesis of independent evolution of TERT like proteins in different eukaryotic lineages (Fig. 1). *TERT* sequences from early-branching metazoans reported to date were from the genomes of *Hydra magnipapillata* and *Nematostella vectensis* [78, 79]. The genomes of the sponge *Amphimedon queenslandica* [80], comb jelly *Mnemiopsis leidyi* [81], myxosporean *Thelohanellus kitauei* [82] and sea anemone *Aiptasia*

pallida [83] have been sequenced and from these genomes we identified new *TERT* sequences for comparative analysis (Fig. 1; Additional file 1: Table S2). While TERT proteins have been identified in many bilaterian phyla [21–25, 34, 46, 52, 55, 84–94], we additionally identified novel TERT sequences that have not been reported elsewhere. These include TERT orthologs from nematodes species of the classes Secernentea, Chromadorea, Enoplea and Rhabditea and multiple orthologs from Insecta, Actinopterygii, and Mammalia (Additional file 1: Table S2). We also identified novel *TERT* sequences from parasitic platyhelminthes from the Trematoda class, from the newly sequenced tardigrade genomes of *Hypsibius dujardini* [95] and *Ramazzottius varieornatus* [96] and the crustacean *Parhyale hawaiensis* [97] (Additional file 1: Table S2).

TERT is absent from the bdelloid rotifer

Despite the ubiquity of TERT amongst metazoans, its presence is not universal. TERT has been reported to be absent from some animal species; one prominent example being the loss of TERT in Diptera before divergence from Siphonaptera [98]. Like in dipterans, we failed to identify *TERT* in the bdelloid rotifer *Adineta vaga*, suggesting the loss of TERT from this group of long-term asexual species [99, 100]. In the absence of TERT and canonical telomere repeats, *Drosophila melanogaster* uses a retrotransposon-based telomere maintenance in which telomere-specific retrotransposons are reverse transcribed onto chromosome ends [101]. Three non-Long Terminal Repeat (LTR) retrotransposons, *HeT-A* [102, 103], *TART* [104, 105] and *TAHRE* [106] are specifically used for telomere maintenance in this species. The bdelloid chromosome ends also have specialised telomere-associated retrotransposons, such as the *Hebe* transposon [107] and sub-telomeric reverse transcriptase (RT)-related genes [108]. *Penelope*-like elements, which lack the endonuclease domain and are located in sub-telomeric regions and are found in two bdelloid rotifer species *A. vaga* and *Philodina roseola* [109]. Although no evidence yet directly links non-LTR transposons to end chromosome maintenance in *A. vaga*, it seems plausible that in the absence of *TERT*, bdelloid rotifers may employ retrotransposon-based mechanisms for telomere maintenance.

The ancestral structure of TERT in metazoans is likely to possess a complete complement of canonical motifs

The TERT protein contains three canonical domains and 11 motifs: 1) a telomerase essential N-terminal (TEN) domain with the GQ motif, 2) a telomerase RNA-binding (TRBD) domain with the CP, QFP and T motifs and 3) a reverse transcriptase (RT) domain with the 1, 2, A, B′, C, D and E motifs (Fig. 1) [9, 110]. Beyond these

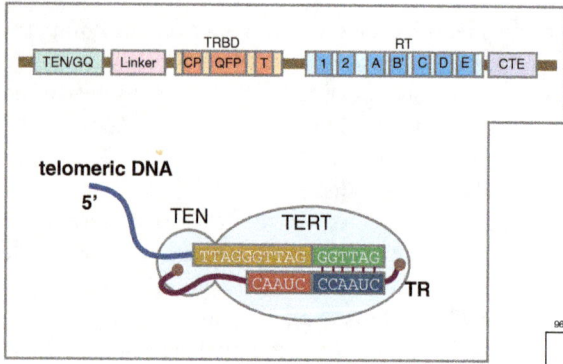

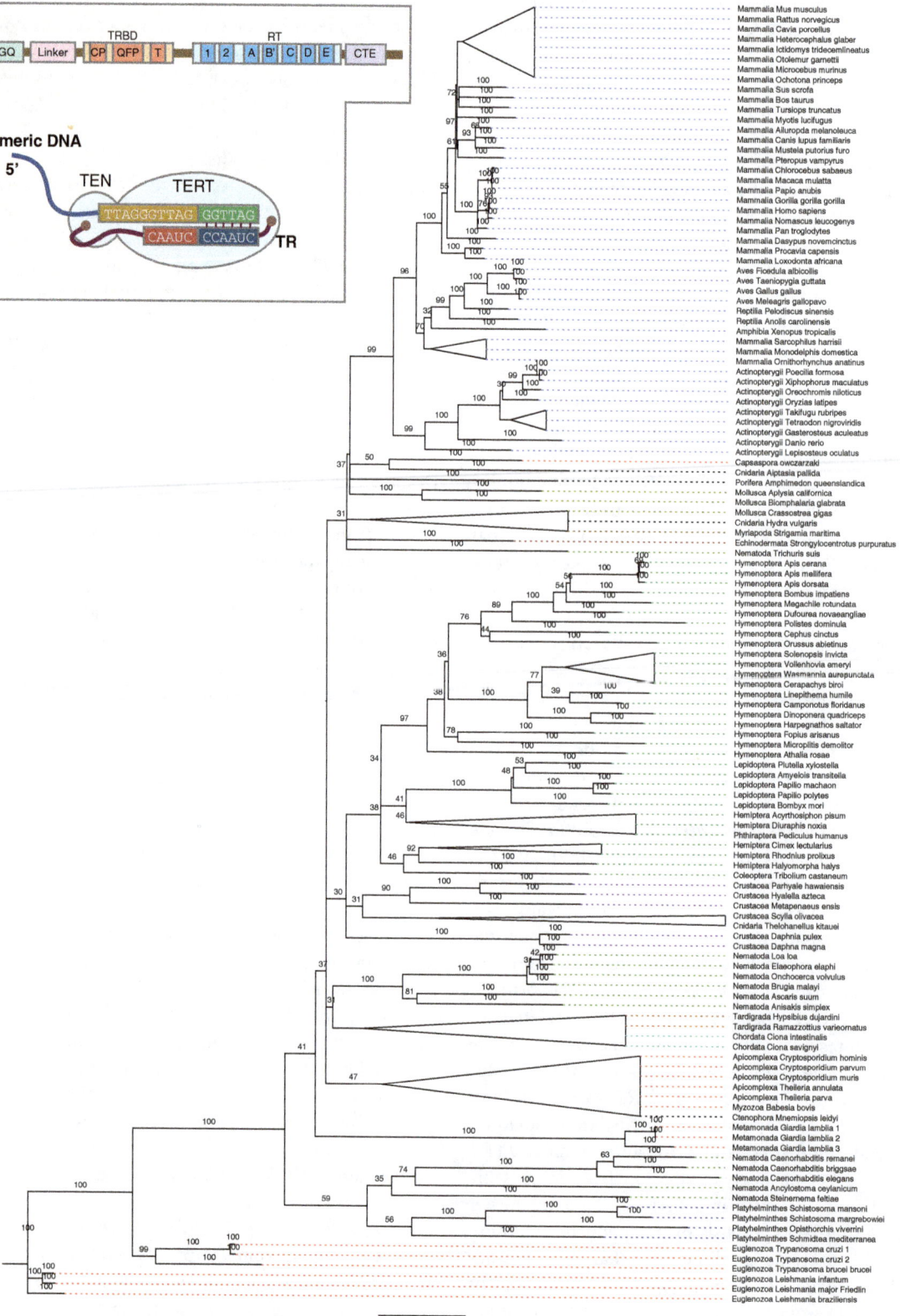

Fig. 1 (See legend on next page.)

(See figure on previous page.)
Fig. 1 Evolution of TERT across the Animal kingdom. Phylogenetic tree of TERT sequences from representative metazoan phyla, early-branching metazoans and unicellular relatives of metazoans constructed using the maximum-likelihood method. Bootstrap support percentages from 1000 replicates are labelled at the nodes. *Scale bar* denotes substitutions per site. Branches that contained less than 30% boostrap replicates were collapsed. Full list of species used and accession numbers can be obtained from Additional file 1: Table S2. Figure inset shows a schematic diagram of the canonical telomerase enzyme bound to the telomeric DNA substrate and the telomerase RNA template. The domain structure of TERT is also shown: the TEN domain with the GQ motif, the TRBD domain with the CP, QFP and T motifs, the RT domain with the 1, 2, A, B', C, D and E motifs, the N-terminal linker and the CTE region

canonical domains and motifs, there is a flexible linker region called the N-terminal linker between the GQ and CP motifs and a C-terminal extension region that starts after the last E motif (Fig. 1) [9]. The T motif allows for correct distinction of TERT family proteins from other prototypical RT proteins and we find that this motif is present in all species used in this study (Fig. 2). To confirm the ancestral metazoan structure of TERT, we performed a cross-species comparison of all TERT proteins (Fig. 2). We inferred that TERT in the metazoan ancestor possesses all 11 canonical motifs outlined above based on the observation that a complete complement of canonical motifs is present in the unicellular filasterean *C. owczarzaki* and the cnidarian *Hydra vulgaris* (Fig. 2). Within Bilateria taxa considered here, only vertebrates and the echinoderm *Strongylocentrotus purpuratus* have retained all 11 motifs

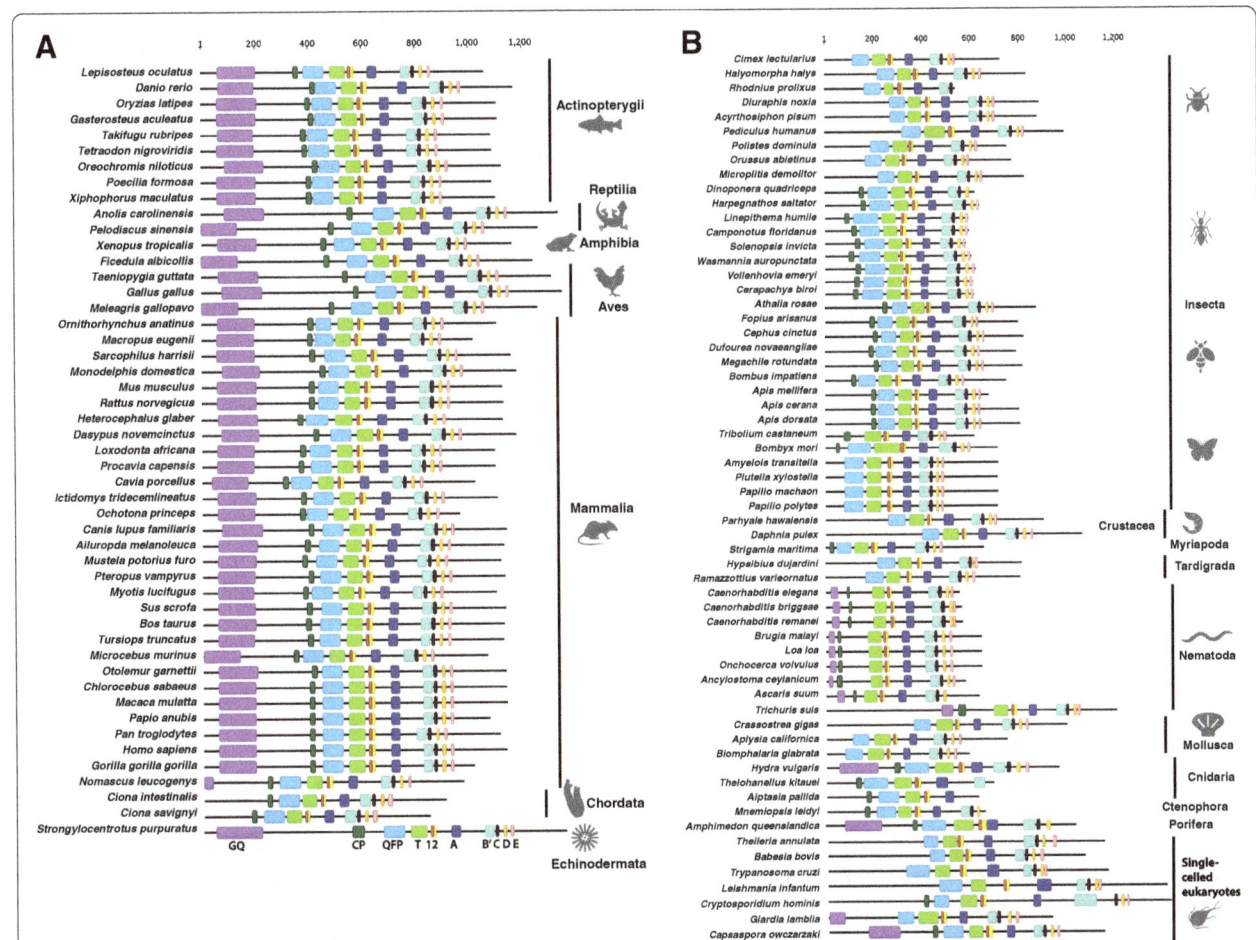

Fig. 2 Taxa-specific modifications of TERT domains. TERT canonical motifs from representative metazoan phyla, early-branching metazoans and unicellular relatives of metazoans were annotated in colour codes: TEN domain GQ motif (*purple*); TRBD domain CP (*dark green*), QFP (*blue*) and T (*light green*) motifs; reverse transcriptase domain motifs 1 (*brown*), 2 (*yellow*), A (*dark blue*), B' (*sky blue*), C (*black*), D (*orange*) and E (*pink*). **a** Deuterostome TERT canonical motifs. **b** Protostome, early-branching metazoans and unicellular eukaryotes canonical motifs. Motif annotations were drawn according to scale based on multiple sequence alignments performed using MAFFT

(Fig. 2), while all protostomes have lost one or more canonical motifs suggesting broad differences in how TERT executes its core function may exist in metazoans.

Taxa-specific divergence of canonical TERT motifs

We next examined evolution of TERT canonical motifs in detail across taxa. Sequence analyses of 11 TERT canonical motifs revealed that the TEN (GQ motif) and TRBD (CP, QFP and T motifs) domains have high sequence conservation across their whole lengths (Additional file 2: Figure S1).

TERT sequence divergence in unicellular relatives of metazoans

Amongst unicellular eukaryotes considered here, only *C. owczarzaki* TERT has all 11 motifs (Fig. 2). An independent study reported that TERT in *Leishmania amazonensis* also contained all 11 motifs [76]. However, from our sequence alignment analyses, we were unable to identify clearly defined GQ motifs in *Leishmania sp.*, *Trypanosoma sp.*, *Theileria sp.*, *Babesia sp.* and *Cryptosporidium sp.* (Additional file 3: Figure S2). The observation of GQ in *L. amazonensis* might be an alignment artefact because from their alignment, only eight out of 244 amino acid residues are found to be in common with other TERTs [76]. We also failed to find the CP motif in this group of unicellular eukaryotes, except for *C. owczarzaki* and *Cryptosporidium sp.* (Additional file 3: Figure S2). It was reported *Giardia lamblia* lacks the T motif [111]. However, this could be due to the use of evolutionarily distant TERT sequences from plants, *Caenorhabditis elegans* and yeast, for the alignments in their earlier study. We retrieved three TERT sequences for the parasitic protozoan *G. lamblia*, obtained from different animal isolates [112]. Sequence alignment revealed that they share only 72% identity. From an alignment of these *G. lamblia* TERTs with other protozoans, we were able to identify a putative T motif, suggesting that it hasn't been lost. The *G. lamblia* T motif, however, has eight extra residues not present in other unicellular eukaryotes (Additional file 3: Figure S2). Apart from *C. owczarzaki* and *G. lamblia*, it appears that the TEN domain has diverged substantially in unicellular eukaryotes with poor conservation to the canonical GQ motif. Within kinetoplastid protozoans from the genera *Leishmania* and *Trypanosoma*, we identified a conserved block of residues, (I/V)QQRVxLQF, between the QFP and T motif. Considering that the divergence time of *Trypanosoma sp.* from other trypanosomatids including *Leishmania sp.* is well over 300 million years ago (m.y.a) [113], it is likely this block of residues in TERT is an example of a lineage specific motif.

TERT sequence divergence in early branching metazoans

Amongst early branching metazoans, TERT sequences from the comb jelly *M. leidyi* and two cnidarians, *A. pallida* and the myxozoan *T. kitauei* appear to be incomplete because their sequences do not extend beyond reverse transcriptase canonical motifs D, B and A respectively (Fig. 2). TERT from the sponge *A. queenslandica* appears to be complete but entirely lacks the E motif. The GQ motif is collectively absent from the comb jelly and cnidarians *T. kitauei* and *A. pallida*, but not from *H. vulgaris*, indicating that multiple independent losses of GQ occurred amongst early branching metazoans (Fig. 2).

TERT sequence divergence in bilaterians

Within Deuterostomia, vertebrates and *Strongylocentrotus purpuratus* have retained all 11 conserved motifs while *Ciona sp.* lack the GQ motif (Fig. 2) [90]. The canonical GQ motif has also been lost from protostomes, except from the nematode lineage (Fig. 2, Additional file 4: Figure S3). In addition to the free-living *Caenorhabditis sp.* [52], we find that other nematodes have also retained the GQ motif (Additional file 4: Figure S3). As previously reported we find that the early branching nematode *Trichinella spiralis* TERT has all 11 canonical TERT motifs [91], so the ancestor of nematode likely possessed all canonical motifs. Studies on human cell lines have shown that within the GQ motif, an additional region known as 'dissociates activities of telomerase' (DAT), is essential for TERT activity [114]. Although nematodes have retained the GQ motif, the DAT region within this motif appears to have lost multiple important residues (Additional file 4: Figure S3). The DAT region was proposed to be involved in other aspects of in vivo telomere elongation because mutations in this region have no direct effect on human TERT multimerization, nuclear targeting or template binding [114]. It appears that either nematodes do not have sensu stricto DAT regions, or that the GQ motif in protostomes, including nematodes, have diverged significantly that sequence conservation with the deuterostome GQ can no longer be detected (Additional file 4: Figure S3). The GQ motif in yeast, human cell lines and *T. thermophila* plays a role in repeat addition processivity [9, 114–117]. It remains to be determined, however, whether the loss or sequence divergence of GQ in protostomes bears any functional implications in species-specific life histories. From the crystal structure of the red flour beetle *Tribolium castaneum* TERT, it was shown that although GQ is absent in this species, remaining motifs (CP and T) in the N-terminal region of TERT not only share high degrees of structural conservation with *T. thermophila* TERT [118, 119] but also conferred in vitro activity [120]. Therefore, even in the absence of a canonical GQ sequence across protostomes (Fig. 2), structural and functional conservation of TERT N-terminal region may persist in these phylogenetic groups. The second TERT canonical motif CP has also been also lost in some protostomes: e.g. in molluscs and tardigrades (Fig. 2). Similarly, most arthropods lack the CP motif, but several exceptions exist: the centipede *Strigamia maritima*, hymenopterans (ants, sawflies, bees but not wasps) and *T. castaneum* [118] have retained the CP motif (Fig. 2; Additional file 5: Figure S4). Other TERT motifs are maintained in metazoan taxa.

We next performed functional divergence analyses on monophyletic clusters from the TERT phylogeny using the DIVERGE programme [121]. A coefficient of functional divergence (θ) value that is significantly different from zero, as indicated by the z-score (MFE z-score) for the model-free method (MFE θ) or the likelihood ratio test (LRT θ) for the maximum likelihood estimate (θ ML), suggests evidence for functional divergence between two clusters. MFE θ and θ ML estimates appeared to be statistically significant for five pairwise comparisons: Euglenozoa/Nematoda (MFE θ = 0.56, *P* value <0.01, θ ML = 0.55, LRT θ = 16.54), Actinopterygii/Euglenozoa (MFE θ = 0.79, *P* value <0.01, θ ML = 0.74, LRT θ = 22.63), Mammalia/Crustacea (MFE θ = 0.54, *P* value <0.01, θ ML = 0.66, LRT θ = 33.55), Mammalia/Euglenozoa (MFE θ = 0.76, *P* value <0.01, θ ML = 0.76, LRT θ = 22.28) and Mammalia/Apicomplexa (MFE θ = 0.61, *P* value <0.01, θ ML = 0.69, LRT θ = 35.25; Additional file 6: Table S1). Future experimental validations will be required to investigate these results in detail.

Identification of novel lineage-specific motifs in TERT

TERT proteins from most organisms possess two highly variable regions that are not part of the 11 canonical motifs: the C-terminal extension (CTE) and the N-terminal linker region [9]. The CTE region starts after the last canonical E motif, and despite poor sequence homology of CTEs between human, *T. castaneum* and *T. thermophila*, it is thought to constitute the RT 'thumb' that is made up of three α-helices crucial in template binding and TERT processivity [118, 122–124]. The mechanistic significance of the CTE has remained elusive until recently, when the CTE was shown to be directly involved in differential binding of DNA and a mutated version of the CTE resulted in defective DNA binding and faster DNA dissociation rates [125].

CTE regions in unicellular relatives of metazoans and early branching metazoans have regions with high conservation

Since the CTE appears to play an important role despite having poor sequence conservation we examined whether this region has evolved to harbour lineage-specific amino acid conservation. We extracted regions after the last canonical E motif from 96 species representing various taxa and investigated these as potential CTEs. Not all species have CTE regions because TERT proteins in some animals do not extend beyond the E motif. We show that unicellular eukaryotes and two early branching metazoan species, *H. vulgaris* and *A. queenslandica* possess CTE regions (Additional file 7: Figure S6A). Alignment of unicellular eukaryotes CTE regions revealed that blocks of conservation exist within different protozoan lineages, although not between unicellular species and human TERT (Additional file 7: Figure S6C). *Crytosporidium sp.* CTE regions are considerably shorter than those of other protozoans. Kinetoplastid protozoans, *Leishmania sp.* and *Trypanosoma sp.*, have a block of conserved residues made up of 48 amino acids at the end of their TERT proteins (Additional file 7: Figure S6C). In other early-branching metazoans, although the parasitic cnidarian *T. kitauei* and the comb jelly *M. leidyi* lack CTE regions, we identified some conserved residues in CTE regions when aligning these regions from *A. queenslandica* and *H.vulgaris* (Additional file 7: Figure S6A).

Taxa specific conservation of CTE regions exist in bilaterians

Within bilaterians, it has been reported that nematodes from the *Caenorhabditis* genus have some of the shortest TERT proteins and appear to have lost their CTE structure [52, 111]. We observe that this is also true for the parasitic roundworm *Ancylostoma ceylanicum* where the C-terminal region beyond the E motif only has 42 amino acid residues (Additional file 7: Figure S6E). We discovered that most parasitic nematodes from the Chromadorea and Secernentea classes possess intact CTE regions (Additional file 7: Figure S6E). In addition, we show that filarial nematodes *B. malayi*, *O. volvulus* and *L. loa* share a RIAVL(R/K)FLKASLLEKYR(M/V motif (Additional file 7: Figure S6E). The observation that a CTE is present in most parasitic nematodes but not the free-living *Caenorhabditis sp.* may reflect adaptation of TERT to their very different life histories; i.e. some parasitic nematodes have long generation times and life expectancies and are able to survive for years in their hosts [126]. CTE structures in Arthropoda varied markedly and we performed separate alignments for insects and crustaceans (Additional file 7: Figure S6D and F). We identified two crustacean specific motifs within the CTE region RL(K/Q)x(I/V) and R(L/F)xAL (Additional file 7: Figure S6F). Within insects, ants lack the CTE completely as TERT proteins within this lineage do not extend beyond the E motif (Additional file 7: Figure S6D). Lepidopterans have additional residues at the extreme ends of their TERT proteins conserved within this lineage only; a lepidopteran-specific CTE contains 32 to 36 residues (Additional file 7:

Figure S6D). Within Deuterostomia, we annotated highly conserved CTE motifs in tetrapods and fishes (Additional file 7: Figure S6G).

Our findings suggest that the ancestor of metazoans possessed a CTE and that while this feature has been largely retained in most animal lineages, it has been lost multiple times. We have demonstrated that these CTE regions can be incredibly diverse across animal phyla, and we show that novel motifs exist in different lineages. It seems likely that these lineage specific motifs in the CTE region would have significant roles, since functional studies in *T. thermophila*, yeast and humans have revealed the importance of CTE for TERT catalytic activity related to enzyme multimerization, processivity and DNA binding [124, 125, 127–129].

Conserved residues within the N-terminal linkers of unicellular relatives of metazoans and early-branching metazoans

We next investigated the N-terminal hypermutable linker (hereafter referred to as the linker) of TERT that is defined as the region between the GQ motif and CP motif. The linker region, although evolutionarily divergent, has been shown to be biologically essential for the function of TERT. In vertebrates, the association between TERT and telomerase RNA (TR) is mediated by the VSR motif in the linker region that binds the activation domain of TR [130, 131]. The puffer fish *Takifugu rubripes* has an additional short alpha-helix upstream of the VSR motif, known as the TFLY motif that is also implicated in binding the template boundary element [119, 132], which is a precise region within the RNA template that is defined by structural elements identified in *T. thermophila*, *Kluyveromyces lactis* and mammals [133–136]. Amongst early-branching metazoans, we found pockets of conservation in the linker region; the motif LxxAIF is present in three out of five early-branching metazoan species (Additional file 8: Figure S7A). Unicellular eukaryotes have very divergent N-terminal linkers between phyla but some conservation within each protozoan lineage, the kinetoplastids (*Leishmania sp.* and *Trypanosoma sp.*) have the ALxR(T/L)(D/N)V(P/S)RxxL motif and the piroplasmids (*Babesia sp.* and *Theileria sp.*) have the LLxGLF motif (Additional file 8: Figure S7C).

Vertebrates have a distinct VL region within their N-terminal linkers

We performed an alignment of the linker regions from representative deuterostome phyla and observed that linker regions have poor conservation and they do not align well. Separately, an alignment of linkers from vertebrates alone revealed that vertebrates share a well-conserved block of 91 amino acid residues directly upstream of the CP motif, which we named the Vertebrata-specific N-terminal linker (VL; Additional file 8: Figure S7F). Within this VL region we noticed that class specific motifs exist. According to the molecular timescale for vertebrate evolution, mammals diverged from birds 310 m.y.a. and marsupials from placental mammals at 173 m.y.a. [137]. Teleosts and tetrapods diverged approximately 450 m.y.a. after the vertebrate genome duplication [138]. Such long evolutionary distances enabled us to distinguish conserved motifs from neutrally evolving sequences. For example, we demonstrated that ray-finned fishes have the FIRTLGFLY and RRxQGxD motif, birds have the NQSL motif and marsupials have the HLFxxKGDP xQQ motif (Additional file 8: Figure S7F). It seems highly likely that these conserved motifs have functional roles that will be important for TERT function in each of these lineages.

Protostomes have truncated N-terminal regions

Extending our analyses to protostomes, we observed that protostomes generally have truncated N-terminal regions. Because arthropods and lophotrochozoans lack the GQ motif altogether, we extracted sequences upstream of the CP motif (for species that have CP motifs) or the QFP motif (for species without the CP motif) as putative linker regions (Additional file 8: Figure S7B, D and E). Insects, including ants, bees and lepidopterans exhibit long stretches of conserved residues (Additional file 8: Figure S7D). Crustaceans from the classes Malacostraca and Branchiopoda share both FPxxHILS and ILxxNxG motifs in their N-terminal linker regions (Additional file 8: Figure S7A). Amongst nematodes, most species with *Trichuris suris* as the exception (Fig. 2), have truncated N-termini upstream of the CP motif. Nevertheless, some conservation exists within this region, for instance nematodes share a IGxxNxxxx(L/V) motif (Additional file 5: Figure S4A). Overall, although the functional significance of the N-terminal linker is unclear beyond vertebrates, it is apparent that the linker sequence can provide additional surface areas to allow conformational flexibility between the N-terminus and the rest of the TERT protein [9]. It is likely that the conserved linker modifications within different lineages confer some but yet unknown essential biological functions.

TERT exhibits conservation of intron-exon structure and pan-metazoan regulation by alternative splicing
Intron-exon structure of TERT in metazoans

TERT genes are the only example of reverse transcriptase-related genes that have defined biological functions and the preservation of exon-intron structure over long evolutionary periods [24, 108]. We wished to assess changes in both intron structure and alternative splicing of *TERT* across animals. We performed cross-phyla comparison of intron positions within the *TERT* gene family. We defined intron positions according to methods described by another study that compared introns across large evolutionary distances [139]. Intron positions were annotated on protein sequences, where data were available, by determining the boundaries of each coding exon. TERT protein sequences with annotated intron positions were aligned and conserved intron positions were defined as shared intron placements at equivalent positions with hTERT that has 15 introns. With the exceptions of *C. owczarzaki* and *Theileria sp.*, *TERT* in unicellular eukaryotes were found to lack introns altogether (Fig. 3). At the base of Eumetazoa, *TERT* from the comb jelly *M. leidyi* and cnidarian *T. kitauei* are also intronless while *H. vulgaris* and *A. queenslandica* have seven and 29 introns respectively. All seven introns in *H. vulgaris* have shared positions with *hTERT* while 14 out of 29 intron positions in *A. queenslandica* are shared with *hTERT*. These data suggest great flexibility in *TERT* intron content early in metazoan evolution (Fig. 3). Within deuterostome taxa, we observed that introns in 41 out of 48 species share

equivalent positions with all 15 *hTERT* introns (Fig. 3a). Many deuterostome *TERTs* have additional introns in positions not conserved with *hTERT*. For example, marsupials have between one to six additional introns in non-conserved positions and the sea urchin *Strongylocentrotus purpuratus* has 11 additional non-conserved introns (Fig. 3a). We observed that *TERTs* in protostomes generally have lower intron numbers (Fig. 3b). Nematodes have nine to 14 introns and they share seven to 10 intron positions with *hTERT* (Fig. 3b). *TERT* in crustaceans are surprisingly intron poor, despite reports that the genome of *Daphnia pulex* has one of the highest intron densities [140]. *D. pulex* and *P. hawaiensis* have only one and three introns respectively, and all introns have shared positions with *hTERT* (Fig. 3b). *TERT* genes in the Hemiptera suborder Sternorrhyncha have undergone unique genomic changes compared to other hemipteran insects; *TERT* in aphids and lepidopterans have lost all introns (Fig. 3b). Within lophotrochozoa, only the pacific oyster is intron rich; the oyster has 27 introns and 14 of these share equivalent positions with *hTERT* (Fig. 3b). Other molluscs are intron poor; *Aplysia*

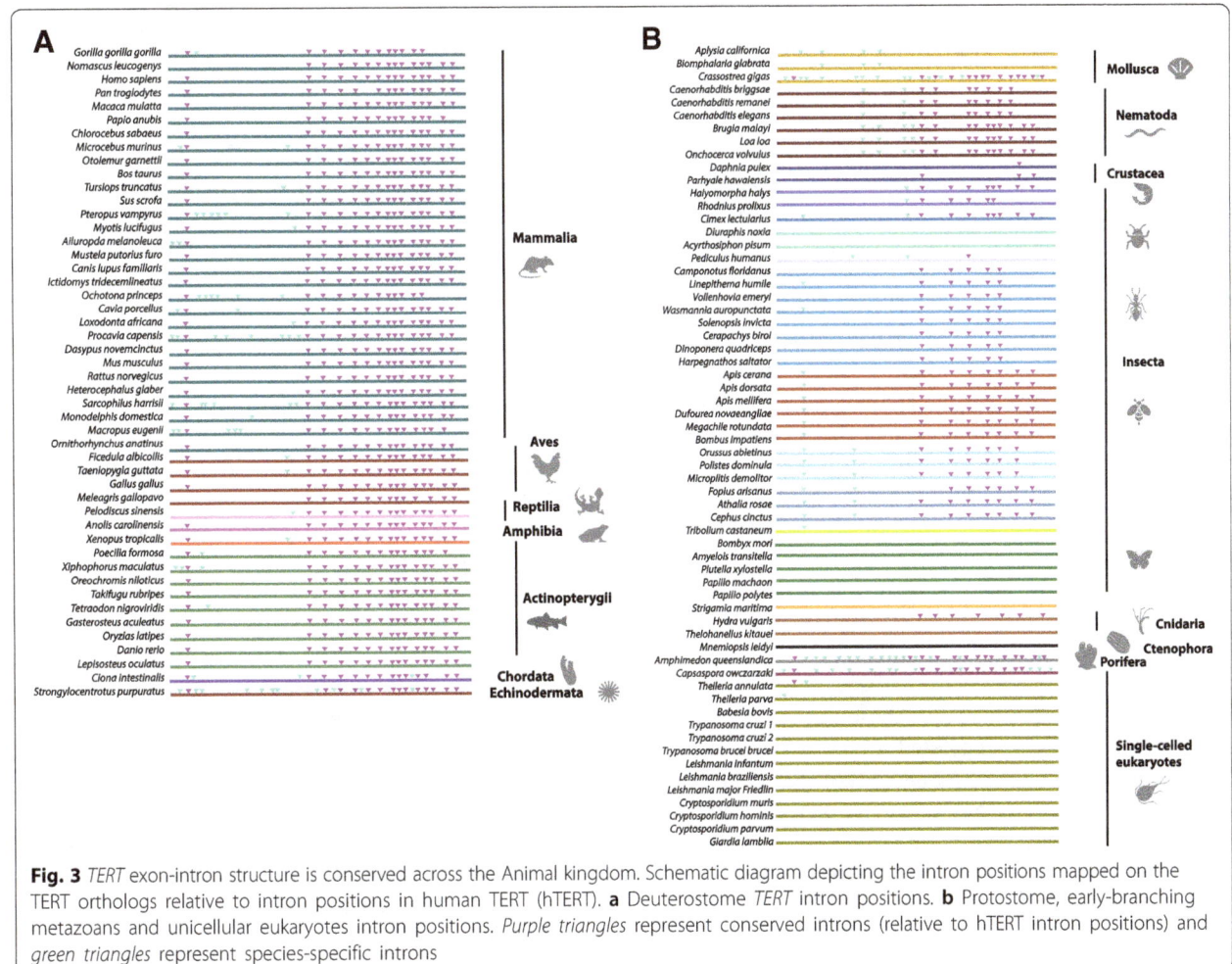

Fig. 3 *TERT* exon-intron structure is conserved across the Animal kingdom. Schematic diagram depicting the intron positions mapped on the TERT orthologs relative to intron positions in human TERT (hTERT). **a** Deuterostome *TERT* intron positions. **b** Protostome, early-branching metazoans and unicellular eukaryotes intron positions. *Purple triangles* represent conserved introns (relative to hTERT intron positions) and *green triangles* represent species-specific introns

californica and *Biomphalaria glabrata* have five and three introns respectively, none of which have positions shared with *hTERT* (Fig. 3b).

Evidence of alternative splicing of TERT in bilaterians

Introns provide the opportunity for creating new gene products via alternative splicing mechanisms. *TERT* alternatively spliced products have been implicated in non-canonical functions in human disease processes, particularly cancer and in adaptation to reproductive mode in planarians flatworms [46]. Alternative splicing has been reported in many other taxa [51–54, 89, 92, 93, 141–144]. We sought to determine whether alternative splicing of TERT is common across animals and if AS variants show any conservation. We probed available genomic and expression datasets for the presence of AS variants. We identified 19 animal species with *TERT* AS variants as evident by computational predictions using RNA-sequencing datasets (Additional file 9: Table S3; Additional file 10). We found AS variants in both protostomes (10 species) and deuterostomes (9 species). From this list, AS variants have been previously described for these five species: *Homo sapiens* [51, 56, 141, 144], *Ornithorhynchus anatinus* [93], *Gallus gallus* [54, 142], *Danio rerio* [89, 143] and *Solenopsis invicta* [92]. We observed extensive patterns of alternative splicing in hymenopteran and hemipteran insects (Additional file 11: Figure S8). Some of the most common splicing events in hymenopterans and hemipterans involved the exclusion of exon 2 (6 out of 9 species) and exon 3 (4 out of 9 species; Additional file 11: Figure S8). All 43 AS variants in insects have retained an open reading frame (ORF; Additional file 12: Figure S9). Of these, 11 have truncated TRBD (motifs CP, QFP and T) domains and nine have truncated RT (motifs 1, 2, A, B′, C, D and E) domains (Additional file 11: Figure S8). The substantial lengths and complete ORFs of many AS variants in insects indicate that they may still be functional at the protein level. In Lophotrochozoa, two AS variants were identified in the mollusc *A. californica* (Additional file 12: Figure S9). Since the combination of exons 2 and 3 only encode 39 amino acid residues, all AS variants in *A. californica* appear to retain a complete set of motifs like the full-length TERT (Additional file 12: Figure S9). Within Arthropoda, aphids and lepidopterans lack any introns and hence alternative splicing (Fig. 3). The short-lived nature of these taxa may preclude the need for highly regulated TERT activity.

Amongst vertebrate species, many AS variants have been previously described but to our knowledge, splicing patterns of *TERTs* in the Amazon molly *Poecilia formosa,* the green spotted puffer fish *Tetraodon nigroviridis*, the wild turkey *Meleagris gallopavo* and the Rhesus macaque *Macaca mulatta* have not been reported (Additional file 13: Figure S10). AS variants

in *hTERT* involved exons 6 to 8 and we witness the splicing events concerning these exons in five out of nine species (3 mammals, the wild turkey and zebrafish) (Additional file 13: Figure S10), suggesting some conservation of splicing across the vertebrates. The exclusion of the unusually large exon 2 (1.3 kb) is a common 5′ splice variant in primate *TERTs* [55] and we observe this splicing event in all vertebrates except the Rhesus macaque (Additional file 13: Figure S10). The lack of observation of the AS variant in Rhesus Macaque may be due to lack of confirmatory data. Overall, many AS variants in mammals and fishes appear to be of substantial lengths, suggesting that they would encode functional proteins given that most of their TRBD and RT motifs are still present (Additional file 9: Table S3). In avians however, many AS variants appear to lack any canonical motifs. The wild turkey has 10 AS variants, eight of these involve the exclusion of multiple exons and/or loss of frame mutations, do not contain any TERT motifs and are therefore likely to be non-coding and be subjected to nonsense-mediated decay (Additional file 13: Figure S10B). This observation of complex splicing patterns generating many short AS variants with premature termination codon has also been reported in chickens [54, 142] (Additional file 13: Figure S10B). It is unknown as to whether these AS variants would participate directly or indirectly in canonical telomerase function or in other non-canonical functions of TERT. But clearly alternative splicing of the TERT protein is a pan-metazoan phenomenon. Given that we do not see broad conservation of splicing patterns (except within the vertebrate lineage) our data suggest, that in combination with the lineage specific sequence evolution we have also described, AS is likely to contribute to lineage specific functions of this core eukaryotic protein.

Free-living planarian species exhibit complex patterns of alternative splicing

While our analysis of alternative splicing on TERT revealed extensive evidence of splicing, data available to assess evolutionary changes of splicing in taxa outside of the vertebrates is relative sparse. A previous study had investigated telomere length, telomerase activity and TERT regulation in the highly regenerative model planarian *Schmidtea mediterranea* [46]. This study found that the asexually reproducing biotype upregulated active full length splice forms of *TERT* during regeneration to sufficient levels to maintain telomere length during stem cell division. Another study in the related species *Dugesia ryukyuensis* demonstrated equivalent telomere length maintenance differences between sexual and asexual strains [145]. In order to use planarians as a study system for understanding how AS could contribute to the evolution of TERT regulation we performed a detailed investigation of *TERT* gene structure, expression and alternative splicing in Tricladida planarians, a group where most, but not all species, are known to be highly regenerative.

Planarian-specific motifs within the N-terminal linkers and CTE regions of TERT

We obtained five new partial *TERT* sequences from published transcriptome sources or previous studies [46, 146–151]. Based on these sequences, we designed putative pan-flatworm degenerate primers and using a combination of degenerate PCR together with 5′ and 3′ RACE techniques, we cloned full-length *TERT* sequences from 11 Tricladida planarian species. These included ten Continenticola species (a clade consisting of freshwater triclads from the family Dendrocoelidae, Dugesiidae and Planariidae) and one Maricola species (a suborder of marine triclads from the family Procerodidae; Fig. 4a & b; Additional file 14: Table S4). We also confirmed both the identity of each species and phylogenetic relationships amongst taxa using 18S rDNA resequencing (Additional file 15: Figure S11; Additional file 14: Table S4) [60–62]. We observed that like other lophotrochozoans (*A. californica*, *Biomphalaria glabata* and *Crassostrea gigas*), TERT in triclad planarians lack the GQ and CP motifs. (Fig. 4c; Additional file 16: Figure S12). We retrieved six partial sequences for other platyhelminthes from the following classes: Trematoda (*Schistosoma mansoni*, *S. margrebowiei*, *Echinostoma caproni; Opisthorchis viverrini*) [152, 153], Cestoda (*Echinococcus multilocularis*) [154] and Turbellaria (*Macrostomum lignano*) [155]. Since these sequences are incomplete, we only used species with intact CTE or N-terminal linker regions for comparison. Although we failed to compare N-termini of parasitic flatworms, we were able to identify planarian-specific N-terminal motifs within this region such as the I(K/D)xKC and PIYK motifs (Fig. 4d). With regards to CTE regions, there is poor conservation between molluscs and platyhelminthes and they require separate alignments. Within platyhelminthes CTE regions, we observed pockets of conservation, e.g. the P(Y/F)KGQMKWIL and LSR(F/Y)(A/S)GxP(Q/R)NA motifs in triclad planarians (Fig. 4e). The VTLKVNNQL and YE(V/L)AL(K/D)SLKxTL motifs are conserved within Dugesiidae (Fig. 4e).

TERT intron-exon structure in planarians and the prevalence of alternative splicing

The observation of multiple PCR products when using primers that span the entire *TERT* transcript in planarians was suggestive of *TERT* alternative splicing (Fig. 4b). We wished to investigate the extent of the regulation by alternative splicing in this group of animals that exist

over short evolutionary time frames. It was previously reported that *TERT* in *S. mediterranea* is alternatively spliced into four variants [46]. Analysis of *TERT* intron-exon boundaries using existing genome assembly datasets for *S. mediterranea* [149] and *Girardia tigrina* (unpublished data) revealed that *Smed-TERT* and *Gtig-TERT* have 16 and 15 exons respectively. Of these, 11 *Smed-TERT* and 10 *Gtig-TERT* introns are conserved with *hTERT* introns where they share the same positions on the TERT protein (Fig. 4f). As planarian *TERTs* shared a high degree of structural similarity and due to the lack of genomic data for the nine remaining planarian species, we based our AS variant analysis on an inferred genomic structure generated from the *Smed-TERT* sequence. From the splicing patterns of cloned AS variants, we could visualise exons that are being excluded and this confirmed that the intron-exon boundaries were accurately inferred based on *Smed-TERT* sequence (Additional file 17: Figure S13; Additional file 18: Figure S14; Additional file 19). AS variants were identified from all seven Dugesiidae species, two Planariidae species and one Dendrocoelidae species Additional file 17: Figure S13; Additional file 18: Figure S14A and B). We did not find any AS variants from the marine planarian *Procerodes plebeja* (Procerodidae), an outgroup to all other species (Additional file 18: Figure S14C). The highest number of AS variants was identified in *G. tigrina* (13 variants) with the lowest being *Dendrocoelum lacteum* (2 variants; Additional file 17: Figure S13; Additional file 18: Figure S14). Two of the shortest AS variants in *G. tigrina* did not contain any TERT motifs (Additional file 17: Figure S13). Dugesiidae species in general had more AS variants compared to Dendrocoelidae and Planariidae species. Five AS variants in *D. ryukyuensis* only have partial QFP motifs and not any other TERT motifs (Additional file 17: Figure S13). Of all 52 AS variants in triclad planarians, the most prevalent pattern is the truncation of the RT domain where 24 out of 52 have partial or missing RT domains (Additional file 17: Figure S13; Additional file 18: Figure S14; Additional file 20: Table S5). We observed 20 out of 52 AS variants containing truncated versions of both TRBD and RT domains and only 7 out of 52 contain a truncated TRBD but intact RT domain (Additional file 20: Table S5; Additional file 17: Figure S13; Additional file 18: Figure S14). Despite sharing conserved positions of introns (as inferred from splicing patterns of each species), there is low degree of conservation in the selection of spliced sites for these species. The most prevalent splicing pattern is the deletion of exons 8 to 10 (represented by orange triangles), which is present in six out of ten species where AS variants were found (Additional file 17: Figure S13; Additional

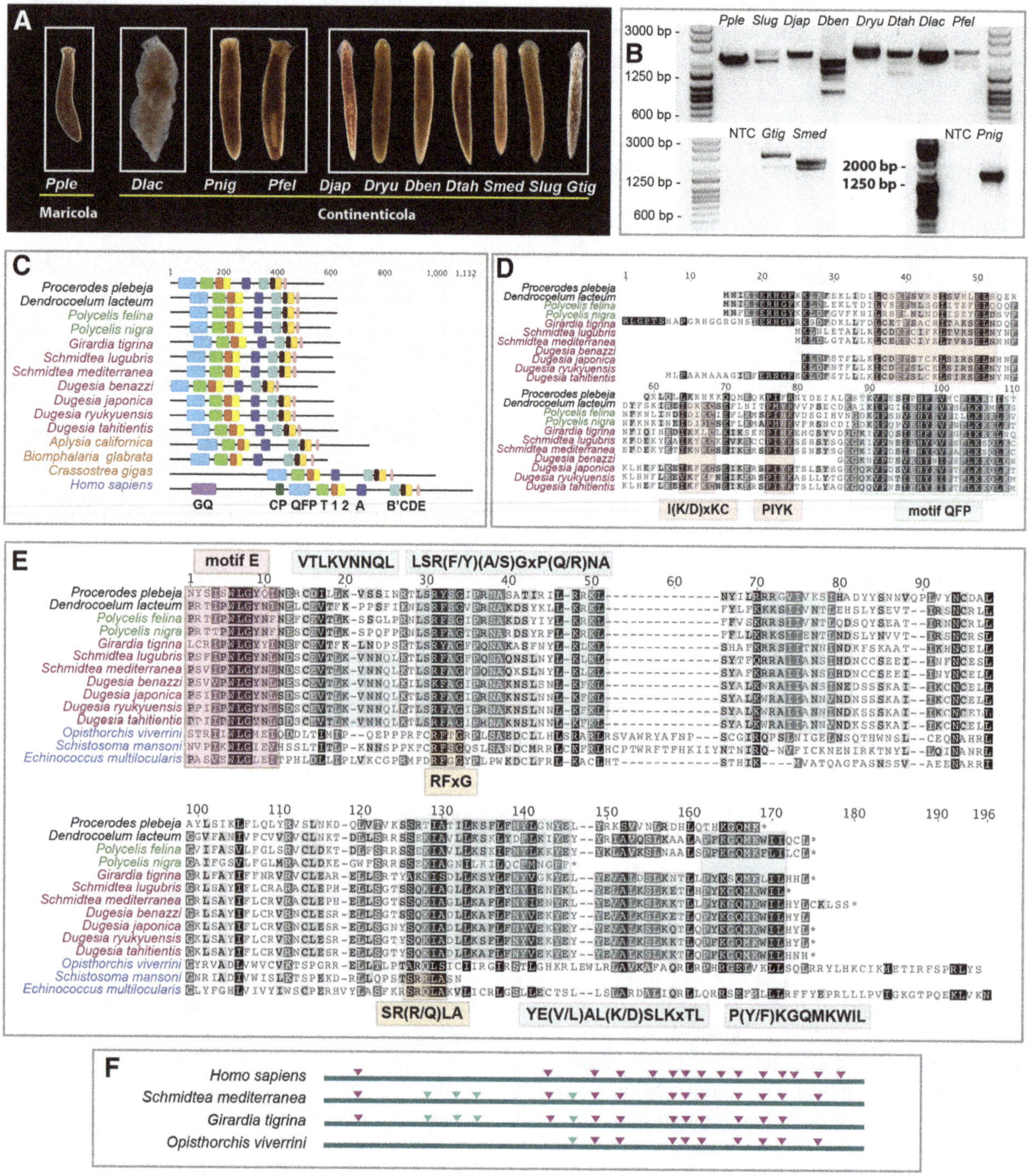

Fig. 4 *TERT* in highly regenerative triclad planarians. **a** The 11 flatworm species used in this study. Abbreviations represent: *Pple (Procerodes plebeja)*, *Dlac (Dendrocoelum lacteum)*, *Pnig (Polycelis nigra)*, *Pfel (Polycelis felina)*, *Djap (Dugesia japonica)*, *Dryu (Dugesia ryukyuensis)*, *Dben (Dugesia Benazzi)*, *Dtah (Dugesia tahitientis)*, *Smed (Schmidtea mediterranea)*, *Slug (Schmidtea lugubris)* and *Gtig (Girardia tigrina)*. **b** Gel image showing full length RT-PCR results of *TERT* transcripts and isoforms from 11 planarian species (NTC represents no cDNA template control for the PCR). **c** Domain structure of planarian TERTs; mollusc TERTs and hTERT were used for comparison. Multiple sequence alignments of the (**d**) N-terminal linker and (**e**) C-terminal extensions from planarians and the trematodes. *Boxes* also denote planarian-specific motifs with their respective descriptions written inside. **f** *TERT* gene structure of *hTERT*, *S. mediterranea*, *G. tigrina* and the liver fluke *Opisthorchis viverrini* (Trematoda Class) are shown. *Purple triangles* indicate conserved intron positions (relative to *hTERT* intron positions) and *green triangles* indicate Platyhelminthes-specific introns. Genomic sequence data is not available for other planarian species. Therefore, putative *TERT* exon-intron boundaries were annotated based on *Smed_TERT* data and confirmed by AS variant analyses of each species

file 18: Figure S14). Deletion of exons 11 and 12 (represented by blue triangles) were identified in five out of ten species. These splicing events would affect the *TERT* RT domain that is encoded by exons 8 to 13 (Additional file 17: Figure S13; Additional file 18: Figure S14). We note that the least amount of *TERT* splicing is observed in the least regenerative species, *P. plebeja* and *D. lacteum* (Additional file 21: Figure S15; Additional file 22: Figure S16) [156]. The prevalence of *TERT* splicing and regenerative capacity is broadly correlative at this stage and future functional studies will be required to ascertain whether any direct linkage between splicing and regenerative capacity exists. Overall our data suggest that AS splicing of TERT is evolving rapidly within this group of animals and that both fine grained study of *TERT* alternative splicing as well as functional study of AS variants are required to understand why *TERT* has evolved so dynamically during animal evolution.

Conclusion

We performed thorough analyses of TERT canonical and non-canonical motifs, gene structure and alternative splicing across representative metazoan species. We show that although the ancestral TERT protein in Metazoa is likely to possess all 11 canonical motifs, the GQ and CP motifs are prone to lineage specific losses (Fig. 2). Beyond the canonical motifs, we demonstrate that the N-terminal linkers and CTE regions of TERTs are highly divergent across phyla. In these regions, we discovered novel motifs that exhibit high levels of sequence conservation over long evolutionary times indicating that they may serve an unknown but important biological function. For example, the CTE regions are distinct within phyla (Additional file 7: Figure S6) and it seems likely that such differences are reflective of life history diversity as growing evidence points to the importance of the CTE for telomerase processivity in vivo [123, 124, 157]. We also demonstrated that the N-terminal linker regions have poor sequence conservation between metazoan phyla (Additional file 8: Figure S7). The N-terminal linkers of TERT proteins have been implicated to mediate a conserved function in enhancing the translocation of the 3′ end of DNA substrate relative to the RNA template [158, 159]. Deleting residues in this region reduces the translocation of DNA substrate and overall processivity [115]. Future studies will be needed to unravel the biological significance of phyla specific sequences and whether they fine tune telomerase processivity to life history strategy.

Analysis on *TERT* intron exon structure revealed that the metazoan ancestor is intron rich and share many intron positions with *TERTs* in the vertebrate lineage (Fig. 3). *TERTs* in protostomes have undergone significant intron loss; lepidopterans and aphids do not have any introns, perhaps due to short lifespans of these taxa. Alternative splicing of TERT has been extensively studied in many species. We report new observations on splicing events for 14 metazoan species and show that the selection of spliced exons is poorly conserved in these animals (Additional file 11: Figure S8; Additional file 12: Figure S9; Additional file 13: Figure S10). Lastly, we investigated *TERT* alternative splicing in Tricladida planarians and demonstrated that splicing is dynamic and rapidly evolving within this group of closely related species (Fig. 4, Additional file 17: Figure S13; Additional file 18: Figure S14). It is likely that splicing in planarian *TERTs* is significant for the highly regenerative and potentially immortal life histories of some of these species and AS variants could possess important non-canonical functions. This study highlights that neither *TERT* structure nor its regulation is static. A future endeavour will be to unravel mechanistic linkages that connect unique sequence evolution and the dynamic regulation of TERT to actual biological roles.

Methods

Data mining of TERT sequences

TERT sequences were retrieved from Ensembl Genome and Ensembl Metazoa [160], NCBI and Uniprot. Sequences were confirmed using reciprocal blastx against the nr database using the NCBI Blast suite. We retrieved tardigrade TERTs from the *Hypsibius dujardini* genome browser (http://badger.bio.ed.ac.uk/H_dujardini/blast/index) and the *Ramazzottius varieornatus gene models* (http://kumamush-i.org/database.html). TERT sequence from the crustacean *Parhyale hawaiensis* was retrieved from the assembled genome [97]. Additional references for other datasets are included in the main text. Complete species names, accession numbers and database sources are listed in Additional file 1: Table S2.

Phylogenetic analyses

Multiple sequence alignment of TERT protein sequences (from motif T to motif E) was performed using MAFFT with the following parameters: scoring matrix BLOSUM62, gap penalty 1.53, offset value 0.123 [161]. Phylogenetic tree was constructed using RAxML [162] employing the WAG matrix and gamma distribution rate model with 1000 bootstrap replicates. A single tree topology was generated based on the best-scoring maximum likelihood tree. The tree Figure was made using FigTree (http://tree.bio.ed.ac.uk/software/figtree/). Branches that contained less than 30% boostrap replicates were collapsed.

Annotation of TERT canonical motifs and non-canonical motifs in C-terminal extension (CTE) and N-terminal linker regions

TERT protein motifs (GQ, CP, QFP, T, 1, 2, A, B′, C, D and E) were annotated based on regional comparisons with hTERT protein annotations in multiple sequence alignments performed using MAFFT with the following parameters: scoring matrix BLOSUM62, gap penalty 1.53, offset value 0.123 [161]. For the alignments of canonical domains, we used published domain annotations from two reference sequences (hTERT and *C. elegans* TERT). For the alignment of TERT domains in unicellular relatives of metazoans (Additional file 3: Figure S2) and insects (Additional file 5: Figure S4), the hTERT sequence was used as a reference while for nematodes, the *C. elegans* TERT sequence was used as a reference (Additional file 4: Figure S3). Global alignments of full-length TERT sequences were first performed on the target taxa to allow the annotation of regions that align well. Regions that do not align well were extracted and local alignments were then performed with the reference sequence. For example, we identified the GQ motif in unicellular relatives of metazoans from a local alignment of the N-terminal region (Additional file 3: Figure S2). For CTE and linker regions, local alignments were performed only on selected taxa for the identification of non-canonical motifs. Fasta files for all alignments performed in this study are included (Additional files 23, 24, 25, 26, 27, 28, 29, 30, 31, 32, 33, 34, 35, 36, 37, 38, 39, 40, 41, 42, 43, 44, 45, 46, 47, 48, 49, 50, 51, 52). Alignment Figures were generated using Geneious (version 7) [163].

Animal culture

All freshwater flatworm strains were cultured at 20 °C using the planarian water formulation. 1X Montjuic salt solution was prepared using milliQ ddH2O with the following composition: 1.6 mM NaCl, 1 mM $CaCl_2$, 1 mM $MgSO_4$, 0.1 mM $MgCl_2$, 0.1 mM KCl, 1.2 mM $NaHCO_3$ [164]. The marine flatworm *Procerodes plebeja* was cultured at 14 °C using a salt water formulation made with Tropic Marin sea salt to a salinity of 28-30 ppm. *Dendrocoelum lacteum* was fed with shrimp while all other flatworms used in this study were fed organic beef liver once a week. All flatworms were starved for 1 week prior to any experimental procedures. The worms were kept in the dark at all times apart from feeding and water changing times.

RNA extraction and reverse transcription in Triclad flatworms

Total RNA was isolated from three animals of each species using the TRIzol reagent (Thermo Fisher Scientific) according to manufacturer's instructions. cDNA was generated from total RNA extract using the QuantiTect reverse transcription kit (Qiagen) containing the RT primer mix (blend of oligo-dT and random hexamers). A genomic DNA elimination step was also included in this kit performed using the gDNA wipeout buffer.

Degenerate polymerase chain reaction (PCR) and 5' 3' RACE for the cloning of *TERT* genes

TERT transcript sequences for *Girardia tigrina* [148], *Dugesia japonica* [147], *D. lacteum,* and *P. nigra* [151] were retrieved using tblastn with *Smed_TERT* as a query from published transcriptome sequences of the respective species. Top hits with the best e-value for each species was used for reciprocal blast against the nr database in NCBI to confirm the identification of *TERT*. For degenerate PCR, first strand cDNA synthesis was performed as mentioned previously. For 5' and 3' RACE, first strand cDNA synthesis was performed on TRIzol extracted total RNA using the SMARTer RACE 5' 3' Kit (Clontech).

For the cloning of *TERT* in all the other *Dugesia sp.* (*Dugesia ryukyuensis, Dugesia tahitientis, Dugesia benazzi*) degenerate PCR primers were designed based on the pairwise alignment of *Smed-TERT* and *Djap-TERT*: DugesiaFD (5'-CACTGGTGYGAATCACCA-3') and DugesiaRD (5'- AAAACATCATCMACRTATTG-3'). PCR products were cloned in the pGEMT-easy vector (Promega) followed by colony PCR with M13F (5'- GTAAAACGACGGCCAGT-3') and M13R (5'-GGAAACAGCTATGACCATG-3') primers. Positive colonies were selected for Sanger sequencing. Once the middle sequence of *TERT* transcripts for all three *Dugesia sp.* were obtained, 5' and 3' RACE was performed on cDNA made with the SMARTer RACE 5' 3' Kit using gene specific primers listed in Additional file 14: Table S4 according to manufacturer's instructions.

For the cloning of *TERT* in *Schmidtea lugubris*, the same set of degenerate PCR primers (DugesiaFD and DugesiaRD) was used for PCR followed by cloning and sequencing. PCR products were cloned in the pGEMT-easy vector (Promega) followed by colony PCR and sequencing as mentioned previously. The 5' end of *Slug_-TERT* transcript was cloned using the following primer set: Dugesia5F (5'-AATYGAGMGWMATGGTTT-3'; this primer was designed based on 5' sequence region in *Dugesia sp.*) and LugR (5'- CTGAAATTTGTGCC ATTG-3'; *Slug-TERT* gene specific primer). Next, 3' RACE was performed with the SMARTer RACE 5' 3' Kit using gene specific primers listed in Additional file 14: Table S4 according to manufacturer's instructions.

For the cloning of *TERT* in *Polycelis felina*, degenerate PCR primers were designed based on pairwise alignment of *Pnig-TERT* and *Polycelis tenuis TERT* (obtained from PlanMine) [151]: PolycelisDF (5'-AATTGGCACM TSTTYCTG-3') and PolycelisDR (5'-GACTCRTAR-CAAYTCTT-3'). PCR products were cloned in the pGEMT-easy vector (Promega) followed by colony PCR and sequencing as mentioned previously. The 5' and 3' end of *Pfel-TERT* was obtained using 5' and 3' RACE with the SMARTer RACE 5' 3' Kit using gene specific primers listed in Additional file 14: Table S4 according to manufacturer's instructions.

For the cloning of *TERT* in *P. plebeja*, degenerate PCR primers were designed based on pairwise alignment of *Pnig_TERT* and *Pfel_TERT*: PpleDF (5′- ATGSARTA TAAAGGATWTAT-3′) and PpleDR (5′-AAYRTCRTC GACATATTG-3′). PCR products were cloned in the pGEMT-easy vector (Promega) followed by colony PCR and sequencing as mentioned previously. The 5′ and 3′ end of *Pple-TERT* was obtained using 5′ and 3′ RACE with the SMARTer RACE 5′ 3′ Kit using gene specific primers listed in Additional file 14: Table S4 according to manufacturer's instructions. All degenerate PCRs were performed using the Advantage 2 Polymerase (Clontech) according to recommended thermal cycling parameters. All RACE products were ran on a gel and gel extraction was performed with the MinElute gel extraction kit (Qiagen) following by cloning and sequencing of the RACE products.

DNA extraction, 18S rDNA PCR and sequencing
High molecular weight genomic DNA (gDNA) was isolated using the phenol-chloroform method followed by ethanol precipitation. PCR was performed according to conditions in Carranza et al. [61] using the Advantage 2 Polymerase (Clontech). List of 18S primer sequences are provided in Additional file 14: Table S4. PCR products were gel extracted using the MinElute gel extraction kit (Qiagen) and eluted in 10uL of molecular grade water. Gel extracted products were sequenced in both directions using 18S nested primers listed in Additional file 14: Table S4. Sequences were aligned with known 18S sequence data available from NCBI (accession numbers provided in Additional file 14: Table S4). 18S rDNA sequences obtained from this study fell into two groups (18S type I and 18S type II).

Cloning of planarian *TERT* AS variants
PCR primers were designed to amplify from the start to end of *TERT* transcript sequences obtained from above. Full list of primers used to amplify *TERT* AS variants are provided in Additional file 14: Table S4. PCR was performed using Phusion Polymerase (Thermo Fisher Scientific) and all PCR products were ran on a gel. In order to cut out gel pieces containing the AS variants, the gel image was overexposed to enable visualization of low abundance variants. Gel regions containing bands were excised and extracted using the MinElute gel extraction kit (Qiagen). A-tailing was subsequently performed with GoTaq Polymerase (Promega) for TA-cloning purposes since Phusion PCR products were blunt ended. PCR products were cloned in the pGEMT-easy vector (Promega) followed by colony PCR with M13F and M13R primers. Colony PCR products were

run on a gel for 2 h using 1.3% agarose in 1X TAE buffer to allow good resolution of bands. Colony PCR gel would display an array of bands with different sizes representing the AS variants. Bands of different sizes were selected for Sanger sequencing. For each species, 25 colonies were selected for sequencing to allow exhaustive identification of *TERT* AS variants.

Sequence analyses of flatworm *TERT* AS variants
Eleven planarian species were used in this study. Planarians *Schmidtea mediterranea*, *S. lugubris*, *G. tigrina*, *D. japonica*, *D. benazzi*, *D. tahitientis*, *P. felina*, *P. nigra* and *D. lacteum* were purchased from Sciento (http://www.sciento.co.uk/). *D. ryukyuensis* was a gift from Dr. *Midori Matsumoto (Keio University, Japan) and P. plebeja* from Dr. Bernhard Egger (Universität *Innsbruck)*. The gene structures (exon-intron boundaries) for *Smed-TERT* and *Gtig-TERT* were identified using blastn against the *S. mediterranea* [149] and *G. tigrina* genome assembly (unpublished data). Both *Smed-TERT* and *Gtig-TERT* have almost identical intron positions, with *Gtig-TERT* having one less exon. Since genome data is not available for all the other flatworm species, we based our analyses on the gene structure of *Smed-TERT* and *Gtig-TERT* and refer to them as inferred exon positions. Further analyses on cloned AS variants for species without genomic data confirmed the accurate positioning of intron-exon boundaries based on the excluded exons. The full-length *TERT* sequences for each species were referred to as 'isoform 1'. Cloned AS variants were mapped to 'isoform 1', which set as a reference sequence for each respective species. From this, we were able to visualize the skipped exons, splice site mutations or retained introns by comparing the AS variant sequences to the reference sequence. All the AS variants including the longest isoform were in silico translated using Geneious (version 7) [163] to the correct open reading frame. Multiple sequence alignments of all AS protein sequences for each respective species were performed using MAFFT with the following parameters: scoring matrix BLOSUM62, gap penalty 1.53, offset value 0.123 [161]. Annotations of TERT canonical motifs were performed according to positions in hTERT; telomerase RNA-binding domain motifs (QFP and T) and canonical reverse transcriptase motifs (1, 2, A, B′, C, D and E) in all AS protein variants. Further information on the flatworm AS variants were provided in Additional file 20: Table S5.

Annotation of *TERT* AS variants in vertebrates, insects and mollusc
The annotation of *TERT* AS variants were performed based on publically available transcriptomic and genomic datasets (Additional file 1: Table S2). To evaluate alternatively spliced exons, we have identified 19 species that contained more than one *TERT* transcripts in the transcriptomes and these additional *TERT* transcripts

were considered as *TERT* AS variants. These *TERT* transcripts were in silico translated according to the correct open reading frame into protein sequences. Subsequent analyses were performed as described in the previous section "Sequence analyses of flatworm *TERT* AS variants". The longest transcripts/isoforms were used as reference sequences and all other AS variants were mapped to the reference using Geneious (version 7) [163]. This allowed the visualization of skipped exons, splice site mutations or retained introns by comparing the AS variant sequences to the reference sequence. Detailed descriptions on splicing patterns are available in Additional file 9: Table S3.

Annotation of intron positions on TERT protein sequences

Methods for the identification of conserved intron positions were adapted from two previous reports [139, 165]. The locations of the exon boundaries were first annotated on TERT transcripts based on their cognate genomic sequences (schematic diagram in Additional file 53: Figure S17). Next, the transcripts were in silico translated to protein sequences while retaining information on the location of each exon boundary (also referred here as intron positions). TERT protein sequences were independently aligned with the hTERT sequence. Pairwise alignments with hTERT were performed to allow the generation of a schematic diagram that is scaled according to the hTERT sequence. From the pairwise alignments, intron positions that overlap with hTERT introns were defined as conserved introns, while intron positions that do not overlap with hTERT introns were defined as species-specific introns (Additional file 53: Figure S17).

Additional files

Additional file 1: Table S2. TERT sequences from representative species and their corresponding accession numbers. (XLSX 57 kb)

Additional file 2: Figure S1. Graphs showing sequence conservation of TERT canonical motifs: TEN domain (GQ), TRBD domain (CP, QFP and T), and RT domain (1, 2, A, B', C, D and E). The y-axes represent frequency of the most-abundant amino acid residue at any given position with a frequency of 1.0 indicating 100%. The x-axes represent the most-abundant amino acid residue at any given position across the length of the motif. (PDF 115 kb)

Additional file 3: Figure S2. Multiple sequence alignments of the (A) TEN domain (GQ motif) and TRBD domain (CP, QFP, T) and (B) RT domain (1, 2, A, B', C, D and E) from unicellular relatives of metazoans. Areas highlighted in coloured boxes delimit the canonical motifs. Boxes also denote species- or class-specific motifs with their respective descriptions written inside. Except for *Capsaspora*, the protozoan orders are represented by colour codes in the figure inset. (PDF 2723 kb)

Additional file 4: Figure S3. Multiple sequence alignments of the (A) GQ and CP motifs and (B) T motif and RT domain (1, 2, A, B', C, D and E) from nematodes. Areas highlighted in coloured boxes delimit the canonical motifs. Boxes also denote Species- or Class-specific motifs with their respective descriptions written inside. Nematode classes are represented by colour codes in the figure inset. (PDF 2291 kb)

Additional file 5: Figure S4. Multiple sequence alignments of the (A) TRBD domain (CP, QFP and T) and (B) and RT domain (1, 2, A, B', C, D and E) from insects. Areas highlighted in coloured boxes delimit the canonical motifs. Boxes also denote Species- or order-specific motifs with their respective descriptions written inside. Insect orders are represented by colour codes in the figure inset. (PDF 3367 kb)

Additional file 6: Table S1. Analyses of functional divergence between gene clusters based on model-free and maximum likelihood estimates. (XLSX 31 kb)

Additional file 7: Figure S6. Multiple sequence alignments of the C-terminal extensions (CTEs) from (A) early-branching metazoans, (B) molluscs, (C) unicellular relatives of metazoans, (D) insects, (E) nematodes, (F) crustaceans and (G) vertebrates. The last TERT canonical motif E is highlighted in pink boxes to illustrate the start of CTE regions. Boxes also denote phylum-, order- or class-specific motifs with their respective descriptions written inside. (PDF 6781 kb)

Additional file 8: Figure S7. Multiple sequence alignments of the N-terminal linker regions from (A) early-branching metazoans, (B) crustaceans, (C) unicellular relatives of metazoans, (D) insects, (E) molluscs and (F) vertebrates. Boxes denote phylum-, order- or class-specific motifs with their respective descriptions written inside. (PDF 2999 kb)

Additional file 9: Table S3. Alternatively spliced variants from vertebrates, insects and mollusc. (XLSX 49 kb)

Additional file 10: Fasta file of TERT alternatively spliced variants from representative metazoan species. (FASTA 184 kb)

Additional file 11: Figure S8. *TERT* alternative splicing in Hymenopteran and Hemipteran insects. *TERT* AS variants data were obtained from Ensembl Genomes or NCBI. These are generated by computational predictions based on RNA sequencing read evidence for (A) hymenopterans and (B) hemipterans. Schematic diagram depicts gene structure and splicing of *TERT*. Orange boxes represent exons conserved with *hTERT*. Blue boxes represent Hymenoptera specific exons while green boxes represent Hemiptera specific exons. Gray boxes represent skipped or deleted exons resulted from alternative splicing events. 'M' denotes splice site mutations, deletions or intron retention. The left margin shows *TERT* gene and AS variant names for each species and the right margin shows descriptive names of *TERT* AS sequences. The numbers in parentheses represent the length of splice site mutations (deletion or intron retention) for the respective AS variants. Schematic diagrams on the far right illustrate the presence or absence of canonical motifs (CP, QFP, T, 1, 2, A, B', C, D and E) on TERT AS protein variants drawn to scale. (PDF 774 kb)

Additional file 12: Figure S9. *TERT* alternative splicing in mollusc. Schematic diagram depicts gene structure and *TERT* AS variants. Yellow boxes represent *Aplysia californica* exons. Gray boxes represent skipped exons resulted from alternative splicing events. 'M' denotes splice site mutations, deletions or intron retention. The left margin shows *TERT* gene and AS variant names for each species and the right margin shows descriptive names of *TERT* AS sequences. The numbers in parentheses represent the length of splice site mutations (deletion or intron retention) for the respective AS variants. Schematic diagrams on the far right illustrate the presence or absence of canonical motifs (QFP, T, 1, 2, A, B', C, D and E) on TERT AS protein variants drawn to scale. (PDF 178 kb)

Additional file 13: Figure S10. *TERT* alternative splicing in vertebrates. AS variants are shown for (A) mammals, (B) birds and (C) fishes. Only the main AS variants for *hTERT* and chicken *TERT* are shown in this diagram. Additional variants have been reported in Hrdličková et al. [93]. The presence of AS variants with skipped exon 2 is conserved across different vertebrate species. AS variants with spliced exons 7 to 8 are conserved in mammals except in *G. gorilla gorilla*. Mammalian *TERT* has 16 exons whereas *TERT*s in non-mammalian vertebrates have anywhere between 14 to 17 exons. *TERT* AS variants data were obtained from Ensembl Genomes or NCBI. These are generated by computational predictions based on RNA sequencing read evidence. Schematic diagram depicts gene structure and splicing of *TERT*. Orange boxes represent exons conserved with *hTERT*. Purple and green boxes represent species-specific exons. Grey boxes represent skipped or deleted exons resulted from alternative splicing events. 'M' denotes splice site mutations, deletions or intron retention. The left margin shows *TERT* gene and AS variant names for each species and the right margin shows descriptive names of *TERT* AS sequences. The numbers in parentheses represent the length of splice site mutations (deletion or intron retention) for the respective AS variants except for platypus (*O. anatinus*) and chicken (*G. gallus*). For these two species, the numbers in parentheses represent the position of deletion in the exon indicated. Schematic diagrams on the far right illustrate the presence or absence of canonical motifs (GQ, CP, QFP, T, 1, 2, A, B', C, D and E) on TERT AS protein variants drawn to scale. (PDF 334 kb)

Additional file 14: Table S4. Primers for 18S rDNA PCR and sequencing, 5' and 3' RACE and amplification of TERT alternatively spliced variants in triclad planarians. (XLSX 53 kb)

Additional file 15: Figure S11. Phylogenetic tree illustrating the positions of planarian type I and type II sequences in the 18S rDNA phylogeny. Sequences of 18S rDNA were cloned from 11 flatworm species in the lab for comparison against published sequences in GenBank. (A) Type I 18S rDNA sequences with species marked in blue as cloned in the lab. (B) Type II 18S rDNA sequences with species marked in brown as cloned in the lab. Asterisks indicate novel 18S rDNA sequences obtained from this study. (PDF 139 kb)

Additional file 16: Figure S12. Alignment of canonical motifs of TERT from 11 planarian species, three molluscs and hTERT. These include TRBD domain (QFP and T motifs) and canonical reverse transcriptase motifs (1, 2, A, B', C, D and E). Clearly defined GQ and CP motifs were not identified from planarian TERTs. (PDF 1901 kb)

Additional file 17: Figure S13. *TERT* in Dugesiidae is alternatively spliced. Comparison of alternatively spliced (AS) variants in seven Dugesiidae species from the genera *Schmidtea*, *Girardia* and *Dugesia*. Structure of AS variants in *S. mediterranea*, *S. lugubris*, *G. tigrina*, *D. japonica*, *D. ryukyuensis*, *D. benazzi* and *D. tahitientis*. Full-length or wild-type *TERT* structure is shown at the top of each set of AS variants. Functional TERT domains (TRBD and RT) and positions of (putative) exons are indicated. Deletions (skipped exons) are denoted by triangles and insertions (retained introns or splice site mutations) are denoted by grey rectangles. 'M' abbreviations represent splice site mutations. Red triangles represent conserved alternatively spliced exons in all Dugesiidae species except *S. mediterranea* and *D. tahitientis*. Blue triangles represent conserved alternatively spliced exons in all Dugesiidae species except *G. tigrina* and *D. japonica*. Asterisks indicate stop codon positions caused by frame shift mutation or retained introns. The left margin shows *TERT* gene and AS variant names for each species and the right margin shows descriptive names of *TERT* AS sequences. Schematic diagrams on the far right illustrate the presence or absence of canonical motifs (QFP, T, 1, 2, A, B', C, D and E) on TERT AS protein variants drawn to scale. (JPEG 1654 kb)

Additional file 18: Figure S14. *TERT* in Planariidae and Dendrocoelidae but not Procerodidae is alternatively spliced. Comparison of AS variants in two Planariidae species from the genus *Polycelis*, one Dendrocoelidae species and one Procerodidae species. Structure of AS variants in (A) Planariidae *P. felina* and *P. nigra*, (B) Dendrocoelidae *D. lacteum* and (C) Procerodidae *P. plebeja*. Genomic sequence data is not available for these flatworm species. Therefore, putative *TERT* exon-intron boundaries were annotated based on *Smed_TERT* sequence and confirmed by AS variant analyses of each species. Full-length or wild-type *TERT* structure is shown at the top of each set of AS variants. Functional TERT domains (TRBD and RT) and positions of (putative) exons are indicated. Deletions (skipped exons) are denoted by triangles and insertions (retained introns or splice site mutations) are denoted by gray rectangles. Red triangles represent alternatively spliced exons conserved with Dugesiidae. Asterisks indicate stop codon positions caused by frame shift mutation or retained introns. The left margin shows *TERT* gene and AS variant names for each species and the right margin shows descriptive names of *TERT* AS sequences. Schematic diagrams on the far right illustrate the presence or absence of canonical motifs (QFP, T, 1, 2, A, B', C, D and E) on TERT AS protein variants drawn to scale. (JPEG 673 kb)

Additional file 19: Fasta file of planarian TERT alternatively spliced variants. (FASTA 85 kb)

Additional file 20: Table S5. TERT alternatively spliced variants and respective features in triclad planarians. (XLSX 47 kb)

Additional file 21: Figure S15. Regenerative capacity in planarians. Transverse sections assaying regenerative abilities along anterior-posterior axis of planarians. Ten worms were used for each species and were amputated into three fragments each (head, trunk and tail). Representative images depict time-course observation of the same animal undergoing cephalic regeneration in (A) trunk and (B) tail fragments. Anterior blastema (unpigmented region) was imaged. Appearances of pigmented eyes were noted and plotted as frequency graphs. Scale bars indicate 0.5 mm. (JPEG 3354 kb)

Additional file 22: Figure S16. Head regeneration frequency graph for (A) trunk fragments and (B) tail fragments of amputated flatworms (*N* = 10 for each species). Cartoon depicts the amputation planes used. The timing of eye appearance in cohorts of 10 fragments was noted. *P. plebeja* failed to regenerate at all and is not shown. (PDF 182 kb)

Additional file 23: Multiple sequence alignment of the DAT and GQ motifs in unicellular relatives of metazoans. (FASTA 737 bytes)

Additional file 24: Multiple sequence alignment of the CP motif in unicellular relatives of metazoans. (FASTA 264 bytes)

Additional file 25: Multiple sequence alignment of the QFP and T motifs in unicellular relatives of metazoans. (FASTA 4 kb)

Additional file 26: Multiple sequence alignment of the motifs 1 and 2 in unicellular relatives of metazoans. (FASTA 2 kb)

Additional file 27: Multiple sequence alignment of the A motif in unicellular relatives of metazoans. (FASTA 998 bytes)

Additional file 28: Multiple sequence alignment of the B' motif in unicellular relatives of metazoans. (FASTA 1 kb)

Additional file 29: Multiple sequence alignment of the C and D motifs in unicellular relatives of metazoans. (FASTA 1 kb)

Additional file 30: Multiple sequence alignment of the E motif in unicellular relatives of metazoans. (FASTA 641 bytes)

Additional file 31: Multiple sequence alignment of the DAT, GQ and CP motifs in nematodes. (FASTA 2 kb)

Additional file 32: Multiple sequence alignment of the T, 1, 2, A, B', C, D and E motifs in nematodes. (FASTA 4 kb)

Additional file 33: Multiple sequence alignment of the CP motif in insects. (FASTA 620 bytes)

Additional file 34: Multiple sequence alignment of the QFP and T motifs in insects. (FASTA 3 kb)

Additional file 35: Multiple sequence alignment of the 1, 2, A, B', C, D and E motifs in insects. (FASTA 6 kb)

Additional file 36: Multiple sequence alignment of the CTE region in early branching metazoans. (FASTA 308 bytes)

Additional file 37: Multiple sequence alignment of the CTE region in molluscs. (FASTA 925 bytes)

Additional file 38: Multiple sequence alignment of the CTE region in unicellular relatives of metazoans. (FASTA 6 kb)

Additional file 39: Multiple sequence alignment of the CTE region in insects. (FASTA 5 kb)

Additional file 40: Multiple sequence alignment of the CTE region in nematodes. (FASTA 1 kb)

Additional file 41: Multiple sequence alignment of the CTE region in crustaceans. (FASTA 1 kb)

Additional file 42: Multiple sequence alignment of the CTE region in deuterostomes. (FASTA 8 kb)

Additional file 43: Multiple sequence alignment of the N-terminal linker region in early branching metazoans. (FASTA 627 bytes)

Additional file 44: Multiple sequence alignment of the N-terminal linker region in crustaceans. (FASTA 714 bytes)

Additional file 45: Multiple sequence alignment of the N-terminal linker region in kinetoplastids. (FASTA 1 kb)

Additional file 46: Multiple sequence alignment of the N-terminal linker region in *C. owczarzaki*. (FASTA 313 bytes)

Additional file 47: Multiple sequence alignment of the N-terminal linker region in piroplasmids. (FASTA 455 bytes)

Additional file 48: Multiple sequence alignment of the N-terminal linker region in ants. (FASTA 1019 bytes)

Additional file 49: Multiple sequence alignment of the N-terminal linker region in bees. (FASTA 954 bytes)

Additional file 50: Multiple sequence alignment of the N-terminal linker region in lepidopterans. (FASTA 345 bytes)

Additional file 51: Multiple sequence alignment of the N-terminal linker region in molluscs. (FASTA 159 bytes)

Additional file 52: Multiple sequence alignment of the N-terminal linker region in vertebrates. (FASTA 3 kb)

Additional file 53: Figure S17. Schematic diagram demonstrating the annotation of intron positions in pairwise alignments with the hTERT sequence as a reference. Amino acid residues are highlighted in black (amino acids that are conserved amongst all sequences), grey (amino acids with similar properties) and light grey (amino acids that are dissimilar). (PDF 134 kb)

Acknowledgements
We thank Dr. Hanae Nodono (Kagoshima University, Japan) for her contribution on TERT identification in planarians. We thank Dr. Midori Matsumoto (Keio University, Japan) for the gift of *Dugesia ryukyuensis*. We thank Dr. Bernhard Egger (Universität Innsbruck) for the gift of *Procerodes plebeja*.

Funding
This work was supported by the Biotechnology and Biological Sciences Research Council (grant number BBK0075641); Human Frontier Science Program Postdoctoral Fellowship (to A.G.L.); University of Oxford Elizabeth Hannah Jenkinson Grant (to A.G.L. and S.S.) and Clarendon Scholarship (to S.S.).

Authors' contributions
AGL and AAA conceived and designed the experiments. AGL, NP-S, AD, GK and SS performed the experiments. AGL and AAA analysed the data. AGL and AAA wrote the paper with help from all the other authors. All authors read and approved the final manuscript.

Competing interests
The authors declare that they have no competing interests.

References
1. Muller HJ. The production of mutations by X-rays. Proc Natl Acad Sci U S A. 1928;14:714–26.
2. Harley CB, Futcher AB, Greider CW. Telomeres shorten during ageing of human fibroblasts. Nature. 1990;345:458–60.
3. De Lange T. Telomere-related genome instability in cancer. Cold Spring Harb Symp Quant Biol. 2005;70:197–204.
4. Watson JD. Origin of concatemeric T7 DNA. Nat New Biol. 1972;239:197–201.
5. Olovnikov AM. A theory of marginotomy. The incomplete copying of template margin in enzymic synthesis of polynucleotides and biological significance of the phenomenon. J Theor Biol. 1973;41:181–90.
6. Hayflick L, Moorhead PS. The serial cultivation of human diploid cell strains. Exp Cell Res. 1961;25:585–621.
7. Harley CB, Villeponteau B. Telomeres and telomerase in aging and cancer. Curr Opin Genet Dev. 1995;5:249–55.
8. Rudolph KL, Chang S, Lee HW, Blasco M, Gottlieb GJ, Greider C, et al. Longevity, stress response, and cancer in aging telomerase-deficient mice. Cell. 1999;96:701–12.
9. Autexier C, Lue NF. The structure and function of telomerase reverse transcriptase. Annu Rev Biochem. 2006;75:493–517.
10. Blackburn EH, Greider CW, Szostak JW. Telomeres and telomerase: the path from maize, Tetrahymena and yeast to human cancer and aging. Nat Med. 2006;12:1133–8.
11. Tomás-Loba A, Flores I, Fernández-Marcos PJ, Cayuela ML, Maraver A, Tejera A, et al. Telomerase reverse transcriptase delays aging in cancer-resistant mice. Cell. 2008;135:609–22.
12. Bernardes de Jesus B, Vera E, Schneeberger K, Tejera AM, Ayuso E, Bosch F, et al. Telomerase gene therapy in adult and old mice delays aging and increases longevity without increasing cancer. EMBO Mol Med. 2012;4:691–704.
13. Counter CM, Avilion AA, LeFeuvre CE, Stewart NG, Greider CW, Harley CB, et al. Telomere shortening associated with chromosome instability is arrested in immortal cells which express telomerase activity. EMBO J. 1992;11:1921–9.
14. Vaziri H, Dragowska W, Allsopp RC, Thomas TE, Harley CB, Lansdorp PM. Evidence for a mitotic clock in human hematopoietic stem cells: loss of telomeric DNA with age. Proc Natl Acad Sci U S A. 1994;91:9857–60.
15. Bryan TM, Englezou A, Gupta J, Bacchetti S, Reddel RR. Telomere elongation in immortal human cells without detectable telomerase activity. EMBO J. 1995;14:4240–8.
16. Blackburn EH, Gall JG. A tandemly repeated sequence at the termini of the extrachromosomal ribosomal RNA genes in Tetrahymena. J Mol Biol. 1978;120:33–53.
17. Szostak JW, Blackburn EH. Cloning yeast telomeres on linear plasmid vectors. Cell. 1982;29:245–55.
18. Greider CW, Blackburn EH. Identification of a specific telomere terminal transferase activity in Tetrahymena extracts. Cell. 1985;43:405–13.
19. Greider CW, Blackburn EH. The telomere terminal transferase of Tetrahymena is a ribonucleoprotein enzyme with two kinds of primer specificity. Cell. 1987;51:887–98.
20. Greider CW, Blackburn EH. A telomeric sequence in the RNA of Tetrahymena telomerase required for telomere repeat synthesis. Nature. 1989;337:331–7.
21. Kilian A, Bowtell DD, Abud HE, Hime GR, Venter DJ, Keese PK, et al. Isolation of a candidate human telomerase catalytic subunit gene, which reveals complex splicing patterns in different cell types. Hum Mol Genet. 1997;6:2011–9.
22. Harrington L, Zhou W, McPhail T, Oulton R, Yeung DS, Mar V, et al. Human telomerase contains evolutionarily conserved catalytic and structural subunits. Genes Dev. 1997;11:3109–15.
23. Meyerson M, Counter CM, Eaton EN, Ellisen LW, Steiner P, Caddle SD, et al. hEST2, the putative human telomerase catalytic subunit Gene, is up-regulated in tumor cells and during immortalization. Cell. 1997;90:785–95.
24. Nakamura TM, Morin GB, Chapman KB, Weinrich SL, Andrews WH, Lingner J, et al. Telomerase catalytic subunit homologs from fission yeast and human. Science. 1997;277:955–9.
25. Lingner J, Hughes TR, Shevchenko A, Mann M, Lundblad V, Cech TR. Reverse transcriptase motifs in the catalytic subunit of telomerase. Science. 1997;276:561–7.
26. Nugent CI, Lundblad V. The telomerase reverse transcriptase: components and regulation. Genes Dev. 1998;12:1073–85.
27. Nakayama J, Tahara H, Tahara E, Saito M, Ito K, Nakamura H, et al. Telomerase activation by hTRT in human normal fibroblasts and hepatocellular carcinomas. Nat Genet. 1998;18:65–8.
28. Rubio MA, Davalos AR, Campisi J. Telomere length mediates the effects of telomerase on the cellular response to genotoxic stress. Exp Cell Res. 2004;298:17–27.
29. Suram A, Herbig U. The replicometer is broken: telomeres activate cellular senescence in response to genotoxic stresses. Aging Cell. 2014;13:780–6.
30. Bischoff C, Petersen HC, Graakjaer J, Andersen-Ranberg K, Vaupel JW, Bohr VA, et al. No association between telomere length and survival among the elderly and oldest old. Epidemiology. 2006;17:190–4.
31. Kimura M, Cherkas LF, Kato BS, Demissie S, Hjelmborg JB, Brimacombe M, et al. Offspring's leukocyte telomere length, paternal age, and telomere elongation in sperm. PLoS Genet. 2008;4:e37.
32. Steenstrup T. Hjelmborg J v B, Kark JD, Christensen K, Aviv a. The telomere lengthening conundrum–artifact or biology? Nucleic Acids Res. 2013;41:e131.
33. Stone RC, Horvath K, Kark JD, Susser E, Tishkoff SA, Aviv A. Telomere length and the cancer–atherosclerosis trade-off. Fisher EMC, editor. PLoS Genet. 2016;12:e1006144.
34. Kim EB, Fang X, Fushan AA, Huang Z, Lobanov AV, Han L, et al. Genome sequencing reveals insights into physiology and longevity of the naked mole rat. Nature. 2011;479:223–7.
35. Haussmann MF, Winkler DW, Huntington CE, Nisbet ICT, Vleck CM. Telomerase activity is maintained throughout the lifespan of long-lived birds. Exp Gerontol. 2007;42:610–8.
36. Blasco MA, Lee H-W, Hande MP, Samper E, Lansdorp PM, DePinho RA, et al. Telomere shortening and tumor formation by mouse cells lacking telomerase RNA. Cell. 1997;91:25–34.
37. Samper E, Flores JM, Blasco MA. Restoration of telomerase activity rescues chromosomal instability and premature aging in Terc−/− mice with short telomeres. EMBO Rep. 2001;2:800–7.

38. Blasco MA. Telomeres and human disease: ageing, cancer and beyond. Nat Rev Genet. 2005;6:611–22.

39. Jaskelioff M, Muller FL, Paik J-H, Thomas E, Jiang S, Adams AC, et al. Telomerase reactivation reverses tissue degeneration in aged telomerase-deficient mice. Nature. 2011;469:102–6.

40. Charnov EL, Berrigan D. Evolution of life history parameters in animals with indeterminate growth, particularly fish. Evol Ecol. 1991;5:63–8.

41. Klapper W, Heidorn K, Kühne K, Parwaresch R, Krupp G. Telomerase activity in "immortal" fish. FEBS Lett. 1998;434:409–12.

42. Reznick D, Ghalambor C, Nunney L. The evolution of senescence in fish. Mech Ageing Dev. 2002;123:773–89.

43. Owen R, Sarkis S, Bodnar A. Developmental pattern of telomerase expression in the sand scallop, Euvola ziczac. Invertebr Biol. 2007;126:40–5.

44. Ebert TA, Russell MP, Gamba G, Bodnar A. Growth, survival, and longevity estimates for the rock-boring sea urchin Echinometra lucunter Lucunter (Echinodermata, Echinoidea) in Bermuda. Bull Mar Sci. 2008;82:381–403.

45. Klapper W, Kühne K, Singh KK, Heidorn K, Parwaresch R, Krupp G. Longevity of lobsters is linked to ubiquitous telomerase expression. FEBS Lett. 1998;439:143–6.

46. Tan TCJ, Rahman R, Jaber-Hijazi F, Felix DA, Chen C, Louis EJ, et al. Telomere maintenance and telomerase activity are differentially regulated in asexual and sexual worms. Proc Natl Acad Sci U S A. 2012;109:4209–14.

47. Smith LL, Coller HA, Roberts JM. Telomerase modulates expression of growth-controlling genes and enhances cell proliferation. Nat Cell Biol. 2003;5:474–9.

48. Parkinson EK, Fitchett C, Cerezer B. Dissecting the non-canonical functions of telomerase. Cytogenet Genome Res. 2008;122:273–80.

49. Choi J, Southworth LK, Sarin KY, Venteicher AS, Ma W, Chang W, et al. TERT promotes epithelial proliferation through transcriptional control of a Myc- and Wnt-related Developmental program. PLoS Genet. 2008;4:e10.

50. Low KC, Tergaonkar V. Telomerase: central regulator of all of the hallmarks of cancer. Trends Biochem Sci. 2013;38:426–34.

51. Ulaner GA, Hu JF, Vu TH, Giudice LC, Hoffman AR. Telomerase activity in human development is regulated by human telomerase reverse transcriptase (hTERT) transcription and by alternate splicing of hTERT transcripts. Cancer Res. 1998;58:4168–72.

52. Meier B, Clejan I, Liu Y, Lowden M, Gartner A, Hodgkin J, et al. Trt-1 is the Caenorhabditis elegans catalytic subunit of telomerase. PLoS Genet. 2006;2:e18.

53. Sýkorová E, Fajkus J. Structure-function relationships in telomerase genes. Biol Cell. 2009;101:375–92.

54. Nehyba J, Bose HR Jr, Hrdličková R. The regulation of telomerase by alternative splicing of TERT. INTECH open access Publisher; 2012.

55. Withers JB, Ashvetiya T, Beemon KL. Exclusion of Exon 2 is a common mRNA splice variant of primate telomerase reverse Transcriptases. Tian B, editor. PLoS One. 2012;7:e48016.

56. Hrdlickova R, Nehyba J, Bose HR. Alternatively spliced telomerase reverse transcriptase variants lacking telomerase activity stimulate cell proliferation. Mol Cell Biol. 2012;32:4283–96.

57. Listerman I, Sun J, Gazzaniga FS, Lukas JL, Blackburn EH. The major reverse transcriptase-incompetent splice variant of the human telomerase protein inhibits telomerase activity but protects from apoptosis. Cancer Res. 2013;73:2817–28.

58. Wong MS, Chen L, Foster C, Kainthla R, Shay JW, Wright WE. Regulation of telomerase alternative splicing: a target for chemotherapy. Cell Rep. 2013;3:1028–35.

59. Wong MS, Wright WE, Shay JW. Alternative splicing regulation of telomerase: a new paradigm? Trends Genet. 2014;30:430–8.

60. Riutort M, Field KG, Raff RA, Baguñà J. 18S rRNA sequences and phylogeny of platyhelminthes. Biochem Syst Ecol. 1993;21:71–7.

61. Carranza S, Giribet G, Ribera C, Baguñà J, Riutort M. Evidence that two types of 18S rDNA coexist in the genome of Dugesia (Schmidtea) mediterranea (Platyhelminthes, Turbellaria, Tricladida). Mol Biol Evol. 1996;13:824–32.

62. Alvarez-Presas M, Baguñà J, Riutort M. Molecular phylogeny of land and freshwater planarians (Tricladida, Platyhelminthes): from freshwater to land and back. Mol Phylogenet Evol. 2008;47:555–68.

63. Lang BF, O'Kelly C, Nerad T, Gray MW, Burger G. The closest unicellular relatives of animals. Curr Biol. 2002;12:1773–8.

64. Ruiz-Trillo I, Roger AJ, Burger G, Gray MW, Lang BF. A Phylogenomic investigation into the origin of Metazoa. Mol Biol Evol. 2008;25:664–72.

65. Shalchian-Tabrizi K, Minge MA, Espelund M, Orr R, Ruden T, Jakobsen KS, et al. Multigene phylogeny of Choanozoa and the origin of animals. Aramayo R, editor. PLoS One. 2008;3:e2098.

66. Torruella G, Derelle R, Paps J, Lang BF, Roger AJ, Shalchian-Tabrizi K, et al. Phylogenetic relationships within the Opisthokonta based on Phylogenomic analyses of conserved single-copy protein domains. Mol Biol Evol. 2012;29:531–44.

67. Suga H, Chen Z, de Mendoza A, Sebé-Pedrós A, Brown MW, Kramer E, et al. The Capsaspora genome reveals a complex unicellular prehistory of animals. Nat Commun. 2013;4:2325.

68. Robertson HM. The choanoflagellate Monosiga Brevicollis karyotype revealed by the genome sequence: telomere-linked helicase genes resemble those of some fungi. Chromosom Res. 2009;17:873–82.

69. Fairclough SR, Chen Z, Kramer E, Zeng Q, Young S, Robertson HM, et al. Premetazoan genome evolution and the regulation of cell differentiation in the choanoflagellate Salpingoeca Rosetta. Genome Biol. 2013;14:R15.

70. Xu P, Widmer G, Wang Y, Ozaki LS, Alves JM, Serrano MG, et al. The genome of Cryptosporidium hominis. Nature. 2004;431:1107–12.

71. Figueiredo LM, Rocha EPC, Mancio-Silva L, Prevost C, Hernandez-Verdun D, Scherf A. The unusually large plasmodium telomerase reverse-transcriptase localizes in a discrete compartment associated with the nucleolus. Nucleic Acids Res. 2005;33:1111–22.

72. Hall N, Karras M, Raine JD, Carlton JM, Kooij TWA, Berriman M, et al. A comprehensive survey of the plasmodium life cycle by genomic, transcriptomic, and proteomic analyses. Science. 2005;307:82–6.

73. Pain A, Renauld H, Berriman M, Murphy L, Yeats CA, Weir W, et al. Genome of the host-cell transforming parasite Theileria annulata compared with T. parva. Science. 2005;309:131–3.

74. Dreesen O, Li B, Cross GAM. Telomere structure and shortening in telomerase-deficient Trypanosoma brucei. Nucleic Acids Res. 2005;33:4536–43.

75. El-Sayed NM, Myler PJ, Blandin G, Berriman M, Crabtree J, Aggarwal G, et al. Comparative genomics of trypanosomatid parasitic protozoa. Science. 2005;309:404–9.

76. Giardini MA, Lira CBB, Conte FF, Camillo LR, de Siqueira Neto JL, Ramos CHI, et al. The putative telomerase reverse transcriptase component of Leishmania amazonensis: gene cloning and characterization. Parasitol Res. 2006;98:447–54.

77. Morrison HG, McArthur AG, Gillin FD, Aley SB, Adam RD, Olsen GJ, et al. Genomic minimalism in the early diverging intestinal parasite Giardia lamblia. Science. 2007;317:1921–6.

78. Chapman JA, Kirkness EF, Simakov O, Hampson SE, Mitros T, Weinmaier T, et al. The dynamic genome of Hydra. Nature. 2010;464:592–6.

79. Steele RE, David CN, Technau U. A genomic view of 500 million years of cnidarian evolution. Trends Genet. 2011;27:7–13.

80. Srivastava M, Simakov O, Chapman J, Fahey B, Gauthier MEA, Mitros T, et al. The Amphimedon Queenslandica genome and the evolution of animal complexity. Nature. 2010;466:720–6.

81. Ryan JF, Pang K, Schnitzler CE, Nguyen A-D, Moreland RT, Simmons DK, et al. The genome of the ctenophore Mnemiopsis leidyi and its implications for cell type evolution. Science. 2013;342:1242592.

82. Yang Y, Xiong J, Zhou Z, Huo F, Miao W, Ran C, et al. The genome of the Myxosporean Thelohanellus kitauei shows adaptations to nutrient acquisition within its fish host. Genome Biol Evol. 2014;6:3182–98.

83. Baumgarten S, Simakov O, Esherick LY, Liew YJ, Lehnert EM, Michell CT, et al. The genome of Aiptasia, a sea anemone model for coral symbiosis. Proc Natl Acad Sci. 2015;112:11893–8.

84. Bryan TM, Sperger JM, Chapman KB, Cech TR. Telomerase reverse transcriptase genes identified in Tetrahymena Thermophila and Oxytricha trifallax. Proc Natl Acad Sci. 1998;95:8479–84.

85. Collins K, Gandhi L. The reverse transcriptase component of the Tetrahymena telomerase ribonucleoprotein complex. Proc Natl Acad Sci. 1998;95:8485–90.

86. Greenberg RA, Allsopp RC, Chin L, Morin GB, DePinho RA. Expression of mouse telomerase reverse transcriptase during development, differentiation and proliferation. Oncogene. 1998;16:1723–30.

87. Martin-Rivera L, Herrera E, Albar JP, Blasco MA. Expression of mouse telomerase catalytic subunit in embryos and adult tissues. Proc Natl Acad Sci. 1998;95:10471–6.

88. O'Reilly M, Teichmann SA, Rhodes D. Telomerases. Curr Opin Struct Biol. 1999;9:56–65.

89. Lau BW-M, Wong AO-L, Tsao GS-W, So K-F, Yip HK-F. Molecular cloning and characterization of the zebrafish (Danio rerio) telomerase catalytic subunit (telomerase reverse transcriptase, TERT). J Mol Neurosci. 2008;34:63–75.

90. Li Y, Yates JA, Chen JJ-L. Identification and characterization of sea squirt

telomerase reverse transcriptase. Gene. 2007;400:16–24.

91. Cai Y, Ai Y, Zhao Q, Li J, Yang G, Gong P, et al. Cloning and characterization of telomerase reverse transcriptase gene in Trichinella spiralis. Parasitol Res. 2012;110:411–7.

92. Wurm Y, Wang J, Riba-Grognuz O, Corona M, Nygaard S, Hunt BG, et al. The genome of the fire ant Solenopsis invicta. Proc Natl Acad Sci U S A. 2011;108:5679–84.

93. Hrdličková R, Nehyba J, Lim SL, Grützner F, Bose HR. Insights into the evolution of mammalian telomerase: platypus TERT shares similarities with genes of birds and other reptiles and localizes on sex chromosomes. BMC Genomics. 2012;13:216.

94. Schumpert C, Nelson J, Kim E, Dudycha JL, Patel RC. Telomerase activity and telomere length in Daphnia. Saretzki G, editor. PLoS One. 2015;10:e0127196.

95. Koutsovoulos G, Kumar S, Laetsch DR, Stevens L, Daub J, Conlon C, et al. No evidence for extensive horizontal gene transfer in the genome of the tardigrade Hypsibius dujardini. Proc Natl Acad Sci. 2016;113:5053–8.

96. Hashimoto T, Horikawa DD, Saito Y, Kuwahara H, Kozuka-Hata H, Shin-I T, et al. Extremotolerant tardigrade genome and improved radiotolerance of human cultured cells by tardigrade-unique protein. Nat Commun. 2016;7:12808.

97. Kao D, Lai AG, Stamataki E, Rosic S, Konstantinides N, Jarvis E, et al. The genome of the crustacean Parhyale hawaiensis, a model for animal development, regeneration, immunity and lignocellulose digestion. elife. 2016;5:1.

98. Mason JM, Randall TA, Capkova Frydrychova R. Telomerase lost? Chromosoma. 2016;125:65–73.

99. Mark Welch D, Meselson M. Evidence for the evolution of bdelloid rotifers without sexual reproduction or genetic exchange. Science. 2000;288:1211–5.

100. Gladyshev EA, Arkhipova IR. Genome structure of Bdelloid rotifers: shaped by asexuality or desiccation? J Hered. 2010;101:S85–93.

101. Pardue M-L, DeBaryshe PG. Retrotransposons provide an evolutionarily robust non-telomerase mechanism to maintain telomeres. Annu Rev Genet. 2003;37:485–511.

102. Biessmann H, Carter SB, Mason JM. Chromosome ends in Drosophila without telomeric DNA sequences. Proc Natl Acad Sci U S A. 1990;87:1758–61.

103. Biessmann H, Champion LE, O'Hair M, Ikenaga K, Kasravi B, Mason JM. Frequent transpositions of Drosophila melanogaster HeT-A transposable elements to receding chromosome ends. EMBO J. 1992;11:4459–69.

104. Casacuberta E, Pardue M-L. Coevolution of the telomeric retrotransposons across Drosophila species. Genetics. 2002;161:1113–24.

105. Casacuberta E, Pardue M-L. Transposon telomeres are widely distributed in the Drosophila genus: TART elements in the virilis group. Proc Natl Acad Sci. 2003;100:3363–8.

106. Abad JP, De Pablos B, Osoegawa K, De Jong PJ, Martín-Gallardo A, Villasante A. TAHRE, a novel telomeric retrotransposon from Drosophila melanogaster, reveals the origin of Drosophila telomeres. Mol Biol Evol. 2004;21:1620–4.

107. Gladyshev EA, Arkhipova IR. A subtelomeric non-LTR retrotransposon Hebe in the bdelloid rotifer Adineta vaga is subject to inactivation by deletions but not 5′ truncations. Mob DNA. 2010;1:12.

108. Gladyshev EA, Arkhipova IR. A widespread class of reverse transcriptase-related cellular genes. Proc Natl Acad Sci U S A. 2011;108:20311–6.

109. Gladyshev EA, Arkhipova IR. Telomere-associated endonuclease-deficient Penelope-like retroelements in diverse eukaryotes. Proc Natl Acad Sci. 2007;104:9352–7.

110. Kelleher C, Teixeira MT, Förstemann K, Lingner J. Telomerase: biochemical considerations for enzyme and substrate. Trends Biochem Sci. 2002;27:572–9.

111. Malik HS, Burke WD, Eickbush TH. Putative telomerase catalytic subunits from Giardia lamblia and Caenorhabditis elegans. Gene. 2000;251:101–8.

112. Adam RD, Dahlstrom EW, Martens CA, Bruno DP, Barbian KD, Ricklefs SM, et al. Genome sequencing of Giardia lamblia genotypes A2 and B isolates (DH and GS) and comparative analysis with the genomes of genotypes A1 and E (WB and pig). Genome Biol Evol. 2013;5:2498–511.

113. Stevens JR, Noyes HA, Schofield CJ, Gibson W. The molecular evolution of Trypanosomatidae. Adv Parasitol. 2001;48:1–53.

114. Armbruster BN, Banik SS, Guo C, Smith AC, Counter CM. N-terminal domains of the human telomerase catalytic subunit required for enzyme activity in vivo. Mol Cell Biol. 2001;21:7775–86.

115. Moriarty TJ, Marie-Egyptienne DT, Autexier C. Functional Organization of Repeat Addition Processivity and DNA synthesis determinants in the human telomerase Multimer. Mol Cell Biol. 2004;24:3720–33.

116. Sekaran VG, Soares J, Jarstfer MB. Structures of telomerase subunits provide functional insights. Biochim Biophys Acta-Proteins Proteomics.

1804;2010:1190–201.

117. Wyatt HDM, West SC, Beattie TL. InTERTpreting telomerase structure and function. Nucleic Acids Res. 2010;38:5609–22.

118. Gillis AJ, Schuller AP, Skordalakes E. Structure of the Tribolium castaneum telomerase catalytic subunit TERT. Nature. 2008;455:633–7.

119. Rouda S, Skordalakes E. Structure of the RNA-binding domain of telomerase: implications for RNA recognition and binding. Structure. 2007;15:1403–12.

120. Mitchell M, Gillis A, Futahashi M, Fujiwara H, Skordalakes E. Structural basis for telomerase catalytic subunit TERT binding to RNA template and telomeric DNA. Nat Struct Mol Biol. 2010;17:513–8.

121. Gu X. Statistical methods for testing functional divergence after gene duplication. Mol Biol Evol. 1999;16:1664–74.

122. Huang H, Chopra R, Verdine GL, Harrison SC. Structure of a covalently trapped catalytic complex of HIV-1 reverse transcriptase: implications for drug resistance. Science. 1998;282:1669–75.

123. Hossain S, Singh S, Lue NF. Functional analysis of the C-terminal extension of telomerase reverse transcriptase. A putative "thumb" domain. J Biol Chem. 2002;277:36174–80.

124. Huard S, Moriarty TJ, Autexier C. The C terminus of the human telomerase reverse transcriptase is a determinant of enzyme processivity. Nucleic Acids Res. 2003;31:4059–70.

125. Tomlinson CG, Holien JK, Mathias JAT, Parker MW, Bryan TM. The C-terminal extension of human telomerase reverse transcriptase is necessary for high affinity binding to telomeric DNA. Biochimie. 2016;128–129:114–21.

126. Morand S. Life-history traits in parasitic nematodes: a comparative approach for the search of invariants. Funct Ecol. 1996;10:210–8.

127. Friedman KL, Cech TR. Essential functions of amino-terminal domains in the yeast telomerase catalytic subunit revealed by selection for viable mutants. Genes Dev. 1999;13:2863–74.

128. Beattie TL, Zhou W, Robinson MO, Harrington L. Functional multimerization of the human telomerase reverse transcriptase. Mol Cell Biol. 2001;21:6151–60.

129. Lai CK, Mitchell JR, Collins K. RNA binding domain of telomerase reverse transcriptase. Mol Cell Biol. 2001;21:990–1000.

130. Moriarty TJ, Huard S, Dupuis S, Autexier C. Functional multimerization of human telomerase requires an RNA interaction domain in the N terminus of the catalytic subunit. Mol Cell Biol. 2002;22:1253–65.

131. Bley CJ, Qi X, Rand DP, Borges CR, Nelson RW, Chen JJ-L. RNA-protein binding interface in the telomerase ribonucleoprotein. Proc Natl Acad Sci U S A. 2011;108:20333–8.

132. Harkisheimer M, Mason M, Shuvaeva E, Skordalakes E. A motif in the vertebrate telomerase N-terminal linker of TERT contributes to RNA binding and telomerase activity and processivity. Structure. 2013;21:1870–8.

133. Autexier C, Greider CW. Boundary elements of the Tetrahymena telomerase RNA template and alignment domains. Genes Dev. 1995;9:2227–39.

134. Tzfati Y, Fulton TB, Roy J, Blackburn EH. Template boundary in a yeast telomerase specified by RNA structure. Science. 2000;288:863–7.

135. Lai CK, Miller MC, Collins K. Template boundary definition in Tetrahymena telomerase. Genes Dev. 2002;16:415–20.

136. Chen J-L, Greider CW. Template boundary definition in mammalian telomerase. Genes Dev. 2003;17:2747–52.

137. Kumar S, Hedges SB. A molecular timescale for vertebrate evolution. Nature. 1998;392:917–20.

138. Hedges SB. The origin and evolution of model organisms. Nat Rev Genet. 2002;3:838–49.

139. Raible F, Tessmar-Raible K, Osoegawa K, Wincker P, Jubin C, Balavoine G, et al. Vertebrate-type intron-rich genes in the marine annelid Platynereis dumerilii. Science. 2005;310:1325–6.

140. Colbourne JK, Pfrender ME, Gilbert D, Thomas WK, Tucker A, Oakley TH, et al. The ecoresponsive genome of Daphnia pulex. Science. 2011;331:555–61.

141. Saebøe-Larssen S, Fossberg E, Gaudernack G. Characterization of novel alternative splicing sites in human telomerase reverse transcriptase (hTERT): analysis of expression and mutual correlation in mRNA isoforms from normal and tumour tissues. BMC Mol Biol. 2006;7:26.

142. Chang H, Delany ME. Complicated RNA splicing of chicken telomerase reverse transcriptase revealed by profiling cells both positive and negative for telomerase activity. Gene. 2006;379:33–9.

143. Imamura S, Uchiyama J, Koshimizu E, Hanai J, Raftopoulou C, Murphey RD, et al. A non-canonical function of Zebrafish telomerase reverse transcriptase is required for Developmental Hematopoiesis. PLoS One. 2008;3:e3364.

144. Ulaner GA, Hu J-F, Vu TH, Oruganti H, Giudice LC, Hoffman AR. Regulation of telomerase by alternate splicing of human telomerase reverse

transcriptase (hTERT) in normal and neoplastic ovary, endometrium and myometrium. Int J Cancer. 2000;85:330–5.

145. Tasaka K, Yokoyama N, Nodono H, Hoshi M, Matsumoto M. Innate sexuality determines the mechanisms of telomere maintenance. Int J Dev Biol. 2013;57:69–72.

146. Robb SMC, Ross E, Sánchez Alvarado A. SmedGD: the Schmidtea mediterranea genome database. Nucleic Acids Res. 2008;36:D599–606.

147. Nishimura O, Hirao Y, Tarui H, Agata K. Comparative transcriptome analysis between planarian Dugesia Japonica and other platyhelminth species. BMC Genomics. 2012;13:289.

148. Wheeler NJ, Agbedanu PN, Kimber MJ, Ribeiro P, Day TA, Zamanian M. Functional analysis of Girardia tigrina transcriptome seeds pipeline for anthelmintic target discovery. Parasit Vectors. 2015;8:34.

149. Robb SMC, Gotting K, Ross E, Sánchez AA. SmedGD 2.0: the Schmidtea mediterranea genome database. Genesis. 2015;53:535–46.

150. Egger B, Lapraz F, Tomiczek B, Müller S, Dessimoz C, Girstmair J, et al. A Transcriptomic-Phylogenomic analysis of the evolutionary relationships of flatworms. Curr Biol. 2015;25:1347–53.

151. Brandl H, Moon H, Vila-Farré M, Liu S-Y, Henry I, Rink JC. PlanMine–a mineable resource of planarian biology and biodiversity. Nucleic Acids Res. 2016;44:D764–73.

152. Young ND, Nagarajan N, Lin SJ, Korhonen PK, Jex AR, Hall RS, et al. The Opisthorchis viverrini genome provides insights into life in the bile duct. Nat Commun. 2014;5:4378.

153. Howe KL, Bolt BJ, Cain S, Chan J, Chen WJ, Davis P, et al. WormBase 2016: expanding to enable helminth genomic research. Nucleic Acids Res. 2016;44:D774–80.

154. Tsai IJ, Zarowiecki M, Holroyd N, Garciarrubio A, Sanchez-Flores A, Brooks KL, et al. The genomes of four tapeworm species reveal adaptations to parasitism. Nature. 2013;496:57–63.

155. Wasik K, Gurtowski J, Zhou X, Ramos OM, Delás MJ, Battistoni G, et al. Genome and transcriptome of the regeneration-competent flatworm, Macrostomum lignano. Proc Natl Acad Sci. 2015;112:12462–7.

156. Liu S-Y, Selck C, Friedrich B, Lutz R, Vila-Farré M, Dahl A, et al. Reactivating head regrowth in a regeneration-deficient planarian species. Nature. 2013;500:81–4.

157. Lue NF. Adding to the ends: what makes telomerase processive and how important is it? BioEssays. 2004;26:955–62.

158. Collins K. Ciliate telomerase biochemistry. Annu Rev Biochem. 1999;68:187–218.

159. Peng Y, Mian IS, Lue NF. Analysis of telomerase processivity: mechanistic similarity to HIV-1 reverse transcriptase and role in telomere maintenance. Mol Cell. 2001;7:1201–11.

160. Aken BL, Ayling S, Barrell D, Clarke L, Curwen V, Fairley S, et al. The Ensembl Gene annotation system. Database (Oxford). 2016;2016:baw093.

161. Katoh K, Asimenos G, Toh H. Multiple alignment of DNA sequences with MAFFT. Methods Mol Biol. 2009;537:39–64.

162. Stamatakis A. RAxML version 8: a tool for phylogenetic analysis and post-analysis of large phylogenies. Bioinformatics. 2014;30:1312–3.

163. Kearse M, Moir R, Wilson A, Stones-Havas S, Cheung M, Sturrock S, et al. Geneious basic: an integrated and extendable desktop software platform for the organization and analysis of sequence data. Bioinformatics. 2012;28:1647–9.

164. González-Estévez C, Felix DA, Smith MD, Paps J, Morley SJ, James V, et al. SMG-1 and mTORC1 act antagonistically to regulate response to injury and growth in planarians. PLoS Genet. 2012;8:e1002619.

165. Lek M, MacArthur DG, Yang N, North KN. Phylogenetic analysis of gene structure and alternative splicing in alpha-actinins. Mol Biol Evol. 2010;27:773–80.

Duplication and concerted evolution of MiSp-encoding genes underlie the material properties of minor ampullate silks of cobweb weaving spiders

Jannelle M. Vienneau-Hathaway[1], Elizabeth R. Brassfield[1], Amanda Kelly Lane[1], Matthew A. Collin[2], Sandra M. Correa-Garhwal[2], Thomas H. Clarke[1,2], Evelyn E. Schwager[3], Jessica E. Garb[3], Cheryl Y. Hayashi[2] and Nadia A. Ayoub[1*]

Abstract

Background: Orb-web weaving spiders and their relatives use multiple types of task-specific silks. The majority of spider silk studies have focused on the ultra-tough dragline silk synthesized in major ampullate glands, but other silk types have impressive material properties. For instance, minor ampullate silks of orb-web weaving spiders are as tough as draglines, due to their higher extensibility despite lower strength. Differences in material properties between silk types result from differences in their component proteins, particularly members of the spidroin (spider fibroin) gene family. However, the extent to which variation in material properties within a single silk type can be explained by variation in spidroin sequences is unknown. Here, we compare the minor ampullate spidroins (MiSp) of orb-weavers and cobweb weavers. Orb-web weavers use minor ampullate silk to form the auxiliary spiral of the orb-web while cobweb weavers use it to wrap prey, suggesting that selection pressures on minor ampullate spidroins (MiSp) may differ between the two groups.

Results: We report complete or nearly complete *MiSp* sequences from five cobweb weaving spider species and measure material properties of minor ampullate silks in a subset of these species. We also compare MiSp sequences and silk properties of our cobweb weavers to published data for orb-web weavers. We demonstrate that all our cobweb weavers possess multiple *MiSp* loci and that one locus is more highly expressed in at least two species. We also find that the proportion of β-spiral-forming amino acid motifs in MiSp positively correlates with minor ampullate silk extensibility across orb-web and cobweb weavers.

Conclusions: MiSp sequences vary dramatically within and among spider species, and have likely been subject to multiple rounds of gene duplication and concerted evolution, which have contributed to the diverse material properties of minor ampullate silks. Our sequences also provide templates for recombinant silk proteins with tailored properties.

Keywords: Theridiidae, *Latrodectus*, *Steatoda*, Silk glands, Gene expression, Silk proteins, Spidroin gene family

* Correspondence: ayoubn@wlu.edu
[1]Department of Biology, Washington and Lee University, Lexington, VA, USA
Full list of author information is available at the end of the article

Background

Orb-web and cobweb weaving spiders in the superfamily Araneoidea use silk fibers and glues for a variety of functions, including prey wrapping, prey capture, egg casings, and web framing [1]. Each task-specific silk is made up of a unique combination of proteins produced in one of seven different gland types [2, 3]. Despite making architecturally distinct prey-capture webs (Fig. 1), orb-web and cobweb weaving araneoid spiders possess homologous gland types, most of which produce silks with homologous functions. For example, major ampullate glands synthesize dragline fibers, tubuliform glands express proteins that form outer egg-casing fibers, and pyriform gland protein products are used to attach other fiber types to substrates by all araneoid spiders [1, 4]. Other gland types are homologous in morphology, but have diverged in some functions, such as minor ampullate glands which produce silk used as scaffolding fibers

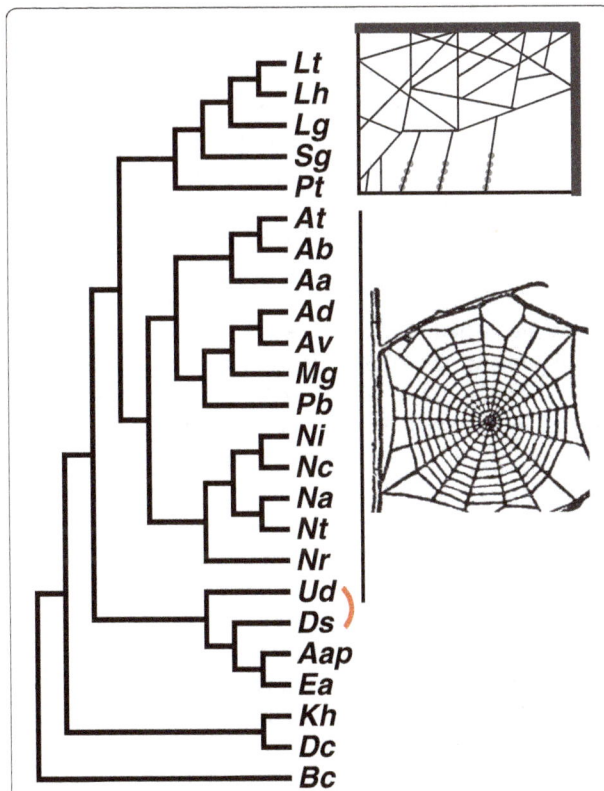

Fig. 1 Relationships among spider species in this study. See Table 1 for species abbreviations and family status. We focused on comparing minor ampullate silks between species that build cobwebs, a disordered three-dimensional web that typically catches walking prey (top graphic) and species that build orb-webs, a two dimensional wagon-wheel shaped web that catches flying prey (bottom graphic, from http://cliparts.co/spiders-web-clip-art). Relationships are based on a compilation of published spider phylogenies (see Methods). The *red* elbow indicates that one study [38] found the relationship shown in the figure, while another [39] switched the placement of the families to which these species belong

by most araneoids, but additionally functions to form auxiliary spirals in the orb-web versus contributing to prey-wrapping in cobweb weavers [5].

Major ampullate silk has been the primary focus of silk research because it is relatively easy to harvest and has tensile strength rivaling that of steel [6, 7]. However, all silk fibers have impressive and unique material properties [8]. For instance, minor ampullate silk of the orb-web weavers *Argiope trifasciata*, *Argiope argentata*, and *Nephila clavipes* have higher extensibility and lower strength than major ampullate silks from the same species, yet are equally tough [8–12]. Additionally, minor and major ampullate silks display different physical properties when wet. Supercontraction in water occurs with *A. trifasciata* and *Nephila inaurata* major ampullate silks, which allows them to become more extensible while maintaining their high strength [13]. However, when minor ampullate silks from the same species are exposed to water, supercontraction does not occur and the mechanical properties do not change [13]. Despite having consistent behaviors within species whether wet or dry, mechanical properties of minor ampullate silks varied more among species than did the properties of major ampullate silks [14, 15]. The consistency of minor ampullate silks' mechanical properties when wet or dry, as well as their increased extensibility relative to draglines and greater variability among species, could allow for different applications of minor ampullate silk that would not be feasible for major ampullate silk. The material properties of minor ampullate silks made by cobweb weavers have yet to be measured, but given the different functions of this silk in cobweb weavers compared to in orb-web weavers, we predict divergence in the mechanical performance of these fibers.

The variation in physical properties between major and minor ampullate silks and among minor ampullate silks of different species likely results from sequence differences among their component proteins, termed spidroins (spider fibroins). Spidroins are extremely large (>200 kDa) and are made almost entirely of a highly repetitive region, which is flanked by short, conserved carboxy (C)- and amino (N)-terminal regions [16–21]. The repetitive region of many spidroins contain numerous subrepeats of short amino acid sequences, called motifs [2]. Motifs found frequently in spidroin repetitive regions include GGX (where X refers to a subset of amino acids), $(GA)_n$ where $n \geq 2$, A_n where $n \geq 4$, and GPG [2, 18]. A_n and $(GA)_n$ form β-sheets, GGX forms 3_1-helices, and GPG forms β-spirals [2, 14, 18, 22–26]. Secondary structures of β-sheets likely confer strength, and β-spirals contribute to elasticity [18].

Major ampullate silks of all araneoid spiders examined thus far are made up of two proteins, MaSp1 and MaSp2, which are distinguished by a high GPG content in the latter but not the former [27, 28]. Similarly, *N. clavipes* minor ampullate silk is made up of two proteins, MiSp1 and MiSp2, although they are not as distinctly different from each other as MaSp1 and MaSp2 [29]. All described minor ampullate spidroins (generalized as MiSp) have consensus repeats that are primarily made up of A_n, $(GA)_n$ and GGX motifs disrupted by serine and threonine-rich "spacer" regions [2, 14, 18, 29, 30]. The functions of the spacer regions are not well understood, but artificially made recombinant proteins composed solely of spacer regions tend to form oligomers and only recombinant MiSp proteins that include at least one spacer, repetitive region, and the C-terminal domain can form fibers [31]. Both MaSp1 and MiSp are predicted to be composed of β-sheets connected by helical regions and amorphous chains [2, 14, 18]. The higher tensile strength of major ampullate silk may result from longer β-sheets made up of A_n repeats, in contrast to shorter β-sheets made up of A_n and $(GA)_n$ repeats in minor ampullate silk [14].

The large size and repetitiveness of spidroins make spidroin-encoding genes extremely difficult to sequence completely. The only full-length sequence of *MiSp* known is for *Araneus ventricosus* [30]. Furthermore, even partial *MiSp* sequences have been characterized from only a few orb-web weavers [29, 30]. The full length sequence shows that MiSp repeat units are not as similar to each other as are the repeats of MaSp1 or MaSp2 repeats; and that *A. ventricosus* MiSp possesses exactly two spacers that interrupt the glycine and alanine rich motifs [30]. The generality of these features for other MiSp-encoding sequences needs to be tested.

The only known sequences of *MiSp* in a cobweb weaver are partial sequences from *Latrodectus hesperus* [5, 32]. The known repetitive regions are similar between *L. hesperus* and orb-web weavers' MiSp, with the main difference being length of spacers [30, 32]. However, because cobweb weavers use minor ampullate silks for prey wrapping, while orb-web weavers use them in the auxiliary spiral of their web [5], we predict different selective pressures on cobweb weaving MiSp that could consequently alter secondary structure and physical properties.

Here, we characterize complete or nearly complete *MiSp* sequences from five cobweb weaving spider species: *L. hesperus*, *L. tredecimguttatus*, *L. geometricus*, *Steatoda grossa*, and *Parasteatoda tepidariorum* (Theridiidae). We also measured mechanical properties of minor ampullate silks in *L. hesperus*, *L. geometricus*, and *S. grossa*. Our goals were to determine the extent of variation in MiSp sequences and minor ampullate silk properties within cobweb weaving spiders, compare this variation to that found in orb-web weaving spiders, and identify any relationship between sequence and material property variation across both groups. We found

evidence for at least two *MiSp* loci in each cobweb weaving species, and that one copy is more highly expressed than the other in minor ampullate glands in *L. hesperus* and *L. geometricus*. Phylogenetic analyses suggested extensive gene duplication and concerted evolution of MiSp-encoding loci contributed to variation in MiSp sequences. Finally, we found significant differences in mechanical properties among minor ampullate silks that were partially explained by variation in MiSp sequences.

Results
Multiple loci encode MiSp in cobweb weavers
We employed a series of molecular experiments to identify MiSp-encoding sequences in our five species of cobweb weavers. Additional file 1: Figure S1 provides an overview of these various methods.

L. hesperus
Shotgun sequencing of an *L. hesperus MiSp*-positive fosmid genomic clone resulted in the assembly of three contiguous sequences (contigs). One ~13 kilobase (kb) contig contained a complete open reading frame (ORF) predicted to encode MiSp (6549 base pairs, bp) and 4689 bp upstream and 2183 bp downstream flanking sequence. Amplification and cloning of *L. hesperus* genomic DNA with primers designed from the 3'UTR of a *MiSp* cDNA and an N-terminal encoding cDNA from this species resulted in an almost complete ORF (5568 bp) and 85 bp of 3'UTR, which was 98.8% identical to the published *MiSp* cDNA (HM752571) over a 983 bp alignment and 97.7% identical to the published *MiSp* N-terminal cDNA (HM752570) over a 903 bp alignment. In contrast to the similarity between the amplified gene and the cDNAs, the fosmid clone and the amplified gene were distinct. Pairwise differences between the fosmid clone and the amplified gene were 4.5% at the N-terminal encoding end (975 bp alignment) and 20.3% at the C-terminal/ 3' UTR end (960 bp alignment). The repetitive regions were too divergent to align. We designated the fosmid clone as *Lh* MiSp variant 1 (*Lh* MiSp_v1) and the amplified gene as *Lh* MiSp variant 2 (*Lh* MiSp_v2; see Table 1 for accession numbers).

Primer sets designed to be specific to *Lh* MiSp_v1 versus *Lh* MiSp_v2 successfully amplified genomic DNA of three single individuals. Variant-specific PCR products were 99.3% identical to the targeted variant, but only 86.3% identical to each other on average at the 616 bp C-terminal/ 3'UTR region. Sequence chromatograms of *Lh* MiSp_v1 and MiSp_v2 amplifications exhibited double peaks in both directions in at least one individual (10 double peaks in one of three individuals amplified with *Lh* MiSp_v1 specific primers and one double peak in all three individuals amplified with *Lh* MiSp_v2

Table 1 Spidroins included in phylogenetic analyses of N- and C-terminal sequences

Spidroin name	Species	Family	N-term accession	C-term accession
Bc fibroin1	Bothriocyrtum californicum	Ctenizidae	HM752562	EU117162
Kh MaSp1	Kukulcania hibernalis	Filistatidae	HM752563	
Dc MaSp	Diguetia canities	Diguetidae	HM752564	HM752565
Dc MaSp-like	Diguetia canities	Diguetidae	HM752566	HM752567
Ds MaSp2	Deinopis spinosa	Deinopidae	HM752568	DQ399328, DQ399329[a]
Aap MaSp	Agelenopsis aperta	Agelenidae	HM752573	AAT08436
Ud MiSp	Uloborus diversus	Uloboridae	HM752574	ABD61597
Mg MiSp	Metepeira grandiosa	Araneidae	HM752575	HM752569
Lh MiSp	Latrodectus hesperus	Theridiidae	HM752570	HM752571
Aap TuSp1	Agelenopsis aperta	Agelenidae	HM752576	HM752572
Aa TuSp1	Argiope argentata	Araneidae	HM752577	AY953071
Ab TuSp1	Argiope bruennichi	Araneidae	AB242144	AB242144
Nct TuSp1	Nephila clavata	Nephilidae	AB218974	AB218973
Lh TuSp1	Latrodectus hesperus	Theridiidae	DQ379383	AY953070
At MaSp2	Argiope trifasciata	Araneidae	DQ059136S1	DQ059136S2
Nc MaSp2	Nephila clavipes	Nephilidae	EU599240	AY654297
Ni MaSp2	Nephila inaurata madagascariensis	Nephilidae	DQ059135	AF350278
Nc MaSp1a	Nephila clavipes	Nephilidae	EU599238	AY654292
Nc MaSp1b	Nephila clavipes	Nephilidae	EU599239	AY654291
Lh MaSp1	Latrodectus hesperus	Theridiidae	EF595246	EF595246
Lh MaSp2	Latrodectus hesperus	Theridiidae	EF595245	EF595245
Lg MaSp1	Latrodectus geometricus	Theridiidae	DQ059133S1[b]	DQ059134
Ea MaSp	Euprosthenops australis	Pisauridae	AM259067	AJ973155
Nc Flag	Nephila clavipes	Nephilidae	AF027972[b]	AF027973
Ni Flag	Nephila inaurata madagascariensis	Nephilidae	AF218623S1	AF218623S2
Av Flag	Araneus ventricosus	Araneidae	AY945306	AY587193
Lg MaSp2	Latrodectus geometricus	Theridiidae	EU177657	AF350275
Ds fibroin1a	Deinopis spinosa	Deinopidae	JX978170	DQ399326
Lh AcSp1	Latrodectus hesperus	Theridiidae	JX978171	
Lg AcSp1	Latrodectus geometricus	Theridiidae	JX978180	JX978181
Lh MiSp_v1	Latrodectus hesperus	Theridiidae	KX584003	
Lh MiSp_v2	Latrodectus hesperus	Theridiidae	KX584020	
Lg MiSp_v1	Latrodectus geometricus	Theridiidae	KX584023	
Lg MiSp_v2	Latrodectus geometricus	Theridiidae	KX584024	
Lt MiSp_v1	Latrodectus tredecimguttatus	Theridiidae	KX584027	
Lt MiSp_v2	Latrodectus tredecimguttatus	Theridiidae	KX584026	
Sg MiSp_pseudo	Steatoda grossa	Theridiidae	KX584019	
Sg MiSp	Steatoda grossa	Theridiidae	KX584035	KX584021
Pt MiSp	Parasteatoda tepidariorum	Theridiidae	KX584055	KX584022
Ncr PySp	Nephilengys cruentata	Nephilidae		GU062417
Nc PySp	Nephila clavipes	Nephilidae		GQ980330
Nc PySp2	Nephila clavipes	Nephilidae		HM020705
Lh PySp	Latrodectus hesperus	Theridiidae		FJ973621
Ud AcSp1	Uloborus diversus	Uloboridae		DQ399333

Table 1 Spidroins included in phylogenetic analyses of N- and C-terminal sequences *(Continued)*

Na MiSp	Nephila antipodiana	Nephilidae	DQ399324
Nc MiSp1	Nephila clavipes	Nephilidae	AF027735
Ncr MiSp	Nephilengys cruentata	Nephilidae	EF638447
Ds MiSp	Deinopis spinosa	Deinopidae	DQ399324
Ad MiSp	Araneus diadematus	Araneidae	U47853
Av MiSp	Araneus ventricosus	Araneidae	JX513956.1
Pb MiSp	Parawixia bistriata	Araneidae	GQ275358
Ad MaSp1	Araneus diadematus	Araneidae	U47854
Ad MaSp2a	Araneus diadematus	Araneidae	U47855
Ad MaSp2b	Araneus diadematus	Araneidae	U47856
Ncr MaSp_like	Nephilengys cruentata	Nephilidae	EF638446
Na MaSp1	Nephila antipodiana	Nephilidae	DQ338461
Ud MaSp1	Uloborus diversus	Uloboridae	DQ399331
Ud MaSp2a	Uloborus diversus	Uloboridae	DQ399334
Ud MaSp2b	Uloborus diversus	Uloboridae	DQ399335
Ncr Flag	Nephilengys cruentata	Nephilidae	EF638444
Ncr TuSp	Nephilengys cruentata	Nephilidae	EF638445
Ds TuSp1	Deinopis spinosa	Deinopidae	AY953073
Ud TuSp1	Uloborus diversus	Uloboridae	AY953072
Na TuSp1	Nephila antipodiana	Nephilidae	DQ089048

[a]DQ399329 contains the *Ds* MaSp2a C-terminal sequence and DQ399328 the *Ds* MaSp2b C-terminal sequence
[b]DQ059133S1 and AF027972 were edited to include corrections outlined by [73]

specific primers). The pattern of double peaks in one individual but not others for the *Lh* MiSp_v1 specific PCR product is consistent with allelic variation for *Lh* MiSp_v1 and confirms that *Lh* MiSp_v1 and *Lh* MiSp_v2 represent separate genomic loci.

L. tredecimguttatus and *L. geometricus*
Cloning and end sequencing of nearly complete *MiSp* PCR products from genomic DNA of single individuals of *L. tredecimguttatus (Lt)* and *L. geometricus (Lg)* resulted in 11 unique *Lg* sequences (11 clones) and 11 unique *Lt* sequences (12 clones). Within each of these species, end sequences clustered into two distinct groups (Fig. 2; Additional file 1: Figure S2). Average pairwise differences within a cluster were low for N-terminal and adjacent repetitive encoding sequences: 0.9% (0–1.2%) and 2.0% (1.2–3.5%) for *Lt*; 0.7% (0–1.6%) and 1.0% (0.2–1.5%) for *Lg*. Average pairwise differences between clusters for the same region were much higher: 6.4% (4.5–7.6%) for *Lt* and 11.6% (5.3–14%) for *Lg*. A similar pattern was found for C-terminal and adjacent repetitive encoding sequences, except divergence among sequences was higher; average pairwise distances within clusters were 0.7% (0–1.5%) and 8% (4.5–15%) for *Lt*, and 2.8 (0–9.1%) and 6.0 (0.2–12.6%) for *Lg*; versus average pairwise distances between clusters were 10.2% (2.3–19.5%) for *Lt* and 52% (42.1–56.6%) for *Lg*.

Because these species are diploid, we expect no more than two alleles per locus within a single individual. Thus each species may possess as many as six loci encoding MiSp (11 variants identified in each species). However, because some PCR or cloning error could produce slight variants among sequences (despite the use of a proofreading enzyme) we conservatively considered the two distinct clusters to represent separate loci and completely sequenced one representative from each cluster (Fig. 2). We designated the completely sequenced clones from *L. tredecimguttatus* as *Lt* MiSp_v1 and *Lt* MiSp_v2, and the completely sequenced clones from *L. geometricus* as *Lg* MiSp_v1 and *Lg* MiSp_v2. Designation of MiSp variants with 1 or 2 is not intended to imply orthology among species, but simply reflects different genomic loci within species.

S. grossa
TOPO cloning and end sequencing of an ~5000 bp *S. grossa* PCR product that was generated with primers targeting complete *MiSp* resulted in ten identical clones from a single individual. When translated, these ~800 bp end sequences aligned well with our *Latrodectus* MiSp sequences, but contained numerous stop codons. This variant is likely a non-functional pseudogene, designated *Sg* MiSp_pseudo.

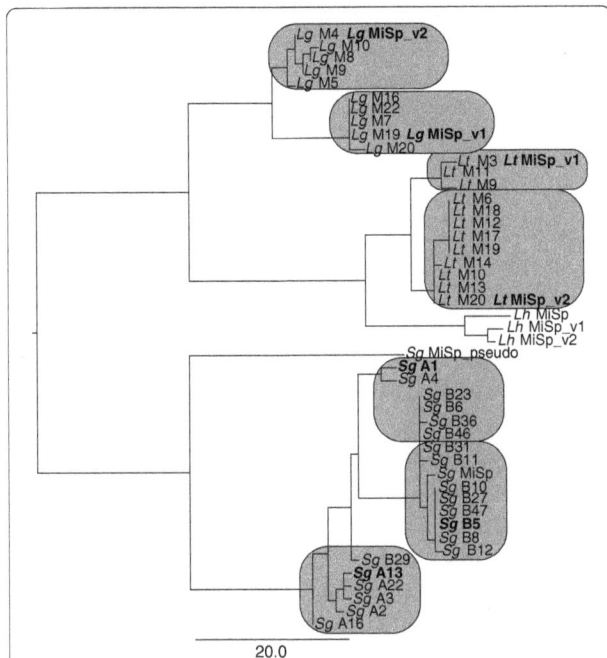

Fig. 2 Neighbor joining tree for N-terminal MiSp encoding sequences demonstrates variation within and among species. Variants are shown for *L. hesperus* (*Lh*) and TOPO clones from *L. tredecimguttatus* (*Lt*), *L. geometricus* (*Lg*), and *S. grossa* (*Sg*). Units are number of substitutions. *Latrodectus* clones are arbitrarily numbered, except that *Lt* M1-7 and *Lg* M1-5 resulted from amplification with primers designed from *L. hesperus*, while higher clone numbers resulted from amplification with species-specific primers. *S. grossa* TOPO clones resulted from two separate PCR reactions, which are indicated here as "A#" and "B#". Completely sequenced clones are indicated by *bolded* names. See Table 1 and Additional file 1: Table S4 for accession numbers. *Boxes* indicate distinct clusters identified in neighbor joining trees based on N-terminal sequences including adjacent repetitive sequences and in C-terminal and adjacent repetitive region encoding sequences (Additional file 1: Figures S2 and S3)

PCR amplification using primers designed from our ~3 kb C-terminal encoding *S. grossa* MiSp cDNA (designated *Sg* MiSp) failed to find any further variants. In contrast, TOPO cloning and sequencing of 20 N-terminal and adjacent repetitive MiSp encoding PCR products for a single *S. grossa* individual resulted in 20 unique sequences, each different from our ~500 bp N-terminal encoding cDNA. The clones (designated by clone numbers) grouped into three clusters (Additional file 1: Figure S3); within a cluster pairwise differences among sequences were <0.9% on average (0.13–1.5%) while between clusters pairwise differences were much greater (average pairwise distance = 10%). Nucleotide differences within a cluster could reflect alleles or multiple loci, but could also be attributable to PCR error because amplification was performed with *Taq* polymerase alone, which does not have proofreading capability, in contrast to the high fidelity polymerases used to amplify the *Latrodectus* species (see Methods). Conservatively, the

three clusters represent three true variants. Thus there is a minimum of two loci that encode MiSp in *S. grossa*.

P. tepidariorum

We found partial length N and C-terminal MiSp-encoding cDNAs from *P. tepidariorum* (designated *Pt* MiSp, see Table 1 for accessions) that were nearly identical (>98%) to the 5' and 3' ends, respectively, of an Augustus-predicted gene model on Scaffold 853 in the publicly available genome. This gene model was predicted to span 10292 bp, but included four gaps (4353 bp of assembled sequence, 5939 bp of unknown). Characterization of a TOPO clone containing an ~2.3 kb PCR product for *P. tepidariorum* MiSp (KX584004) resulted in a sequence that was 100% identical to the 104 bp of overlap with the C-terminal encoding region of Scaffold 853 and filled 2236 bp of the last gap. We refer to the combination of our TOPO clone with Scaffold 853 as *Pt* MiSp variant 1 (*Pt* MiSp_v1). The fully sequenced TOPO clone included a 1175 bp long intron with canonical donor and acceptor sites identified by aligning with our C-terminal encoding cDNA using SPIDEY [33]. We also found BLASTN alignments of this 2.3 kb TOPO clone to three other scaffolds (954, 3280, and 3758) that may represent additional MiSp-encoding loci. The C-terminal encoding region of our TOPO clone aligned with 91% identity on Scaffold 954. The intron sequence aligned with 95% identity to Scaffold 3280 and 99% identity to Scaffold 3758. The repeat region flanking the intron (including 100 bases of intron) aligned with 89% identity to Scaffold 3280 and 90% identity to Scaffold 3758. The presence of two C-terminal encoding sequences that differ by 9% and are placed on separate scaffolds suggests that at least two loci encode MiSp in *P. tepidariorum*.

MiSp loci are expressed at different rates in cobweb weaving spiders

In *L. geometricus* and *L. hesperus*, both completely sequenced loci (*MiSp_v1* and *MiSp_v2*) are highly transcribed in minor ampullate glands (2.4–44.7% of all reads in minor ampullate gland RNA-seq libraries aligned to *MiSp_v1* or *MiSp_v2*) and transcribed at low levels in other tissue types (e.g. <1% of all reads in major ampullate gland RNA-seq libraries aligned to *MiSp_v1* or *MiSp_v2*, Table 2). *MiSp_v2* transcripts are more abundant than *MiSp_v1* in both *L. geometricus* and in *L. hesperus* minor ampullate glands (Table 2). The ratio was consistent for all tissue types in *L. hesperus*, however, ratios were much more variable in *L. geometricus* between libraries of minor ampullate glands and even more so among tissues (Table 2).

Table 2 *MiSp* transcript abundance based on read counts in minor ampullate, major ampullate, and total silk gland RNA-seq libraries

RNA-seq Library	L. geometricus					L. hesperus				
	Total reads[a]	MiSp_v2	MiSp_v1	Ratio[c]	Ratio 3'[d]	Total reads[b]	MiSp_v2	MiSp_v1	Ratio[c]	Ratio 3'[d]
Minor 1	4832818	275164	57912	4.8	2.6	11679960	5289870	1487650	3.6	3.6
Minor 2	20640831	3496061	564983	6.2	4.2	11148437	4919531	1443331	3.4	3.4
Total Silk	8101748	61016	72774	0.8	0.9	15093424	572185	183028	3.1	4.3
Major 1	13978630	1943	4493	0.4	0.3	6116968	68	24	2.8	2.7
Major 2	13636997	69	9	7.7	5.5	9574596	315	85	3.7	4.4

See Table 1 for accession numbers
[a] Raw RNA-seq reads are available on NCBI's SRA database: SRR1539569
[b] Raw RNA-seq reads are available on NCBI's SRA database: SRR1539570
[c] Ratio of *MiSp_v2* to *MiSp_v1* based on counts of reads aligned to the entire sequences
[d] Ratio of *MiSp_v2* to *MiSp_v1* based on counts of reads aligned to the 3'-most 500 bases of coding sequences. See Methods for rationale

Variable genomic structure of MiSp

All *MiSp* genes characterized from *Latrodectus* lacked introns interrupting coding sequence. While only *L. hesperus MiSp_v1* included both the start and stop codons, the other *MiSp* sequences missed no more than 300 bp of the N or C-terminal coding regions on either end. Our *P. tepidariorum MiSp* genomic TOPO clone that includes the C-terminal encoding region has an intron of 1175 bp. For comparison, orb-weaver *A. ventricosus MiSp* has a large, single intron of 5628 bp [30].

Inspection of the *L. hesperus* fosmid genomic clone revealed a TATA box approximately 50 bases upstream of the full-length *L. hesperus* MiSp coding region. The nucleotide motif CACG found in all previously characterized spidroin genes was also found in *L. hesperus* MiSp, 11 bases upstream of the TATA box. The *P. tepidariorum* sequence on Scaffold 853 in the i5k genome possessed a TATA box 63 bases upstream of the MiSp coding region and a CACG motif 10 bases upstream of the TATA box. The *A. ventricosus* sequence has a TATA box 63 bases upstream of the full-length MiSp coding region, and a CACG motif 30 bases upstream of the TATA box [30]. Polyadenylation signals were found 110 bases downstream of the stop codon for *L. hesperus MiSp_v1*, 242 bases downstream of the stop for *P. tepidariorum MiSp_v1*, and 82 bases downstream of the stop for *A. ventricosus MiSp* [30].

MiSp amino acid variation within and among species

Cobweb weaver MiSp length is highly variable within and among species (815 aa to 2751 aa, Additional file 1: Figure S4). For comparison, the completely sequenced *L. hesperus* MaSp1 is 3129 aa and MaSp2 is 3779 aa, and *A. ventricosus* MiSp is 1,766 aa [30]. Glycine, alanine, and serine are the three most abundant amino acids for all MiSp variants, showing similar composition percentages to *L. hesperus* MaSp1 and MaSp2. Serine is found in both the spacer regions and the repeat regions, while glycine and alanine are predominantly found in the repeat region only. Proline is found in very low abundance

for both *L. hesperus* MiSp variants, both *L. tredecimguttatus* MiSp variants, *L. geometricus* MiSp_v2, and *P. tepidariorum* MiSp_v1, but is more abundant in *L. geometricus* MiSp_v1 and is the fourth most abundant amino acid in *S. grossa* MiSp, surpassing *L. hesperus* MaSp2 proline composition (Additional file 1: Figure S4). Codon usage is variable among MiSp loci and species, but there is an overall bias toward A or T in the third position of the codon, especially for alanine and glycine (Additional file 1: Table S1).

Despite gross similarities in MiSp amino acid composition among species, arrangement of amino acids into motifs is more variable. *L. hesperus* MiSp variants, *L. tredecimguttatus* MiSp variants, *L. geometricus* MiSp_v2, and *P. tepidariorum* MiSp_v1 contain mostly (GA)ₙ and GGX motifs, with GPG being nonexistent. However, GPG is prevalent in both *L. geometricus* MiSp_v1 and *S. grossa* MiSp (Fig. 3). In contrast to MaSp1 and MaSp2, which have ensemble repeat units ~30–50 aa long that are nearly identical, there are multiple shorter, more variable ensemble repeats that describe MiSp sequences (Additional file 1: Table S2). Tandem repeats range in length from 7 to 569 aa, repeated in the protein anywhere from 2 to 28 times, and with amino acid sequence identity ranging from 81% to 97% (Additional file 1: Table S2). Modularity of ensemble repeats is highly variable among and within species, with *L. hesperus* MiSp_v1, *L. tredecimguttatus* MiSp_v2, *L. geometricus* MiSp_v1, and *S. grossa* MiSp having a higher order repeat unit (>160 aa repeated at least once) and *L. hesperus* MiSp_v2, *L. tredecimguttatus* MiSp_v1, *L. geometricus* MiSp_v2, and *P. tepidariorum* MiSp_v1 having multiple short repetitive units only (24 to 30 aa repeated >10 times).

The number and arrangement of spacers is also highly variable among cobweb weaver MiSp. Some MiSp variants lack spacers while others have up to seven spacers. Spacers are varied in their distribution within a monomer, ranging from 100 aa to 400 aa away from each other in MiSp proteins with more than one spacer

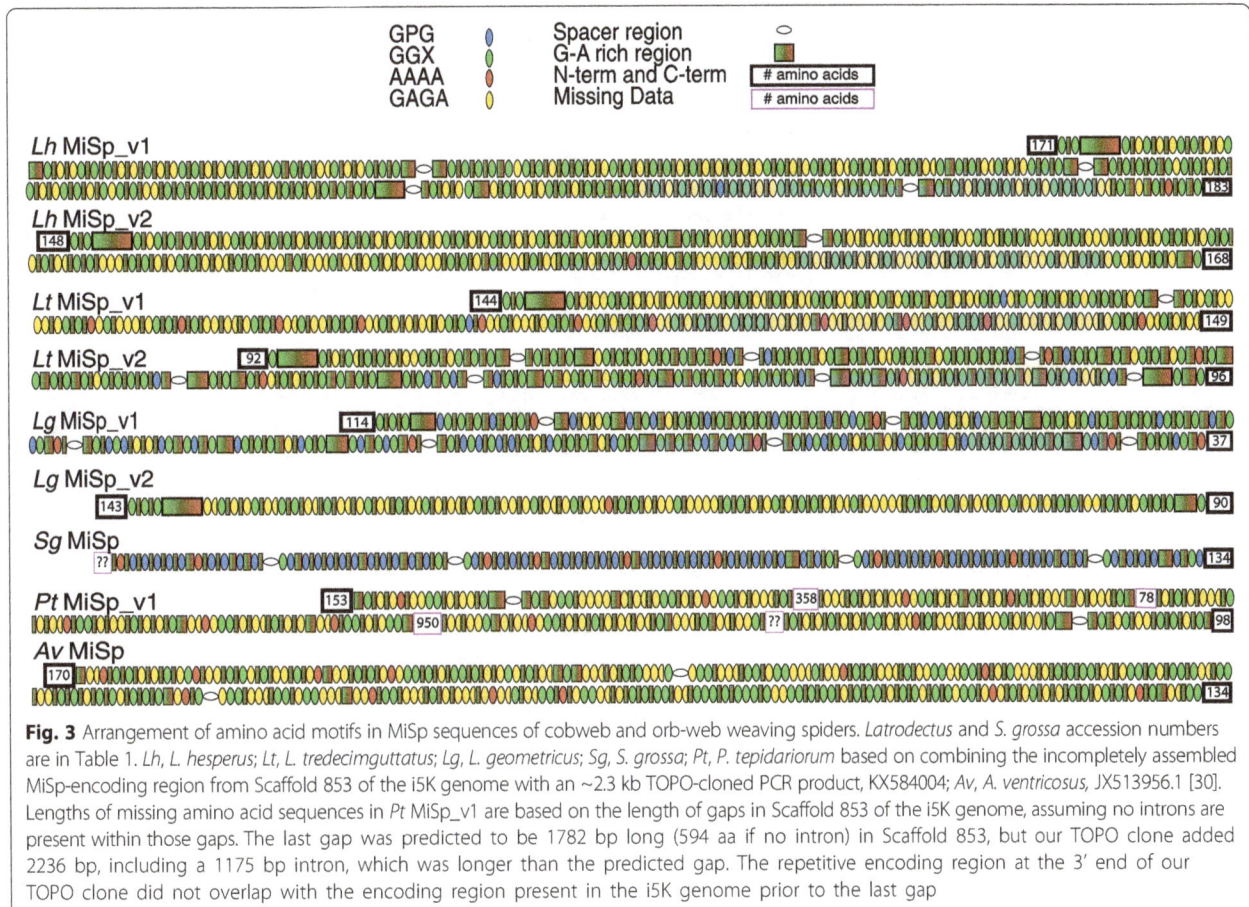

Fig. 3 Arrangement of amino acid motifs in MiSp sequences of cobweb and orb-web weaving spiders. *Latrodectus* and *S. grossa* accession numbers are in Table 1. *Lh, L. hesperus; Lt, L. tredecimguttatus; Lg, L. geometricus; Sg, S. grossa; Pt, P. tepidariorum* based on combining the incompletely assembled MiSp-encoding region from Scaffold 853 of the i5K genome with an ~2.3 kb TOPO-cloned PCR product, KX584004; *Av, A. ventricosus*, JX513956.1 [30]. Lengths of missing amino acid sequences in *Pt* MiSp_v1 are based on the length of gaps in Scaffold 853 of the i5K genome, assuming no introns are present within those gaps. The last gap was predicted to be 1782 bp long (594 aa if no intron) in Scaffold 853, but our TOPO clone added 2236 bp, including a 1175 bp intron, which was longer than the predicted gap. The repetitive encoding region at the 3' end of our TOPO clone did not overlap with the encoding region present in the i5K genome prior to the last gap

(Fig. 3). Spacers are easily distinguished from the repetitive motifs in their amino acid composition (rich in threonine, serine, and valine), near perfect repetition within a MiSp protein, and similar length across species (Fig. 4). Furthermore, with the exception of *S. grossa* MiSp, spacers are much more hydrophilic than the repetitive region (Additional file 1: Figure S5).

Material properties of minor ampullate spidroins correlate with MiSp sequence variation

We found significant differences in each of four tensile properties of minor ampullate silk fibers measured for three cobweb weaving species (Fig. 5). These properties were tensile strength, the amount of force applied to break a single silk fiber, accounting for the diameter of the fiber; extensibility, the proportion of the original fiber length that the fiber was extended at breakage; stiffness or Young's modulus, the initial slope of the force-extension curve (initial resistance to force); and toughness, the energy required to break the fiber (see Methods for details). Post-hoc Tukey's test revealed that minor ampullate silk from *L. geometricus* had the lowest tensile strength (Fig. 5). Minor ampullate silk from *S.*

grossa was significantly more extensible and less stiff than the other two species (Fig. 5).

Minor ampullate silks from the cobweb weavers are generally weaker, less stiff, and less tough than orb-web weavers, but more extensible (Table 3). We found a positive, but non-significant, relationship between the proportion of the GPG motif in MiSp and the extensibility of minor ampullate silk fibers among the five species with both published MiSp sequences and material properties (adjusted $R^2 = 0.53$, $p = 0.10$; Additional file 1: Figure S6). However, when accounting for phylogenetic relationships and evolutionary distances among species using phylogenetic independent contrasts, we found a significant positive relationship between GPG proportion and extensibility (adjusted $R^2 = 0.92$, $p = 0.006$), due to the extensive change in GPG and extensibility on the relatively short *S. grossa* branch (see Additional file 1: Figure S6 for the effect of varying branch lengths). This pattern of higher GPG content contributing to extensibility parallels patterns found for major ampullate silks, although for major ampullate silks higher GPG content results from higher expression of MaSp2, rather than sequence differences among orthologs [2, 14, 18, 22–25].

```
Cobweb Weaver Spacers
Lh MiSp_v1 spacer 1     SSTGYTERQNEVVTTVTTTRQEIAD--RRQ
Lh MiSp_v1 spacer 2     SSTGYTERQNEVVTTVTTTRQEIAD--RRQ
Lh MiSp_v1 spacer 3     ASTGYTERQNEVTTTVSTTRKETAD--RRQ
Lh MiSp_v1 spacer 4     SSTGYAGRQNEVITTVTTTRQETADYANKQ
Lh MiSp_v2 spacer 1     SSTGYTERQNEVVTTVTTTRQETADYAQKQ
Lt MiSp_v1 spacer 1     SSTGYTERQNEVVTTVTTTRQESADYARKQ
Lt MiSp_v2 spacer 1     SSTGYTERQNEVVTTVTTTRQETAD--RRQ
Lt MiSp_v2 spacer 2     SSTGYTERQNEVVTSVTSTRQETAD--RRQ
Lt MiSp_v2 spacer 3     SSTGYTERQNEVITTVTTTRQETAD--QRQ
Lt MiSp_v2 spacer 4     SATGYTERQNEVVTTVTTTRQETAD--RRQ
Lt MiSp_v2 spacer 5     SSTGYTERQNEVVTTVTSTRQETAD--RRQ
Lt MiSp_v2 spacer 6     SATGYTERQNEVVTTVTTTRQETAD--RRQ
Lt MiSp_v2 spacer 7     SSTGYTERQNEVVTTVTSTRQETAD--RRQ
Lg MiSp_v1 spacer 1     ---GYTQRQNEVITTVSTTRQETADYGQKQ
Lg MiSp_v1 spacer 2     ---GYTQKQNEVITTVSTTRQKIADYGQKQ
Lg MiSp_v1 spacer 3     ---GYTQKQNEVITTVSTTRQKIADYGQKQ
Lg MiSp_v1 spacer 4     ---GYTQRQNEVITTVSTTRQKTADYGQKQ
Lg MiSp_v1 spacer 5     ---GYTQRQNEVITTVSTTRQKTADYGQKQ
Lg MiSp_v1 spacer 6     ---GYIQKQNEVITTVSTTRQETADYGRKQ
Sg MiSp    spacer 1     RSTTSTSAATASATTVNNGPQI--------
Sg MiSp    spacer 2     RSTTSTSAATASATTVDIGPQV--------
Sg MiSp    spacer 3     RSTTSTSAATASATTVDNGPQF--------
Sg MiSp    spacer 4     WSTTSTTAATASATTVNNGPQI--------
Sg B5      spacer 1     RST--TSATTTSTTTVVNGPQV--------
Sg A1      spacer 1     RSTTSTSAATASATTVNNGPQI--------
Sg A13     spacer 1     RSTTSTSAATASATTVDNGPQF--------
Pt MiSp    spacer N     -----TSSQVSQTQNVVTTTQTENMI----
Pt MiSp    spacer C     -----TASQVSQTQNVVTTTQTENMI----

Orb-web Weaver Spacers
Av  MiSp   spacer      TGNRAGDAFAQVFSQNVINSG---VITSTTVTKNSAQAAASSMVST
Ad  MiSp   spacer      ??????????????????????????????????????????????
Aar MiSp   spacer      ???????AFAQAFSQNMLSSG---VISSETITTSSAQAAASTMATT
Na  MiSp   spacer      -GSAAGNAFAQSLSSNLLSSGDFVQMISSTTSTDQAVSVATSVAQN
Nc  MiSp1  spacer      -GSSAGNAFAQSLSSNLLSSGDFVQMISSTTSTDHAVSVATSVAQN
Ncr MiSp   spacer      -GSAAGNAFAQSLSSNLLSSGDFVQMISTTTSTDQAVSVATSVAQN
Ds  MiSp   spacer      GATAATTAFAQSMSSALASSPSFSSLFSSGLQTQDAVSASVSVAQT

Av  MiSp    AAKSLGLDENTARSMANAMSSYAAAMAKSFRNSDEFIRNMSYQMGRMLSNAGAINESTA
Ad  MiSp    ???????????????????SSYAAAMAASTRNS-DFIRNMSYQMGRLLSNAGAITESTA
Aar MiSp    AAKGMGLDDATANYLSNALASMAASMAQSSKNQADFIQSMSYAMGKLLSDAGAITESSA
Na  MiSp    VGNQLGLDANAMSLLGAVSGYVSTLGNAISDASAYANAISSAIGNVLANSGSISESTAS
Nc  MiSp1   VGSQLGLDANAMNNLLGAVSGYVSTLGNAISDASAYANALSSAIGNVLANSGSISESTA
Ncr MiSp    VGNQLGLDANAMNNLLAAVGGYVSSLGGAVADAAAYANAISSAIGNVLANTGSINESTA
Ds  MiSp    LANQVGLDNSATANLLQLVQRYVSSVG-AYADAVAYANAISKALGSVLANTGQITTSTA

Av  MiSp    SAAASSASSTVTETVRTYGPAAIFS-----------------
Ad  MiSp    SSAASSASSTVTESIRTYGPAAIFS-----------------
Aar MiSp    SSAASSASSTVTETIRTYGPASIFS-----------------
Na  MiSp    SAASSAASSVTTTLTSYGPAVFYAPTSSA-------------
Nc  MiSp1   SSAASSAASSVTTTLTSYGPAVFYAPSASS------------
Ncr MiSp    SSAASSAASSVTTTLTSYGPAVFY------------------
Ds  MiSp    YSTANSFAQTVTTYINNVSPRVLFSPPIGILPVPSGEAAPRE
```

Fig. 4 Comparison of theridiid (*top*) and orb-web weaver (*bottom*) MiSp spacers. Theridiid MiSp spacers are aligned for *L. hesperus* (*Lh*), *L. tredecimguttatus* (*Lt*), *L. geometricus* (*Lg*), *S. grossa* (*Sg*), and *P. tepidariorum* (*Pt*). Orb-web weaver MiSp spacers aligned for *A. ventricosus* (*Av*), *A. diadematus* (*Ad*), *Argiope argentata* (*Aar*), *Nephila antipodiana* (*Na*), *N. clavipes* (*Nc*), *Nephilengys cruentata* (*Ncr*), and *Deinopis spinosa* (*Ds*) following [30]. See Table 1 and Additional file 1: Table S4 for accession numbers. Note that the sequence of spacers in the *Pt* MiSp N and C-terminal encoding cDNAs shown here are identical to spacers found in *Pt* MiSp_v1

Phylogenetic relationships of spidroins among orb-web and cobweb weaving species suggest multiple duplication events

Spidroin N- and C-terminal domains and encoding nucleotides demonstrate MiSp sequences found in eight species form a monophyletic group that is usually recovered as sister to a clade of MaSp1 and MaSp2 sequences (Fig. 6, Additional file 1: Figure S7). Theridiid MiSp consistently group together with strong support (Fig. 6, Additional file 1: Figure S7). Within theridiids, MiSp sequences from *Latrodectus* species group together, *S. grossa* MiSp variants form a sister clade to *Latrodectus* MiSp variants, and *P. tepidariorum* MiSp is sister to *Latrodectus* and *Steatoda* MiSp variants. Within *Latrodectus, L. geometricus* MiSp variants form a sister clade to *L. hesperus and L. tredecimguttatus* MiSp variants; these patterns are congruent with species relationships [34, 35] (Fig. 6). However, MiSp variants group together within a species, suggesting independent duplication events within each species as inferred by reconciling our gene trees with the species trees in Fig. 1 (Fig. 6). Alternatively, this pattern could be explained by concerted evolution of terminal-encoding regions within each species, which we detail in the Discussion.

Outside of Theridiidae, relationships among MiSp sequences are more ambiguous and do not necessarily

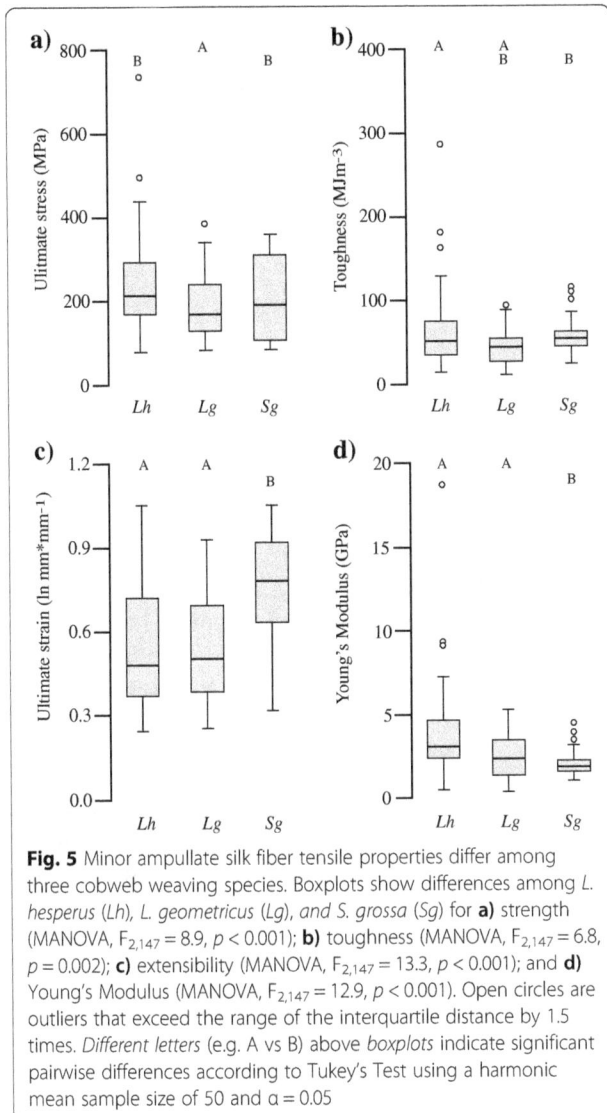

Fig. 5 Minor ampullate silk fiber tensile properties differ among three cobweb weaving species. Boxplots show differences among *L. hesperus* (*Lh*), *L. geometricus* (*Lg*), and *S. grossa* (*Sg*) for **a)** strength (MANOVA, $F_{2,147} = 8.9$, $p < 0.001$); **b)** toughness (MANOVA, $F_{2,147} = 6.8$, $p = 0.002$); **c)** extensibility (MANOVA, $F_{2,147} = 13.3$, $p < 0.001$); and **d)** Young's Modulus (MANOVA, $F_{2,147} = 12.9$, $p < 0.001$). Open circles are outliers that exceed the range of the interquartile distance by 1.5 times. *Different letters* (e.g. A vs B) above *boxplots* indicate significant pairwise differences according to Tukey's Test using a harmonic mean sample size of 50 and $\alpha = 0.05$

reflect species relationships. For instance, molecular and morphological characters strongly support monophyly of the Araneoidea (represented here by Theridiidae, Nephilidae, and Araneidae) to the exclusion of Deinopidae and Uloboridae [36–39]. In contrast, N- and C-terminal domains sometimes place deinopid and uloborid MiSp sister to theridiid MiSp, and sometimes place araneid MiSp sequences sister to theridiid MiSp sequences (Fig. 6, Additional file 1: Figure S7). The former placement is consistent with an ancient duplication of a MiSp-encoding locus in the common ancestor of Araneoidea, Deinopidae, and Uloboridae (Fig. 6). C-terminal domains and encoding nucleotides for MiSp from 14 species generally show the same patterns within Theridiidae, but theridiid MiSp sequences tend to group with nephilid and deinopoid MiSp sequences to the exclusion of araneid MiSp sequences, consistent with multiple ancient

duplication events (Fig. 7, Additional file 1: Figure S8). Relationships among other spidroins are consistent with previous findings that gland-specific spidroins tend to group together, with the exception of major ampullate spidroins [17, 32, 40]. Overall, we inferred 24–30 duplication events within the spidroin gene family, depending on which C-terminal gene tree was used for reconciliation (Additional file 1: Figure S8).

Discussion

Our characterization of complete or almost complete encoding sequences for the minor ampullate spidroin (MiSp) in multiple cobweb weaving spider species (Theridiidae) demonstrate the presence of at least two *MiSp* loci in each species. The substantial variation of MiSp sequences within and among species and its relationship to variation in material properties has multiple implications for molecular evolution, spider ecology, and biomimetic applications through recombinant DNA technology.

Two MiSp-encoding loci have also been documented in the golden orb-weaver spider, *Nephila clavipes* (Nephilidae), and therefore the presence of multiple MiSp-encoding loci in spider genomes most likely dates minimally back to the common ancestor of theridiids and nephilids (e.g. Araneoidea). Our reconciliation of gene trees with species trees suggests even older duplication events (Fig. 7). However, the maintenance of two loci appears to involve complex molecular evolutionary processes including intergenic concerted evolution of the loci within species, and multiple gains and losses of individual *MiSp* copies. Although our reconciliation analyses inferred independent duplication events within each of our cobweb weaving species, the grouping of *MiSp* loci within species based on the terminal domains (Figs. 6 and 7), could also result from intergenic concerted evolution of the N- and C-terminal encoding regions. We favor the latter hypothesis because of the dramatic differences in the repetitive region within some species (Fig. 3). Concerted evolution of the N and C-terminal encoding regions could occur through non-homologous recombination between the loci, facilitated by their similar sequences, as proposed for the major ampullate spidroin paralogs, *MaSp1* and *MaSp2*, of multiple species [2, 32, 41–43]. Within theridiids, the rate of concerted evolution appears to be faster than speciation, since relationships among each species' pair of *MiSp* loci reflect species relationships (Figs. 6 and 7). Intergenic concerted evolution of terminal-encoding regions could be favored by selection because these regions are involved in assembly of multiple MiSp monomers into polymers and the conversion of the protein complex from a liquid to a solid [31, 44]. Highly similar terminal regions may be necessary for polymers to form.

Table 3 Properties of minor ampullate silk for cobweb and orb-web weaving species

Species	Diameter (μm)	Young's Modulus (GPa)	Strength (MPa)	Extensibility (ln mm/mm)	Toughness (J/cm³)	GPG (%)
Latrodectus hesperus	1.12 ± 0.47	3.9 ± 2.9	245.4 ± 120.5	0.57 ± 0.02	66.7 ± 46.4	0.07 [g]
Latrodectus geometricus	1.09 ± 0.29	2.6 ± 1.3	174.4 ± 17.6	0.54 ± 0.02	43.6 ± 20.1	3.48 [h]
Steatoda grossa	1.07 ± 0.26	2.1 ± 0.7	251.3 ± 106.2	0.74 ± 0.02	57.3 ± 20.7	12.79 [i]
Nephila clavipes [a]	2.5 ± 0.2	3.0 ± 0.6	346 ± 53	0.30 ± 0.15		0 [j]
Nephila clavipes [b]	2.5		960 ± 50	0.25 ± 0.03		0 [j]
Nephila inaurata [c]	1.8 ± 0.1	11.2 ± 0.7	1500 ± 200	0.46 ± 0.05	300 ± 50	
Argiope trifasciata [d,e]	0.69 ± 0.10	8.9 ± 0.05	751.5 ± 40	0.44 ± 0.04	150 ± 12	
Argiope trifasciata [c]	1.8 ± 0.1	10 ± 0.4	1040 ± 6	0.45 ± 0.02	240 ± 20	
Argiope argentata [e]	1.1 ± 0.7	10.6 ± 1.2	923 ± 154	0.33 ± 0.03	137 ± 22	
Araneus diadematus [f]				0.29		0 [k]
Araneus gemmoides [b]	2.0		1400 ± 100	0.22 ± 0.07		

Material properties collected from [a][10], [b][15], [c][13], [d][9], [e][8], [f][74]
Proportion of MiSp that is made up of the motif GPG was averaged across sequences described for each species: [g] *Lh* MiSp_v1, MiSp_v2; [h] *Lg* MiSp_v1, MiSp_v2; [i] *Sg* MiSp: C-terminal encoding cDNA (KX584021); [j] AF027735; [k] U47853

Outside of theridiids, relationships among MiSp N and C-terminal domains are not congruent with species relationships. The low posterior probabilities and bootstrap support for these incongruous relationships suggest there is limited phylogenetic signal retained in the MiSp C-termini at the distant time scale of divergence of theridiids from other araneoid families (~170 million years ago, [36, 38, 39]). However, the consistent grouping of *Nephila* and theridiid MiSp C-terminal domains to the exclusion of araneid MiSp C-terminal domains suggests that nephilid and theridiid MiSp sequences are derived from a different ancient copy than the araneid MiSp sequences, as supported by our reconciliation analyses (Figs. 6 and 7). The loss of functional *MiSp* loci through multiple single nucleotide mutations is further supported by the presence of a *MiSp* pseudogene in the *S. grossa* genome. Many additional losses were inferred by our reconciliation analyses (Additional file 1: Figure S8), but because of incomplete spidroin sampling for most species we do not feel confident in estimating the extent of spidroin gene loss.

We found extensive variation in the MiSp repetitive regions between loci within a genome and among species (Fig. 3). MiSp variants in *L. hesperus* and *L. tredecimguttatus*, *L. geometricus* MiSp_v2, *P. tepidariorum* MiSp_v1, *Nephila* MiSp 1 & 2, and araneid MiSp sequences are similar in terms of amino acid motif composition. *S. grossa* MiSp and *L. geometricus* MiSp_v1 are especially divergent, with the difference between variants within *L. geometricus* being extremely striking (Fig. 3). The divergent repeats have an increased proline content, which likely occurred independently in *S. grossa* MiSp and *L. geometricus* MiSp_v1, based on gene tree relationships (Figs. 6 & 7). The mutation of alanine or glutamine codons to a proline codon requires only a single base change and the GPG motif is frequently found as GPGA

or GPGQ, indicating that it could have evolved from $(GA)_n$ or $(GQ)_n$ motifs, which are the most common motifs in all other MiSp sequences (Fig. 3). Intragenic concerted evolution could rapidly proliferate these mutations throughout a single gene as has been proposed for other spidroin paralogs [16, 17, 45, 46]. The near identity of spacer sequences within each of our cobweb weaver MiSp variants (Fig. 4) supports the hypothesis that intragenic concerted evolution is common, as also found in orb-web weaver MiSp [30, 31]. However, intragenic concerted evolution must be offset by some other molecular processes, since the remainder of the repetitive regions are not as homogenized as the spacers (Fig. 3). The simple amino acid motifs found in MiSp such as $(GA)_n$ are encoded by microsatellite-like sequences and could thus be prone to high rates of slipped strand mispairing. It is also possible that excessive homogenization of MiSp repetitive regions adversely affects its function and is eliminated by selection.

The high proportion of the GPG amino acid motif in *S. grossa* MiSp correlates with higher extensibility in *S. grossa* minor ampullate silk fibers (Fig. 5c, Table 3). GPG content is probably not the only predictor of extensibility, however, since *Latrodectus* minor ampullate fibers were more extensible than orb-web weavers' even though GPG content for *L. hesperus* is similar to the orb-web weavers. Furthermore, we did not find a significant difference in extensibility between *L. hesperus* and *L. geometricus* minor ampullate silk fibers despite *L. geometricus* MiSp_v1 having a relatively high percentage of GPG motifs (Fig. 3). The *MiSp_v2*: *MiSp_v1* transcript ratio in *L. geometricus* suggests that the GPG-rich MiSp_v1 is in lower abundance in *L. geometricus* (Table 2), which could explain why there is not a significant increase in extensibility of its minor ampullate silk fibers in comparison to *L. hesperus*. Cobweb weaving

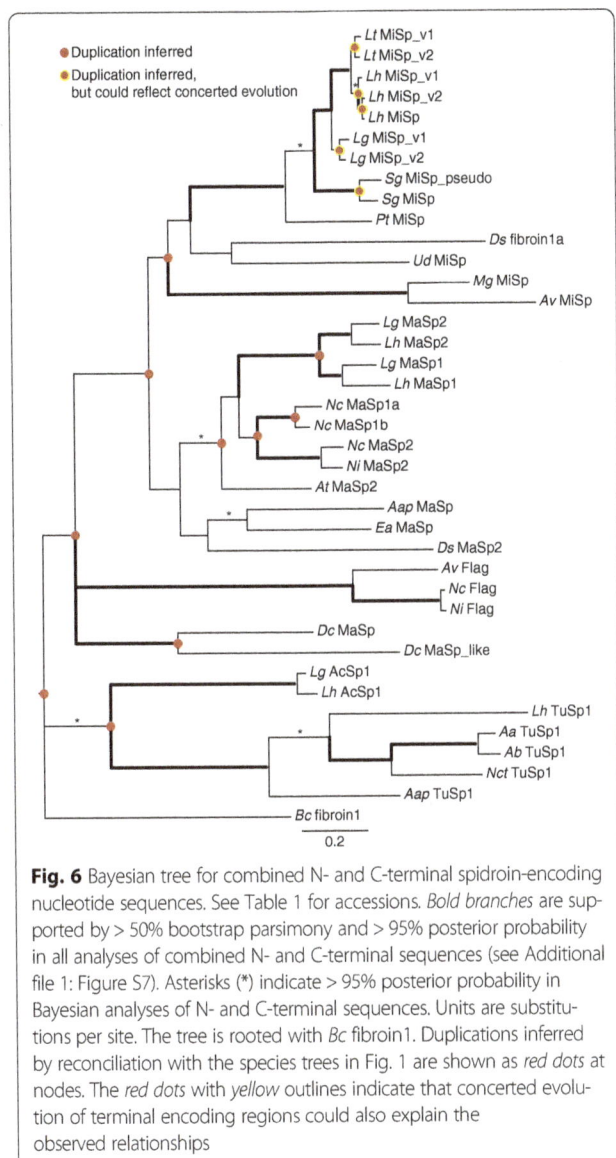

Fig. 6 Bayesian tree for combined N- and C-terminal spidroin-encoding nucleotide sequences. See Table 1 for accessions. *Bold branches* are supported by > 50% bootstrap parsimony and > 95% posterior probability in all analyses of combined N- and C-terminal sequences (see Additional file 1: Figure S7). Asterisks (*) indicate > 95% posterior probability in Bayesian analyses of N- and C-terminal sequences. Units are substitutions per site. The tree is rooted with *Bc* fibroin1. Duplications inferred by reconciliation with the species trees in Fig. 1 are shown as *red dots* at nodes. The *red dots* with *yellow* outlines indicate that concerted evolution of terminal encoding regions could also explain the observed relationships

spiders use MiSp in their prey wrapping silks [5], and evolution of extensibility-conferring motifs could potentially allow for capturing different prey types by *S. grossa* compared to other cobweb weavers, although little is known about the ecology or diet of this species. It is also possible that *L. geometricus* modulates expression of its two MiSp-encoding loci in response to prey availability. Experimentally manipulating prey type has been shown to affect proline content of major ampullate fibers in *Nephila*, potentially as a result of plastic changes in relative expression levels of the proline-poor MaSp1 and the proline-rich MaSp2 [47, 48]. Variation among individuals in *MaSp1* and *MaSp2* transcript abundance has also been demonstrated for black widows [49]. Analogous experiments have not been done for minor ampullate fibers. However, the variation in *MiSp_v2*: *MiSp_v1* ratios

that we found between *L. geometricus* individuals (Table 2) suggests plasticity of expression is possible.

Our six complete or nearly complete single exon *MiSp*-containing clones can also serve as templates for recombinant proteins. Due to the dramatic differences in length and amino acid content of some of the proteins encoded by these loci, it may be possible to spin artificial fibers with custom-made properties. For instance, the GPG-containing *L. geometricus* MiSp_v1 could be used to make more extensible fibers, while the longer alanine-rich *L. hesperus* MiSp_v1 may make stiffer fibers.

Conclusions

We found that intragenic concerted evolution within MiSp-encoding genes likely led to rapid proliferation of proline replacements for alanine or glutamine in MiSp protein sequences independently in at least two species. For one species, the proliferation of proline coincides with higher extensibility of minor ampullate silks. This could allow cobweb weavers to access new prey types or the ability to modulate mechanical properties of their silks through altering expression levels of *MiSp* gene copies. Our multiple nearly complete *MiSp* sequences also provide various templates for tailored biomimetic applications through recombinant DNA technologies.

Methods
Taxon sampling
Individual *L. hesperus* were collected in Riverside, California and Tucson, Arizona. *L. geometricus* individuals were collected in San Diego, California. *L. tredecimguttatus* and *S. grossa* were obtained from SpiderPharm (Yarnell, Arizona). *P. tepidariorum* were obtained from a laboratory culture founded with spiders collected near Cologne, Germany and purchased from SpiderPharm.

Sequencing
Genomic library clone
We screened an *L. hesperus* genomic library with PCR for clones containing *MiSp* (see [16] for library construction and screening protocols). Primers used in screening were designed from the C-terminal encoding region of a partial-length *L. hesperus* *MiSp* cDNA (HM752571; primers listed in Additional file 1: Table S3). An ~49 kb *MiSp* containing clone was shotgun sequenced and assembled to 8× coverage by Qiagen (Hilden, Germany).

PCR amplification and cloning of Latrodectus genomic MiSp
We amplified full-length *L. hesperus* *MiSp* from genomic DNA using primers designed from the 3' UTR of the C-terminal encoding *MiSp* cDNA (HM752571) and the N-terminal region of another partial *MiSp* cDNA (HM752570). We amplified genomic *MiSp* from *L.*

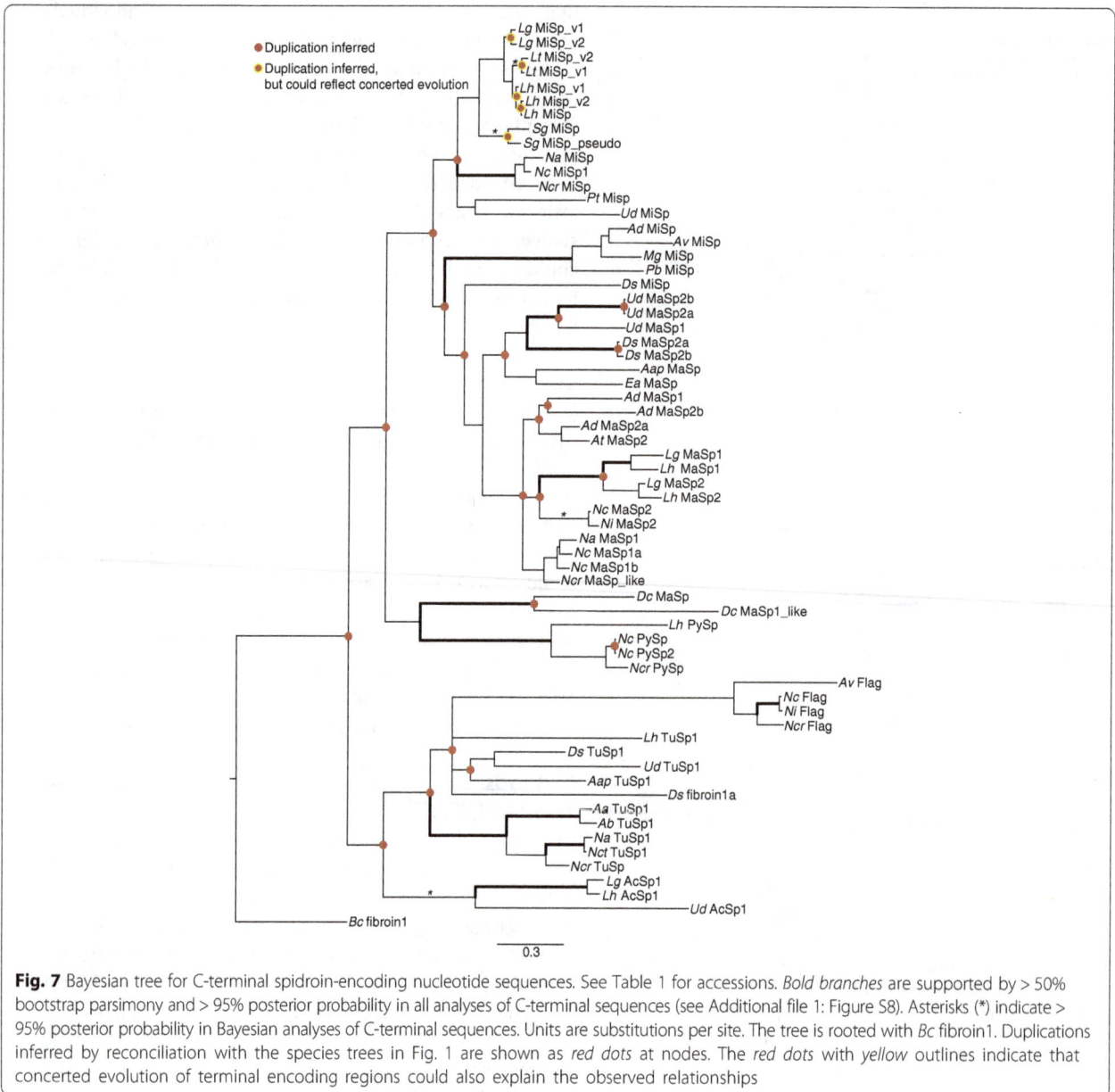

Fig. 7 Bayesian tree for C-terminal spidroin-encoding nucleotide sequences. See Table 1 for accessions. *Bold branches* are supported by > 50% bootstrap parsimony and > 95% posterior probability in all analyses of C-terminal sequences (see Additional file 1: Figure S8). Asterisks (*) indicate > 95% posterior probability in Bayesian analyses of C-terminal sequences. Units are substitutions per site. The tree is rooted with *Bc* fibroin1. Duplications inferred by reconciliation with the species trees in Fig. 1 are shown as *red dots* at nodes. The *red dots* with *yellow* outlines indicate that concerted evolution of terminal encoding regions could also explain the observed relationships

tredecimguttatus and *L. geometricus* first using primers designed from *L. hesperus* MiSp N and C-terminal cDNAs and then with internal species-specific primers (primers in Additional file 1: Table S3). Genomic PCR was performed with 1 unit of AccuPrime™ *Taq* DNA Polymerase High Fidelity (Invitrogen, Carlsbad, CA) according to manufacturer's instructions. Cycling conditions were 40 cycles of 94 °C for 30 s, 50–60 °C (depending on primer pair) for 45 s, and 68 °C for 10 min.

PCR products were gel-excised and cloned using the TOPO®-TA Cloning® Kit (Invitrogen). Clones were screened by PCR amplification of MiSp N and/or C-terminal encoding region, and *MiSp* positive clones were sequenced with Sp6, T7, M13F, or M13R universal primers. End sequencing of TOPO clones indicated that, within both *L. geometricus* and *L. tredecimguttatus*, clones clustered into two distinct groups (Additional file 1: Figure S2). For both *L. tredecimguttatus* and *L. geometricus*, the longest clones representing each cluster were completely sequenced. End sequences and restriction digest patterns of *L. hesperus* TOPO clones were very similar. Thus, one clone was chosen for complete sequencing. Clones chosen for complete sequencing were subjected to random transposon insertion with the GPS-1 Genome Priming System (New England Biolabs, Ipswich, MA). Position of transposon insertion was mapped with restriction digests and clones were sequenced from both ends of the transposon. Sequences were assembled with Sequencher v. 4.9 (GeneCodes).

Steatoda grossa and P. tepidariorum cDNA libraries

We constructed cDNA libraries from *S. grossa* total silk glands and *P. tepidariorum* major and minor ampullate silk glands following procedures detailed in Garb et al. (2010). In brief, total RNA was extracted with TRIzol® (Invitrogen) and the RNeasy Mini Kit (Qiagen, Valencia, CA). mRNA was isolated with oligo-(dT)-tagged magnetic beads (Invitrogen). cDNA was synthesized with SuperScript®III (Invitrogen), fractionated by size with ChromaSpin 1000 columns (Clontech, Mountain View, CA), blunt-end ligated into pZErO™-2 (Invitrogen), and electroporated into TOP10 *Escherichia coli* cells (Invitrogen). Approximately 2000 colonies per library were arrayed and stored at −80 °C. The libraries were screened by gel electrophoresis of plasmid DNA [50]. Plasmids with inserts >500 bp were sequenced with T7 or SP6 universal primers. We identified silk protein-encoding sequences by conceptual translations and comparisons to published sequences with BLASTX [51].

An *S. grossa* cDNA clone (designated *Sg* MiSp) with a 3 kb insert containing the C-terminal encoding portion of MiSp was completely sequenced by random insertion of transposons with the EZ-Tn5 < Tet > kit (Epicentre). Transposon mapping, sequencing, and assembly were performed as for TOPO clones above.

Amplification and cloning of S. grossa and P. tepidariorum MiSp

We attempted to amplify complete MiSp encoding sequences from *S. grossa* genomic DNA using primers designed from conserved regions of *Latrodectus* MiSp N-terminal encoding sequences and from the C-terminal encoding region of the *S. grossa MiSp* cDNA (Additional file 1: Table S3). *S. grossa* genomic DNA was amplified with 1 unit of Phusion® High-Fidelity DNA Polymerase (New England Biolabs), 200 μM each dNTP, 0.5 μM each primer, 1X Phusion® GC buffer, and 3% DMSO. Cycling conditions were 40 cycles of 98 °C for 5 s, 58 °C for 15 s, and 72 °C for 2.5 min. PCR products were gel-excised and cloned using the TOPO®-Blunt Cloning® Kit (Invitrogen). End sequencing of these clones revealed numerous premature stop codons and thus clones were not completely sequenced.

In order to obtain longer N-terminal MiSp encoding sequences from *S. grossa*, we amplified *S. grossa* genomic DNA using forward primers designed from the conserved region of *Latrodectus* MiSp N-terminal encoding sequences or from the *S. grossa* N-terminal *MiSp* cDNA, and a reverse primer designed from the repetitive region of the *S. grossa* C-terminal cDNA (Additional file 1: Table S3). Genomic DNA was amplified with 1 unit

GoTaq Polymerase (Promega), 1X GoTaq Buffer, 200 μM each dNTP, 0.5 μM each primer, and 1.5 mM MgCl₂ with 45 cycles of 30s at 94 °C, 45 s at 54.2 °C or 61 °C, and 1.5 or 2 min at 72 °C. PCR products from the two amplifications were separately gel-excised and cloned using the TOPO®-TA XL PCR Cloning® Kit (Invitrogen). TOPO clones from each reaction were amplified with M13 forward and reverse primers and sequenced with T7 or SP6 universal primers.

A *MiSp* gene was identified in the *P. tepidariorum* genome by BLASTN searches of our N- and C-terminal encoding MiSp cDNA sequences against the genome assembly [52] generated by the i5K initiative [53]. We found alignments of both cDNAs to a region on Scaffold 853 that included an Augustus-predicted gene model (aug3.g7951.t3). Using the genomic sequence, we designed a reverse primer in the C-terminal encoding region and a forward primer in the adjacent repetitive region. Genomic DNA and cDNA were amplified as for *S. grossa* N-terminal encoding region but with 40 cycles of 30s at 94 °C, 45 s at 60.1 °C and 1 min at 72 °C. PCR products were gel-excised and sequenced. Genomic PCR products had sizes ranging from 250 bp to 2500 bp, while cDNA PCR products had sizes ranging from 250 bp to 1200 bp. Those bands that were too large to directly sequence were TOPO cloned using the TOPO-TA PCR Cloning Kit (Invitrogen) and completely sequenced by random insertion of transposons as for *S. grossa* TOPO clones above. Completed TOPO clones were compared to the *P. tepidariorum* genome by BLASTN.

Amino acid sequence characteristics

Amino acid composition and codon usage were calculated with codonW [54]. Kyte-Doolittle hydropathy plots [55] with a window size of seven were generated online [56]. Amino acid motifs GGX, A_n ($n \geq 4$), $(GA)_n$ ($n \geq 2$), and GPG were identified by word searches in Microsoft Word. Spacer regions were defined manually as regions that were not in the N or C-termini and were low in alanine and glycine and were manually aligned. Two authors (JVH and CYH) independently assigned spacer regions. Tandem repeats were identified by XSTREAM v1.4.8 using default parameters for moderate repeat degeneracy and high significance [57].

Transcript abundance estimation

Transcript abundance was estimated by aligning raw RNA-seq reads [58] (SRR1539569 and SRR1539569 for *L. hesperus* and *L. geometricus* respectively) to our full or almost full-length *MiSp* paralogs found in *L. hesperus* (*Lh MiSp_v1 and MiSp_v2*) and *L. geometricus* (*Lg MiSp_v1 and MiSp_v2*) using Bowtie 2 with default parameters, which searches for multiple alignments for each read but reports the one best alignment [59]. RNA-seq reads were generated for mRNA isolated from major and minor ampullate glands of ~30 adult females and from all the silk glands ("total silk") of a single adult female for both *L. hesperus* and *L. geometricus*. SAMtools v1.1 was used to report the number of sequence reads matching our reference *MiSp* sequences [60]. To compare transcript abundance between loci within the same library, we first compared the total counts of reads aligned to the entire reference sequence. We did not correct for gene length because many of our libraries were heavily 3'-prime biased. However, to ensure that differences in gene length did not dramatically affect our estimates of the ratio of one MiSp locus to another, we also compared the counts that aligned to the last 500 bases of coding sequence (Table 2).

Relationships among spidroin paralogs

We added our theridiid and *A. ventricosus* [30] MiSp N and C-terminal coding sequences to an alignment of 29 spidroin termini analyzed previously (see [17]). N-terminal sequences were not available for many spidroins and thus we added 25 additional spidroin sequences to the C-terminal alignment to more comprehensively represent spidroin gene family diversity in our analyses (Table 1). Additional C-termini of spidroin paralogs were chosen for species for which MiSp has been characterized. The expanded amino acid alignment was manually edited and used to guide the nucleotide alignment in SeaView v. 4.2.7 [61, 62].

We conducted heuristic searches for maximum parsimony (MP) and maximum likelihood (ML) trees based on amino acid and nucleotide alignments in PAUP* v4.0b10 using tree bisection reconnection branch swapping and 1000 (MP) or 10 (ML) replicates of random stepwise addition of taxa [63]. Support for clades recovered in MP analyses was evaluated with 1000 bootstrap pseudoreplicates and 10 random addition sequences per pseudoreplicate.

Bayesian analyses were carried out with MRBAYES v.3.1.2 [64, 65]. Optimal models of evolution were determined for nucleotide sequences with JMODELTEST v0.1.1 [66] and for protein sequences with PROTTEST v2.4 [67] for N and C-termini separately. Combined analysis of nucleotides employed a model partitioned by N and C-termini. Combined analysis of amino acids employed a mixed model, which allowed estimation of the optimal model of protein evolution during the Bayesian analysis. Default priors and Metropolis coupled, Markov-chain, Monte Carlo sampling procedures were executed for two independent runs of 10 million generations each, sampled every 1000th generation. Convergence was assessed every 1000th generation and the posterior distribution was considered adequately sampled if the standard deviation of split frequencies of these two runs was below 0.01. The mygalomorph spidroin, *Bothriocyrtum californicum* fibroin1 (*Bc* fibroin1), was set as the outgroup in all analyses.

Gene tree – species tree reconciliation

We inferred gene duplication events by reconciling each of our gene trees with two species trees (Fig. 1) using Notung v. 2.8.1.7 [68]. Species trees were based on published phylogenies as follows. Relationships among families were based on Garrison et al. [38] and Dimitrov et al. [39]. These studies agreed on all family-level relationships except the placement of Deinopidae and Uloboridae. The former placed Deinopidae sister to the RTA-clade (represented here by *E. australis* and *A. aperta*) with Uloboridae sister to Deinopidae plus the RTA-clade. The latter switched the placement of Uloboridae and Deinopidae. For relationships among theridiid species we used Liu et al. [35] and Garb et al. [34]. Relationships among nephilid species were based on Kuntner et al. [69]. Relationships among araneid genera were based on Dimitrov et al. [39] with relationships for *Argiope* species from Cheng and Kuntner [70].

Mechanical properties of minor ampullate fibers

We obtained minor ampullate silk from *L. hesperus*, *L. geometricus*, and *S. grossa* adult females within one week of being received or collected. Individual spiders were anesthetized with CO_2 for 2–5 min. We then secured the spider ventral side up to a stereo microscope stage using Scotch* tape, exposing the spinnerets. We manually pulled silk emanating from the minor ampullate spigots. Single fibers were taped to m-shaped cardstock (2.54 × 7.62 cm), with a single silk fiber suspended across each of the two rectangular notches (1 cm wide × 1.5 cm deep) in a collection card. Fibers were secured onto the cards using cyanoacrylate glue. We collected four to ten minor ampullate fibers from each of seven (*L. hesperus* and *S. grossa*) or eight (*L. geometricus*) individuals per species.

We measured silk fiber diameters using polarized light microscopy (as described in [6]). Tensile tests were conducted with a NanoBionix tensile tester (formerly MTS Systems Corp., Oak Ridge, TN; currently Keysight Technologies, Santa Rosa, CA) [6]. In brief, each m-shaped card was cut in half, such that only one fiber segment was tested at one time. Each silk sample was mounted onto the NanoBionix and extended at a rate of 1% per second until failure. We defined strength as the true stress of the fiber at breakage. True stress is the force applied to the fiber divided by its cross-sectional area calculated from the original cross-sectional area assuming constant volume of the fiber during extension [71]. We defined extensibility as the true strain at breakage.

True strain is the natural log of the ratio of the instantaneous length of the fiber to the original gage length of the fiber. We also measured stiffness or Young's modulus, the initial slope of the true stress-true strain curve, and toughness, the area under the true stress-true strain curve. Force-extension data were visualized using Testworks 4.0 software (MTS Systems Corp.). We tested for differences among species for the four tensile properties using multivariate analysis of variance (MANOVA). We used Tukey's test for post-hoc pairwise comparisons. All statistical tests were carried out with SPSS 13.0 (SPSS Inc, Chicago, IL).

We compiled published tensile data for minor ampullate silks and fit the relationship between proportion of GPG amino acid motifs and extensibility to linear models implemented in R for each species for which at least one MiSp sequence and minor ampullate fiber tensile data were available. Minor ampullate silk was assumed to be composed only of MiSp and for each species all available MiSp motif percentages were averaged to estimate composition. To account for phylogenetic relationships and evolutionary distances among species we calculated phylogenetic independent contrasts using APE [72] implemented in R. We used estimates of divergence time from three published molecular clock studies [35, 38, 39] for branch lengths (see Additional file 1: Figure S6).

Acknowledgements
We thank Stephen Richards and colleagues at Baylor College of Medicine's Human Genome Sequencing Center for access to the *P. tepidariorum* genome. We also thank Alistair McGregor, Mario Stanke, and colleagues for annotating this genome. Students in Genetics Lab (BIOL 221) at Washington and Lee University assisted with collecting sequence data.

Funding
This work was supported by the National Science Foundation (IOS-0951086 to NAA, IOS-0951061 to CYH), National Institutes of Health (F32 GM78875-1A to NAA; 1F32GM083661-01 and 1R15GM097714-01 to JEG), Army Research Office (W911NF-06-1-0455 and W911NF-11-1-0299 to CYH), and Washington and Lee University through Lenfest Summer Fellowships to NAA and Summer Research Scholarships to JVH and AKL. ERB was supported in part by a grant to Washington and Lee University from the Howard Hughes Medical Institute through the Precollege and Undergraduate Science Education Program (52007570).

Availability of data and materials
Sequences generated for this study are available in GenBank (Accessions: KX584003 - KX584055). Accessions for previously published sequences are reported in the manuscript text and in Table 1.

Authors' contributions
CYH and NAA conceived the study. JMV, ERB, JEG, EES, and NAA generated sequence data. AKL, MAC, SMC, and CYH measured tensile properties of minor ampullate silk fibers. JMV, ERB, MAC, THC, and NAA analyzed the data. JMV and NAA wrote the manuscript. All authors edited the manuscript and read and approved the final manuscript.

Competing interests
The authors declare that they have no competing interests.

Author details
[1]Department of Biology, Washington and Lee University, Lexington, VA, USA.
[2]Department of Biology, University of California, Riverside, CA, USA.
[3]Department of Biological Sciences, University of Massachusetts, Lowell, MA, USA.

References
1. Foelix RF. Biology of Spiders. 3rd ed. New York: Oxford University Press; 2011.
2. Gatesy J, Hayashi C, Motriuk D, Woods J, Lewis R. Extreme diversity, conservation, and convergence of spider silk fibroin sequences. Science. 2001;291:2603–5.
3. Guerette PA, Ginzinger DG, Weber BH, Gosline JM. Silk properties determined by gland-specific expression of a spider fibroin gene family. Science. 1996;272:112–5.
4. Vollrath F. Spider webs and silks. Sci Am. 1992;266:70–6.
5. Mattina CL, Reza R, Hu X, Falick AM, Vasanthavada K, McNary S, et al. Spider minor ampullate silk proteins are constituents of prey wrapping silk in the cob weaver *Latrodectus hesperus†*. Biochemistry (Mosc). 2008;47:4692–700.
6. Blackledge TA, Summers AP, Hayashi CY. Gumfooted lines in black widow cobwebs and the mechanical properties of spider capture silk. Zoology. 2005;108:41–6.
7. Gosline JM, DeMont ME, Denny MW. The structure and properties of spider silk. Endeavour. 1986;10:37–43.
8. Blackledge TA, Hayashi CY. Silken toolkits: biomechanics of silk fibers spun by the orb web spider *Argiope argentata* (Fabricius 1775). J Exp Biol. 2006; 209:2452–61.
9. Hayashi CY, Blackledge TA, Lewis RV. Molecular and mechanical characterization of aciniform silk: Uniformity of iterated sequence modules in a novel member of the spider silk fibroin gene family. Mol Biol Evol. 2004;21:1950–9.
10. Liivak O, Flores A, Lewis R, Jelinski LW. Conformation of the polyalanine repeats in minor ampullate gland silk of the spider *Nephila clavipes*. Macromolecules. 1997;30:7127–30.
11. Pérez-Rigueiro J, Elices M, Llorca J, Viney C. Tensile properties of *Argiope trifasciata* drag line silk obtained from the spider's web. J Appl Polym Sci. 2001;82:2245–51.
12. Swanson BO, Blackledge TA, Beltrán J, Hayashi CY. Variation in the material properties of spider dragline silk across species. Appl Phys A. 2006;82:213–8.
13. Guinea GV, Elices M, Plaza GR, Perea GB, Daza R, Riekel C, et al. Minor ampullate silks from *Nephila* and *Argiope* spiders: tensile properties and microstructural characterization. Biomacromolecules. 2012;13:2087–98.
14. Papadopoulos P, Ene R, Weidner I, Kremer F. Similarities in the structural organization of major and minor ampullate spider silk. Macromol Rapid Commun. 2009;30:851–7.
15. Stauffer SL, Coguill SL, Lewis RV. Comparison of physical properties of three silks from *Nephila clavipes* and *Araneus gemmoides*. J Arachnol. 1994;5–11.
16. Ayoub NA, Garb JE, Tinghitella RM, Collin MA, Hayashi CY. Blueprint for a high-performance biomaterial: full-length spider dragline silk genes. PLoS One. 2007;2, e514.
17. Ayoub NA, Garb JE, Kuelbs A, Hayashi CY. Ancient properties of spider silks revealed by the complete gene sequence of the prey-wrapping silk protein (AcSp1). Mol Biol Evol. 2013;30:589–601.
18. Hayashi CY, Shipley NH, Lewis RV. Hypotheses that correlate the sequence, structure, and mechanical properties of spider silk proteins. Int J Biol Macromol. 1999;24:271–5.
19. Motriuk-Smith D, Smith A, Hayashi CY, Lewis RV. Analysis of the conserved N-terminal domains in major ampullate spider silk proteins. Biomacromolecules. 2005;6:3152–9.
20. Sponner A, Schlott B, Vollrath F, Unger E, Grosse F, Weisshart K. Characterization of the protein components of *Nephila clavipes* dragline silk. Biochemistry (Mosc). 2005;44:4727–36.
21. Vasanthavada K, Hu X, Falick AM, La Mattina C, Moore AM, Jones PR, et al. Aciniform spidroin, a constituent of egg case sacs and wrapping silk fibers from the black widow spider *Latrodectus hesperus*. J Biol Chem. 2007;282: 35088–97.
22. Holland GP, Jenkins JE, Creager MS, Lewis RV, Yarger JL. Quantifying the fraction of glycine and alanine in β-sheet and helical conformations in spider dragline silk using solid-state NMR. Chem Commun. 2008;43:5568–70.

23. Holland GP, Creager MS, Jenkins JE, Lewis RV, Yarger JL. Determining secondary structure in spider dragline silk by carbon- carbon correlation solid-state NMR spectroscopy. J Am Chem Soc. 2008;130:9871–7.

24. Jenkins JE, Creager MS, Lewis RV, Holland GP, Yarger JL. Quantitative correlation between the protein primary sequences and secondary structures in spider dragline silks. Biomacromolecules. 2009;11:192–200.

25. Jenkins JE, Creager MS, Butler EB, Lewis RV, Yarger JL, Holland GP. Solid-state NMR evidence for elastin-like β-turn structure in spider dragline silk. Chem Commun. 2010;46:6714–6.

26. van Beek JD, Hesst S, Vollrath F, Meier BH. The molecular structure of spider dragline silk: Folding and orientation of the protein backbone. Proc Natl Acad Sci U S A. 2002;99:10266–71.

27. Hinman MB, Lewis RV. Isolation of a clone encoding a second dragline silk fibroin. Nephila clavipes dragline silk is a two-protein fiber. J Biol Chem. 1992;267:19320–4.

28. Xu M, Lewis RV. Structure of a protein superfiber: spider dragline silk. Proc Natl Acad Sci. 1990;87:7120–4.

29. Colgin MA, Lewis RV. Spider minor ampullate silk proteins contain new repetitive sequences and highly conserved non-silk-like "spacer regions.". Protein Sci. 1998;7:667–72.

30. Chen G, Liu X, Zhang Y, Lin S, Yang Z, Johansson J, et al. Full-length minor ampullate spidroin gene sequence. PloS One. 2012;7, e52293.

31. Gao Z, Lin Z, Huang W, Lai CC, Fan J, Yang D. Structural characterization of minor ampullate spidroin domains and their distinct roles in fibroin solubility and fiber formation. PLoS ONE. 2013;8, e56142.

32. Garb JE, Ayoub NA, Hayashi CY. Untangling spider silk evolution with spidroin terminal domains. BMC Evol Biol. 2010;10:243.

33. Wheelan SJ, Church DM, Ostell JM. Spidey: A tool for mRNA-to-genomic alignments. Genome Res. 2001;11:1952–7.

34. Garb JE, González A, Gillespie RG. The black widow spider genus Latrodectus (Araneae: Theridiidae): Phylogeny, biogeography, and invasion history. Mol Phylogenet Evol. 2004;31:1127–42.

35. Liu J, May-Collado LJ, Pekár S, Agnarsson I. A revised and dated phylogeny of cobweb spiders (Araneae, Araneoidea, Theridiidae): a predatory cretaceous lineage diversifying in the era of the ants (Hymenoptera, Formicidae). Mol Phylogenet Evol. 2015;94:658–75.

36. Bond JE, Garrison NL, Hamilton CA, Godwin RL, Hedin M, Agnarsson I. Phylogenomics resolves a spider backbone phylogeny and rejects a prevailing paradigm for orb web evolution. Curr Biol. 2014;24:1765–71.

37. Coddington JA. Phylogeny and classification of spiders. In: Ubick D, Paquin P, Cushing PE, Roth V, editors. Spiders of North America: and Identification Manual. 2005. p. 18-24.

38. Garrison NL, Rodriguez J, Agnarsson I, Coddington JA, Griswold CE, Hamilton CA, et al. Spider phylogenomics: untangling the Spider Tree of Life. PeerJ. 2016;4, e1719.

39. Dimitrov D, Benavides LR, Arnedo MA, Giribet G, Griswold CE, Scharff N, et al. Rounding up the usual suspects: a standard target-gene approach for resolving the interfamilial phylogenetic relationships of ecribellate orb-weaving spiders with a new family-rank classification (Araneae, Araneoidea). Cladistics. 2016;online early.

40. Starrett J, Garb JE, Kuelbs A, Azubuike UO, Hayashi CY. Early events in the evolution of spider silk genes. PLoS One. 2012;7, e38084.

41. Ayoub NA, Hayashi CY. Multiple recombining loci encode MaSp1, the primary constituent of dragline silk, in widow spiders (Latrodectus: Theridiidae). Mol Biol Evol. 2008;25:277–86.

42. Beckwitt R, Arcidiacono S, Stote R. Evolution of repetitive proteins: spider silks from Nephila clavipes (Tetragnathidae) and Araneus bicentenarius (Araneidae). Insect Biochem Mol Biol. 1998;28:121–30.

43. Rising A, Johansson J, Larson G, Bongcam-Rudloff E, Engström W, Hjälm G. Major ampullate spidroins from Euprosthenops australis: multiplicity at protein, mRNA and gene levels. Insect Mol Biol. 2007;16:551–61.

44. Ittah S, Cohen S, Garty S, Cohn D, Gat U. An essential role for the C-terminal domain of a dragline spider silk protein in directing fiber formation. Biomacromolecules. 2006;7:1790–5.

45. Chaw RC, Zhao Y, Wei J, Ayoub NA, Allen R, Atrushi K, et al. Intragenic homogenization and multiple copies of prey-wrapping silk genes in Argiope garden spiders. BMC Evol Biol. 2014;14:31.

46. Hayashi CY, Lewis RV. Molecular architecture and evolution of a modular spider silk protein gene. Science. 2000;287:1477–9.

47. Blamires SJ, Chao I-C, Tso I-M. Prey type, vibrations and handling interactively influence spider silk expression. J Exp Biol. 2010;213:3906–10.

48. Tso I-M, Wu H-C, Hwang I-R. Giant wood spider Nephila pilipes alters silk protein in response to prey variation. J Exp Biol. 2005;208:1053–61.

49. Lane AK, Hayashi CY, Whitworth GB, Ayoub NA. Complex gene expression in the dragline silk producing glands of the Western black widow (Latrodectus hesperus). BMC Genomics. 2013;14:1.

50. Beuken E, Vink C, Bruggeman CA. Enhanced efficiency of cloning FACS®-sorted mammalian cells. BioTechniques. 1998;24:750–2.

51. Altschul SF, Gish W, Miller W, Myers EW, Lipman DJ. Basic local alignment search tool. J Mol Biol. 1990;215:403–10.

52. BLAST Query | i5k - App [Internet]. [cited 2014 Jun 3]. Available from: https://i5k.nal.usda.gov/webapp/blast/.

53. i5K Consortium. The i5K Initiative: advancing arthropod genomics for knowledge, human health, agriculture, and the environment. J Hered. 2013; 104:595–600.

54. Peden J. codonW [Internet]. [cited 2015 Jul 7]. Available from: http://codonw.sourceforge.net/.

55. Kyte J, Doolittle RF. A simple method for displaying the hydropathic character of a protein. J Mol Biol. 1982;157:105–32.

56. Kyte-Doolittle entry form [Internet]. [cited 2015 Jul 6]. Available from: http://gcat.davidson.edu/rakarnik/kyte-doolittle.htm.

57. Newman AM, Cooper JB. XSTREAM: a practical algorithm for identification and architecture modeling of tandem repeats in protein sequences. BMC Bioinformatics. 2007;8:382.

58. Clarke TH, Garb JE, Hayashi CY, Arensburger P, Ayoub NA. Spider transcriptomes identify ancient large-scale gene duplication event potentially important in silk gland evolution. Genome Biol Evol. 2015;7: 1856–70.

59. Langmead B, Salzberg SL. Fast gapped-read alignment with Bowtie 2. Nat Methods. 2012;9:357–9.

60. Li H, Handsaker B, Wysoker A, Fennell T, Ruan J, Homer N, et al. The sequence alignment/map format and SAMtools. Bioinformatics. 2009;25: 2078–9.

61. Galtier N, Gouy M, Gautier C. SEAVIEW and PHYLO_WIN: two graphic tools for sequence alignment and molecular phylogeny. Comput Appl Biosci CABIOS. 1996;12:543–8.

62. Gouy M, Guindon S, Gascuel O. SeaView version 4: a multiplatform graphical user interface for sequence alignment and phylogenetic tree building. Mol Biol Evol. 2010;27:221–4.

63. Swofford DL. PAUP* version 4.0 b10. Sunderland: Sinauer; 2002.

64. Huelsenbeck JP, Ronquist F. MRBAYES: Bayesian inference of phylogenetic trees. Bioinformatics. 2001;17:754–5.

65. Ronquist F, Huelsenbeck JP. MrBayes 3: Bayesian phylogenetic inference under mixed models. Bioinformatics. 2003;19:1572–4.

66. Posada D. jModelTest: phylogenetic model averaging. Mol Biol Evol. 2008; 25:1253–6.

67. Abascal F, Zardoya R, Posada D. ProtTest: selection of best-fit models of protein evolution. Bioinformatics. 2005;21:2104–5.

68. Durand D, Halldórsson BV, Vernot B. A hybrid micro-macroevolutionary approach to gene tree reconstruction. J Comput Biol. 2006;13:320–35.

69. Kuntner M, Arnedo MA, Trontelj P, Lokovšek T, Agnarsson I. A molecular phylogeny of nephilid spiders: evolutionary history of a model lineage. Mol Phylogenet Evol. 2013;69:961–79.

70. Cheng R-C, Kuntner M. Phylogeny suggests nondirectional and isometric evolution of sexual size dimorphism in argiopine spiders: nondirectional and isometric evolution of SSD. Evolution. 2014;68:2861–72.

71. Vollrath F, Madsen B, Shao Z. The effect of spinning conditions on the mechanics of a spider's dragline silk. Proc R Soc Lond B Biol Sci. 2001;268: 2339–46.

72. Paradis E, Claude J, Strimmer K. APE: Analyses of Phylogenetics and Evolution in R language. Bioinformatics. 2004;20:289–90.

73. Rising A, Hjälm G, Engström W, Johansson J. N-terminal nonrepetitive domain common to dragline, flagelliform, and cylindriform spider silk proteins. Biomacromolecules. 2006;7:3120–4.

74. Work RW. Dimensions, birefrigences, and force-elongation behavior of major and minor ampullate silk fibers from orb-web-spinning spiders - the effects of wetting on these properties. Text Res J. 1977;47:650–62.

Permissions

The contributors of this book come from diverse backgrounds, making this book a truly international effort. This book will bring forth new frontiers with its revolutionizing research information and detailed analysis of the nascent developments around the world.

We would like to thank all the contributing authors for lending their expertise to make the book truly unique. They have played a crucial role in the development of this book. Without their invaluable contributions this book wouldn't have been possible. They have made vital efforts to compile up to date information on the varied aspects of this subject to make this book a valuable addition to the collection of many professionals and students.

This book was conceptualized with the vision of imparting up-to-date information and advanced data in this field. To ensure the same, a matchless editorial board was set up. Every individual on the board went through rigorous rounds of assessment to prove their worth. After which they invested a large part of their time researching and compiling the most relevant data for our readers.

The editorial board has been involved in producing this book since its inception. They have spent rigorous hours researching and exploring the diverse topics which have resulted in the successful publishing of this book. They have passed on their knowledge of decades through this book. To expedite this challenging task, the publisher supported the team at every step. A small team of assistant editors was also appointed to further simplify the editing procedure and attain best results for the readers.

Apart from the editorial board, the designing team has also invested a significant amount of their time in understanding the subject and creating the most relevant covers. They scrutinized every image to scout for the most suitable representation of the subject and create an appropriate cover for the book.

The publishing team has been an ardent support to the editorial, designing and production team. Their endless efforts to recruit the best for this project, has resulted in the accomplishment of this book. They are a veteran in the field of academics and their pool of knowledge is as vast as their experience in printing. Their expertise and guidance has proved useful at every step. Their uncompromising quality standards have made this book an exceptional effort. Their encouragement from time to time has been an inspiration for everyone.

The publisher and the editorial board hope that this book will prove to be a valuable piece of knowledge for researchers, students, practitioners and scholars across the globe.

List of Contributors

Ute Radespiel
Institute of Zoology, University of Veterinary Medicine Hannover, Buenteweg 17, 30559 Hannover, Germany

Philipp Hohenbrink
Institute of Zoology, University of Veterinary Medicine Hannover, Buenteweg 17, 30559 Hannover, Germany
Department of Zoology, University of Cambridge, Downing St, Cambridge CB2 3EJ, UK

Nicholas I. Mundy
Department of Zoology, University of Cambridge, Downing St, Cambridge CB2 3EJ, UK
Vera van NoortCentre of Microbial and Plant Genetics, KU Leuven, Leuven, Belgium

Stijn Wittouck
Centre of Microbial and Plant Genetics, KU Leuven, Leuven, Belgium
Department of Bioscience Engineering, University of Antwerp, Antwerp, Belgium

Carlos G. Schrago
Departamento de Genética, Laboratório de Biologia Evolutiva Teórica e Aplicada, Universidade Federal do Rio de Janeiro, Rio de Janeiro, RJ, Brazil

Damien S. Waits, Scott R. Santos and Kenneth M. Halanych
Department of Biological Sciences, Molette Biology Laboratory for Environmental and Climate Change Studies, Auburn University, Auburn, AL 36849, USA

Elisa M. Costa-Paiva
Departamento de Genética, Laboratório de Biologia Evolutiva Teórica e Aplicada, Universidade Federal do Rio de Janeiro, Rio de Janeiro, RJ, Brazil
Department of Biological Sciences, Molette Biology Laboratory for Environmental and Climate Change Studies, Auburn University, Auburn, AL 36849, USA

Nathan V. Whelan
Department of Biological Sciences, Molette Biology Laboratory for Environmental and Climate Change Studies, Auburn University, Auburn, AL 36849, USA
Warm Springs Fish Technology Center, U.S. Fish and Wildlife Service, 5308 Spring ST, Warm Springs, GA 31830, USA

Roland Butôt
Endless Forms group, Naturalis Biodiversity Center, Vondellaan 55, 2332 AA Leiden, The Netherlands

Anita Dirks-Mulder
Endless Forms group, Naturalis Biodiversity Center, Vondellaan 55, 2332 AA Leiden, The Netherlands
Faculty of Science and Technology, University of Applied Sciences Leiden, Zernikedreef 11, 2333 CK Leiden, The Netherlands

Peter van Schaik, Jan Willem P. M. Wijnands, Roel van den Berg, Louie Krol, Sadhana Doebar and Kelly van Kooperen
Faculty of Science and Technology, University of Applied Sciences Leiden, Zernikedreef 11, 2333 CK Leiden, The Netherlands

Elena M. Kramer
Department of Organismic and Evolutionary Biology, Harvard University, 16 Divinity Ave, Cambridge, MA 02138, USA

Hugo de Boer
Endless Forms group, Naturalis Biodiversity Center, Vondellaan 55, 2332 AA Leiden, The Netherlands
The Natural History Museum, University of Oslo, Blindern, 0318 Oslo, Norway
Department of Organismal Biology, Evolutionary Biology Centre, Uppsala University, Norbyvägen 18D, Uppsala SE-75236, Sweden

Erik F. Smets
Endless Forms group, Naturalis Biodiversity Center, Vondellaan 55, 2332 AA Leiden, The Netherlands
Ecology, Evolution and Biodiversity Conservation cluster, KU Leuven, Kasteelpark Arenberg 31, 3001 Leuven, Belgium

Rutger A. Vos
Endless Forms group, Naturalis Biodiversity Center, Vondellaan 55, 2332 AA Leiden, The Netherlands
Institute for Biodiversity and Ecosystem Dynamics, University of Amsterdam, Science Park 904, 1098 XH Amsterdam, The Netherlands

Alexander Vrijdaghs
Ecology, Evolution and Biodiversity Conservation cluster, KU Leuven, Kasteelpark Arenberg 31, 3001 Leuven, Belgium

Barbara Gravendeel
Endless Forms group, Naturalis Biodiversity Center, Vondellaan 55, 2332 AA Leiden, The Netherlands
Faculty of Science and Technology, University of Applied Sciences Leiden, Zernikedreef 11, 2333 CK Leiden, The Netherlands
Institute Biology Leiden, Leiden University, Sylviusweg 72, 2333 BE Leiden, The Netherlands

Ana Caroline Paiva Gandara
Instituto de Bioquímica Médica Leopoldo de Meis, Universidade Federal do Rio de Janeiro, Rio de Janeiro, Brazil

André Torres
Laboratório de Biologia Computacional e Sistemas, Instituto Oswaldo Cruz, Fiocruz, Rio de Janeiro, Brazil

Ana Cristina Bahia
Instituto de Biofísica, Universidade Federal do Rio de Janeiro, Rio de Janeiro, Brazil

Pedro L. Oliveira
Instituto de Bioquímica Médica Leopoldo de Meis, Universidade Federal do Rio de Janeiro, Rio de Janeiro, Brazil
Instituto Nacional de Ciência e Tecnologia em Entomologia Molecular – INCT-EM, Rio de Janeiro, Brazil

Renata Schama
Laboratório de Biologia Computacional e Sistemas, Instituto Oswaldo Cruz, Fiocruz, Rio de Janeiro, Brazil
Instituto Nacional de Ciência e Tecnologia em Entomologia Molecular – INCT-EM, Rio de Janeiro, Brazil

Kira S. Makarova, Yuri I. Wolf and Eugene V. Koonin
National Center for Biotechnology Information, National Library of Medicine, National Institutes of Health, Bethesda, MD 20894, USA
Department of Division of Genetics and Physiology, Department of Biology, University of Turku, Turku, Finland

Pere Puigbò
National Center for Biotechnology Information, National Library of Medicine, National Institutes of Health, Bethesda, MD 20894, USA
Division of Genetics and Physiology, Department of Biology, University of Turku, Turku, Finland

David M. Kristensen
National Center for Biotechnology Information, National Library of Medicine, National Institutes of Health, Bethesda, MD 20894, USA
Department of Biomedical Engineering, University of Iowa, Iowa City, IA, USA

Cecilio Valadez-Cano and Luis Delaye
Departamento de Ingeniería Genética, Centro de Investigación y de Estudios Avanzados del Instituto Politécnico Nacional, Unidad Irapuato, Km. 9.6 Libramiento Norte Carr. Irapuato-León, 36821 Guanajuato, Irapuato, Mexico

Roberto Olivares-Hernández
Departamento de Procesos y Tecnología, Universidad Autónoma Metropolitana-Cuajimalpa, Av. Vasco de Quiroga 4871, Santa Fe, Del. Cuajimalpa, C.P. 05348 Ciudad de Mexico, México, Mexico

Osbaldo Resendis-Antonio
Human Systems Biology Laboratory, Coordinación de la Investigación Científica-Red de Apoyo a la Investigación (RAI), UNAM, México City, Mexico
Instituto Nacional de Medicina Genómica (INMEGEN), 14610 México City, Mexico

Alexander DeLuna
Unidad de Genómica Avanzada (Langebio), Centro de Investigación y de Estudios Avanzados del IPN, Guanajuato, Irapuato, Mexico

Christina Nagler
Functional morphology group, Department of Biology II, Ludwig-Maximilians-University, Großhaderner Strasse 2, 82152 Planegg-Martinsried, Germany

Matúš Hyžný
Geology and Palaeontology, Faculty of Natural Sciences, Comenius University, Mlynská dolina, Ilkovičova 6, 84215 Bratislava, Slovakia

Joachim T. Haug
Functional morphology group, Department of Biology II, Ludwig-Maximilians-University, Großhaderner Strasse 2, 82152 Planegg-Martinsried, Germany
GeoBio-Center, Richard-Wagner Strasse 10, 80333 Munich, Germany

Pierre Pontarotti
Aix Marseille Université, CNRS, Centrale Marseille, I2M UMR 7373, Evolution Biologique et Modélisation, 3 place Victor Hugo, Marseille 13331, France

Didier Raoult and Vicky Merhej
Aix Marseille Univ, CNRS, IRD, INSERM, AP-HM URMITE, IHU –Méditerranée Infection, 19-21 Boulevard Jean Moulin, Marseille 13005, France

Sandrine Pinos
Aix Marseille Université, CNRS, Centrale Marseille, I2M UMR 7373, Evolution Biologique et Modélisation, 3 place Victor Hugo, Marseille 13331, France
Aix Marseille Univ, CNRS, IRD, INSERM, AP-HM URMITE, IHU –Méditerranée Infection, 19-21 Boulevard Jean Moulin, Marseille 13005, France

Haiqing Wang and Quanwen Dou
Key Laboratory of Adaptation and Evolution of Plateau Biota, Northwest Institute of Plateau Biology, Chinese Academy of Sciences, Xining 810008, China
University of Chinese Academy of Sciences, Beijing 100049, China

Feng Yu, Yanyan Zhao and Ruijuan Liu
Key Laboratory of Adaptation and Evolution of Plateau Biota, Northwest Institute of Plateau Biology, Chinese Academy of Sciences, Xining 810008, China
University of Chinese Academy of Sciences, Beijing 100049, China

Jiangli Dong and Tao Wang
State Key Laboratory of Agrobiotechnology, College of Biological Sciences, China Agricultural University, Beijing 100193, China

Marie Marchal and David R. Johnson
Department of Environmental Microbiology, Eawag, Überlandstrasse 133, 8600 Dübendorf, Switzerland
Department of Environmental Systems Science, ETH Zürich, 8092 Zürich, Switzerland

Felix Goldschmidt, Selina N. Derksen-Müller and Martin Ackermann
Department of Environmental Microbiology, Eawag, Überlandstrasse 133, 8600 Dübendorf, Switzerland
Department of Environmental Systems Science, ETH Zürich, 8092 Zürich, Switzerland

Sven Panke
Department of Biosystems Science and Engineering, ETH Zürich, 4058 Basel, Switzerland

Astrid Schuster
Department of Earth- and Environmental Sciences, Palaeontology and Geobiology, Ludwig Maximilians-Universität München, Richard-Wagner-Str. 10, 80333 Munich, Germany

Jose V. Lopez
Halmos College of Natural Sciences and Oceanography, Nova Southeastern University, Dania Beach, FL 33004, USA

Leontine E. Becking
Marine Animal Ecology, Wageningen University and Research Centre, Wageningen, The Netherlands
Naturalis Biodiversity Center, Marine Zoology Department, Leiden, The Netherlands

Michelle Kelly
National Centre for Aquatic Biodiversity and Biosecurity, National Institute of Water and Atmospheric Research, Newmarket, Auckland, New Zealand

Shirley A. Pomponi
Harbor Branch Oceanographic Institute-Florida Atlantic University, 5600 U.S. 1 North, Ft Pierce, FL 34946, USA

Gert Wörheide
Department of Earth- and Environmental Sciences, Palaeontology and Geobiology, Ludwig-Maximilians-Universität München, Richard-Wagner-Str. 10, 80333 Munich, Germany
SNSB - Bavarian State Collections of Palaeontology and Geology, Richard-Wagner Str. 10, 80333 Munich, Germany
GeoBio-CenterLMU, Ludwig-Maximilians-Universität München, Richard-Wagner Str. 10, 80333 Munich, Germany

Dirk Erpenbeck
Department of Earth and Environmental Sciences, Palaeontology and Geobiology, Ludwig-Maximilians-Universität München, Richard-Wagner-Str. 10, 80333 Munich, Germany GeoBio-CenterLMU, Ludwig-Maximilians-Universität München, Richard-Wagner Str. 10, 80333 Munich, Germany

Paco Cárdenas
Department of Medicinal Chemistry, Division of Pharmacognosy, BioMedical Center, Uppsala University, Husargatan 3, 75123 Uppsala, Sweden

Alvina G. Lai, Natalia Pouchkina-Stantcheva, Alessia Di Donfrancesco, Gerda Kildisiute, Sounak Sahu and A. Aziz Aboobaker
Department of Zoology, University of Oxford, South Parks Road, Oxford OX1 3PS, UK

Jannelle M. Vienneau-Hathaway, Elizabeth R. Brassfield, Amanda Kelly Lane and Nadia A. Ayoub
Department of Biology, Washington and Lee University, Lexington, VA, USA

Matthew A. Collin, Sandra M. Correa-Garhwal and Cheryl Y. Hayashi
Department of Biology, University of California, Riverside, CA, USA

Thomas H. Clarke
Department of Biology, Washington and Lee University, Lexington, VA, USA
Department of Biology, University of California, Riverside, CA, USA

Evelyn E. Schwager and Jessica E. Garb
Department of Biological Sciences, University of Massachusetts, Lowell, MA, USA

Index

A

Amino Acid Composition, 203, 209
Archaeal Genomes, 69-70, 80
Arthropod Phylogeny, 53-54, 61
Aspartic Acid Residue, 24

B

Balancing Selection, 1, 4, 11

C

Chromatophore, 82-97
Chromatophore Model (ICV265), 89, 91, 96
Computing Bootstrap Trees, 21

D

Deceptive Pollination, 35-36, 50
Depalmitoylating Enzymatic Activity, 14
Dna Substrate, 179, 187
Drosophila Melanogaster, 15, 17, 53-54, 57, 62-63, 66-67, 194

E

Erycina Pusilla, 35, 37, 51
Eukaryotes, 12-13, 15-16, 18-19, 53, 61, 63-64, 86, 97-98, 119, 124, 177, 179-183, 194
Euteleostomi, 16-17
Experimental Evolution, 140, 144, 146-148

F

False Discovery Rate, 4, 11
Free-living Cyanobacteria, 82, 86, 88, 90, 94-96

G

Gene Duplication, 8, 24, 30, 38, 51, 61, 67, 69, 71, 196, 198, 210, 212
Gene Ontology, 34
Genetic Diversity, 1, 4-5, 8-9, 138, 151
Genome Degradation, 82
Group I Introns, 154-155, 157, 161, 163, 165, 167-171, 173-174
Gut Homeostasis, 64

H

Hemerythrin, 24-25, 32-34
Hemocyanins, 24, 33-34
Hemoglobins, 24-25, 30

Histone Deacetylase, 17, 22
Horizontal Gene Transfer, 69, 78, 80-81, 83, 98, 125, 127, 172, 194
Hr Gene Tree, 24, 26

I

Isopods, 100-101, 103, 107-113

L

Lysosomes, 13, 19

M

Main Olfactory Epithelium, 8, 11
Mcdonald-kreitman Tests, 5, 8
Membrane Biogenesis, 114, 123, 153
Metabolic Evolution, 83-84, 96
Microcebus Murinus, 1-2, 10-11, 178-179
Mitochondrial Dna, 10-11, 138, 172-174
Molecular Evolution, 10, 33-34, 66-67, 174, 194, 205
Mutualistic Interactions, 83, 140-142
Myohemerythrins, 25

N

Neurohemerythrin, 24-25, 33-34
Non-binary Species Tree, 21
Nox4 Genes, 53, 57, 63
Nox Gene, 53-54, 60, 64, 66

O

Orchid Code, 37-38, 48, 51
Ovohemerythrin, 24, 33-34
Oxidoreductases, 84

P

Palmitoyl Fatty Acid, 12
Palmitoylation, 12-13, 15, 19, 22-23
Palmitoylation Writer, 12, 15, 19
Parasitism, 100-101, 109, 111-112, 124, 195
Phylogenetic Signal, 1, 55, 113, 206
Phylogeography, 2
Polyploidy, 128-129, 139
Positive Selection, 1-4, 8-10, 95
Protein Palmitoyl Thioesterases, 13
Protein Palmitoylation, 12-13, 23
Protein Post-translational Modifications, 12, 22

Purifying Selection, 1-2, 8-10, 82, 92, 95

R

Reactive Oxygen Species, 53, 66-67

Retrotransposons, 177, 194

Rna Template, 31, 175, 179, 182, 187, 194

S

Saccharomyces Cerevisiae, 13, 15, 22, 88

Sequences Encoding Putative, 55, 64

Signal Transduction, 67, 117, 123-124

Species Tree, 15, 17-18, 21-22, 71, 75, 79, 115-116, 125, 210

Spidroin Gene Family, 196, 210

Stamen-derived Origin, 35

T

Teleost, 19

Telomerase Reverse Transcriptase, 175-176, 192-194

Tetrapods, 182

Toll-like Receptors, 10

Transcription, 51, 65, 67, 117, 123-124, 153, 188, 193

Transferred Sequences Insertion, 115

V

Vertical Gene Transfer, 156

Virulence, 115, 124-125, 152-153

Vomeronasal Receptors, 1-2, 11

W

Wilcoxon Matched Pairs Test, 4